BOWLING GREEN STATE UNIVERSITY
DISCARDED
LIBRARY

CONTROLLED GROWTH OF NANOMATERIALS

Lide Zhang Xiaosheng Fang Changhui Ye
Chinese Academy of Sciences, China

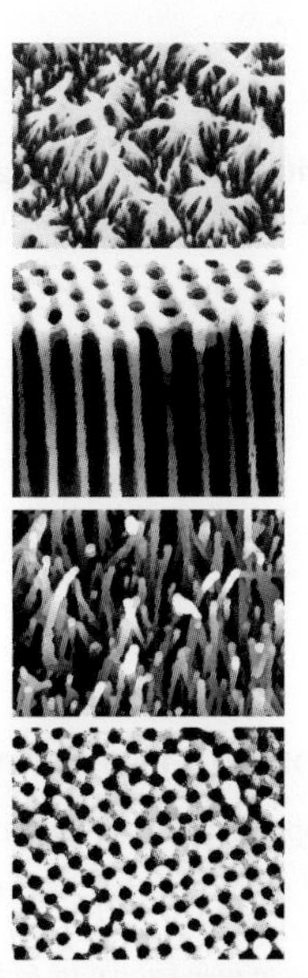

CONTROLLED GROWTH OF NANOMATERIALS

World Scientific

NEW JERSEY • LONDON • SINGAPORE • BEIJING • SHANGHAI • HONG KONG • TAIPEI • CHENNAI

Published by

World Scientific Publishing Co. Pte. Ltd.
5 Toh Tuck Link, Singapore 596224
USA office: 27 Warren Street, Suite 401-402, Hackensack, NJ 07601
UK office: 57 Shelton Street, Covent Garden, London WC2H 9HE

British Library Cataloguing-in-Publication Data
A catalogue record for this book is available from the British Library.

CONTROLLED GROWTH OF NANOMATERIALS

ISBN-13 978-981-256-728-4
ISBN-10 981-256-728-3

Typeset by Stallion Press
Email: enquiries@stallionpress.com

Printed by FuIsland Offset Printing (S) Pte Ltd, Singapore

Contents

Chapter 1

Introduction

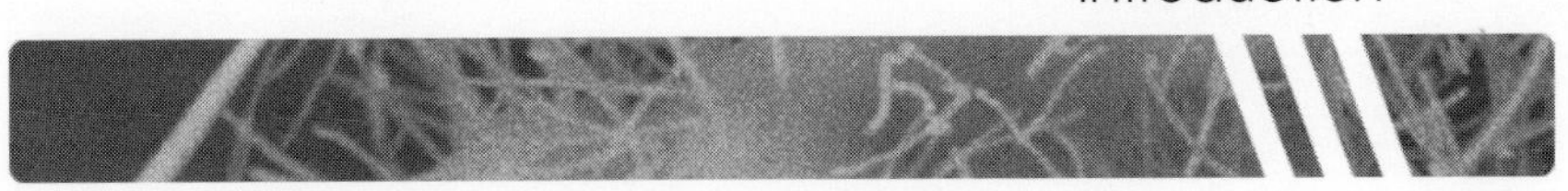

Chapter 1

Introduction

In the past decades, nanoscience and nanotechnology has been making significant progress, and the effect of nanoscience and nanotechnology on every field has been acknowledged in the world. Therefore, in the 21st century, their strategic position has already been established. The study of nanomaterials and nanostructures is one field with the earliest starting that obtained rich achievements in nanoscience and nanotechnology. Nanomaterials and nanostructures play a very important supporting role for applications of nanoscience and nanotechnology in the field of fabrication, such as information and techniques, energy sources, environments, health and medical treatments. All countries place the study of nanomaterials and nanostructures in a very important position. While the United States arranges the development of the front fields of nanoscience and nanotechnology, such as nanoelectronic technologies and devices, nano- and microfabrication techniques, nanobiotechnology, nanomedicines and diagnosis techniques, nanoenvironmental monitoring and treatment techniques, the continuation of the in depth study of nanomaterials and nanostructures is placed in an extremely important position. In the investment for the study of nanoscience and nanotechnology, the investment for nanomaterials and nanostructures occupies 49%. This sufficiently indicates the guiding role and the importance of the study of nanomaterials and nanostructures. Now, the motivational power of the study of nanomaterials and nanostructures are mainly national strategy requirements and enhancements of the national competitive ability in the high science and technology field. In addition, the in depth study of nanomaterials and nanostructures is an important source for producing new principles, new techniques and new methods, and potentially leading to breakthroughs in great scientific problems. At the same time, the nanomaterial market is also a native power for the development of nanomaterials. It will stimulate and promote the development of nanomaterial and nanostructures.

The high technique field requiring nanomaterials with unique properties is the important motivational power for drawing the nanomaterial development to new depths. In the IT industry, the fabrication of ultramicrochips will inevitably drive the sizes of CUP, transistors, high-density storage, field emission displays, ultraspeed calculation and logical devices towards the nanometer sizes. This makes the requirement of nanomaterials with unique properties more urgent. When the line width made by lithography becomes the order of nanomaterials and the CMOS circuits decrease to the ultramicro sizes, the requirement of nanomaterials and nanostructures with stable properties becomes more urgent. Fabricating the nanosized optoeletronic devices is likely to push the IT industry development to a new level. The study of nanomaterials and nanostructures will be faced with a strict challenge as a result.

The high sensitivity, high resolution nanofluid biodetectors, virus and bacterium detection and diagnosis, cell restoration, rapid diagnosis of the diseased position, carriers of target treatment, early rapid detection of toxic gases and burnable gases, rapid diagnosis of explosive molecules, etc., fields are developing at a high speed. This will make nanomaterials and nanostructures become inevitably one of the mainstream materials, and this also proposes a new challenge for the study of the unique properties of nanomaterials and nanostructures.

The rapid progress of global industries and the promotional requirements of traditional industries will encourage nanomaterials and nanostructures to play an important role in our competition. This will inevitably raise the fabricating technique and applied technique of nanomaterials and the study of optimization of structures and properties of nanomaterials to a new level.

Nanomaterials and nanostructures are established as the knowledge frame of an important branch of learning. This will promote the study of nanomaterials and nanostructures to develop deeply and expansively. Although the preparation science of nanomaterials made remarkable progress, such as the synthesis of a series of materials with new structures and new properties which has never been seen in the traditional material field, and the discovery of some special physical properties and proposal of some new concepts, the field of nanomaterial as a new branch of learning did not develop perfectly. A description of the new and unique physical and chemical phenomena which were discovered, still stops at the explanation step when using the traditional theory. In a strict sense, the relatively independent new theory of nanomaterials and nanostructures has yet to be established. The research of new laws is not systematical. The new effects appeared in the nanofield, such as the quantum confinement effect and coulomb blockade effect etc., are also established according to the traditional quantum theory. Different from the thermodynamics of infinite systems, for establishment of the thermodynamics of nanomaterials systems, controlled growth dynamics of nanomaterials, new lows of matter and energy transport of nanosystems, new principles and new methods of regulation and control of structures and properties etc., all have expansive knowledge innovation spaces. The research of nanomaterials and nanostructures at the new step is expanded around the above-mentioned scientific problems. It will inevitably provide the basis for designing nanomaterials with special structural control of the properties of nanomaterials and nanostructures. Therefore, an establishment of the self-scientific frame of nanomaterials is a significant and extremely difficult assignment. It must be established on the basis for a deeper study of nanomaterials and nanostructures. Now, our important assignment is to realize "three transitions" based on the nanomaterial research, and push nanomaterials to develop towards a newer more profound level. One is that the preparation realizes scientifically the transition from random growth to controlled growth, and the growth dynamics of nanomaterials, nanostructures and the thermodynamics of the confinement systems are studied. The second one is the transition from the exploration of random new phenomena to the more profound exploration of new laws. The conditions for unique property appearance, the change laws of properties and the factors which affect the property stability are systematically studied, so that the properties become controllable. The third transition is that of the research focus of gravity from preparation and synthesis as the main focus to the study of the relation between nanomaterials and nanostructures. The study of the law and process of nanostructure development and the effect on properties, and the investigation of the relation of preparation processes, nanostructures and properties, theoretically provide the theoretical basis for designing nanostructures with the required properties and obtaining the required properties and structures through the preparation methods. When the sample size reaches the order of nanometers, this does not imply the appearance of new properties. The new properties occurring in single element nanomaterials and nanostructures certainly do not satisfy the applied requirements. Therefore, through heterogeneous doping, composite and heterogeneous nanostructure designing, the largest limit excavation of nano effects and optimalization, modulation and control of properties are realized. All of these problems are the most challenging problems in the research field of nanomaterials and nanostructures, and are also the important market of the deep research of nanomaterials and nanostructures.

In the past decades, significant progress has already been made in the field of zero- (0D) and two-dimensional nanomaterials and nanostructures, since the discovery of pure carbon nanotubes,[1] and one-dimensional (1D) nanomaterials have stimulated great interest. Due to their potential interests to the understanding of physical concepts and for applications in constructing nanoscale electronic and optoelectronic devices,[2–8] they are the ideal systems for investigating the dependence of electrical transport, optical and mechanical properties on size and dimensionally and are expected to play important roles as both interconnects and functional components in the fabrication of nanoscale electronic and optoelectronic devices.[9] Up to now, 1D nanomaterials and nanostructures, such as nanotubes,[10–12] nanorods,[13–15] nanowires,[16–18] nanobelts or nanoribbons,[19–21] nanocables,[22–25] and nanosheets,[26–28] have been successfully synthesized by a variety of methods.

During the past decade, many methods have been developed to synthesize 1D nanomaterials and nanostructures. Overall, they can be categorized into two major approaches based on the reaction media that were used during the preparation: solution and gas phase-based process. Solution-based approaches include template-directed synthesis,[29–33] solution-liquid-solid method,[34] and solvothermal chemical synthesis.[35–38] Gas phase-based process for 1D nanomaterials and nanostructures mainly include four growth mechanisms. One is a well-accepted so-called vapor-liquid-solid (VLS) process of nanowire and nanowhisker growth proposed by Wagner and Ellis.[39] According to this mechanism, the anisotropic crystal growth is promoted by the presence of the liquid alloy/solid interface, and the detailed description about the VLS growth mechanism can be found in the literature by Yang.[40] It is conceivable that the size and diameter of the as-synthesized products can be controlled by selecting different size catalysts. For example, monodisperse silicon nanowires were synthesized by exploiting well-defined gold nanoclusters as catalysts for 1D growth via a VLS mechanism in the Lieber group.[41] Transmission electron microscopy (TEM) studies of the materials grown from 5, 10, 20, and 30 nm nanocluster catalysts showed that the nanowires had mean diameters of 6, 12, 20, and 31 nm, respectively, and were thus well-defined by the nanocluster sizes. The positions of 1D nanostructures can be controlled by the initial position of Au or other catalyst clusters or thin films. In addition, by applying the conventional epitaxial crystal growth technique into the VLS process, it is possible to achieve precise orientation control during the nanowire growth. Second is the vapor-solid (vs) growth process. In this process, vapor is first generated by evaporation, chemical reduction or gaseous reaction. The vapor is subsequently transported by a carrying gas, and condensed onto a substrate. Some metal, oxide and other nanomaterials and nanostructures have been successfully synthesized by the VS process.[29,42–44] It is generally believed that growth temperature and gas-phase supersaturation determine the growth rate of surface planes and the final morphology of the crystals, with other experimental parameters playing minor roles in the VS process. Third is oxide-assisted growth, which is a new nanowire growth route proposed by Lee[45] This synthesis technique, in which oxides instead of metals play an important role in the nucleation and growth of nanowires, is capable of producing a series of nanowires. Fourth is the combination of the anodic alumina membrane (AAM) and chemical vapor deposition (CVD). For example, Zhang reported that large-scale single crystalline GaN nanowires in anodic alumina membrane were achieved through a gas reaction of Ga_2O vapor with a constant flowing ammonia atmosphere at 1273 K.[46] Ordered ZnO nanowire arrays have also been fabricated via this method by Lee[47]

Exploration of nanomaterials and nanostructures that exhibit functionality is the key to nanotechnology. Among the 1D nanomaterials and nanostructures, functional oxides are the fundamental ingredients of smart systems, because the physical and chemical properties of the oxides can be tuned and controlled through adjusting cation valence state and anion deficiency. The structures of functional oxides are very diverse and colorful, and there are endless new phenomena and applications. Such unique characteristics have made oxides the most diverse class of materials, with properties covering almost all aspects of condensed matter physics and solid state chemistry in areas including semiconductivity, superconductivity, ferroelectricity, magnetism, and piezoelectricity.[48] Up to now, much attention has been paid to the preparation of binary oxide nanomaterials and nanostructures.[49]

An important issue in the self-organized growth and application of 1D nanomaterials and nanostructures is how to control the composition, size, morphology, position, orientation and crystallinity etc. in an effective and controllable way. The significance of the controllability manifests in both the chemistry of small-size material synthesis and the realization of their applications. A significant challenge for the chemical synthesis is how to rationally control the nanostructure assembly so that their size, dimensionality, interfaces, and ultimately, their two-dimensional and three-dimensional superstructures can be tailor-made towards desired functionality.[40] Controllable synthesis of nanowires and nanobelts only started to emerge recently.

This book will introduce the new methods and principles about the controlled growth of a nanomaterial system in detail, including a simple nanomaterial system, an ordered nanostructure system and a complicated nanostructure ordered system, the essential conditions for the controlled growth of nanostructures with different morphologies, size, compositions, and microstructures, and we also discuss the dynamics of controlled growth and thermodynamic characteristic of the two-dimensional nanorestricted system. This book also introduces some novel synthesis methods for nanomaterials and nanostructures, such as hierarchical growth and some developing template synthesis methods etc. combined with the application of nanomaterials and nanostructures. This book introduces correlative novel properties and property control arising from nanomaterials and nanostructures. This book also reviews the developing trend of nanomaterials and nanostructures.

Bibliography

1. S. Iijima, *Nature* **354**, 56 (1991).
2. J. T. Hu, J. T. W. Odom, and C. M. Lieber, *Acc. Chem. Res.* **32**, 435 (1999).
3. Z. R. Dai, Z. W. Pan, and Z. L. Wang, *Adv. Funct. Mater.* **13**, 9 (2003).
4. M. Law, J. Goldberger, and P. D. Yang, *Annu. Rev. Mater. Res.* **34**, 83 (2004).
5. L. Miao, S. Tanemura, S. Toh, K. Kaneko, and M. Tanemuram, *J. Mater. Sci. Technol.* **20**, 59 (2004).
6. L. Li, Y. Zhang, Y. W. Yang, X. H. Huang, G. H. Li, and L. D. Zhang, *Appl. Phys. Lett.* **87**, 031912 (2005).
7. G. M. Whitesides, *Small* **1**, 172 (2005).
8. X. S. Fang, C. H. Ye, L. D. Zhang, and T. Xie, *Adv. Mater.* **17**, 1661 (2005).
9. Y. N. Xia, P. D.Yang, Y. G. Sun, Y. Y. Wu, B. Mayers, B. Gates, Y. D. Yin, F. Kim, and H. Q. Yan, *Adv. Mater.* **15**, 353 (2003).
10. P. M. Ajayan, *Chem. Rev.* **99**, 1787 (1999).
11. C. H. Ye, G. W. Meng, Z. Jiang, Y. H. Wang, G. Z. Wang, and L. D. Zhang, *J. Am. Chem. Soc.* **124**, 15180 (2002).
12. G. S. Wu, L. D. Zhang, B. C. Cheng, T. Xie, and X. Y. Yuan, *J. Am. Chem. Soc.* **126**, 5976 (2004).

13. J. Zhang, Z. Q. Li, and J. Xu, *J. Mater. Sci. Technol.* **21**, 128 (2005).
14. S. H. Yu, B. Liu, M. S. Mo, J. H. Huang, X. M. Liu, and Y. T. Qian, *Adv. Funct. Mater.* **13**, 639 (2003).
15. H. L. Zhu, D. R. Yang, and H. Zhang, *J. Mater. Sci. Technol.* **21**, 609 (2005).
16. A. M. Morales and C. M. Lieber, *Science* **279**, 208 (1998).
17. X. Y. Zhang, L. D. Zhang, G. W. Meng, G. H. Li, N. Y. Jin-Phillipp, and F. Phillipp, *Adv. Mater.* **13**, 1238 (2001).
18. B. S. Xu, P. D. Han, J. Liang, Y. Yu, H. Q. Bao, and X. G. Liu, *J. Mater. Sci. Technol.* **20**, 681 (2004).
19. Z. W. Pan, Z. R. Dai, and Z. L.Wang, *Science* **291**, 1947 (2001).
20. Y. W. Wang, L. D. Zhang, G. W. Meng, C. H. Liang, G. Z. Wang, and S. H. Sun, *Chem. Comm.* 2632 (2001).
21. J. S. Wang, J. Q. Sun, Y. Bao, and X. F. Bian, *J. Mater. Sci. Technol.* **19**, 489 (2003).
22. Y. Zhang, K. Suenaga, C. Colliex, and S. Iijima, *Science* **281**, 973 (1998).
23. G. W. Meng, L. D. Zhang, C. M. Mo, S. Y. Zhang, Y. Qin, S. P. Feng, and H. J. Li, *J. Mater. Res.* **13**, 2533 (1998).
24. Q. Li, L. Z. Yao, G. W. Jiang, C. G. Jin, W. F. Liu, W. L. Cai, and Z. Yao, *J. Mater. Sci. Technol.* **20**, 684 (2004).
25. Y. Li, C. H. Ye, X. S. Fang, L. Yang, Y. H. Xiao, and L. D. Zhang, *Nanotech.* **16**, 501 (2005).
26. S. H. Yu and M.Yoshimura, *Adv. Mater.* **14**, 296 (2002).
27. X. S. Fang, C. H. Ye, X. S. Peng, Y. H. Wang, Y. C. Wu, and L. D. Zhang, *J. Cryst. Growth* **263**, 263 (2004).
28. B. Q. Cao, W. P. Cai, Y. Li, F. Q. Sun, and L. D. Zhang, *Nanotech.* **16**, 1734 (2005).
29. C. R. Martin, *Science* **266**, 1961 (1994).
30. H. Masuda and K. Fukuda, *Science* **268**, 1466 (1995).
31. L. Li, Y. Zhang, G. H. Li, and L. D. Zhang, *Chem. Phys. Lett.* **378**, 244 (2003).
32. L. Li, Y. H. Xiao, Y. W. Yang, X. H. Huang, G. H. Li, and L. D. Zhang, *Chem. Lett.* **34**, 930 (2005).

33. M. Z. Wu, L. Z. Yao, W. L. Cai, G. W. Jiang, X. G. Li, and Z. Yao, *J. Mater. Sci. Technol.* **20**, 11 (2004).

34. J. D. Holmes, K. P. Johnston, R. C. Doty, and B. A. Korgel, *Science* **287**, 1471 (2000).

35. Y. D. Li, H. W. Liao, Y. Ding, Y. T. Qian, L. Yang, and G. E. Zhou, *Chem. Mater.* **10**, 2301 (1998).

36. Y. G. Sun and Y. N. Xia, *Science* **298**, 2176 (2002).

37. W. T. Yao, S. H. Yu, L. Pan, J. Li, Q. S. Wu, L. Zhang, and H. Jiang, *Small* **1**, 320 (2005).

38. X. Wang, J. Zhuang, Q. Peng, and Y. D. Li, *Nature* **437**, 121 (2005).

39. R. S. Wagner and W. C. Ellis, *Appl. Phys. Lett.* **4**, 89 (1964).

40. P. D.Yang, Y. Y. Wu, and R. Fan, *Inter. J. Nanosoci.* **1**, 1 (2002).

41. Y. Cui, L. J. Lauhon, M. S. Gudiksen, J. F. Wang, and C. M. Lieber, *Appl. Phys. Lett.* **78**, 2214 (2001).

42. P. D. Yang and C. M. Lieber, *J. Mater. Res.* **12**, 2981 (1997).

43. X. S. Fang, C. H. Ye, L. D. Zhang, Y. Li, and Z. D. Xiao, *Chem. Lett.* **34**, 436 (2005).

44. Y. W. Wang, L. D. Zhang, C. H. Liang, G. W. Wang, and X. S. Peng, *Chem. Phys. Lett.* **357**, 314 (2002).

45. R. Q. Zhang, Y. Lifshitz, and S. T. Lee, *Adv. Mater.* **15**, 635 (20035).

46. G. S. Cheng, L. D. Zhang, Y. Zhu, G. T. Fei, L. Li, C. M. Mo, and Y. Q. Mao, *Appl. Phys. Lett.* **75**, 2455 (1999).

47. C. H. Liu, J. A. Zapien, Y. Yao, X. Meng, C. S. Lee, S. S. Fan, Y. Lifshitz, and S. T. Lee, *Adv. Mater.* **15**, 838 (2003).

48. Z. L. Wang, *Annu. Rev. Phys. Chem.* **55**, 159 (2004).

49. C. N. R. Rao, F. L. Deepak, G. Gundiah, and A. Govindaraj, *Prog. Solid State Chem.* **31**, 5 (2003).

Chapter 2

Controlled Growth of Nanowires and Nanobelts

- Oxides nanowires and nanobelts
- Sulfides nanowires and nanobelts
- Doping of nanowires and nanobelts

Chapter 2

Controlled Growth of Nanowires and Nanobelts

2.1 Introduction

The discovery of carbon nanotubes[1] has resulted in intensive investigations to study the novel physical properties of one-dimensional nanoscale materials, and their potential applications in constructing nanoscale electric and optoelectronic devices.[2–7] Nanowires, nanobelts, and nanoribbons, are a new class of quasi-one-dimensional materials and they are ideal systems for investigating the dependence of electrical transport and mechanical properties on dimensionality and size confinement.[8] They are expected to play an important role both in interconnects and in functional components in the fabrication of nanoscale electric and optoelectronic devices.

Nanowires are defined as having an isotropic nanocrystal with large aspect ratios between length and diameter. Generally, they would have diameters in the range of 1–200 nm and lengths up to several tens of micrometers.[9] Nanobelts or nanoribbons are quasi-one-dimensional structurally controlled nanomaterial that have well-defined chemical composition, crystallographic structure, and surfaces such as growth direction, top/bottom surface, and side surfaces.[10,11] Up to now, inorganic nanowires and nanobelts of elements, oxides, nitrides, carbides and chalcogenides, have been generated by employing various strategies, including vapor phase techniques and solution-based approaches, such as chemical vapor deposition (CVD), precursor decomposition, as well as solvothermal, hydrothermal and carbothermal methods.[5]

An important issue in the study and application of nanowires and nanobelts is how we can control the composition, size, morphology and crystallinity etc. in an effective and controllable way. Controllable synthesis is very important to nanoscale science, which denotes self-organized growth or fabrication of nanomaterials and nanostructures with desirable composition, size, morphology and structure. The significance of the controllability manifests in both the chemistry of small-size material synthesis and the realization of their applications. However, the controllable synthesis of nanowires and nanobelts have only started to emerge recently.

In this chapter, we present some progress in the controlled growth of nanowires and nanobelts, including oxides (ZnO, SnO_2, In_2O_3, MgO, and Al_2O_3) and sulfides (ZnS and CdS) and the doping of some nanowires and nanobelts.

2.2 Oxides nanowires and nanobelts

2.2.1. ZnO

Zinc oxide (ZnO), is a wide-band-gap semiconductor (E_g = 3.37 eV at room temperature) with a large exciton binding energy of 60 meV that is suitable for short wavelength optoelectronic applications. Nanostructured ZnO materials have received broad attention due to their distinguished performance in electronics, optics and photonics. ZnO is also a key technological material. The lack of a center of symmetry in wurtzite, combined with large electromechanical coupling, results in strong piezoelectric and pyroelectric properties and the consequent use of ZnO in mechanical actuators and piezoelectric sensors.[12] The morphology of ZnO has been proven to be the richest among inorganic semiconductors. Up to now, ZnO nanoarrays, nanorods, nanowires, nanobelts, nanotubes, nanorings, nanohelices, and nanosprings have been synthesized via a variety of techniques, such as template-assisted growth, vapor-liquid-solid methods, chemical vapor deposition, and solution-based synthesis etc.[13–30] Here, we mainly introduce the systematic investigation of the morphology evolution characteristics of ZnO nanostructures from dense rods to dense nanoplatelets, nanoplatelet flowers, dense nanobelt flowers, and nanowire flowers in an evaporation-physical transport-condensation approach.[31]

The commercial ZnO and graphite powders mixed in a molar ratio of 4 : 1 were placed in a ceramic boat that was inserted into the central region of a ceramic tube in a conventional horizontal tube furnace, where the temperature, pressure, and evaporation time were controlled. Several pieces of Si wafers were placed downstream in the tube side-by-side, and another Si wafer was put above the source material horizontally, as the deposition substrate. The system was rapidly heated to 1050°C in 12 minutes with the flow rate of the carrier gas (Ar) at 20 standard cubic centimeter per minute (sccm) and kept at this temperature for 90 minutes. Various ZnO nanostructures were deposited on Si substrates in a temperature range of 750–1050°C.

Figure 2.1 shows the typical scanning electron microscopy (SEM) images of as-synthesized products. The products are mainly formed on the downstream side of the source material. We also found some products on the substrate above the source material, as shown in Fig. 2.1(a), where the structure forms a dense film-like layer with areas covering over 1 cm^2. The high magnification image in Fig. 2.1(b) reveals that the film is composed of high density ZnO rods with diameter of several micrometers and length of tens of micrometers. The rods form arrays with nearly parallel orientation perpendicular to the substrate. It is noteworthy that the rods are faceted, both for the side planes and for the growing tips. On the substrates' downstream side of the source material, with the increase in distance between the source material and the substrate, the morphologies of ZnO nanostructures vary from dense nanoplatelets to nanoplatelet flowers, dense nanobelt flowers, and nanowire flowers with each morphology happening in a wide region and in high yield. The SEM images are shown from Figs. 2.1(c) to 2.1(j).

Through the use of the crystal growth theory, the determining factors for the formation of different nanostructural morphologies were found to be the gas-phase supersaturation and the surface energy of the growing surface planes. The growth theories and technologies of the nanoscale structure have some differences from the perfect large size crystals for films. For example, the growth of bulk size perfect crystals is always carried out under a very low gas-phase supersaturation. Under this circumstance, the crystal grows at near thermal equilibrium conditions and the shape of the crystal is well-defined by Wulff's theorem and reflects the internal crystallography of the crystal itself.[32,33] In an effort to grow nanoscale structures, such as nanoplatelets, nanobelts, nanowires, and nanorods, the supersaturation is generally much larger than unity and the crystal grows under conditions very different from thermal equilibrium. Under such a circumstance, growth kinetics takes a part or even determines the growth behavior of the surface planes and the final morphology of the structure.

Similar to what Brenner[34] and Nam[35] proposed, the forced diffusion of vapor by a carrier gas is more important than the molecular diffusion driven by a concentration gradient for the low Reynolds number of our system. Under this advection condition, there generally exists a maximum in ZnO vapor concentration and the supersaturation downstream from the source material, and as the flow rate of the Ar carrier gas increases, the maximum shifts further downstream, as qualitatively illustrated in Fig. 2.2. The

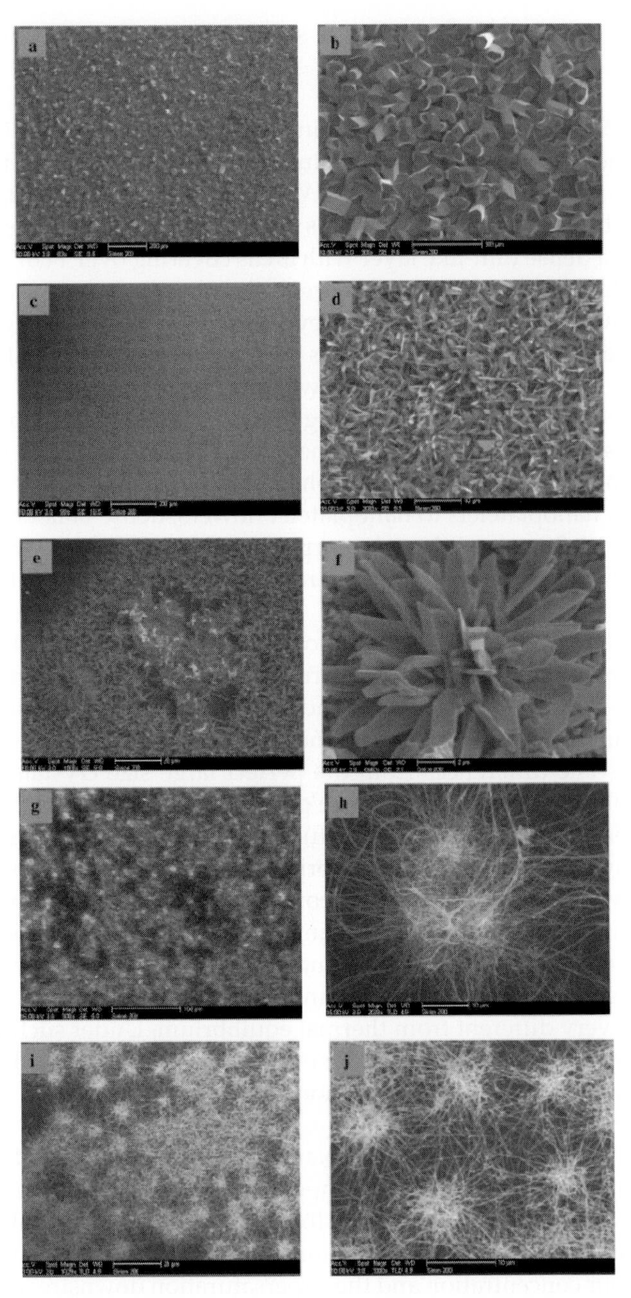

Fig. 2.1. Typical SEM images of as-synthesized products ZnO structures: (a) dense film-like rods, (c) dense film-like nanoplatelets, (e) flower-like nanoplatelets, (g) nanobelts, and (i) nanowires. The corresponding high magnification views are displayed in (b), (d), (f), (h), and (j), respectively.[31]

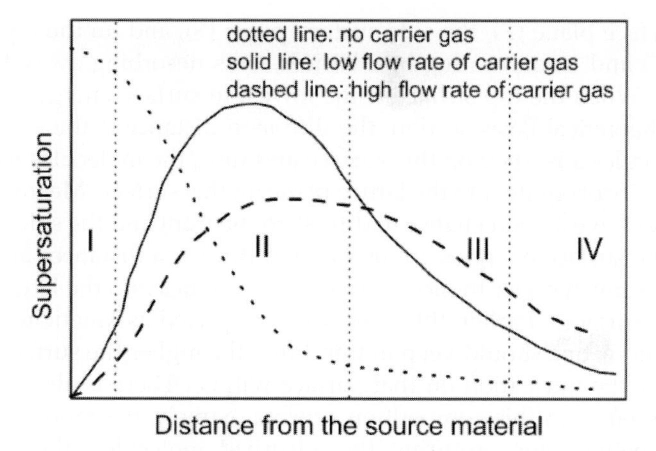

Fig. 2.2. Qualitative supersaturation profile of ZnO vapor in the reactor under a different flow rate of carrier gas along the distance downstream from the source material.[31]

regions I, II, III, and IV correspond to microrods, nanoplatelets, nanobelts, and nanowires, respectively, with a low flow rate of carrier gas.

To highlight the roles of supersaturation and surface energy during the formation of various ZnO structures, we choose the growth of nanoplatelets in region II as an example. The supersaturation is the highest in region II, as shown in Fig. 2.2. The growth takes place under circumstances far from thermal equilibrium, and the kinetics plays a more significant role in determining the morphology of the structure. As illustrated in Fig. 2.3, when a molecule is adsorbed on a surface plane, for example, the top surface of a plate-shaped structure, it will undergo four processes: diffusing to the

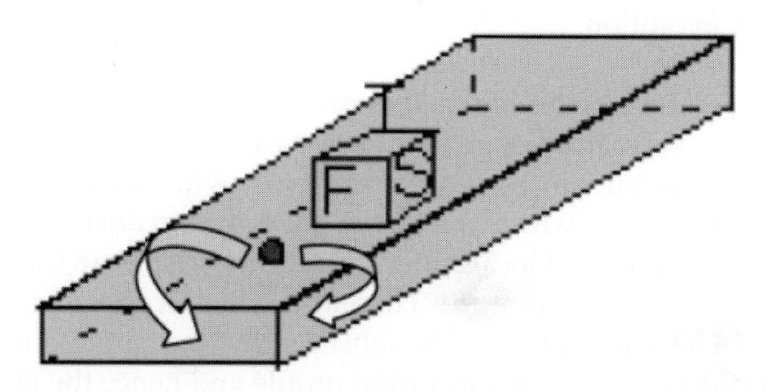

Fig. 2.3. Schematic drawing illustrating the competition for capturing the adsorbed molecules.[31]

front surface plane (*F*), the side surface plane (*S*), and on the top surface plane (*T*) and incorporating into lattices, thus desorbing away from the surfaces. When the top surface is the lowest in surface energy, according to the Theoretical Basis section, the diffusion distance is the smallest for the molecules adsorbed on this surface and thus, the molecules have little chance to incorporate into the lattice plane on this surface. Meanwhile, the molecules have a good chance to diffuse to the front and the side surfaces, where the surface energies are higher and diffusion distances are longer. Thus, it is much easier to incorporate the molecules into the lattice planes on these surfaces. Earlier, this case was interpreted as kinetic roughness. Furthermore, one should keep in mind that the higher the surface energy, the easier the nucleation on that surface will be. There is also a positive feedback effect in this competition process, namely, the more aggressive the competition for capturing the adsorbed molecules, the faster the growth rate for the front and side surfaces and then the larger the surface area for the top surface. Thus, the larger the area for the top surface, the larger the collection number of impinged molecules on this surface there will be to support the growth of the front and side surfaces. Failure to capture the adsorbed molecules leads to the much slower growth of the top surface. The aggressive characteristics of the front and the side surfaces similarly determine their proportion in the final morphology. When these two surfaces are similar in surface energy or belong to the same family of lattice planes, the competitive ability of these surfaces is also similar. Then, the morphology may take the form of a nanoplatelet structure.

We also found that other experimental parameters such as the temperature at the source and the substrate, the temperature difference and the distance between the source and the substrate, the heating rate of the furnace, the gas flow rate, the ceramic tube diameter, and the starting material, are all correlated with supersaturation and impose an effect on the morphology evolution.

The temperature at the source material determines the vapor pressure of it, and the temperature dependent pressure profile of ZnO is shown in Fig. 2.4.[36] The pressure increases exponentially with heating temperature. When the vapor moves within the carrier gas downstream to the substrates, there will be a build-up of local supersaturation higher than that in the case with lower heating temperature. Therefore, the morphology will change accordingly. The temperature of the substrate and the temperature gradient similarly influence the supersaturation profile and hence the morphology. Distance from the source material and the substrate implies the selection of a local supersaturation and temperature. The heating rate may have an

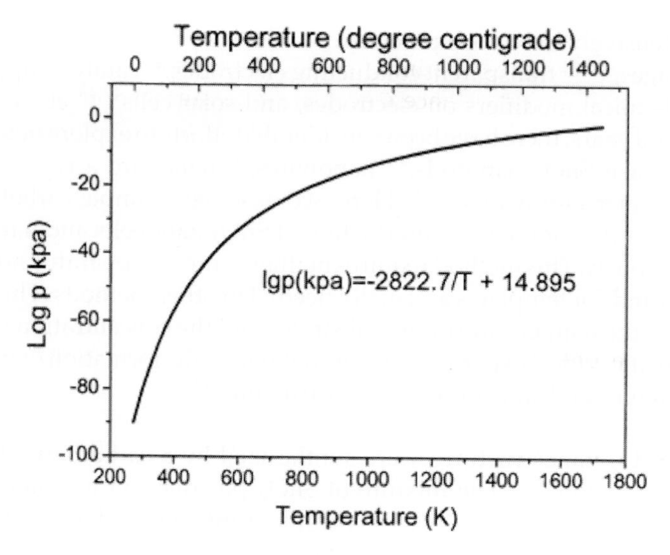

Fig. 2.4. Pressure profile as a function of temperature for ZnO.[31]

influence on the initial nucleation process: a higher heating rate makes the homogeneous nucleation and low dispersity of the morphology possible, while a lower heating rate generally causes high dispersity.

The gas flow rate and the inner diameter of the ceramic tube have a fundamental influence on the supersaturation profile through the interaction with the Reynolds number. In Fig. 2.2, the supersaturation profile under a different gas flow rate has been displayed. With a higher gas flow rate and smaller inner diameter of the ceramic tube, the maximum supersaturation will move downstream from the source material and the morphology of the structure at a fixed position on the substrate will change accordingly.

Materials with nanometer-scaled sizes have been well-known to show a decrease in melting temperature.[37,38] Therefore, the use of nanomaterial as the source corresponds to elevating the heating temperature of the source material and similarly changes the supersaturation profile and the final morphology of the structure.

2.2.2. SnO_2

Tin dioxide (SnO_2), an important *n*-type semiconductor with a wide band gap ($E_g = 3.6\,eV$, at 300 K), is a key functional material that has been extensively used for optoelectronic devices,[39,40] gas sensors detecting leakages,[41–44] transparent conducting electrodes,[45] catalyst supports,[46] electrochemical modifiers on electrodes, and solar cells[47,48] etc. Over the last several years, there have been considerable efforts to explore new routes to synthesize SnO_2 nanorods,[49] nanowires,[50] nanowire arrays,[51,52] and nanobelts or nanoribbons.[53–59] Here, we describe a simple carbothermal reduction route for the mass production of SnO_2 nanobelts and nanowires simultaneously. This method requires neither complex apparatus nor metal catalysts and/or templates as usually needed in other methods. The results show that the temperature of the substrates and the concentration of oxide vapor are the critical experimental parameters for the formation of different morphologies and sizes of SnO_2 nanostructures.[60]

The SnO_2 nanostructures were synthesized by a carbothermal reduction of SnO_2 powder. The mixture of SnO_2 powder and graphite (molar ratio 1:2) was ground to ensure complete mixing, and was placed in an alumina boat. During the experiment, a constant flow of Ar was maintained at flow rate of 60 sccm. The reaction temperature was about 1150°C and kept at this temperature for 3 h. During the experiment, the temperature at any point between the source material and the tube's downstream end was measured *in situ* by a sheathed thermocouple, allowing us to readily measure the temperature of the product formation.

The general morphologies of the products taken from different deposition temperatures are shown in Figs. 2.5(a) to 2.5(d). It is visible from the SEM images that the as-synthesized products consist of nanobelts and nanowires. Figure 2.5(e) shows schematically the formation temperatures of these morphologies. SnO_2 nanobelts grew mainly in the temperature range 1000–850°C [zones I (1000–950°C) and II (950–850°C)]. The typical widths of the nanobelts (zone I) are in the range 1–3 μm, with lengths ranging from several tens to several hundreds of micrometres, and some even up to several millimetres (Fig. 2.5(a)), while the widths of the nanobelts formed in zone II are in the range 50–500 nm, as shown in Fig. 2.5(b). SnO_2 nanowires grew mainly in the temperature range 850–600°C [zone III (850–750°C) and IV (750–600°C)]. The diameters of SnO_2 nanowires (zone III) normally range from 100 to 300 nm and their lengths are several tens of micrometres (Fig. 2.5(c)). while the diameters of the SnO_2 nanowires formed in zone IV are 30–150 nm and their lengths are several micrometres, and some SnO_2 nanoparticles can also be found in this zone (Fig. 2.5(d)). From the experimental results, we found that the widths of the SnO_2 nanobelts and the diameters of SnO_2 nanowires were temperature

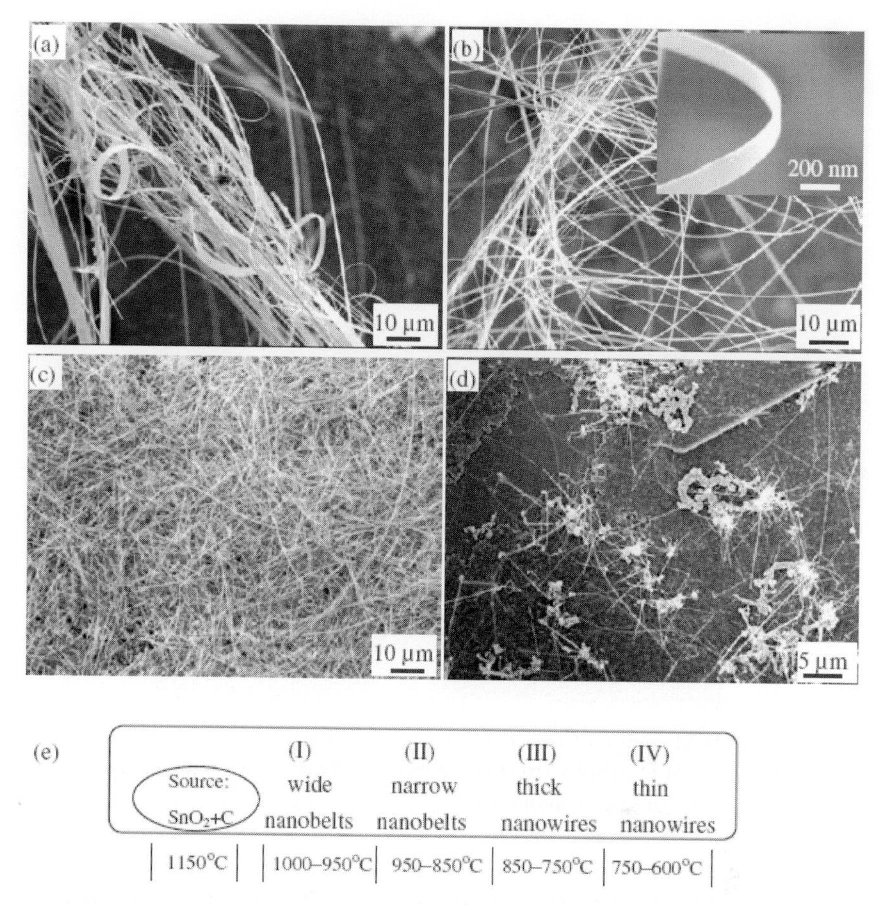

Fig. 2.5. (a)–(d) The general SEM images of the products taken from four different deposition temperatures zones, (e) schematic diagram showing the formation temperatures of these morphologies.[60]

dependent. Low growth temperatures resulted in small widths and diameters of SnO_2 nanobelts and nanowires.

Figure 2.6(a) shows a transmission electron microscopy (TEM) bright field image of a nanobelt with a width of 100 nm (formed in zone II). The nanobelt has a uniform width along its entire length. The corresponding selected area electron diffraction (SAED) pattern (inset in Fig. 2.6(a)) indicates that the SnO_2 nanobelt grows along the [101] direction. To further verify the morphological characteristics of the SnO_2 nanobelts, a cross-section TEM image taken from a nanobelt is given in Fig. 2.6(b), which exhibits a

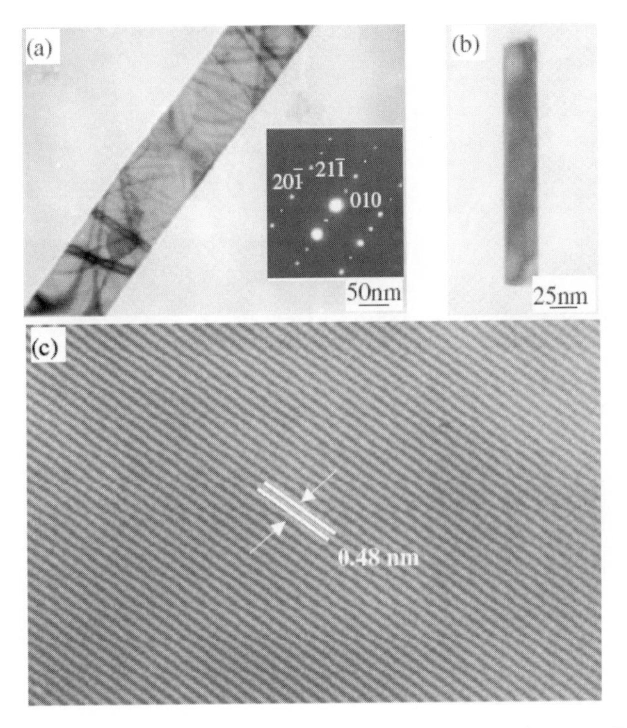

Fig. 2.6. (a) TEM image of one narrow SnO_2 nanobelt, with inset showing the corresponding SAED pattern, (b) cross-section TEM image of a SnO_2 nanobelt, and (c) HRTEM image of a SnO_2 nanobelt.[60]

rectangle-like cross-section. The typical thickness and width-to-thickness ratios of the SnO_2 nanobelts are in the ranges 10–30 nm and 5–10 nm, respectively. A high-resolution transmission electron microscopy (HRTEM) image (Fig. 2.6(c)) reveals that the nanobelts are single crystalline and defect free.

We believe that the temperature of the substrates and the concentration of oxide vapor are the critical experimental parameters for the formation of different morphologies of SnO_2 nanostructures. The growth mechanism of the SnO_2 nanostructures can be explained on the basis of a vapor–solid (VS) mechanism[61] since there are no droplets present at the tips of the nanostructures. To understand the growth mechanism of these SnO_2 nanostructures, the following reactions should be considered:

$$2SnO_2(s) + C(s) = 2SnO(v) + CO_2(v), \tag{2.1}$$

$$SnO(v) = SnO_2(s) + Sn(l), \tag{2.2}$$

$$2Sn(l) + O_2(v) = 2SnO(v). \tag{2.3}$$

At high temperature (1150°C), SnO_2 is reduced to SnO vapor by reaction (2.1). It is well-known that SnO is metastable and will decompose into SnO_2 and Sn.[54] The decomposition of SnO will result in the precipitation of SnO_2 nanoparticles, which are carried by the flowing Ar gas and are directly deposited on the surfaces of the substrates. Then the nanoparticles act as the nucleation sites for the growth of SnO_2 nanostructures. In the 1960s, Si needle-like and ribbon structures were synthesized by VS reactions[62] and some favorable growth directions were suggested to be responsible for the ribbon structure growth.[63] The theoretical and most stable crystal habit of SnO_2 was a tetragon elongated along the *c*-axis. Accordingly, SnO_2 nanostructures have optimized and secondly optimized growth directions. At higher temperatures and higher concentrations, the single crystal grew along the two optimized directions simultaneously. Moreover, the growth along the optimized direction was faster than along the secondly optimized direction. Thus, the SnO_2 nanobelts were formed. At lower temperatures and lower concentrations, the single crystal grew along one of the two optimized directions to form SnO_2 nanowires. The slight differences in temperature and concentration will influence the widths and diameters of the nanobelts and nanowires, respectively. In this case, once the initial nucleation starts, the crystal grows in epitaxial ways, which results in the preferential orientation of SnO_2 lattice planes and the formation of SnO_2 nanostructures in the end.

2.2.3. In_2O_3

Indium oxide (In_2O_3) is an important transparent conducting oxide material (TCO). It is an insulator in its stoichiometric form, whereas in its non-stoichiometric form it behaves as a highly conducting semiconductor with a wide direct band gap ($E_g \approx 3.6\,eV$).[64,65] In_2O_3 has been widely applied in the microelectronic field as window heaters, solar cells, flat-panel display materials, optical and electric devices, liquid crystal devices, and gas detectors, etc. owing to its high electric conductance, high transparency to visible light, and the strong interaction between certain poisonous gas molecules and the In_2O_3 surface.[66–70] Recently, much effort has been paid to the preparation of various In_2O_3 nanostructures, including nanofibers,[71] nanowires,[72–76] nanobelts or nanoribbons,[77,78] nanotubes filled with metal Indium (In),[79] single-crystalline pyramids,[80] etc. Here, morphology-controllable single-crystalline In_2O_3 octahedrons and nanowires were synthesized by manipulating appropriate external conditions in vapor environments.[81]

For the growth of In_2O_3 octahedrons and nanowires, In particles (purity: 99.99%, average particle size: 1 mm) or In_2O_3 powders (purity: 99.99%) loaded in an alumina boat were put into the center of the alumina tube, which was placed in a horizontal tube furnace. The system was respectively heated to 950°C, 1050°C, 1200°C, 1300°C, and 1350°C each time in about 10 minutes. The reaction durations are all 40 minutes under a constant flow of Ar (purity: 99.99%) at a flow rate of 50 sccm. After the furnace was cooled back to room temperature naturally, the gray-white powder-like products were collected on the Si substrates in the vicinity of the central heating zone when metal In was the precursor, and the temperature differences between the central heating zones and the downstream deposition zones were about 50–100°C. The gray-white wool-like products were collected downstream on the alumina tube surface when the heating temperatures were about 1350°C and In_2O_3 powders were the precursors and the temperature differences between the central heating zones and the deposition zones were about 400–500°C.

Powder X-ray diffraction (XRD) measurements show that all as-synthesized products are a body-centered cubic (bcc) structural In_2O_3 with a lattice constant of a = 10.11 Å (JCPD 89-4595). Figures 2.7(a) to 2.7(d) are the SEM images of the products, obtained when In was used as the raw material and the reaction temperature was 950°C, 1050°C, 1200°C, 1300°C, and 1350°C, respectively. When the reaction duration was elongated to 2 hours at 1200°C, the SEM images of the products were shown in Fig. 2.7(e). Figure 2.7(f) is the SEM image of the product obtained only when In_2O_3 powders were heated to at least 1350°C. All SEM observations reveal that the products consist of nano- and micro-sized In_2O_3 octahedrons and nanowires.

The growth mechanisms of In_2O_3 nanostructures with different morphologies by different source materials were described in Fig. 2.8. We think vapor pressures, supersaturation ratios, and surface energy are three important factors to the modulation of different morphologies. For the precursor of metal In, the saturated vapor pressures of liquid In are much higher than those of the solid state In_2O_3 at the same temperatures. Consequently, this condition leads to a high supersaturation ratio of the oxidized In vapor even in the vicinity of central heating zones. Although the difference of surface energies among the [110], [100], [111] facets can lead to their different growth rates, the high supersaturation ratio makes the effect of the surface energy difference on the growth very small. As a result, the crystal growth rates perpendicular to the [111], [100], and [110] facets become quite close, so that the octahedron growth model can be realized easily. It

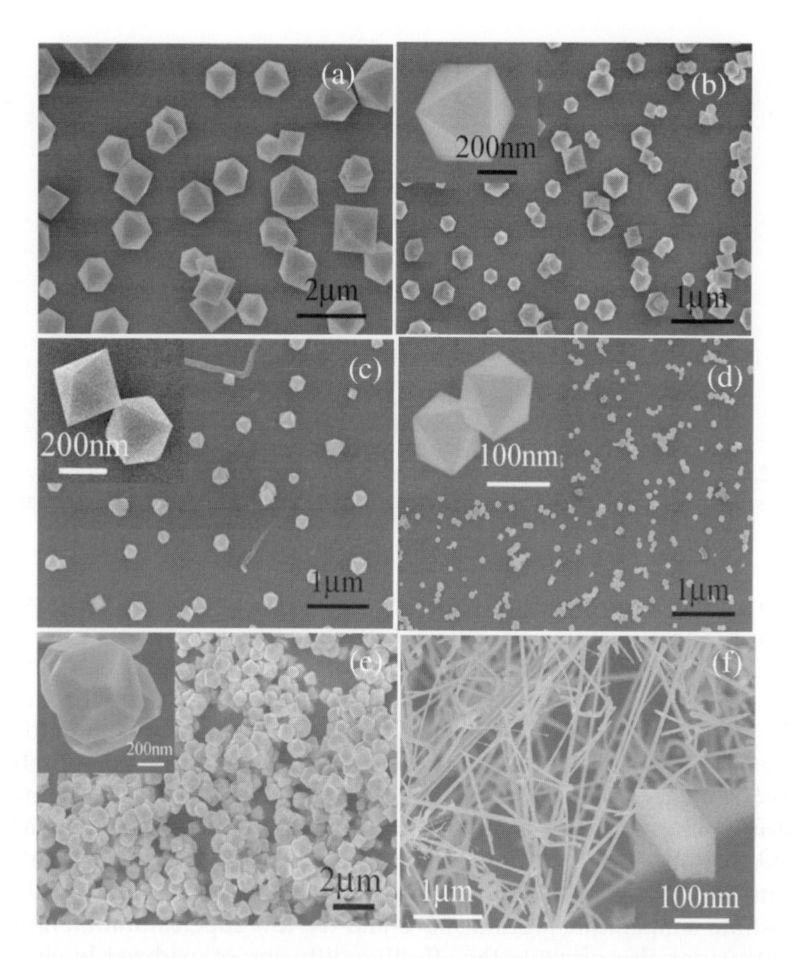

Fig. 2.7. SEM images of In_2O_3 octahedrons and nanowires: (a)–(d) typical In_2O_3 octahedrons with reaction temperatures of 950°C, 1050°C, 1200°C, and 1300°C respectively. Reaction durations are all 40 minutes, (e) In_2O_3 octahedrons with reaction temperature of 1200°C and reaction time of 2 hours, (f) In_2O_3 nanowires with In_2O_3 powders were heated to at least 1350°C. Inset shows the rectangular-like cross-section of one nanowire.[81]

is relatively difficult to obtain In_2O_3 nanowires through directly heating In because most vapor contributes to the nucleation and growth of In_2O_3 octahedrons in the vicinity of central heating zones. When a relatively larger quantity of In precursor is employed, some nanowires may grow at the downstream because some oxidized In does not feed into the growth of octahedrons completely.

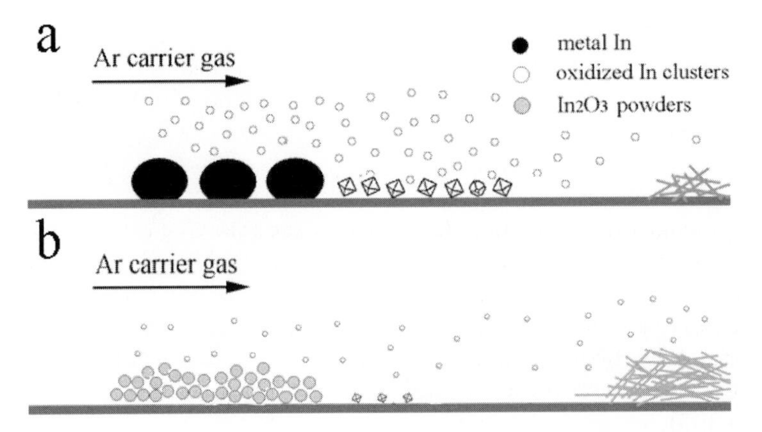

Fig. 2.8. Possible growth processes of octahedrons and nanowires obtained by specific source materials: (a) metal In as source material, and (b) In_2O_3 powders as source material.[81]

For the precursor of In_2O_3 powders, however, there is little product when the temperatures are below 1350°C owing to its excessively low vapor pressure. The saturated vapor pressure of solid state In_2O_3 around 1350°C is also relatively low. In this case, most In_2O_3 vapor will be transported to the deposition zone by the carrier gas: the temperature difference between the central heating zone and the deposition zone is about 400–500°C. In_2O_3 nanowires are observed in this zone. This growth mechanism can be explained as follows: under such a remarkable temperature difference, the surface energy difference among various low index crystallographic planes becomes considerably obvious. On the other hand, the substrate temperature around 900°C and the low supersaturation ratio of In_2O_3 vapor also promote the effective diffusion of oxidized In clusters to high surface energy or high defect density planes, along the surface of formed In_2O_3 nanowires.[82,83] The coordination of the two processes promotes one-dimensional incubation and continuous growth. Therefore, this kind of In_2O_3 nanowires, an extremely anisotropic nanostructure, can be formed in the relatively low supersaturation ratio of In_2O_3 vapor. Occasionally, a little In_2O_3 vapor nucleates in the vicinity of the central heating zone, and then turns into octahedrons.

2.2.4. MgO

Magnesium oxide (MgO) is a typical wide band gap insulator showing high-secondary electron emission (SEE) yield and MgO nanostructures have drawn the special attention due to their unique capability to pin the magnetic flux lines within a high-temperature superconductor (HTSC).[61] The inertness of MgO towards YBa_2Cu_3Ox and $Bi_2Sr_2Ca_2Ox$ superconductors was established.[84] The incorporation of MgO whiskers or fibers into BSCCO-2212 leads to high-Tc composite fibers with textured microstructure.[85] MgO has been investigated as a substrate material for the epitaxial growth of thin films with desirable electrical or magnetic properties.[86–88] The use of MgO as a passivation layer in high-electron-mobility transistors, a substrate for carbon nanotube growth, and a thin film capacitor, is also known.[89] As a high melting point and low heat capacity material, MgO fibers are more appropriate for insulation applications than dense firebrick refractories. MgO has been widely used in protective cases for thermocouples and capillary tubes, high-temperature furnace linings and crucibles for the fusion of certain metals, such as aluminum, copper, and silver etc.

In the past years, various MgO nanostructures have been synthesized. For example, MgO fibers have been prepared using two sol-gel routes.[90] MgO fishbone or fern-like nanostructures have been generated by selective Co-catalyzed growth.[91] MgO nanowires were synthesized by oxide-assisted catalytic growth and a vapor-phase precursor method using MgB_2 powders.[92,93] MgO nanobelts were fabricated by a simple CVD process using $MgCl_3$ as a starting material and using infrared irradiation, the evaporation of Mg_3N_2 precursor, and heating Mg in Ar/O_2 flow.[94–97] Several MgO morphologies including linear nanobelts and two- and three-dimensional entities have been produced by a thermal treatment.[98] MgO and Ga-filled MgO nanotubes which can serve as a unique wide-temperature range nanothermometer owing to a linear thermal expansion of the embedded liquid gallium fillings have also been synthesized by a simple CVD process.[99,100] Here, we describe the controlled synthesis of several types of well-defined MgO nanostructures, including orthogonally-branched nanostructures, nanocubes or microcubes, and straight nanowires, which were achieved simply by heating magnesium (Mg) powders with different oxygen partial pressures.[101] We have also investigated the nucleation and growth process of the novel flower-like MgO nanostructures by a simple chemical route.[102]

2.2.4.1. *Controlled growth of MgO nanostructures*

Controlled growth of MgO nanostructures was carried out in a high-temperature horizontal tubular furnace. The Mg powders (99.99 wt. %, 1 g) in a ceramic boat were placed in the central heating zone of the alumina tube. During the experiment, the reaction temperature and time were about 950°C and 10 minutes. High purity argon (Ar) mixed with different volumes of high purity oxygen (oxygen quantity ratio: 1–10%) was maintained at a flow rate of 50 sccm. Some wool-like or powder-like products were collected for different oxygen partial pressures. Figure 2.9 shows the FESEM images of the synthesized well-defined MgO nanostructures, i.e. orthogonally branched nanostructures (Figs. 2.9(a) and 2.9(b)), nanocubes or microcubes (Fig. 2.9(c)), and straight nanowires (Fig. 2.9(d)) under different oxygen volumes of 10%, 4–5%, and 1–2%, mixed in Ar carrier gas, respectively. It is visible from the FESEM images that the formation of different types of MgO nanostructures can be controlled by altering the

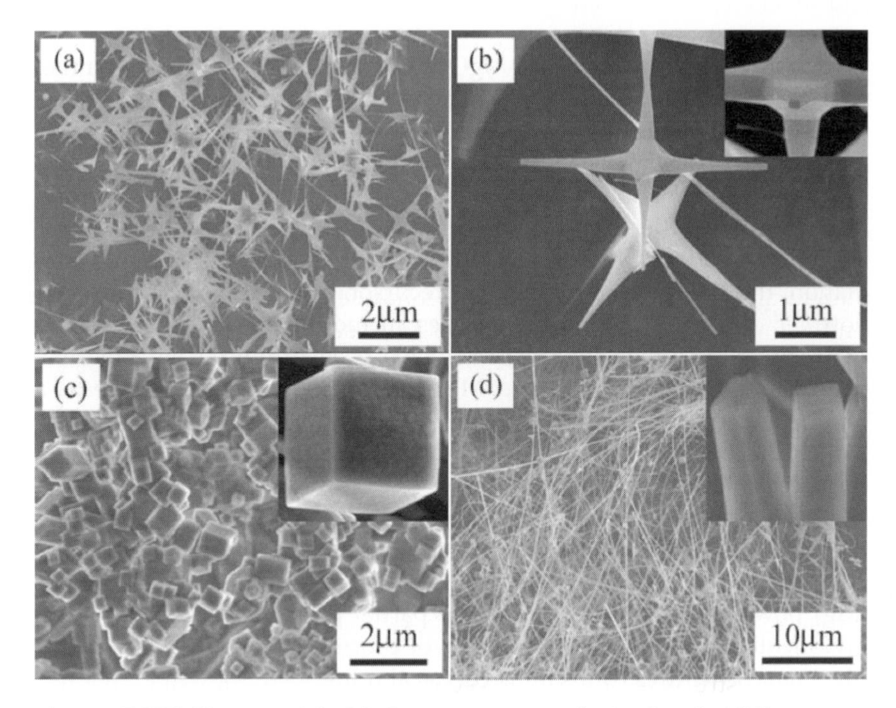

Fig. 2.9. FESEM images of the MgO nanostructures obtained under different oxygen partial pressures. (a), (b) Low- and high-magnification images of branched nanostructures. Inset shows the rectangle-shaped end of one branch, (c) MgO nanocubes or microcubes, and inset shows an enlarged image of a typical cube, (d) MgO nanowires, and inset shows a closed-up view near the ends of the rectangle-shaped MgO nanowires.[101]

oxygen partial pressure during the synthesis. Further observation reveals that more than 90% of the structures in the final products are the desired nanostructures in each case. SAED and HRTEM analysis reveal that all the synthesized MgO nanostructures are cubic single-crystalline enclosed by low-index [100] facets.[101]

Through a mass of experiments, we found that different supersaturation ratios, relatively high substrate temperatures, and surface defects in certain crystallographic planes cooperatively play important effects on determining the product morphologies. During the high temperature vapor reaction, Mg vapor was sufficiently oxidized to MgO clusters by oxygen. Therefore, supersaturation ratios can be adjusted through changing oxygen partial pressures in the Ar carrier gas. During the reaction, MgO clusters are transported onto the Si substrate in the low temperature zone (about 850–900°C), then are deposited and nucleated. Considering the surface energy minimization, the MgO nuclei may be close to the cubic shape. Subsequently, formed MgO clusters transported by the carrier gas will be adsorbed on the surfaces of nuclei. Three activities about these adsorbed MgO clusters may occur[103]: (i) Desorption and evaporation directly or after migrating along the surface; (ii) Incorporation into the crystalline lattice at initial adsorption sites; (iii) Surface diffusion and migration to high energy or high defect density sites and incorporation into the crystalline lattice there. The probability among these three activities is affected by intrinsic structural characteristics and external controlled conditions. Brief schemes on the formation of the three types of MgO nanostructures are shown in Fig. 2.10.

2.2.4.2. *Direct observation of the growth process of MgO nanoflowers*

Direct observation of the growth process of nanostructures is very important for the comprehension of the growth mechanism, and the design and application of nanostructures. By changing the growth parameters such as the reaction time, the nucleation and growth process of novel flower-like MgO, nanostructures were observed directly, and the growth model was established based on the experimental results. We used the high purity (99.9%) magnesium powders and distilled water as starting source materials. A series of growth experiments were performed with different heating time durations of 5, 10, 20, and 30 minutes, with other parameters essentially the same. For example, the reaction temperature was kept at 950°C and the Ar flow rate was kept at 100 sccm.[102]

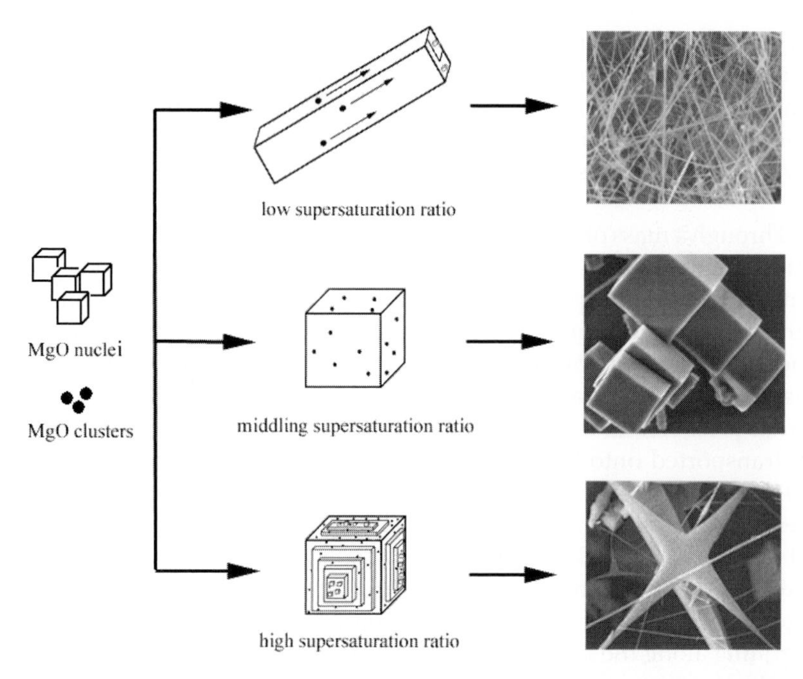

Fig. 2.10. Schematic illustration of three MgO nanostructures formed under corresponding controlled conditions. The small black dots represent adsorbed MgO clusters.[101]

An X-ray diffraction (XRD) spectrum (Fig. 2.11) taken from the as-synthesized products after the different total heating time, demonstrates that the synthesized products vary from pure Mg to the combination of Mg and MgO, and finally to pure MgO, with the increasing of the total heating time. Figure 2.12 shows the nucleation and growth process of novel flower-like MgO nanostructures which composed of four stages. Initially, Mg particles were formed on the Si substrate. This was followed by the formation of clusters of MgO as nucleation centers on the magnesium melt surface and the nucleation of short MgO nanofibers, then the growth of the MgO nanofibers, and MgO nanoflowers finally formed. A growth model is proposed for the MgO nanoflowers grown on Si wafers, as is shown in Fig. 2.13. The first step is that Mg powders were evaporated in a higher temperature zone, and directly deposited on the Si substrate in a lower temperature region of about 850°C (Fig. 2.12(a)). The second step is the formation of the clusters of MgO as nucleation centers on the melt magnesium surface (Fig. 2.12(b)) through the following reaction[104]:

$$Mg + H_2O \rightarrow MgO + H_2. \quad (2.4)$$

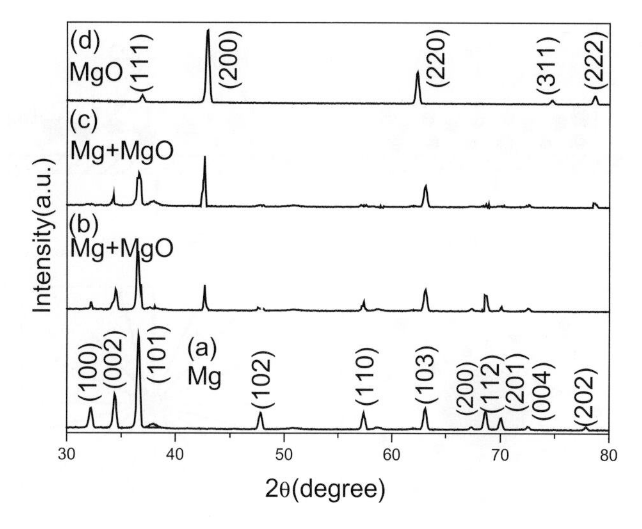

Fig. 2.11. XRD patterns of the as-synthesized products after the different total heating time. (a) 5 minutes, (b) 10 minutes, (c) 20 minutes, (d) 30 minutes.[102]

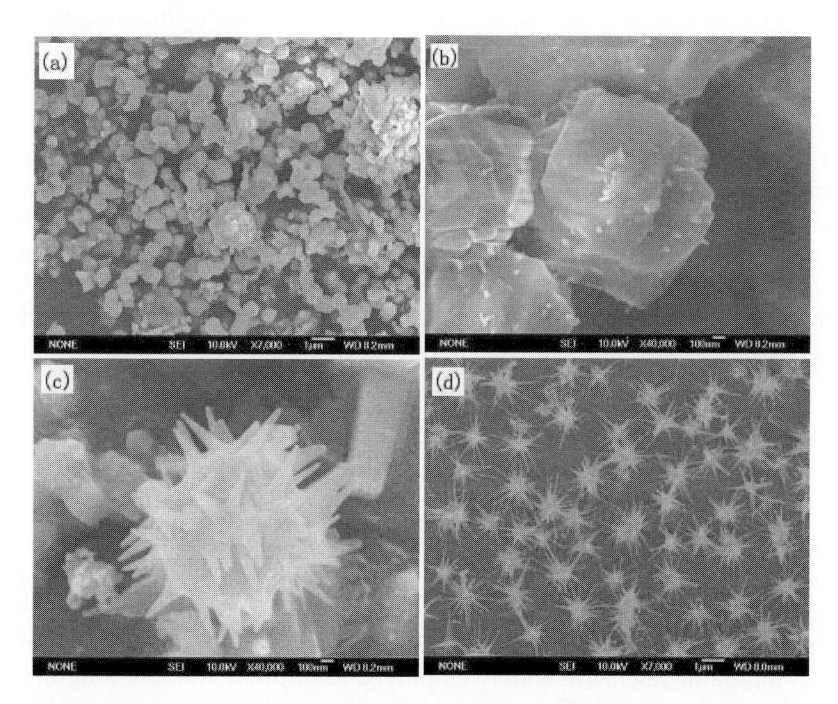

Fig. 2.12. FESEM images of the MgO nanoflowers obtained under different total heating time, which shows the nucleation and growth process of novel MgO nanoflowers.[102]

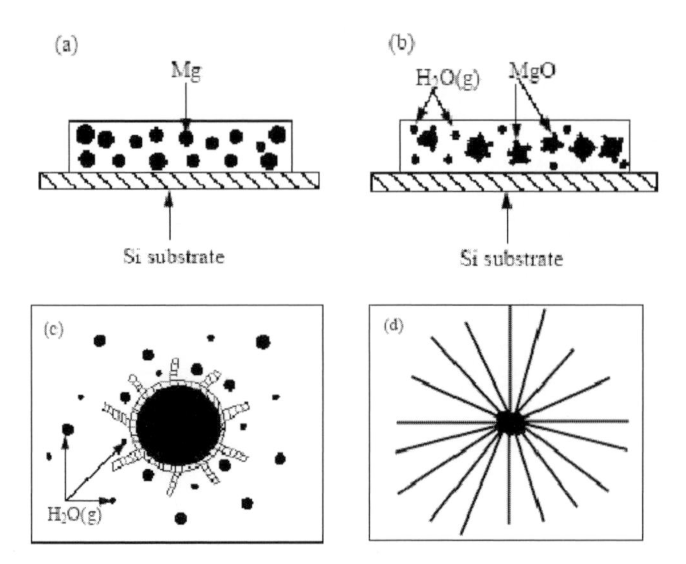

Fig. 2.13. Proposed model for the growth of novel MgO nanoflowers including four stages: (a) Mg particles formed on the Si substrate, (b) the formation of clusters of MgO as nucleation centers on the magnesium melt surface, (c) the nucleation of short MgO nanofibers, and (d) the formation of the MgO nanoflowers.[102]

The third step is the nucleation of short MgO nanofibers and the growth of the MgO nanofibers (Fig. 2.12(c)). When MgO on the magnesium melt surface became supersaturated, it had to precipitate out of the surface in a certain form. This is the driving force for nanofibers growth. The last step is the formation of the MgO nanoflowers, as shown in Fig. 2.12(d). MgO nanofibers will be formed radially through reaction (2.4). On the basis of the growth processes of MgO nanoflowers and experimental observations, the growth of MgO nanoflowers in this work via a VS process is possible.

2.2.5. Al_2O_3

Owing to their brittleness, ceramics have been regarded as materials of modest performance, especially under tension or bending conditions. In contrast to metals or polymers, however, the thermal stability of ceramics above 700°C makes them suitable materials for high-temperature applications. The microstructural design is critical in order to obtain reliable ceramic materials. Such materials are usually called "advanced ceramics". Inorganic fibers are widely employed to improve fracture toughness in advanced ceramics. The use of Al_2O_3 fibers, whiskers, and nanostructures as strengtheners in high-temperature composites is of great interest owing to their high elastic modulus and their thermal and chemical stability.[105,106] Recently, aluminum oxide (Al_2O_3) nanostructures, such as nanotubes, nanorods, nanowires, and nanobelts have especially attracted much attention because of their special properties of high-elastic modulus, thermal and chemical stability, optical characteristics, high dielectric constant, very low permeability, and high thermal conductivity etc. Al_2O_3 nanostructures are most commonly used as adsorbents, catalysts, catalyst supports, and in potential applications especially in high-temperature composite materials and nanodevices. Several methods for preparing these Al_2O_3 nanostructures have been developed, including the coating of CNTs with aluminium isopropoxide by the sol-gel method,[107] electrochemical anodizing of Al films or etching porous alumina membranes,[108,111] heating of partially hydrolyzed $AlCl_3$ powders,[112] coating and filling of multi-walled carbon nanotubes (MWNTs) with atomic-layer deposition (ALD),[113] catalyst-assisted vapor–liquid–solid deposition,[114–117] catalyst-free vapor–solid deposition,[118] chemical etching methods,[119] and the hydrothermal method[120] etc.

Here, we describe the controlled synthesis of Al_2O_3 nanowires and nanobelts by thermal evaporation of Al pieces and SiO_2 nanoparticles at 1200°C and 1150°C, respectively.[121] In addition, α-Al_2O_3 nanobelts and nanosheets with different morphologies and sizes have been successfully synthesized by a simple chemical route from H_2O and Al in an argon atmosphere at different substrate temperatures.[122]

α-Al_2O_3 nanowires and nanobelts were synthesized by using silica nanoparticles and Al pieces without any other catalyst at 1200°C and 1150°C, respectively. SEM observations reveal that the as-synthesized nanowires with typical diameters in the range of 20–70 nm, and lengths from 15–25 μm, and the nanobelts with the width of 0.1–1 μm, thickness of 10–50 nm, and length of several to tens of micrometers (Figs. 2.14(a) and 2.14(b)). HRTEM and SAED observations reveal that the as-synthesized nanowires and nanobelts are single-crystalline (Figs. 2.14(e) and 2.14(f)).

The vapor–liquid–solid (VLS) and vapor–solid (VS) crystal growth mechanism have been widely used for the growth of nanowires and nanobelts.[123,124] To understand the growth mechanism of these Al_2O_3 nanowires and nanobelts, the following reactions may be involved:

$$4Al(g) + SiO_2(s) \rightarrow 2Al_2O(g) + Si(s, l), \tag{2.5}$$

$$2Al(g) + SiO_2 \rightarrow Al_2O(g) + SiO(g), \tag{2.6}$$

$$Al_2O(g) + 2SiO(g) \rightarrow Al_2O_3(s) + Si(l). \tag{2.7}$$

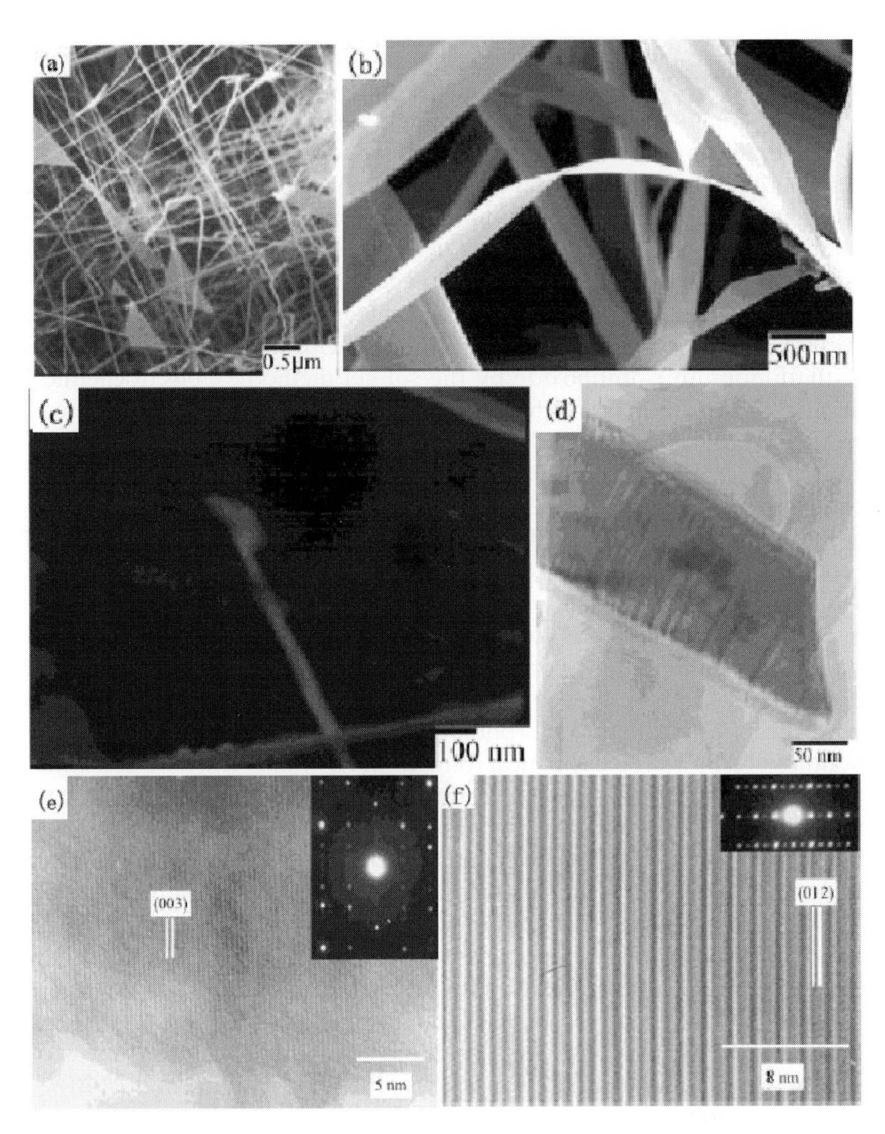

Fig. 2.14. (a) and (b) SEM images showing that the as-synthesized products are wire-like and belt-like structures, (c) and (d) high-magnified SEM image and a typical TEM image further discovered the wire-like and belt-like structures, (e) and (f) HRTEM images of a single Al_2O_3 nanowire and nanobelt and the SAED pattern (the inset in it) reveals the single-crystalline structure.[121]

For nanowires, it is noted that there exists a nanoparticle at the end of the nanowire (Fig. 2.14(c)), which represents strong evidence for a growth process dominated by a VLS mechanism.[123] In the Si-SiO system, a eutectic alloy/mixture can be generated.[125] The compositions needed for the liquid can appear and the gases (Al(*g*), Al_2O(*g*), and SiO(*g*)) can reach and come into the liquid. When the Al_2O_3 is formed and becomes supersaturated in the liquid Si-SiO, Al_2O_3 nanowires will be produced. As for the nanobelts predominately found at the particular locations where the Al pieces were in contact with the SiO_2 nanoparticle bed, it is likely that special features at this interface promote nanobelt formation. At this interface, the yield ratio of vapors should be higher than that for the formation of nanowires, resulting in the formation of nanobelts. In addition, for the nanobelts without nanoparticles at their ends, it is difficult to elucidate how VLS deposition can form a crystal much wider than the drop diameter, and why drops do not appear consistently. Therefore, the growth of Al_2O_3 nanobelts in this work via a VS process is possible. This is similar to previous results reported for the Al_2O_3 micro-ribbon.[106]

Temperature-controlled growth of α-Al_2O_3 nanobelts and nanosheets was achieved by a simple chemical route from H_2O and ultrahigh purity (99.999%) aluminum (Al) in an Ar atmosphere (50 sccm), with the system was rapidly heated to 1350°C in 12 minutes and kept at this temperature for 2 h. A long alumina plate (6 cm in length and 16 mm in width) was placed to act as the deposition substrate and the substrate temperature was from 1100 to 1300°C. After the system was cooled down to room temperature, four distinct zones of white, wool-like products were formed along the whole alumina plate. The as-synthesized products' thicknesses decreased from about 2 mm to 50 μm gradually with an increase in temperature from 1100 to 1300°C. Figure 2.15 shows the SEM images of the as-synthesized Al_2O_3 nanostructures which formed in four distinctive temperature zones: zone I (1100–1150°C), zone II (1150–1200°C), zone III (1200–1250°C), and zone IV (1250–1300°C). The results reveal that zones I and II consisted of α-Al_2O_3 nanobelts with the typical widths ranging from 10 nm to 500 nm, however, zones III and IV consisted of α-Al_2O_3 nanosheets with the typical widths varying from 1 μm–10 μm.

The morphologies of the Al_2O_3 nanobelts and nanosheets are determined by a combination of growth kinetics and thermodynamics. Thermodynamics requires the formation of crystals with well-faceted surface

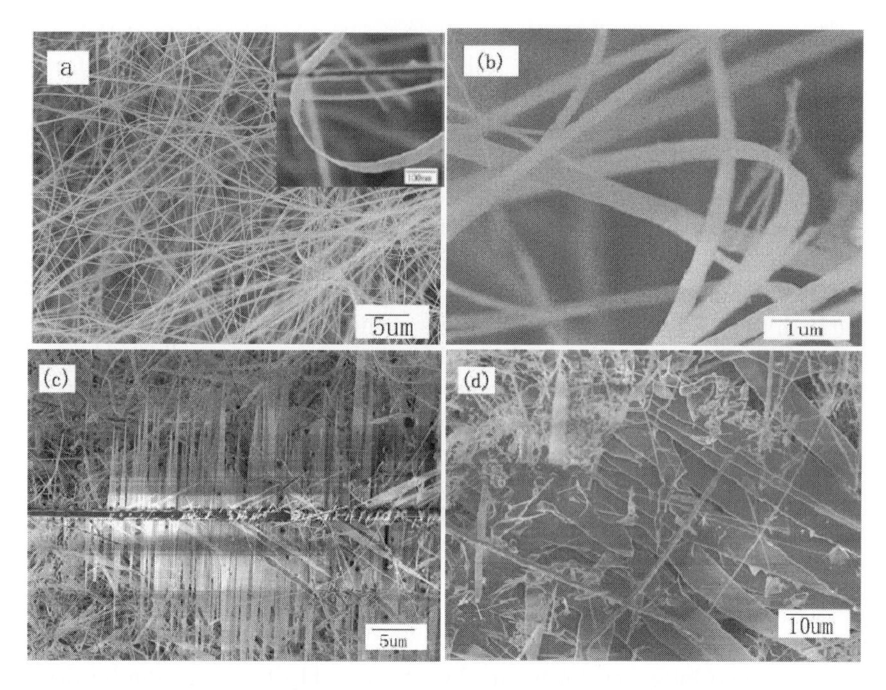

Fig. 2.15. SEM images of Al_2O_3 nanobelts and nanosheets formed in four distinctive temperature zones: (a) zone I (1100–1150°C), (b) zone II (1150–1200°C), (c) zone III (1200–1250°C), and (d) zone IV (1250–1300°C).[122]

planes with low surface energy which is strengthened by kinetics where the growth rate is higher for high-energy surfaces and lower for low energy ones, and only surfaces with low energy could survive in the final crystals. However, for the growth of low-dimensional nanostructures such as Al_2O_3 nanobelts and nanosheets, the difference in the growth rate for varied planes could be enhanced and a real thermodynamic equilibrium crystal shape would not be reached. It implies that the end surfaces of the nanobelts may have a relatively higher surface energy than the top surfaces and the side surfaces, which may lead to a fastest growth rate normal to the end surfaces, medium to the side ones, and slowest to the top ones of the nanobelts, forming a belt-like structure. The low-energy surface defines the planes that can be present in the final product from the thermodynamic point of view, while the growth kinetic determinates the growth direction and the morphology of the product.

2.3 Sulfides nanowires and nanobelts

2.3.1. ZnS

During the last decade, a considerable effort has been spent on the preparation and investigation of the group of wide band gap II–VI nanoscale semiconductors due to their vital optoelectronic applications for laser light-emitting diodes and optical devices based on electronic and optical properties. Zinc sulfide (ZnS), an important semiconductor compound of II-VI group, has a wide band gap energy of 3.7 eV at 300 K. It has been used in electroluminescence,[126] nonlinear optical devices,[127] sensors and lasers,[128] and as an LED when doped.[129] ZnS is also used as catalysts for the photo-oxidation and photo-reduction of organic groups.[130] There is already a lot of progress in the synthesis and characterization of ZnS nanostructures, including nanowires,[131–133] nanobelts or nanoribbons,[134–138] nanosheets,[139–141] nanocables,[142,143] nanotubes,[144] and nanobelt networks[145] using various methods. Recently, the large-scale production of wurtzite ZnS nanowires and nanoribbons has been achieved by thermal evaporation of ZnS powder on Si substrate in the presence of Au catalyst. Morphologies of the ZnS nanostructures were controlled by varying the temperature, as well as the flow rate of the Ar gas.[146]

In this section, ZnS nanostructures with different morphologies, sizes, and microstructures were synthesized by the evaporation of ZnS nanopowders simultaneously in bulk quantities.[147]

Several ZnS nanostructures, such as nanorods, nanowires, nanobelts, and nanosheets, simultaneously in bulk quantities were first reported by Fang. using the evaporation of ZnS nanopowders. Self-made ZnS nanopowders were placed in an alumina boat and then put into the center of the alumina tube. One Si wafer (25 mm in length and 15 mm in width) was placed downstream in the alumina tube 8 mm away from the source material, and the other Si wafer (50 mm in length and 15 mm in width), which was deposited with a layer of Au (about 40 nm) film using a vacuum thermal evaporator, was placed subsequently to act as the deposition substrate for the material growth. The system was heated to 1100°C for 30 minutes under a constant flow of Ar (80 sccm). Before the experiment started, a movable thermocouple was inserted into the tube to measure the temperature distribution along the tube.[147]

On the basis of the SEM images and the curve diagram of temperature and deposition distance, four distinctive temperature zones were identified. From 850 to 1050°C, the morphologies of the ZnS nanostructures changed from nanorods (zone I, 850–900°C) to nanowires (zone II, 900–950°C), then nanobelts (zone III, 950–1000°C), and finally nanosheets (zone IV, 1000–1050°C).

The detailed structural characterizations were performed. The SEM, TEM, HRTEM, SAED, and energy dispersive X-ray spectrum EDX results of the as-synthesized ZnS nanostructures formed from zone I to zone IV, are shown in Figs. 2.16 to 2.19. On the Si wafer deposited with a layer of Au film, namely in zone I and zone II, the morphologies of the products changed from rod-like to wire-like with the growth temperature increasing, and the diameter and length of ZnS nanostructures increased gradually. ZnS nanobelts and nanosheets were found in zones III and IV, and the substrate Si wafer did not deposit other matter. The characteristic results reveal that the nanorods are large-scale and have diameters ranging from 50 to 100 nm and lengths of up to tens of micrometers, the nanowires have typical diameters in the range of 100–500 nm and lengths of up to several tens of micrometers, the nanobelts have widths ranging from 100–500 nm and width-to-thickness ratios of 10–20, and the nanosheets have widths of about 1–10 μm across and thicknesses of several tens of nanometers. It is noted that a few nanoparticles were observed at the top of the nanorods and nanowires, but not for nanobelts and nanosheets. HRTEM

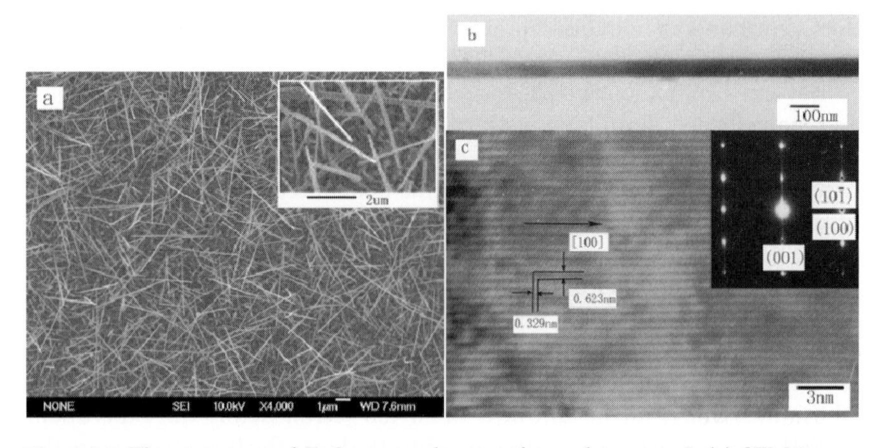

Fig. 2.16. The structure of ZnS nanorods were formed in zone I: (a) SEM image of the as-synthesized ZnS nanorods. The inset shows the high-magnification SEM image of the ZnS nanorods showing a few nanoparticles at the top of the nanorods, (b) and (c) the representative TEM, HRTEM, and SAED images of a single ZnS nanorod.[147]

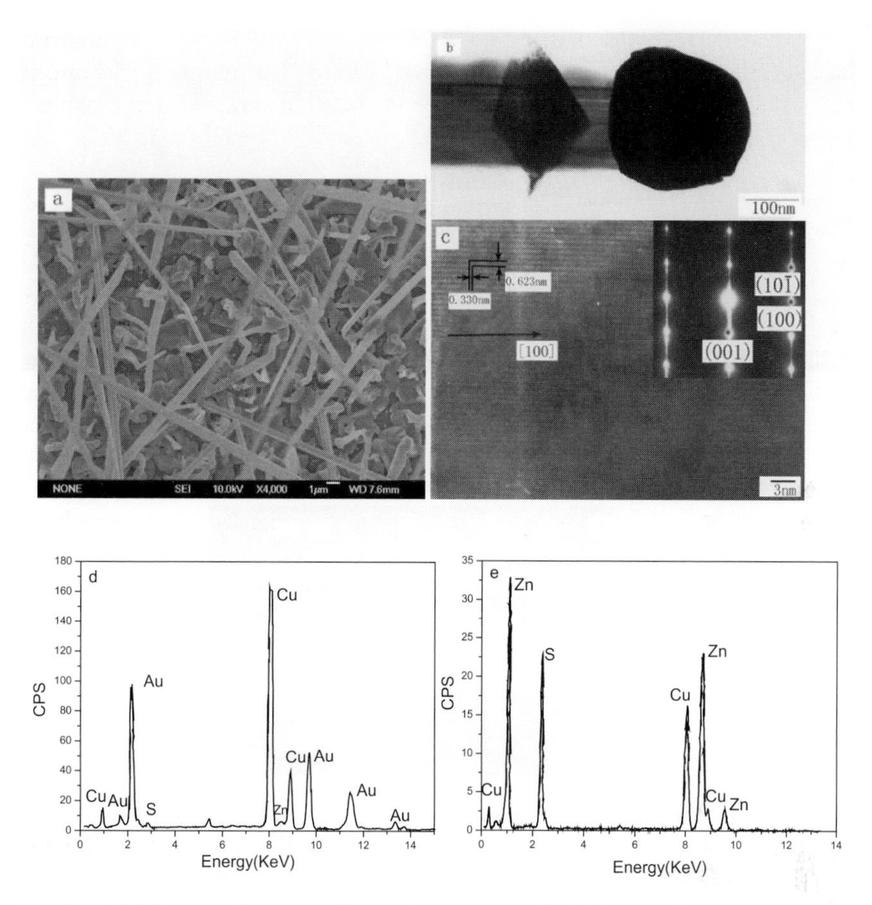

Fig. 2.17. ZnS nanowires were formed in zone II: (a)–(c) SEM, TEM, HRTEM, and SAED images of the as-synthesized ZnS nanowires, (d) and (e) EDX spectra taken from the tip area and stem of the nanowire shown in Fig. 2.3(b). The Cu signal came from the TEM grid used for supporting the sample.[147]

observations reveal that all of the ZnS nanostructures are single-crystalline, and most of the nanorods and nanowires grew along [100] whereas most of the nanobelts and nanosheets grew along [001].

The above results show that ZnS nanostructures with different morphologies, sizes, and microstructures were formed over a narrow temperature range between 850 and 1050°C. All results suggest that the temperature distribution inside the tube furnace and catalyst play a dominant role in the formation of the ZnS nanostructures. Within a certain temperature range, products with a specific morphology can be obtained, therefore it

Fig. 2.18. ZnS nanobelts were formed in zone III: Low-magnification (a) and magnified (b) SEM images of the as-synthesized large-scale ZnS nanobelts, (c) and (d) TEM, HRTEM, and SAED images of an individual ZnS nanobelt.[147]

may be possible to obtain ZnS nanostructures with a specific morphology by controlling the reaction temperature and catalyst.

Regarding the growth of ZnS nanorods and nanowires, we think that the ZnS vapor is rapidly generated at relatively high temperatures by the evaporation of ZnS nanopowders, transported to and reacted with the Au liquid to form alloy droplets. As ZnS in the droplets become supersaturated, ZnS nanorods or nanowires will be formed via a vapor–liquid–solid (VLS) growth process. However, the larger alloy droplets tend to form on the substrate of higher temperature. Therefore, the larger diameters and lengths of ZnS nanostructures were formed in higher temperatures which is similar to the reports of Lee.[148] and Wang.[149] which mentioned that the

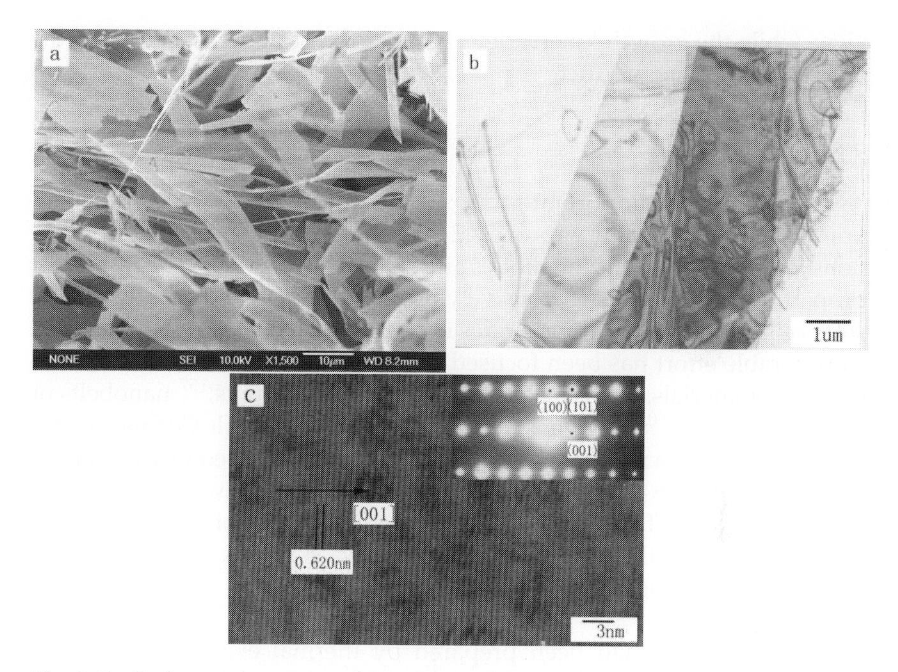

Fig. 2.19. ZnS nanosheets were formed in zone IV: (a)–(c) SEM, TEM, HRTEM, and SAED images of the as-synthesized ZnS nanosheets.[147]

variation of the diameter was due to the variation of the diameter of the droplet nucleated at different temperatures.

The growth of ZnS nanobelts and nanosheets may be a vapor–solid (VS) growth process since no particles can be seen at the tips of the ZnS nanobelts and nanosheets. This is in agreement with the experimental detail when no catalyst was being used in zone III and zone IV. Because catalyst-assisted techniques and template-confined synthesis are not used, ZnS vapor can be solidified again on the precursor's surface when ZnS nanopowders are evaporated into vapors during the heating. The growth mechanism of ZnS nanobelts and nanosheets is similar to that of SnO_2 nanoribbons. Wang demonstrated that the end surfaces of the nanoribbons may have a relatively higher surface energy than the wide surfaces and the side surfaces, which may lead to a faster growth rate normal to the end surfaces of the nanoribbons, forming a ribbon structure.[53] The morphologies of the nanobelts and nanosheets may be determined by a combination of growth kinetics and thermodynamics. The low-energy surface defines the planes that can be present in the final product from the thermodynamic point of view, while the growth kinetic determinates the growth direction and the morphology of the product.

2.3.2. CdS

As one of the most important group II–VI semiconductors, cadmium sulfide (CdS) with a band gap of 2.42 eV has vital optoelectronic applications for laser light-emitting diodes and optical devices based on nonlinear properties.[150] CdS is also extensively used for photoelectric conversion in solar cells and light-emitting diodes in flat panel displays.[151–153] Recently, considerable effort has been focused on the synthesis of one-dimensional CdS nanomaterials, such as nanowires,[154–156] nanoarrays,[157] nanobelts or nanoribbons,[158–160] and nanotubes.[161–163] Here, large-scale CdS nanowires were achieved by a new, simple, and low cost process based on the thermal evaporation of CdS powders under controlled conditions with the presence of Au catalyst.[164] The nucleation and growth process of CdS nanowires were observed for the first time in a typical vapor–solid synthetic route by thermal evaporating of CdS nanosized powders.[165]

CdS nanowires had been prepared by thermal evaporation of CdS powders (99.99%) under controlled conditions in which Au film (about 40 nm) coated on the Si substrate was used as catalyst. Prior to heating, the system was flushed with high-purity Ar for 1 h to eliminate O_2. Next, under a constant flow of Ar (100 sccm), the furnace was rapidly heated to 800°C and kept for 2 h. It was observed that yellow sponge-like products had appeared on the surface of the Si substrate.

The typical SEM image of CdS nanowires prepared using this method is shown in Fig. 2.20(a) with typical lengths of the nanowires in the range of several micrometers to several tens of micrometers. XRD measurement (Fig. 2.20(b)) shows that the products are wurtzite (hexagonal) structured CdS, which is consistent with the standard values for bulk CdS.

The detailed structure of individual CdS nanowire is characterized by TEM. Figure 2.21(a) is the typical TEM image of CdS nanowires which shows that the nanowire is straight with uniform diameters and is typically terminated with a nanoparticle at its end. The nanowire has a high aspect ratio, and its length is about several micrometers. The nanoparticle at the tip of the nanowire generally appears darker and has high contract compared with the nanowire stem. EDX measurements made on the nanoparticle and the nanowire stem indicate that the nanoparticle is composed of Au, Cd, and S (shown in Fig. 2.21(b)) and that the nanowire stem is only

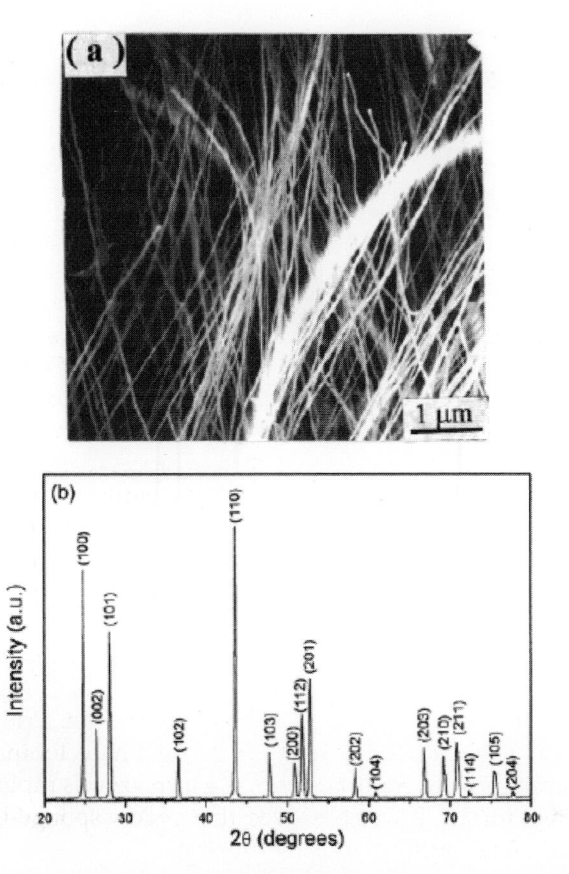

Fig. 2.20. (a) Typical SEM image of bulk CdS nanowires. Large quantities of the nanowires were distributed homogeneously on the Si substrate, (b) XRD pattern taken on bulk CdS nanowires. The numbers above the peaks correspond to the *(hkl)* values of the wurtzite CdS structure.[164]

composed of Cd and S (shown in Fig. 2.21(c)). EDX analysis of the nanowire demonstrates that a 1 : 1 Cd/S composition within experimental error is consistent with stoichiometric CdS. The SAED pattern and HRTEM image (Fig. 2.21(d)) reveal that the CdS nanowires are structurally uniform and single-crystalline, indicating that the nanowire growth occurs along the [131] direction.

For the research of the growth process of CdS nanowires, CdS nanowires were synthesized by thermal evaporation of self-made CdS nanopowders.[165] A series of growth experiments were performed with

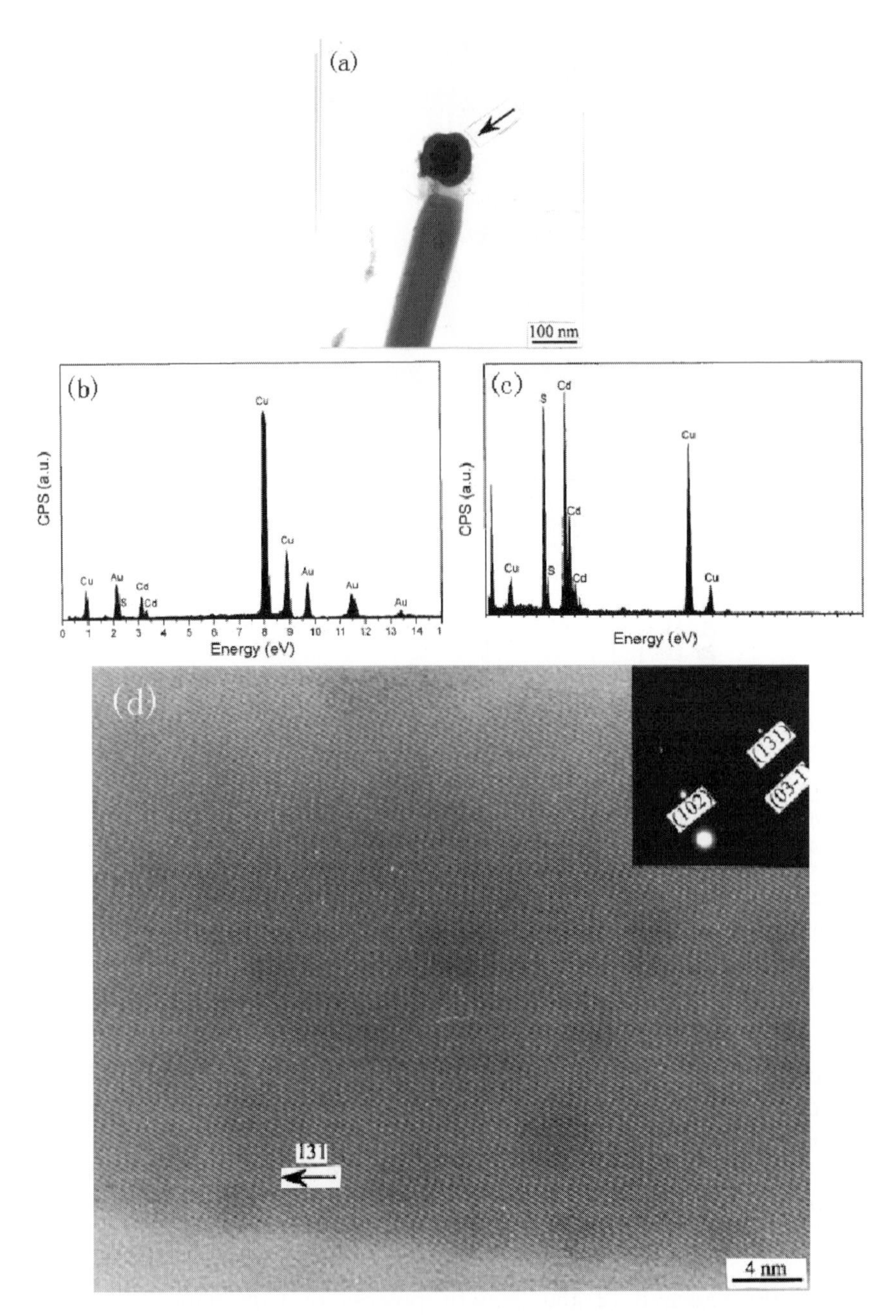

Fig. 2.21. (a) Typical TEM image of a single CdS nanowire with an Au/CdS alloy cluster at its tip, (b) and (c) EDX spectra taken from the nanoparticle (b) and nanowire stem (c), respectively. (d) HRTEM image of CdS nanowire showing that the nanowire is single-crystalline and free from dislocation and defects.[164]

different heating time durations of 15, 30, 45, 60, and 90 minutes and with other parameters, essentially the same. The reaction temperature was about 900°C and the flow rate of Ar was about 80 sccm under a pressure of 250 Torr.

Figures 2.22 and 2.23 show the series SEM images of CdS nanowires synthesized with different heating time durations of 15, 30, 45, 60, and 90 minutes. The results show that three steps are involved in the growth of CdS nanowires: the growth of CdS spheroids, the nucleation of short CdS nanorods, and the growth of long CdS nanowires. The product development was strongly related to the heating duration. With the extension of the heating process, CdS short nanorods nucleated from the cusps of matrix

Fig. 2.22. (a) Low-magnification and (b) high-magnification SEM images of the CdS amorphous particles produced in the first 15-minute heating process, (c) low-magnification and (d) high-magnification SEM images taken after the 30-minute heating process.[165]

Fig. 2.23. SEM images of CdS nanowires. Images were taken after a heating time of (a) 45 minutes, when the nanowires are typically several micrometers in length; (b) 60 minutes, when the nanowires are tens of micrometers in length; and (c) 90 minutes, when the nanowires are hundreds of micrometers in length. The thickness of the nanowires does not change significantly during the heating process.[165]

particles, which were then followed by the growth of long nanowires. Optimization of the heating duration is important in the nanowire synthesis. From this, we learn that the nucleation of the crystalline phase out of an amorphous matrix generally initiates from near the center of the matrix, where temperature can be retained near the melting point longer than at the surface of the particle. The cusps on the surface of the matrix particles go deep into the matrix, therefore, the nucleation of crystalline nanorods from the cusps inside the matrix particle is reasonable.

2.4 Doping of nanowires and nanobelts

Recently, doping of the nanostructures has become an important issue for more diverse range of applications, because its electrical, optical, and magnetic properties can be modified by doping.[167] For example, the modification of ZnO properties by impurity incorporation is currently another important issue. It leads to possible applications in ultraviolet optoelectronics and spin electronics. For example, S-doped ZnO nanostructures have been synthesized via a simple one-step catalyst-free thermal evaporation process on a large scale.[168] Optical property studies of Ga-, In-, and Sn-doped ZnO nanowires synthesized via thermal evaporation have emerged.[169] Here, we describe the doping of ZnO and ZnS nanostructures.

2.4.1. S-doped ZnO nanowires

S-doped ZnO nanowires were synthesized by vapor phase growth using ZnS nanoparticles as starting materials without the presence of a catalyst. The ZnS powder was put in the front of an alumina boat, which was covered with a quartz plate to maintain a higher ZnS vapor pressure. Prior to heating, the system was flushed with high-purity Ar for 1 h to eliminate O_2. Then, it was placed under a constant Ar flow of 100 sccm for 90 minutes, with the reaction temperature of about 900°C. Subsequently, oxygen was introduced into the quartz tube and kept for 10, 60, 90, 100, and 120 minutes under a constant flow of 15 sccm. The time of the oxygen stream flow is the key factor for the quantity of sulfur in ZnO nanowires. After the system was cooled down to room temperature, a large piece of light yellow, wool-like material was found on the inner wall downstream end of the quartz tube.[170]

Figure 2.24(a) shows an SEM image of the as-synthesized products when the time of oxygen flow is 10 minutes. The results show that uniform nanowires are formed with a high yield and with lengths of several tens of micrometers. The XRD pattern shown in Fig. 2.24(b) indicates that all relatively sharp diffraction peaks can be perfectly indexed to the high crystallinity of the hexagonal structure of ZnO and ZnS. Figures 2.25(a) and 2.25(b) give TEM images of several such nanowires, indicating that most nanowires have diameters of about 80 nm. Figures 2.25(c) to 2.25(f) show the results of the EDX measurements, which were made on a different individual nanowire corresponding to 60, 90, 100, and 120 minutes of oxidation time, respectively. The EDX pattern indicates that the as-synthesized nanowire is mainly composed of Zn and O, with a small amount of S, which

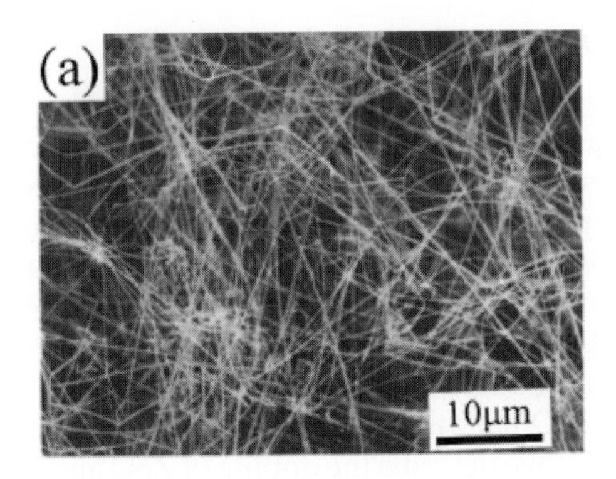

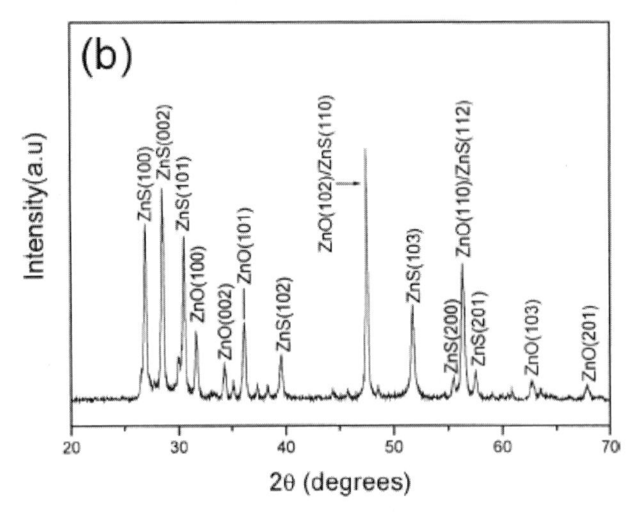

Fig. 2.24. (a) SEM image of the as-synthesized product when the time of oxygen flow is 10 minutes, (b) the XRD pattern of the as-synthesized product.[170]

is determined by the time of oxidation. When the time of oxygen stream lasted 120 minutes, the pure ZnO nanowires were obtained.

2.4.2. Ce-doped ZnO nanostructures

The Ce-doped ZnO nanostructure growth was conducted in a tube furnace by an *in situ* chemical vapor reaction method using amorphous Zn–Ce–C–O composite powders. The powders were obtained by the sol-gel method.[171,172] The source material was placed in a ceramic crucible covered with a ceramic lid, and put into a box furnace. The furnace was heated

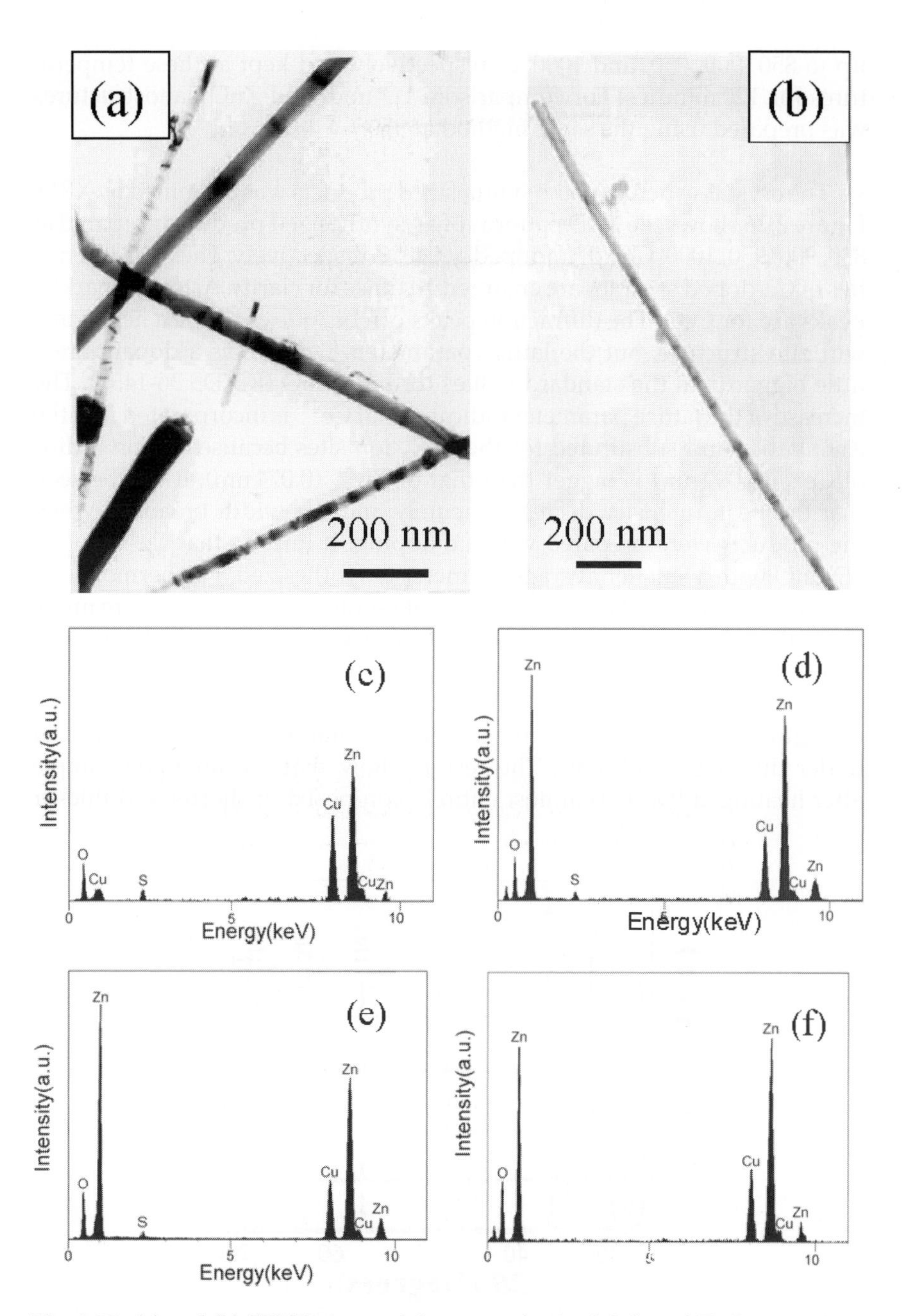

Fig. 2.25. (a) and (b) TEM images of the as-synthesized S-doped ZnO nanowires, (c)–(f) EDX spectra of individual nanowire correspond to 60, 90, 100, and 120 minutes of oxidation time, and the corresponding content of sulfur is 5.40%, 1.47%, 1.41%, and 0%, respectively.[170]

up to 850, 900, 950, and 1000°C, respectively and kept at these temperatures for 120 minutes. For comparison, 1D undoped-ZnO nanostructures was prepared using the same method at 950°C.

The crystal structure of the synthesized products was examined by XRD. Figure 2.26 shows the XRD patterns of as-synthesized products prepared at 850, 900, 950, 1000°C and, undoped at 950°C respectively. The XRD intensities of Ce-doped samples are enlarged 10 times for clarity. Asterisk-marked peaks are for CeO. The diffraction peaks can be indexed to be a hexagonal wurtzite structure, but the lattice parameters with Ce as a dopant are a little bigger than the standard values for bulk ZnO (JCPDS 36-1451). The increase of the lattice parameter indicates that Ce^{4+} is incorporated into the ZnO lattice and substituted for the Zn^{2+} ion sites because the ion radius of Ce^{4+} (0.092 nm) is bigger than that of Zn^{2+} (0.074 nm). It can be seen that the peak intensity decreases acutely and the width broadens when the products were prepared with Ce doping. It implies that Ce^{4+}-doped 1D ZnO with a smaller average diameter is synthesized. Furthermore, the intensities of the XRD peaks of the sample synthesized at 1000°C are much lower than those of others, and weak peaks of CeO appear, which indicates that the Ce ions isolates from the ZnO matrices.

Figure 2.27 shows SEM images of ZnO nanostructures synthesized under different conditions. The results show that the undoped sample after heating at 950°C is almost entirely composed of shorter and thicker

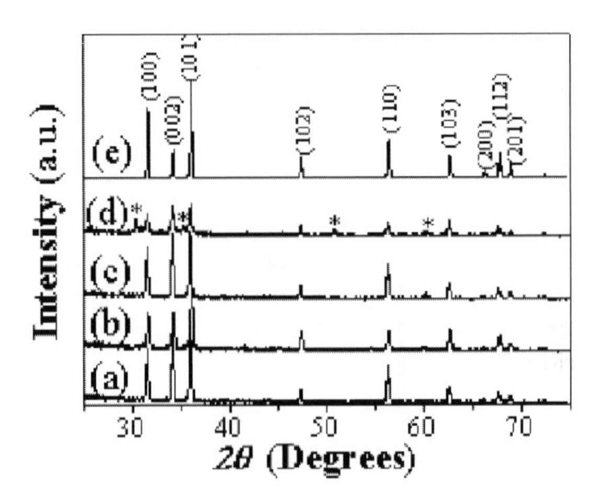

Fig. 2.26. XRD patterns of as-synthesized products prepared at 850, 900, 950, 1000°C and, undoped at 950°C respectively.[171]

Fig. 2.27. SEM images of ZnO nanostructures synthesized under different conditions. (a) Undoped sample synthesized at 950°C; (b), (c) and (d) were doped by Ce and prepared at 850, 950 and 1000°C.[171]

nanobelts (Fig. 2.27(a)). When the heating temperatures are below 950°C, no 1D ZnO is found in the undoped sample. Nanobelts are predominant in the doped-sample prepared at 850°C (Fig. 27(b)). The nanobelts vary in width from several tens to hundreds of nanometers, and the thickness from several to several tens of nanometers. However, the products prepared at 950 and 1000°C mainly present wire-like nanostructures. The length of nanobelts and nanowires both reach several hundreds of micrometers. These results demonstrate that it is possible to control the crystal size and morphology of 1D ZnO by annealing the amorphous precursor with Ce^{4+}-doping at various temperatures.

2.4.3. Sn-doped ZnO nanobelts

Sn-doped ZnO nanobelts were synthesized by thermal evaporation of 99.99% pure zinc powders under controlled conditions with a tin

(Sn)-coated silicon wafer which used as growth substrate placed downstream in the alumina. The system temperature was ramped to 700°C in 5 minutes and kept at this temperature for 30 minutes under different flowing atmospheres, such as air (no flow), dry and humid argon (Ar) flow, and argon/oxygen (O_2) gas mixture. During the heating process, the flow rate was kept at 100 sccm, water was introduced into the reaction by placing an alumina crucible with about 50 mL of distilled water in the upstream side of the alumina tube for humid Ar flow, and the ratio of the gases is 5 : 1 for the Ar/O_2 gas mixture.[173]

Here, we only show the representative results of the as-synthesized samples which were achieved under Ar flow. A typical XRD pattern is shown in Fig. 2.28(a). There is no diffraction signal originating from tin and its compounds in the XRD data and the diffraction peaks can be indexed to a hexagonal wurtzite-structured ZnO which means that the impurity

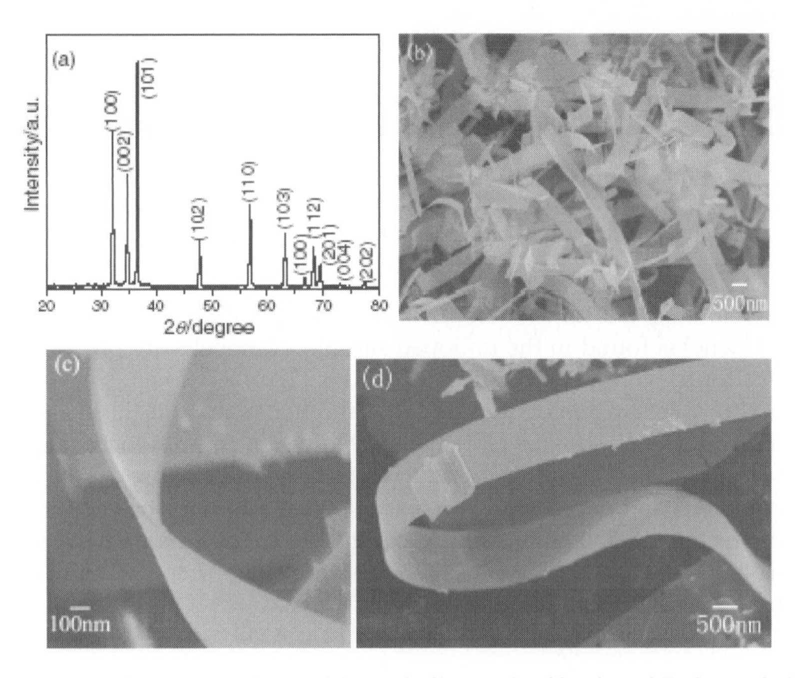

Fig. 2.28. (a) XRD pattern obtained from a bulk sample of Sn-doped ZnO nanobelts, (b) a low-magnification SEM image, (c) and (d) high-magnification SEM images of a Sn-doped ZnO nanobelt, revealing the thin and wide shape characteristics of the nanobelt.[173]

does not change the wurtzite structure of ZnO. Figures 2.28(b) to 2.28(d) are the typical SEM images of synthesized Sn-doped ZnO nanobelts. The low magnification image shows that uniform nanobelts are formed with high field. The high magnification images indicate that the nanobelt has the width of about 800–1500 nm, the thickness of about 20–60 nm, and has the length of up to several tens of microns.

The detailed structure of the individual nanobelt is characterized by TEM. Figure 2.29(a) shows an individual Sn-doped ZnO nanobelt with a width of about 1400 nm and the nanobelt has a uniform width along its entire length. The HRTEM image and its correponding fast Fourier transform (FFT) studies (Fig. 2.29(d)) show that the nanobelt has a hexagonal wurtize structure and the nanobelt grows along $[2\bar{1}\bar{1}0]$ (the *a*-axis), with its top/bottom surface $\pm$ [0001] and the side surfaces $\pm$ $[01\bar{1}0]$ (inset in Fig. 2.29(a)). Through the experiment, we find that most of the nanobelts are single-crystalline without dislocations and defects (Fig. 2.29(d)), and their geometrical shapes is uniform. However, planar defects are observed occasionally and dislocations are rarely seen. Figure 2.29(b) shows the TEM image and the corresponding HRTEM image (inset) of one Sn-doped ZnO nanobelt with a uniform distribution of planar defects along the entire nanobelt. EDX identifies that the composition of the nanobelts is Zn, O and Sn (Fig. 2.29(c)). The Sn content is about 5 atom%. The Cu peaks come from the TEM grid. Naturally, in the ZnO growth, the highest growth rate is along the *c*-axis and the large facets are usually $[01\bar{1}0]$ and $[2\bar{1}\bar{1}0]$.[10,174] Recently, Wang reported that the controlled synthesis of free-standing ZnO nanobelts whose surfaces are dominated by the large polar surfaces, and the nanobelts, grow along the *a*-axis by controlling growth kinetics.[175] Here, we find that it is also possible to change the growth behavior of ZnO nanobelts by doping. In addition, photoluminescence measurements show a blue shift in the near band edge (NBE) emission spectrum compared to a bulk single crystal of ZnO, and the intensity ratio of the NBE to the green emission of these Sn-doped ZnO nanobelts can be controlled by altering the flowing atmosphere.[173]

2.4.4. Mn-doped ZnS nanobelts

Mn-doped ZnS nanobelts have been prepared through a thermal evaporation method starting with a mixture of zinc acetylacetonate [$Zn(acac)_2$] and manganese acetylacetonate [$Mn(acac)_2$] taken in the required molar ratios corresponding to 0%, 1%, 3% and 5% doping. In the experiment, a constant flow of H_2S gas at 100 sccm was maintained. The system was

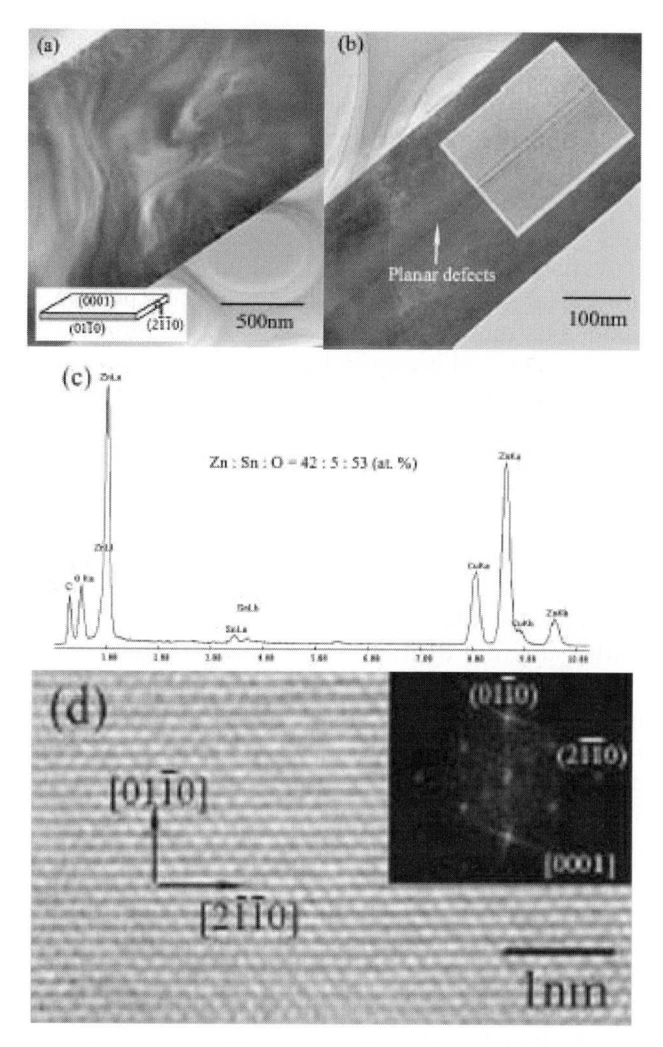

Fig. 2.29. (a) and (d) a low-magnification TEM image and the corresponding HRTEM image and its FFT recorded from a typical unselected Sn-doped ZnO nanobelt. The inset in (a) is the structure model of the Sn-doped ZnO nanobelts, (b) a uniform distribution of planar defects along the entire nanobelt is observed occasionally, (c) EDX data shows 5 atom% Sn.[173]

rapidly heated to the melting point of $Zn(acac)_2$ (138°C) from room temperature in 2 minutes and held at this temperature for 60 minutes. It was then heated to the melting point of $Mn(acac)_2$ (260°C) and maintained at the temperature for 60 minutes. The temperature was then raised to

900°C and kept at that temperature for 120 minutes. After the system was cooled down to room temperature, a large piece of white, cotton-like product was found on the inner wall downstream end of the quartz tube.[176]

XRD measurements showed that the as-synthesized product is a hexagonal wurtzite structure, with the *c* parameter increasing from 6.257 to 6.265 Å while the increase in Mn dopant concentration was from 0% to 5%, and the *a* parameter remained essentially constant around 3.821 Å, which showed that Mn was present in the substitution in ZnS. The results of energy dispersive X-ray analysis (shown in Table 2.1) carried out on several samples and on different regions of the nanobelts, showed the molar fraction of element Mn to be close to those expected from the starting compositions.

Figure 2.30 shows the typical SEM images of the as-synthesized products, which show that uniform nanobelts are formed with a high yield and with lengths of several tens of micrometers. A representative high-magnification SEM image (Fig. 2.30(b)) of the as-synthesized nanobelts reveals that its geometrical shape is belt-like, and its thickness is about 10 nm. The TEM observation (Fig. 2.30(c)) reveals that each nanobelt has a uniform width along its entire length, and the typical widths of the nanobelts are in the range 40–200 nm. Figure 2.30(d) is a typical HRTEM image of a single nanobelt, which clearly reveals the (001) atomic planes with an interplanar spacing of about 0.620 nm along the length of the nanobelt and the (100) atomic planes with an interplanar spacing of about 0.323 nm along the width of the nanobelt. The SAED pattern (inset) also reveals that the ZnS nanobelts are single-crystal hexagonal wurtzite structure, growing along the [001] direction without defect or dislocation.

Table 2.1. Molar fraction (%) of element Mn in starting materials and in as-synthesized products[176].

Starting materials	Products
0	0
1	1.03
3	2.98
5	5.04

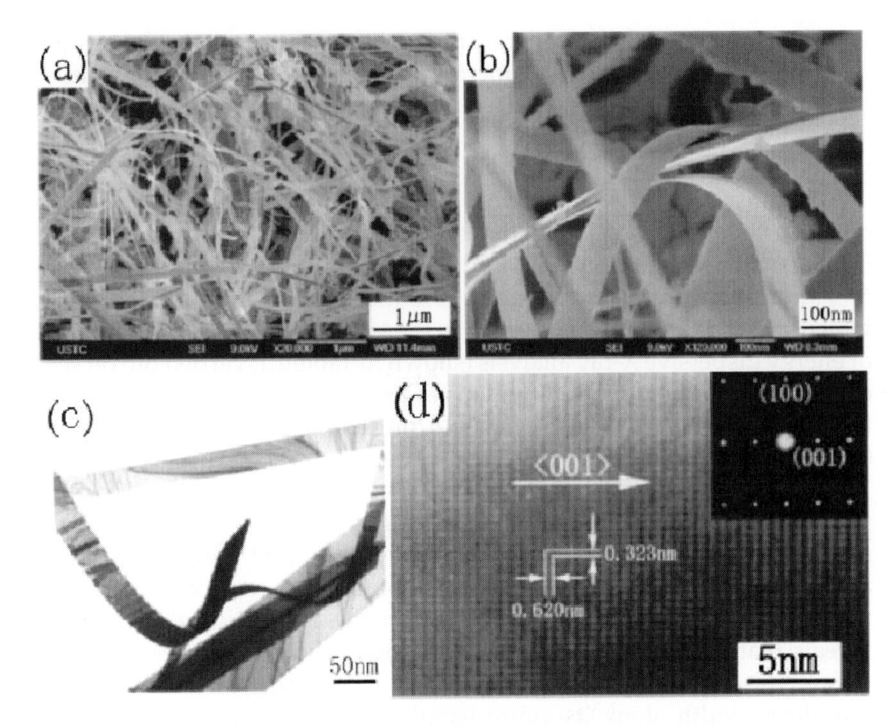

Fig. 2.30. (a) Low magnification and (b) high magnification SEM images of the as-synthesized product, (c) typical TEM image of Mn-doped ZnS nanobelts, (d) HRTEM and SAED pattern (inset) of 5% Mn-doped ZnS nanobelt.[176]

Although we mainly discussed the controlled growth of some oxides, sulfides and doping of nanowires and nanobelts, other nanowires and nanobelts, such as Si,[177–179] Zn,[180] Pd,[181] SiO_x and GeO_x,[182] Sb_2O_3,[183] Fe_2O_3,[184] ZnSe,[185] GaN,[186,187] and so on, have also been reported.

Bibliography

1. S. Iijima, *Nature* **354**, 56 (1991).
2. C. M. Lieber, *Solid. State. Comm.* **107**, 607 (1998).
3. J. T. Hu, T. W. Odom, and C. M. Lieber, *Acc. Chem. Res.* **32**, 435 (1999).
4. Y. N. Xia, P. D. Yang, Y. G. Sun, Y. Y. Wu, B. Mayers, B. Gates, Y. D. Yin, F. Kim, and H. Q. Yan, *Adv. Mater.* **15**, 353 (2003).
5. C. N. R. Rao, F. L. Deepak, G. Gundiah, and A. Govindaraj, *Prog. Solid State Chem.* **31**, 5 (2003).

6. M. Law, J. Goldberger, and P. D. Yang, *Annu. Rev. Mater. Res.* **34**, 83 (2004).
7. A. Kolmakov and M. Moskovits, *Annu. Rev. Mater. Res.* **34**, 151 (2004).
8. Z. L. Wang (Ed.), *Nanowires and Nanobelts*, Kluwer Academic, New York, 2003.
9. P. D. Yang, Y. Y. Wu, and R. Fan, *Inter. J. Nanosci.* **1**, 1 (2002).
10. Z. W. Pan, Z. R. Dai, and Z. L. Wang, *Science* **291**, 1947 (2001).
11. Z. L. Wang, *Annu. Rev. Phys. Chem.* **55**, 159 (2004).
12. Z. L. Wang, *J. Phys.: Condens. Matter* **16**, R829 (2004).
13. P. D. Yang, H. Q. Yan, S. Mao, R. Russo, J. Johnson, R. Saykally, N. Morris, J. Pham, R. He, and H. J. Choi, *Adv. Funct. Mater.* **12**, 323 (2002).
14. Z. Y. Fan, and J. G. Lu, *J. Nanosci. Nanotech.* **5**, 1561 (2005).
15. Y. Li, G. W. Meng, L. D. Zhang and F. Phillipp, *Appl. Phys. Lett.* **76**, 2011 (2000).
16. M. H. Huang, S. Mao, H. Feick, H. Q. Yan, Y. Y. Wu, H. King, E. Weber, R. Russo, and Y. P. Yang, *Science* **292**, 1897 (2001).
17. Y. W. Wang, L. D. Zhang, G. Z. Wang, X. S. Peng, Z. Q. Chu, and C. H. Liang, *J. Cryst. Growth* **234**, 171 (2001).
18. M. H. Huang, Y. Y. Wu, H. N. Feick, N. Tran, E. Weber, and Y. P. Yang, *Adv. Mater.* **13**, 113 (2001).
19. M. J. Zheng, L. D. Zhang, G. H. Li, and W. Z. Shen, *Chem. Phys. Lett.* **363**, 123 (2002).
20. L. Vayssieres, *Adv. Mater.* **15**, 464 (2003).
21. C. H. Liu, J. A. Zapien, Y. Yao, X. Meng, C. S. Lee, S. S. Fan, Y. Lifshitz, and S. T. Lee, *Adv. Mater.* **15**, 838 (2003).
22. X. Y. Kong, Y. Ding, R. Yang, and Z. L. Wang, *Science* **303**, 1348 (2004).
23. P. X. Gao, Y. Ding, W. J. Mai, W. L. Hughes, C. S. Lao, and Z. L. Wang, *Science* **309**, 1700 (2005).
24. P. X. Gao and Z. L. Wang, *Small* **1**, 945 (2005).
25. W. Z. Xu, Z. Z. Ye, D. W. Ma, H. M. Lu, L. P. Zhu, B. H. Zhao, X. D. Yang, and Z. Y. Xu, *Appl. Phys. Lett.* **87**, 093110 (2005).

26. B. Q. Cao, W. P. Cai, G. T. Duan, Y. Li, F. Q. Sun, F. Lu, and L. D. Zhang, *Nanotech.* **16**, 7341 (2005).

27. Q. Wei, G. W. Meng, X. H. An, Y. F. Hao, and L. D. Zhang, *Nanotech.* **16**, 2561 (2005).

28. G. S. Wu, X. Y. Yuan, Y. Li, L. Yang, Y. H. Xiao, and L. D. Zhang, *Solid. State. Comm.* **134**, 485 (2005).

29. Y. H. Xiao, L. Li, Y. Li, M. Fang, and L. D. Zhang, *Nanotech.* **16**, 671 (2005).

30. L. Yang, G. Z. Wang, C. J. Tang, H. Q. Wang, and L. D. Zhang, *Chem. Phys. Lett.* **409**, 337 (2005).

31. C. H. Ye, X. S. Fang, Y. F. Hao, X. M Teng, and L. D. Zhang, *J. Phys. Chem.* **B109**, 19758 (2005).

32. C. Herring, *Phys. Rev.* **82**, 87 (1951).

33. E. Kaldis, *J. Cryst. Growth* **5**, 376 (1969).

34. S. S. Brenner, *Acta. Metall.* **4**, 62 (1956).

35. C. Y. Nam, D. Tham, and J. E. Fischer, *Appl. Phys. Lett.* **85**, 5676 (2004).

36. S. Xiang and T. Cao, *Inorganic Chemistry Series (Scientific Press)* **6**, 717 (2000), in chinese.

37. M. Schmidt, R. Kusche, B. von Issendorff, and H. Haberland, *Nature* **393**, 238 (1998).

38. C. L. Cleveland, W. D. Luedtke, and U. Landman, *Phys. Rev. Lett.* **81**, 2036 (1998).

39. A. Aoki and H. Sasakura, *Jpn. J. Appl. Phys.* **9**, 582 (1970).

40. C. Tatsuyama and S. Ichimura, *Jpn. J. Appl. Phys.* **15**, 843 (1976).

41. N. Yamazoe, *Sens. Actuat.* **B5**, 7 (1991).

42. C. Nayral, E. Viala, V. Colliere, P. Fau, F. Senocq, A. Maisonnat, and B. Chaudret, *Appl. Surf. Sci.* **164**, 219 (2000).

43. M. Law, H. Kind, B. Messer, F. Kim, and P. D. Yang, *Angew. Chem. Int. End.* **41**, 2405 (2002).

44. A. Maiti, J. A. Rodriguez, M. Law, P. Kung, J. R. McKinney, and P. D. Yang, *Nano. Lett.* **3**, 1025 (2003).

45. Y. S. He, J. C. Campbell, R. C. Murphy, M. F. Arendt, and J. S. Swinnea, *J. Mater. Res.* **8**, 3131 (1993).

46. D. Z. Wang, S. L. Wen, J. Chen, S.Y. Zhang, and F. Q. Li, *Phys. Rev.* **B49**, 14282 (1994).

47. P. H. Wei, G. B. Li, S. Y. Zhao, and L. R. Chen, *J. Electrochem. Soc.* **146**, 3536 (1999).

48. H. S. Varol and A. Hinsch, *Sol. Energy Mater. Sol. Cells* **40**, 27 (1996).

49. D. F. Zhang, L. D. Sun, J. L. Yin, and C. H. Yan, *Adv. Mater.* **15**, 1022 (2003).

50. S. Mathur, S. Barth, H. Shen, J. C. Pyun, and U. Werner, *Small* **1**, 713 (2005).

51. M. J. Zheng, G. H. Li, X. Y. Zhang, S. Y. Huang, Y. Lei, and L. D. Zhang, *Chem. Mater.* **13**, 3859 (2001).

52. A. Kolmakov, Y. X. Zhang, G. S. Cheng, and M. Moskovits, *Adv. Mater.* **15**, 997 (2003).

53. Z. R. Dai, Z. W. Pan, and Z. L. Wang, *Solid State Comm.* **118**, 351 (2001).

54. J. Q. Hu, X. L. Ma, N. G. Shang, Z. Y. Xie, N. B. Wong, C. S. Lee, and S. T. Lee, *J. Phys. Chem.* **B106**, 3823 (2002).

55. S. H. Sun, G. W. Meng, Y. W. Wang, T. Gao, M. G. Zhang, Y. T. Tian, X. S. Peng, and L. D. Zhang, *Appl. Phys.* **A76**, 287 (2003).

56. X. S. Peng, L. D. Zhang, G. W. Meng, Y. T. Tian,Y. Lin, B. Y. Geng, and S. H. Sun, *J. Appl. Phys.* **93**, 1760 (2003).

57. S. H. Sun, G. W. Meng, G. X. Zhang, T. Gao, B. Y. Geng, L. D. Zhang, and J. Zuo, *Chem. Phys. Lett.* **376**, 103 (2003).

58. J. Zhang, F. H. Jiang, and L. D. Zhang, *J. Phys. D: Appl. Phys.* **36**, L21 (2003).

59. C. H. Ye, X. S. Fang, Y. H. Wang, T. Xie, A. W. Zhao, and L. D. Zhang, *Chem. Lett.* **33**, 54 (2004).

60. S. H. Sun, G. W. Meng, M. G. Zhang, X. H. An, G. S. Wu, and L. D. Zhang, *J. Phys. D: Appl. Phys.* **37**, 409 (2004).

61. P. D. Yang and C. M. Lieber, *Science* **273**, 1836 (1996).

62. E. S. Greiner, J. A. Gutowski, and W. C. Ellis, *J. Appl. Phys.* **32**, 2489 (1961).

63. R. S. Wagner and R. G. Treuting, *J. Appl. Phys.* **32**, 2490 (1961).

64. X. Li, M. W. Wanlass, T. A. Gessert, K. A. Emery, and T. J. Coutts, *Appl. Phys. Lett.* **54**, 2674 (1989).

65. L. Dai, X. L. Chen, J. K. Jian, M. He, T. Zhou, and B. Q. Hu, *Appl. Phys.* **A75**, 687 (2002).

66. C. G. Granqvist, *Appl. Phys.* **A57**, 19 (1993).

67. I. Hamburg and C. G. Granqvist, *J. Appl. Phys.* **60**, R123 (1986).

68. D. S. Ginley and C. Bright, *Mater. Res. Soc. Bull.* **25**, 15 (2000).

69. M. Emziane and R. Le Ny, *Mater. Res. Bull.* **35**, 1849 (2000).

70. C. Li, D. H. Zhang, X. L. Liu, S. Han, T. Tang, J. Han, and C. W. Zhou, *Appl. Phys. Lett.* **82**, 1613 (2003).

71. C. H. Liang, G. W. Meng, Y. Lei, F. Phillipp, and L. D. Zhang, *Adv. Mater.* **13**, 1330 (2001).

72. M. J. Zheng, L. D. Zhang, G. H. Li, X. Y. Zhang, and X. F. Wang, *Appl. Phys. Lett.* **79**, 839 (2001).

73. M. J. Zheng, L. D. Zhang, X. Y. Zhang, J. Zhang, and G. H. Li, *Chem. Phys. Lett.* **334**, 298 (2001).

74. H. Q. Cao, X. Q. Qiu, Y. Liang, Q. M. Zhu, and M. J. Zhao, *Appl. Phys. Lett.* **83**, 761 (2003).

75. C. Li, D. H. Zhang, S. Han, X. L. Liu, T. Tang, and C. W. Zhou, *Adv. Mater.* **15**, 143 (2003).

76. X. S. Peng, G. W. Meng, J. Zhang, X. F. Wang, Y. W. Wang, C. Z. Wang, and L. D. Zhang, *J. Mater. Chem.* **12**, 1602 (2002).

77. X. Y. Kong and Z. L. Wang, *Solid State Commun.* **128**, 1 (2003).

78. J. S. Jeong, J. Y. Lee, C. J. Lee, S. J. An, and G. C. Yi, *Chem. Phys. Lett.* **384**, 246 (2004).

79. Y. B. Li, Y. Bando, and D. Golberg, *Adv. Mater.* **15**, 581 (2003).

80. H. B. Jia, Y. Zhang, X. H. Chen, J. Shu, X. H. Luo, Z. S. Zhang, and D. P. Yu, *Appl. Phys. Lett.* **82**, 4146 (2003).

81. Y. F. Hao, G. W. Meng, C. H. Ye, and L. D. Zhang, *Cryst. Growth Des.* **5**, 1617 (2005).

82. G. W. Sears, *Acta Met.* **4**, 361 (1955).

83. G. W. Sears, *Acta Met.* **4**, 367 (1955).

84. D. R. Watson, M. Chen, D. M. Glowacka, N. Adamopoulos, B. Soylu, B. A. Glowacki, and J. E. Evetts, *IEEE Trans. Appl. Supercond.* **5**, 801 (1995).

85. B. Soylu, N. Adamopoulos, D. M. Glowacka, and J. E. Evetts, *Appl. Phys. Lett.* **60**, 22 (1992).

86. A. R. Smith, H. A. H. Al-Brithen, D. C. Ingram, and D. Gall, *J. Appl. Phys.* **90**, 1809 (2001).

87. H. Yang, H. Al-Brithen, A. R. Smith, J. A. Borchers, R. L. Cappelletti, and M. D. Vaudin, *Appl. Phys. Lett.* **78**, 3860 (2001).

88. K. A. Shaw, E. Lochner, and D. M. Lind, *J. Appl. Phys.* **87**, 1727 (2000).

89. B. Q. Wei, R. Vajtai, Z. J. Zhang, G. Ramanath, and P. M. Ajayan, *J. Nanosci. Nanotech.* **1**, 35 (2001).

90. G. Kordas, *J. Mater. Chem.* **10**, 1157 (2000).

91. Y. Q. Zhu, Y. Z. Jin, H. W. Kroto, and D. R. M. Walton, *J. Mater. Chem.* **14**, 685 (2004).

92. C. C. Tang, Y. Bando, and T. Sato, *J. Phys. Chem.* **B106**, 7449 (2002).

93. Y. D.Yin, G. T. Zhang, and Y. N. Xia, *Adv. Funct. Mater.* **12**, 293 (2002).

94. J. Zhang, L. Zhang, X. Peng, and X. Wang, *Appl. Phys.* **A73**, 773 (2001).

95. Y. B. Li, Y. Bando, and T. Sato, *Chem. Phys. Lett.* **359**, 141 (2002).

96. R. Ma and Y. Bando, *Chem. Phys. Lett.* **370**, 770 (2003).

97. J. Zhang and L. D. Zhang, *Chem. Phys. Lett.* **363**, 293 (2002).

98. K. L. Klug and V. P. Dravid, *Appl. Phys. Lett.* **81**, 1687 (2002).

99. Y. B. Li, Y. Bando, D. Golberg, and Z. W. Liu, *Appl. Phys. Lett.* **83**, 999 (2003).

100. J. H. Zhan, Y. Bando, J. Q. Hu, and D. Golberg, *Inorg. Chem.* **43**, 2462 (2004).

101. Y. F. Hao, G. W. Meng, C. H. Ye, X. R. Zhang, and L. D. Zhang, *J. Phys. Chem.* **B109**, 11204 (2005).

102. X. S. Fang, C. H. Ye, L. D. Zhang, J. X. Zhang, J. W. Zhao, and P. Yan, *Small* **1**, 422 (2005).

103. J. W. Mullin, *Crystallization*, 3rd ed., Butterworth-Heinemann, Oxford, UK, 1993.

104. B. Zhang, Y. Y. Wu, P. D. Yang, and J. Liu, *Adv. Mater.* **14**, 122 (2002).

105. G. Das, *Ceram. Eng. Sci. Proc.* **16**, 977 (1995).

106. V. Valcárcel, A. Pérez, M. Cyrklaff, and F. Guitián, *Adv. Mater.* **10**, 1370 (1998).

107. B. C. Satishkumar, A. Govindaraj, E. M. Vogl, L. Basumallick, and C. N. R. Rao, *J. Mater. Res.* **12**, 604 (1997).

108. S. S. Berdonosov, S. B. Baronov, Y. V. Kuzmicheva, D. G. Berdonosova, and I. V. Melikhov, *Inorg. Mater.* **37**, 1037 (2001).

109. Z. L. Xiao, C. Y. Han, U. Welp, H. H. Wang, W. K. Kwok, G. A. Willing, J. M. Hiller, R. E. Cook, D. J. Miller, and G. W. Crabtree, *Nano. Lett.* **2**, 1293 (2002).

110. Y. Pang, G. Meng, L. Zhang, W. Shan, C. Zhang, X. Gao, A. Zhao, and Y. Mao, *J. Solid State Electrochem.* **7**, 344 (2003).

111. Y. T. Tian, G. W. Meng, T. Gao, S. H. Sun, T. Xie, X. S. Peng, C. H. Ye, and L. D. Zhang, *Nanotech.* **15**, 189 (2004).

112. L. Pu, X. M. Bao, J. P. Zou, and D. Feng, *Angew. Chem. Int. Ed.* **40**, 1490 (2001).

113. J. S. Leea, B. Mina, K. Choa, S. Kima, J. Parkb, Y. T. Leeb, N. S. Kimb, M. S. Leec, S. O. Parkc, and J. T. Moonc, *J. Cryst. Growth* **254**, 443 (2003).

114. C. C. Tang, S. S. Fan, P. Li, Chapelle, Lamy de la M., and H. Y. Dang, *J. Cryst. Growth* **224**, 117 (2001).

115. J. Zhou, S. Z. Deng, J. Chen, J. C. She, and N. S. Xu, *Chem. Phys. Lett.* **365**, 505 (2002).

116. C. N. R. Rao, G. Gundiah, F. L. Deepak, A. Govindaraj, and A. K. Cheetham, *J. Mater. Chem.* **14**, 440 (2004).

117. X. S. Fang, C. H. Ye, X. X. Xu, T. Xie, Y. C. Wu, and L. D. Zhang, *J. Phys.: Condens. Matter* **16**, 4157 (2004).

118. X. S. Fang, C. H. Ye, L. D. Zhang, and T. Xie, *Adv. Mater.* **17**, 1661 (2005).

119. Z. H. Yuan, H. Huang, and S. S. Fan, *Adv. Mater.* **14**, 303 (2002).

120. H. C. Lee, H. J. Kim, S. H. Chung, K. H. Lee, H. C. Lee, and J. S. Lee, *J. Am. Chem. Soc.* **125**, 2882 (2003).

121. X. S. Peng, L. D. Zhang, G. W. Meng, X. F. Wang, Y. W. Wang, C. Z. Wang, and G. S. Wu, *J. Phys. Chem.* **B106**, 11163 (2002).

122. X. S. Fang, C. H. Ye, X. S. Peng, Y. H. Wang, Y. C. Wu, and L. D. Zhang, *J. Mater. Chem.* **13**, 3040 (2003).

123. R. S. Wagner and W. C. Ellis, *Appl. Phys. Lett.* **4**, 89 (1964).

124. P. D. Yang and C. M. Lieber, *J. Mater. Res.* **12**, 2981 (1997).

125. R. B. Sosman, *Trans. Br. Ceram. Soc.* **54**, 655 (1955).

126. E. Schlam, *Proc. IEEE.* **61**, 894 (1973).

127. J. Xu and W. Ji, *J. Mater. Sci. Lett.* **18**, 115 (1999).

128. T. V. Prevenslik, *J. Lumin.* **87**, 1210 (2000).

129. L. Sun, C. Liu, C. Liao, and C. Yan, *J. Mater. Chem.* **9**, 1655 (1999).

130. Y. Wada, H. Yin, T. Kitamura, and S. Yanagida, *Chem. Commun.* 2683 (1998).

131. X. Jiang, Y. Xie, J. Lu, L. Zhu, W. Hei, and Y. T. Qian, *Chem. Mater.* **13**, 1213 (2001).

132. Y. W. Wang, L. D. Zhang, C. H. Liang, G. W. Wang, and X. S. Peng, *Chem. Phys. Lett.* **357**, 314 (2002).

133. Y. Jiang, X. M. Meng, J. Liu, Z. R. Hong, C. S. Lee, and S. T. Lee, *Adv. Mater.* **15**, 1195 (2003).

134. C. Ma, M. Moore, J. Li, and Z. L. Wang, *Adv. Mater.* **15**, 228 (2003).

135. Y. Jiang, X. M. Meng, J. Liu, Z. Y. Xie, C. S. Lee, and S. T. Lee, *Adv. Mater.* **15**, 323 (2003).

136. Y. C. Zhu, Y. Bando, and D. F. Xue, *Appl. Phys. Lett.* **82**, 1769 (2003).

137. Q. Li and C. Wang, *Appl. Phys. Lett.* **83**, 359 (2003).

138. C. H. Ye, X. S. Fang, G. H. Li, and L. D. Zhang, *Appl. Phys. Lett.* **85**, 3035 (2004).

139. S. H. Yu and M. Yoshimura, *Adv. Mater.* **14**, 296 (2002).

140. X. S. Fang, C. H. Ye, X. S. Peng, Y. H. Wang, Y. C. Wu, and L. D. Zhang, *J. Cryst. Growth* **263**, 263 (2004).

141. C. H. Liang, Y. Shimizu, T. Sasaki, H. Umehara, and N. Koshizaki, *J. Phys. Chem.* **B108**, 9728 (2004).

142. X. Wang, P. Gao, J. Li, C. J. Summers, and Z. L. Wang, *Adv. Mater.* **14**, 1732 (2002).

143. Y. Li, C. H. Ye, X. S. Fang, L. Yang, Y. H. Xiao, and L. D. Zhang, *Nanotech.* **16**, 501 (2005).

144. Y. C. Zhu, Y. Bando, and Y. Uemura, *Chem. Commun.* 836 (2003).

145. P. A. Hu, Y. Q. Liu, L. Fu, L. C. Cao, and D. B. Zhu, *J. Phys. Chem.* **B108**, 936 (2004).

146. S. Kar and S. Chaudhuri, *J. Phys. Chem.* **B109**, 3298 (2005).

147. X. S. Fang, C. H. Ye, X. S. Peng, Y. H. Wang, Y. C. Wu, and L. D. Zhang, *Adv. Funct. Mater.* **15**, 63 (2005).

148. H. Y. Peng, Z. W. Pan, L. Xu, X. H. Fan, N. Wang, C. S. Lee, and S. T. Lee, *Adv. Mater.* **13**, 317 (2001).

149. Z. W. Pan, Z. R. Dai, L. Xu, S. T. Lee, and Z. L. Wang, *J. Phys. Chem.* **B05**, 2507 (2001).

150. X. F. Duan, Y. Huang, R. Argarawal, and C. M. Limber, *Nature* **421**, 24 (2002).

151. I. P. Mcclean and C. B. Thomas, *Semicond. Sci. Technol.* **7**, 1394 (1992).

152. X. Peng, M. C. Schlamp, A. V. Kadavanich, and A. P. Alivisatos, *J. Am. Chem. Soc.* **119**, 7019 (1997).

153. L. S. H. Ichia, J. L. Wasserman, and I. Willner, *Adv. Mater.* **14**, 1323 (2002).

154. J. H. Zhan, X. G. Yang, D. W. Wang, S. D. Li, Y. Xie, Y. Xia, and Y. T. Qian, *Adv. Mater.* **12**, 1348 (2000).

155. H. Q. Cao, Y. Xu, J. M. Hong, H. B. Liu, G. Yin, B. L. Li, C. Y. Tie, and Z. Xu, *Adv. Mater.* **13**, 1393 (2001).

156. Y. T. Chen, Y. Guo, L. B. Kong, and H. L. Li, *Chem. Lett.* **31**, 602 (2002).

157. F. Gao, Q. Y. Lu, and D. Y. Zhao, *Adv. Mater.* **15**, 739 (2003).

158. L. F. Dong, J. Jiao, M. Coulter, and L. Love, *Chem. Phys. Lett.* **376**, 653 (2003).

159. J. Zhang, F. H. Jiang, and L. D. Zhang, *J. Phys. Chem.* **B108**, 7002 (2004).

160. Y. K. Liu, J. A. Zapien, C. Y. Geng, Y. Y. Shan, C. S. Lee, Y. Lifshitz, and S. T. Lee, *Appl. Phys. Lett.* **85**, 3241 (2004).

161. Y. J. Xiong, Y. Xie, J. Yang, R. Zhang, C. Z. Wu, and G. Du, *J. Mater. Chem.* **12**, 3712 (2002).

162. S. M. Zhou, Y. S. Feng, and L. D. Zhang, *Eur. J. Inorg. Chem.* 1794 (2003).

163. X. P. Shen, A. H. Yuan, F. Wang, J. M. Hong, and Z. Xu, *Solid State Comm.* **133**, 19 (2005).

164. Y. W. Wang, G. W. Meng, L. D. Zhang, C. H. Liang, and J. Zhang, *Chem. Mater.* **14**, 1773 (2002).

165. C. H. Ye, G. W. Meng, Y. H. Wang, Z. Jiang, and L. D. Zhang, *J. Phys. Chem.* **B106**, 10338 (2002).

166. X. F. Duan and C. M. Lieber, *Adv. Mater.* **12**, 298 (2000).

167. S. Y. Bae, H. W. Seo, and J. H. Park, *J. Phys. Chem.* **B108**, 5206 (2004).

168. G. Z. Shen, J. H. Cho, J. K. Yoo, G. C. Yi, and C. J. Lee, *J. Phys. Chem.* **B109**, 5491 (2005).

169. S. Y. Bae, C. W. Na, J. H. Kang, and J. H. Park, *J. Phys. Chem.* **B109**, 2526 (2005).

170. B. Y. Geng, G. Z. Wang, Z. Jiang, T. Xie, S. H. Sun, G. W. Meng, and L. D. Zhang, *Appl. Phys. Lett.* **82**, 4791 (2003).

171. B. C. Cheng, Y. H. Xiao, G. S. Wu, and L. D. Zhang, *Adv. Funct. Mater.* **14**, 913 (2004).

172. B. C. Cheng, Y. H. Xiao, G. S. Wu, and L. D. Zhang, *Appl. Phys. Lett.* **84**, 416 (2004).

173. X. S. Fang, C. H. Ye, L. D. Zhang, Y. Li, and Z. D. Xiao, *Chem. Lett.* **34**, 436 (2005).

174. Z. R. Tian, J. A. Voigt, J. Liu, B. Mckenzie, and M. J. Mcdermott, *J. Am. Chem. Soc.* **124**, 12954 (2002).

175. X. Y. Kong and Z. L. Wang, *Appl. Phys. Lett.* **84**, 975 (2004).

176. B. Y. Geng, G. Z. Wang, Z. Jiang, T. Xie, S. H. Sun, G. W. Meng, and L. D. Zhang, *Appl. Phys. Lett.* **82**, 4791 (2003).

177. Y. Q. Chen, K. Zhang, B. Miao, B. Wang, and J. G. Hou, *Chem. Phys. Lett.* **358**, 396 (2002).

178. A. I. Hochbaum, R. Fan, R. R. He, and P. D. Yang, *Nano. Lett.* **5**, 457 (2005).

179. S. P. Ge, K. L. Jiang, X. X. Lu, Y. F. Chen, R. M. Wang, and S. S. Fan, *Adv. Mater.* **17**, 56 (2005).

180. X. S. Peng, L. D. Zhang, G. W. Meng, X. Y. Yuan, Y. Lin, and Y. T. Tian, *J. Phys. D: Appl. Phys.* **36**, L35 (2003).

181. M. A. Bangar, K. Ramanathan, M. Yun, C. Lee, C. Hangarter, and N. V. Myung, *Chem. Mater.* **16**, 4955 (2004).

182. J. Q. Hu, Y. Jiang, X. M. Meng, C. S. Lee, and S. T. Lee, *Small* **1**, 429 (2005).

183. Y. X. Zhang, G. H. Li, J. Zhang, and L. D. Zhang, *Nanotech.* **15**, 762 (2004).

184. X. G. Wen, S. H. Wang, Y. Ding, Z. L. Wang, and S. H. Yang, *J. Phys. Chem.* **B109**, 215 (2005).

185. A. Colli, S. Hofmann, A. C. Ferrari, F. Martelli, S. Rubini, C. Ducati, A. Franciosi, and J. Robertson, *Nanotech.* **16**, S139 (2005).

186. J. Y. Li, C. G. Lu, B. Maynor, S. M. Huang, and J. Liu, *Chem. Mater.* **16**, 1633 (2004).

187. G. Kipshidze, B. Yavich, A. Chandolu, J. Yun, V. Kuryatkov, I. Ahmad, D. Aurongzeb, M. Holtz, and H. Temkin, *Appl. Phys. Lett.* **86**, 033104 (2005).

Chapter 3

Design and Synthesis of One-Dimensional Heterostructures

Chapter 3

Design and Synthesis of One-Dimensional Heterostructures

3.1 Introduction

Heterostructure is a well-developed concept in semiconductor discipline. To form a heterostructure, materials of different kinds should form a sharp interface. According to the form of the interface and the arrangement of the materials forming the interface, heterostructures may be planar (two-dimensional), spherical (zero-dimensional), or cylindrical (one-dimensional). Thin film heterostructures are widely used, and are the core technology of today's electronic industry. Heterostructures of two-dimensional morphology or thin films are the earliest ones in the concept. The second form is the zero-dimensional structure. The concepts of core-shell structure, core-multiple shell structure, and quantum dot quantum well structure all belong to this family. The zero-dimensional heterostructures have been proposed for more than two decades, and have been extensively investigated.[1–5] The one-dimensional heterostructures have the shortest history. However they are developing rather rapidly, and have attracted much research attention in recent years.

One-dimensional nanomaterials, including nanowires and nanotubes, are the research focus in nanomaterials in recent years owing to their rich physical and chemical properties and wide range of applications including optical, optoelectronic, magnetic, and sensing performances. This has been detailed in several review papers.[6–9] The assembly of the one-dimensional building blocks into complex heterostructures makes it possible for us to tune the energy band structure as desired and realize the complicated functionality of the structure. A sharp interface as thin as atomic thickness could act as an effective barrier to confine the transportation of the charge carriers. Therefore, nanodevices could be realized in the near future by using nanoheterostructures. The progress of the design and synthesis of semiconductor nanowire heterostructures has been reviewed recently by Lauhon.[10]

In this chapter, the design and synthesis of one-dimensional heterostructures of various forms are described in detail. The development in this field through the last decade is reviewed.

3.2 Synthesis of one-dimensional heterostructures

One-dimensional heterostructure, according to the arrangement of the composing materials, can be classified into three main categories: (1) coaxial core/shell structure (also known as nanocable) and biaxial nanowires, (2) heterojunction and superlattice nanowire structures, (3) complex branch structure (also known as hierarchical structure if the stem and the branches are in different sizes). In this section, one-dimensional heterostructures will be discussed according to the above classification.

3.2.1. Coaxial core/shell structure (nanocable) and biaxial nanowires

This category is the best investigated one. A lot of such materials have been designed and fabricated. The first member in this category was reported in 1997 by a French research group.[11] In the following year, a Japanese research group reported a similar structure, and explicitly named it nanocable.[12] Also in that year, Meng in our laboratory reported the β-SiC/SiO_2 core/shell structure.[13] In Fig. 3.1, the structure is exhibited, which consists of a crystalline β-SiC core wrapped by amorphous silica.

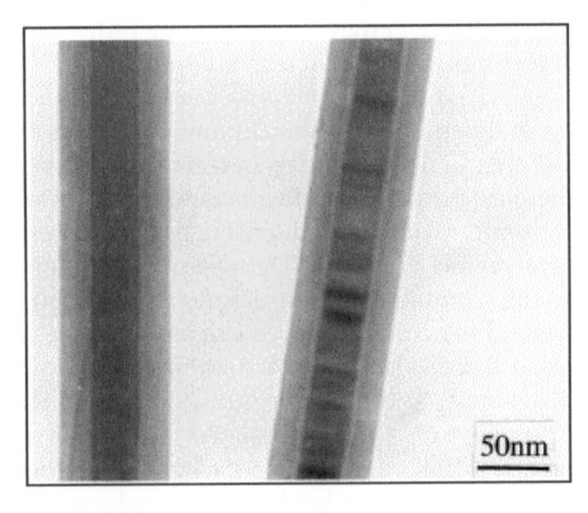

Fig. 3.1. TEM image of the β-SiC/SiO_2 core/shell structure.[13]

The name 'nanocable' was devised by the inspiration of the coaxial electric cables, where a conductive metal core was wrapped by an insulating polymer shell. We added sugar when preparing SiO_2 sols in the sol-gel process. The sugar in the nanochannels transformed to carbon nanoparticles at 700°C. When heated at 1600°C, SiC nucleated on the carbon nanoparticles by carbothermal reactions, then large-scale single crystalline β-SiC nanowires grew by the vapor–solid mechanism because of the multiple nucleation sites and the separated confinement of the channels. In the cooling process, the remaining SiO_2 vapor condensed on the surface of the β-SiC nanowires, and formed SiO_2 insulating layers with thickness of about 30 nm. Thus, single crystalline β-SiC/SiO_2 insulator nanocables were obtained.

Following this direction, Lauhon designed Si/Ge core/shell and core/multiple shell structures using Au as catalyst and by controlling the growth process of the core and the shell.[14] In Fig. 3.2(a), the schematic illustration of the formation of the core/shell and core/multiple shell structures is shown. Figures 3.2(b) and 3.2(d) are TEM images of i-Si/SiO_x/p-Si and Ge/Si core/shell structures, and Figs. 3.2(c) and 3.2(e) are corresponding HRTEM images, respectively.

Lin reported the synthesis of GaP/GaN and GaN/GaP core/shell structures by nitriding GaP nanowires and phosphiding GaN nanowires, respectively.[15] TEM and HRTEM images of the GaP/GaN core/shell structure are shown in Fig. 3.3. The sharp interface between GaP and GaN is demonstrated.

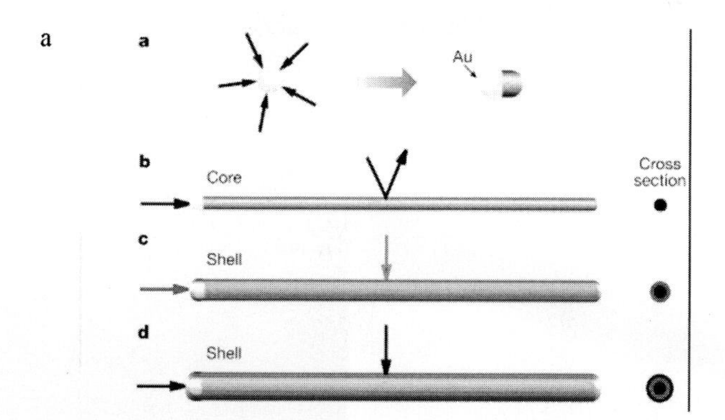

Fig. 3.2. (a) Schematic illustration of the formation of the core/shell and core/multiple shell structures. (b) and (d) TEM images of i-Si/SiO_x/p-Si and Ge/Si core/shell structures. (c) and (e) corresponding HRTEM images, respectively.[14]

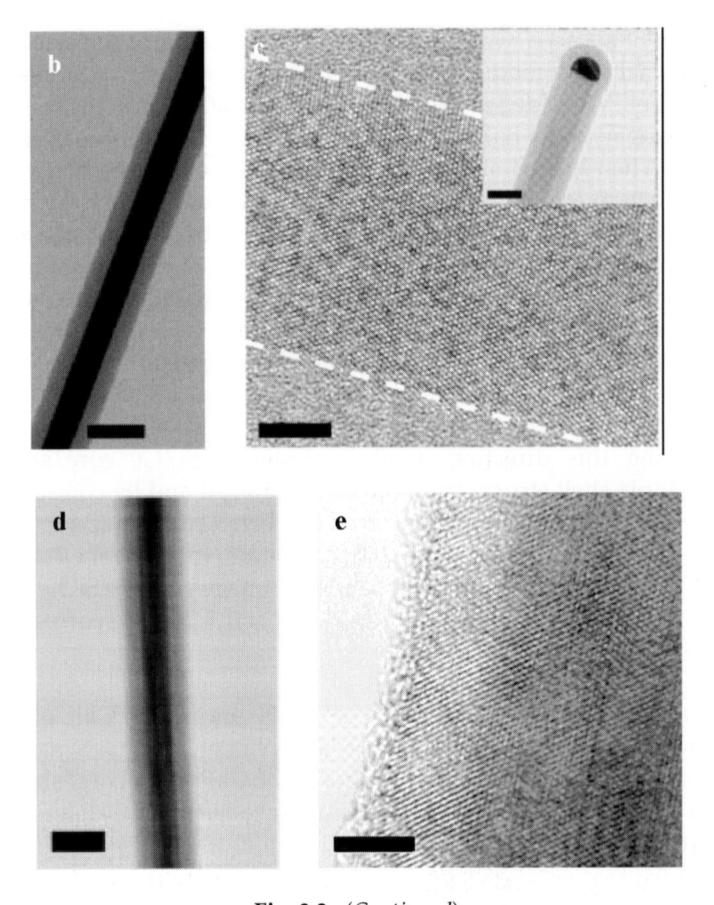

Fig. 3.2. (*Continued*)

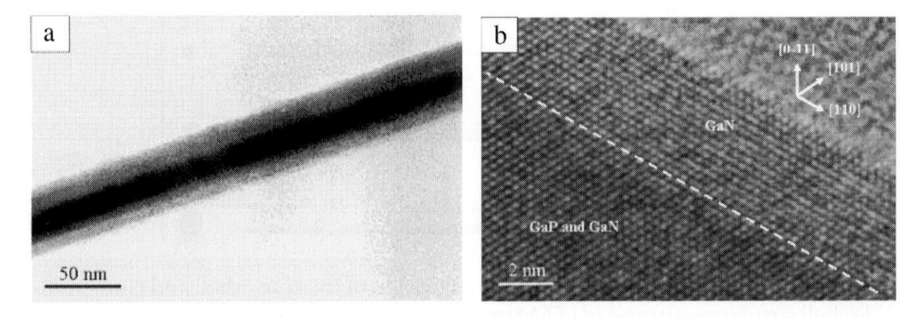

Fig. 3.3. (a) TEM and (b) HRTEM images of the GaP/GaN core/shell structure.[15]

Geng in our laboratory synthesized $CdSe/SiO_2$ nanocables by heating the mixture of Cd and Se in the presence of Si.[16] Partially-filled silica nanotube with CdSe is shown in Fig. 3.4(a), and the corresponding HRTEM image of the nanocable is shown in Fig. 3.4(b).

Li in our laboratory synthesized ZnS/SiO_2 nanocables by heating ZnS powders in the presence of Si.[17] TEM and HRTEM images of ZnS/SiO_2 nanocables are shown in Figs. 3.5(a) and 3.5(b), respectively. It is interesting to note that after evaporating the ZnS core, the silica nanotubes with hexagonal cross-section have been observed and are shown in Figs. 3.5(c) and 3.5(d). These observations imply that the nanocables were formed by templating ZnS nanowires.

Many core/shell structures have been synthesized, for example Si/CdSe,[18] β-Ga_2O_3/TiO_2,[19] ZnS/SiO_2,[20] GaP/C and $GaP/SiO_x/C$,[21] Ga_2O_3/C,[22] ZnS/SiC,[23] GaP/Ga_2O_3,[24] n-GaN/InGaN/p-GaN,[25] CdSe/CdS/ZnS,[26] GaN/BN,[27,28] ZnO/SiO_2,[29] Zn/ZnO,[30] GaN/BCN,[31] $Ga/Ga_2O_3/ZnO$,[32] β-Ga_2O_3/BN,[33] Zn/ZnS,[34] ZnO/ZnS,[35] Ga_2O_3/ZnO,[36] and so on.

Wang synthesized β-SiC/SiO_x biaxial nanowires by heating SiO and graphite powders.[37] TEM and HRTEM images of the biaxial nanowires are shown in Figs. 3.6(a) and 3.6(b), respectively.

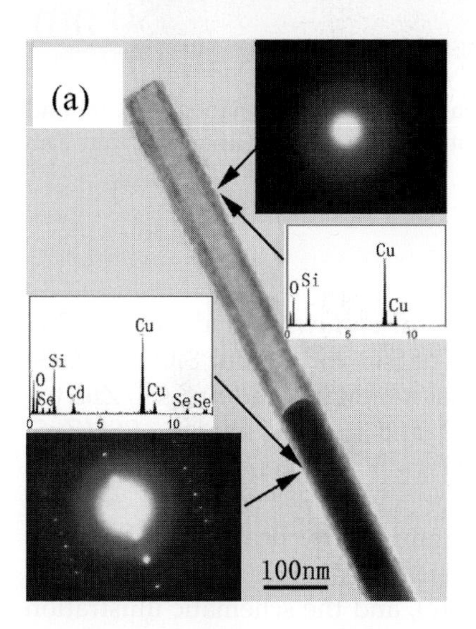

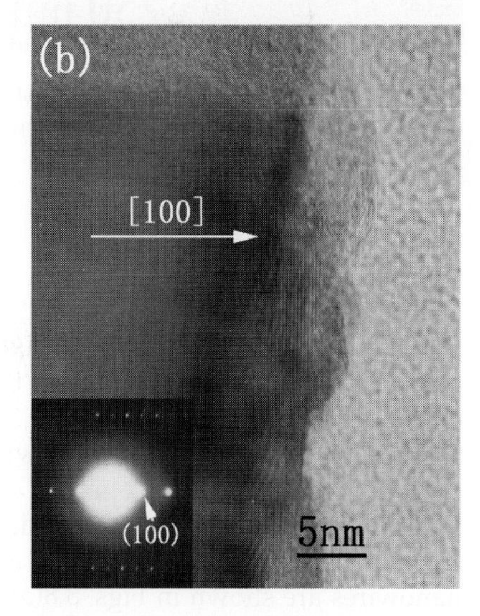

Fig. 3.4. (a) TEM and (b) HRTEM images of $CdSe/SiO_2$ nanocables.[16]

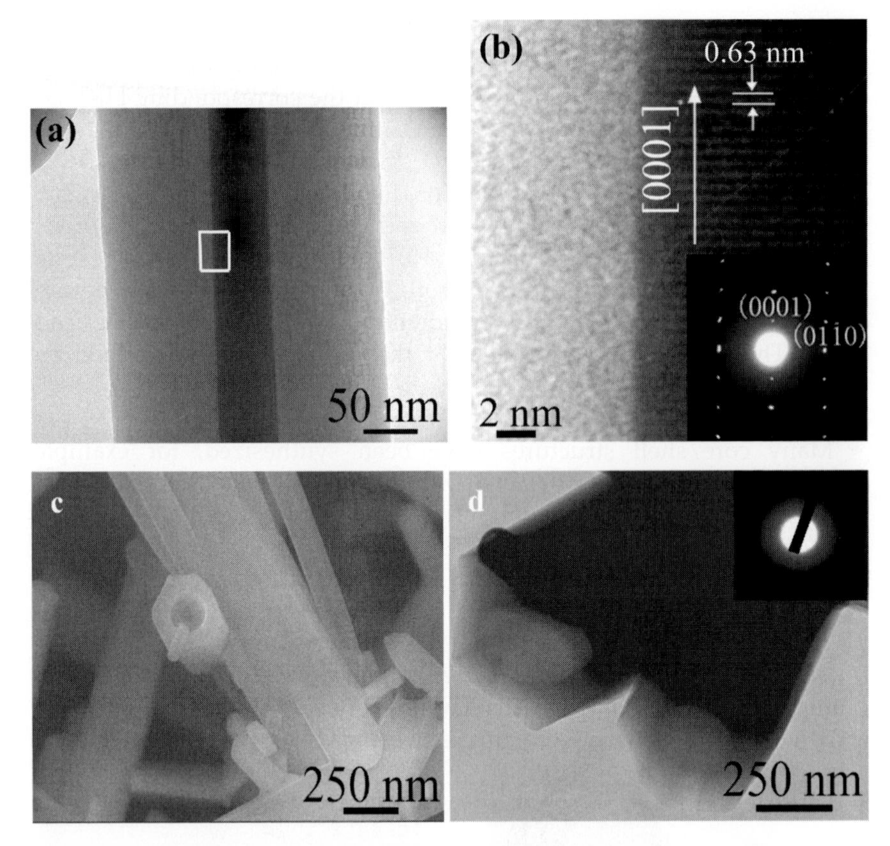

Fig. 3.5. (a) and (b) TEM and HRTEM images of ZnS/SiO_2 nanocables. (c) and (d) SEM and TEM images of silica nanotubes left after evaporating the ZnS core.[17]

Hu synthesized Si/ZnS, Si/ZnSe biaxial nanowires and ZnS/Si/ZnS triaxial nanowires by heating a mixture of SiO and ZnS or SiO and ZnSe powders.[38] TEM images of Si/ZnS biaxial nanowires and ZnS/Si/ZnS triaxial nanowires are shown in Fig. 3.7(a), and HRTEM images of Si/ZnS, Si/ZnSe biaxial nanowires are shown in Figs. 3.7(b) and 3.7(c), respectively.

Zhan synthesized ZnS/Si biaxial nanowires by heating a mixture of ZnS, Si, and SnS powders.[39] TEM and HRTEM images of ZnS/Si biaxial nanowires are shown in Figs. 3.8(a)–3.8(c), and the schematic illustration of the interface model is displayed in Fig. 3.8(d).

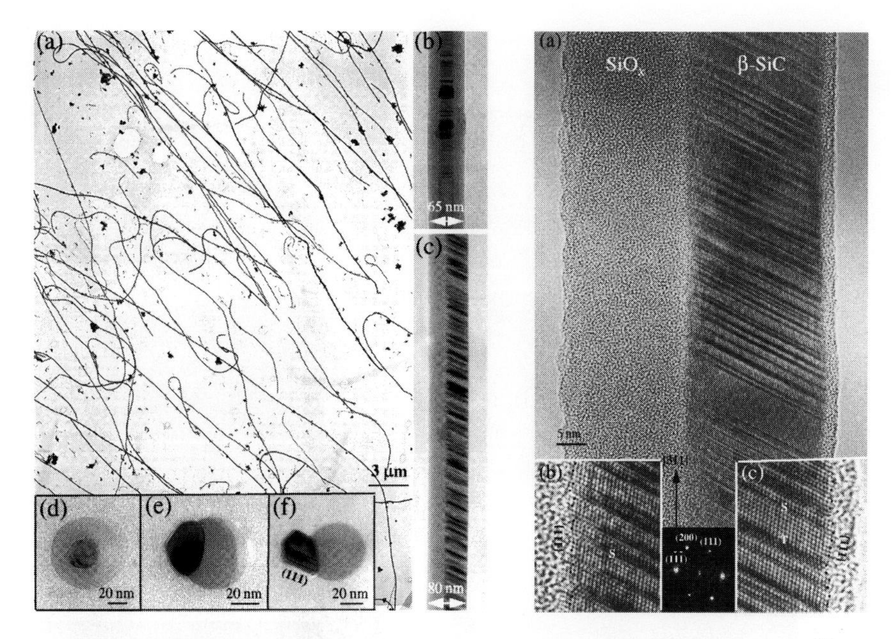

Fig. 3.6. (a) TEM and (b) HRTEM images of β-SiC/SiO_x biaxial structure.[37]

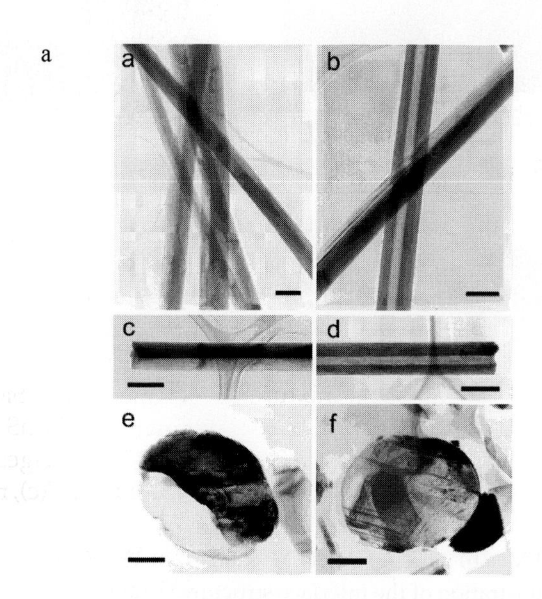

Fig. 3.7. (a) TEM images Si/ZnS biaxial nanowires and ZnS/Si/ZnS triaxial nanowires. (b) and (c) HRTEM images of Si/ZnS, Si/ZnSe biaxial nanowires.[38]

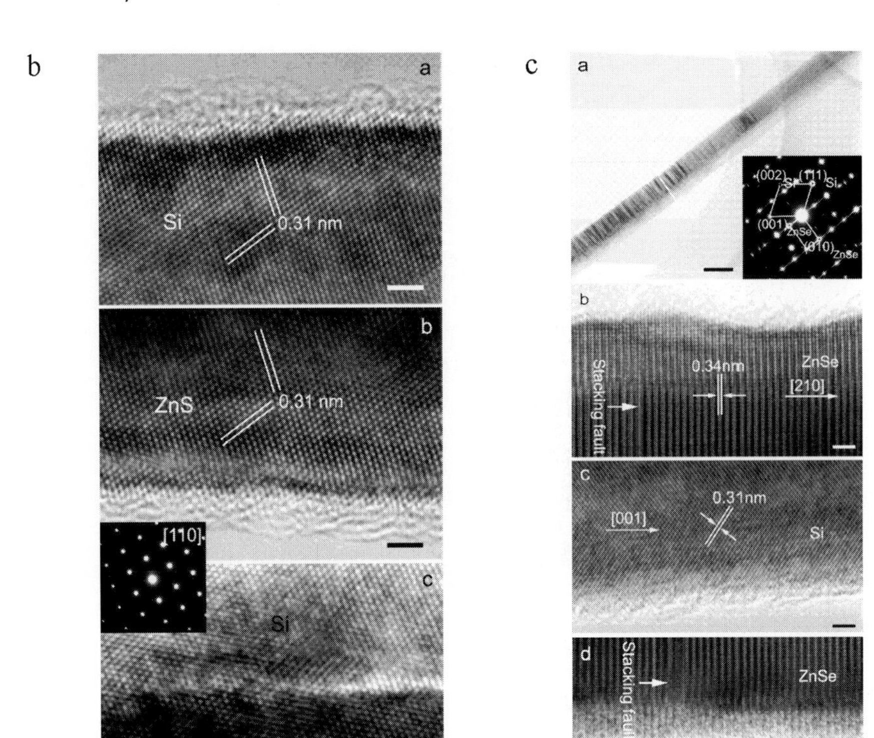

Fig. 3.7. (*Continued*)

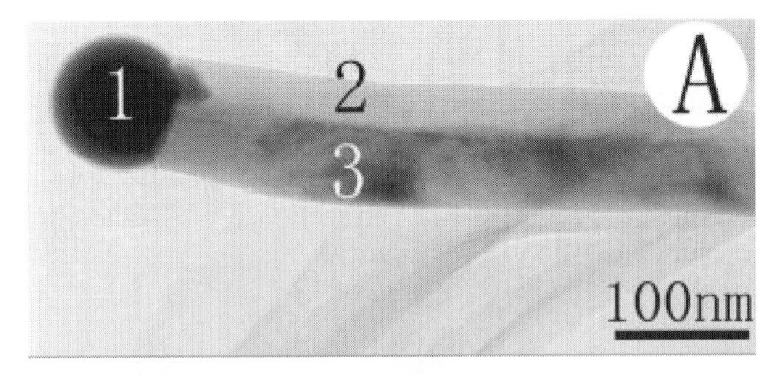

Fig. 3.8. (a) TEM, (b) and (c) HRTEM images of ZnS/Si biaxial nanowires. (d) Schematic illustration of the interface structure.[39]

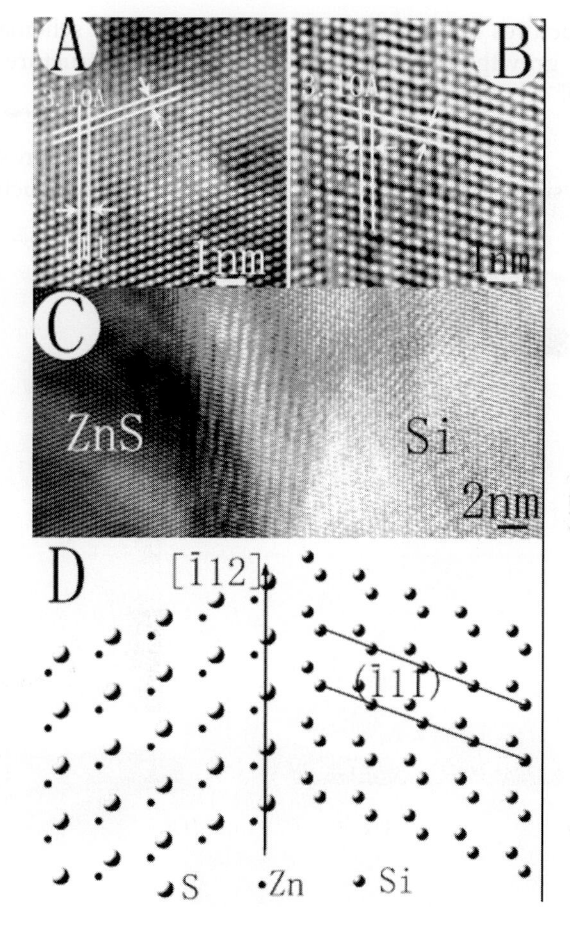

Fig. 3.8. (*Continued*)

3.2.2. Heterojunction and superlattice nanowire structure

Nanowires which are composed of two sections with a heterojunction are called heterojunction nanowire structure. They are nanowires composed of alternatively arranged sections of two different materials.

Hu synthesized the carbon nanotube and the Si nanowire heterojunction structure by Fe catalyzed growth of carbon nanotubes at the end of Si nanowires or growth of Si nanowires at the end of carbon nanotubes.[40] This is the first successful example in the attempt to make structures in this category. TEM images of Si nanowire/carbon nanotube heterojunction structures are shown in Fig. 3.9(a), and the schematic illustrations of the growth process of the heterojunction structure are shown in Fig. 3.9(b).

Ohlsson synthesized InAs/GaAs nanowire heterostructures.[41] Park synthesized the ZnO/ZnMgO nanowire heterostructure by the

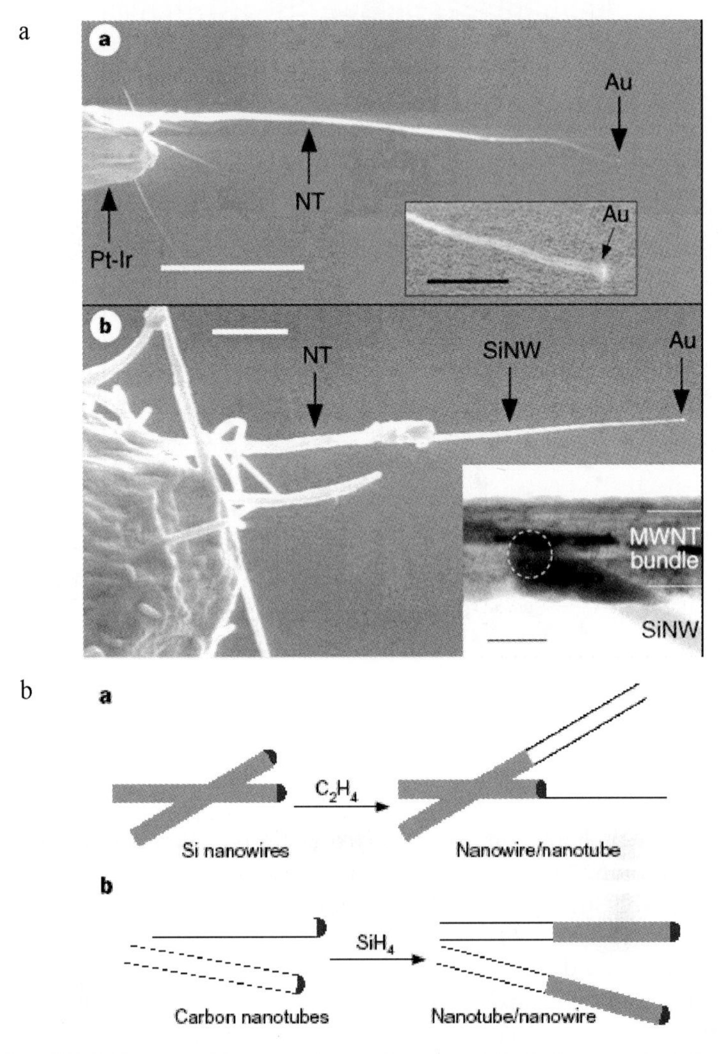

Fig. 3.9. (a) TEM images of Si nanowire/carbon nanotube heterojunction structure. (b) Schematic illustrations of the growth process of the heterojunction structure.[40]

metal-organic vapor phase epitaxy method.[42] The schematic illustration and the SEM image of the heterostructure are shown in Fig. 3.10.

Svensson synthesized GaP/GaAsP/GaP double heterostructure nanowires by the metal-organic vapor phase epitaxy method.[43] TEM images of the structure are shown in Fig. 3.11.

Hu synthesized the silica sheathed Ga/ZnS heterojunction structure by heating a mixture of ZnS, Ga_2O_3, and SiO powders.[44] TEM and HRTEM images of the heterojunction structure are shown in Figs. 3.12(a) and 3.12(b), respectively.

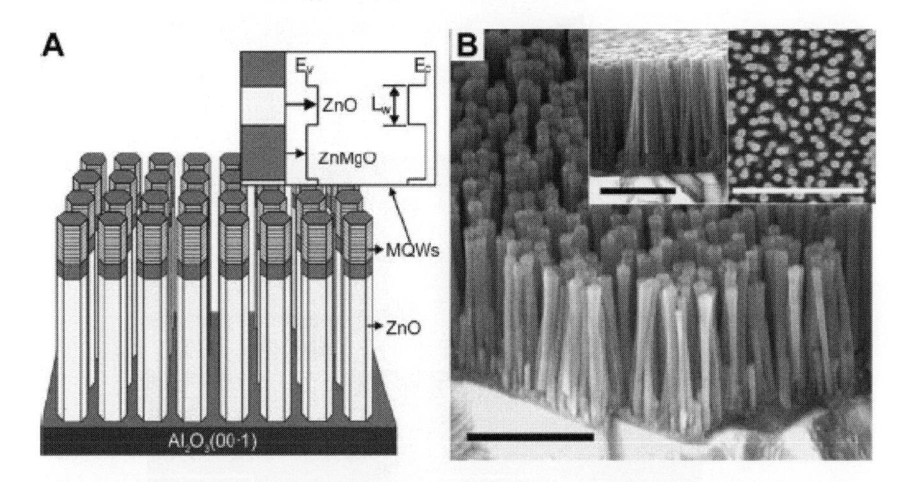

Fig. 3.10. (a) The schematic illustration and (b) the SEM image of the ZnO/ZnMgO nanowire heterostructure.[42]

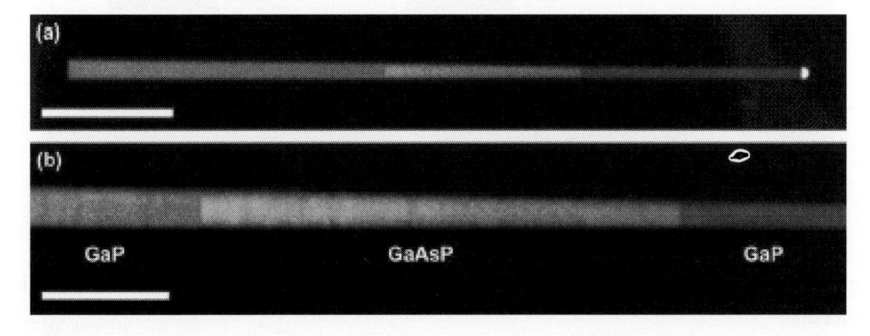

Fig. 3.11. TEM images of the GaP/GaAsP/GaP double heterostructure.[43]

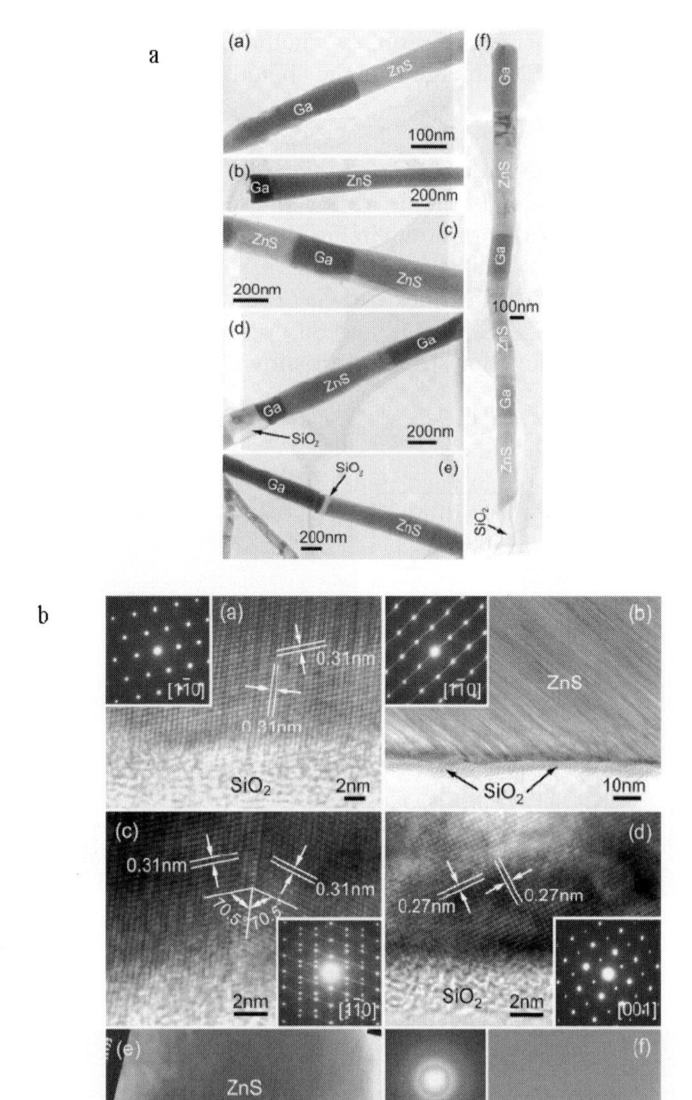

Fig. 3.12. (a) TEM and (b) HRTEM images of the Ga/ZnS heterojunction structure.[44]

Wu synthesized Si/SiGe superlattice nanowires by sequential deposition of Si and SiGe alloy.[45] The TEM image of the superlattice structure is shown in Fig. 3.13.

Björk synthesized InP/InAs superlattice nanowires by sequential deposition of InP and InAs sections.[46] The TEM image of the superlattice structure is shown in Fig. 3.14(a). The SAED pattern is shown in 3.14(b), where the splitting of the diffraction spots indicates the presence of different materials. In Figs. 3.14(c) and 3.14(d), the inverse Fourier transform images of the superlattice structure show clearly the alternating sections of InAs and InP in the nanowires.

Gudiksen synthesized GaP/GaAs superlattice nanowires by alternating the deposition of GaP and GaAs.[47] TEM images and elemental mapping of the structure are shown in Fig. 3.15(a), where the sharp interface between the two sections are clearly shown. The schematic illustration of the growth of the superlattice nanowires is shown in Fig. 3.15(b).

Wu synthesized NiSi/Si superlattice nanowires by selective alloying of preformed Si nanowires.[48] The schematic illustration of the formation process of the superlattice nanowires is shown in Fig. 3.16(a). The dark field optical image, TEM and HRTEM images of the structure are shown in Figs. 3.16(b) to 3.16(d).

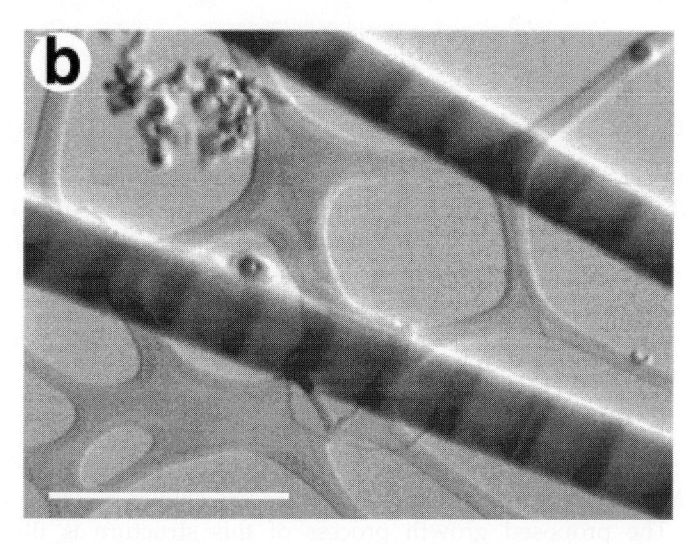

Fig. 3.13. TEM image of Si/SiGe superlattice nanowires.[45]

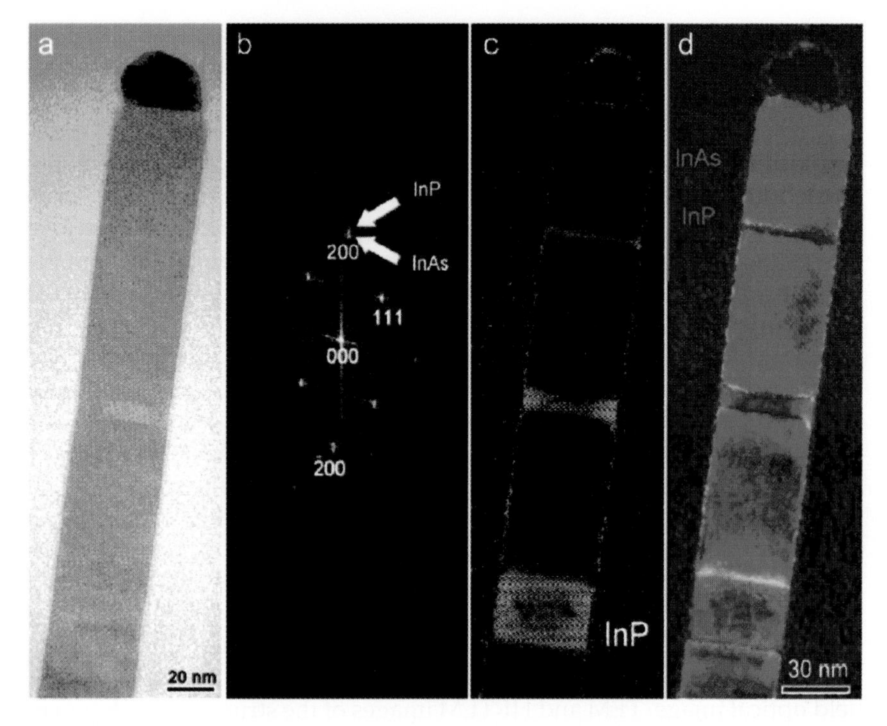

Fig. 3.14. (a) TEM image of the InP/InAs superlattice structure, (b) SAED pattern, (c) and (d), the inverse Fourier transform images of the superlattice structure.[46]

3.2.3. Complex branch structure (hierarchical structure)

The simplest nanostructures in this category are the multipod structures, where the branches stem from a single core. Examples include CdSe,[49] CdS,[50,51] ZnO,[52,53] MnS,[54] CdTe,[55] and so on. A more complex structure is the so-called nanoflower. However, this structure is nothing more in concept than in the former case. Typical examples include ZnO,[56] MgO,[57] GaP,[58] SiC,[59] SiO_x,[60] and so on. These structures will not be described in this section.

Zhu synthesized the Mg_2SiO_3 fishbone structure by heating a mixture of Si and Mg in the presence of Al_2O_3 and Co powders.[61,62] TEM and SEM images of the fishbone structure are shown in Figs. 3.17(a)–3.17(d). The proposed growth process of this structure is illustrated in Fig. 3.17(e).

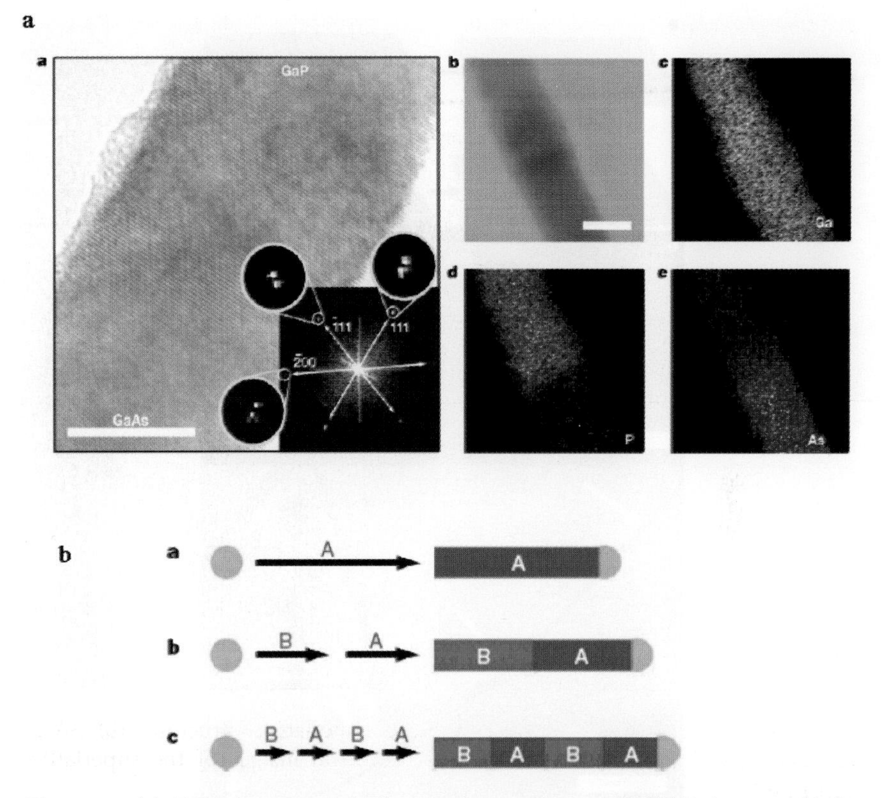

Fig. 3.15. (a) TEM images and elemental mapping of GaP/GaAs superlattice nanowires. (b) Schematic illustration of the growth of the structure.[47]

Yan synthesized the ZnO nanocomb structure by heating Zn powder in the presence of oxygen.[63] SEM and TEM images of the structure are shown in Fig. 3.18. From the HRTEM images and SAED patterns, the nanocomb is a single crystal entity.

Lao synthesized ZnO hierarchical nanostructures by heating a mixture of ZnO, In_2O_3, and graphite powders.[64–66] SEM and TEM images and structural models of 2-, 4-, and 6-fold symmetric ZnO hierarchical nanostructures are shown in Fig. 3.19.

Gao synthesized ZnO nanopropellers by heating ZnO, SnO_2, and graphite powders.[67] SEM images of the nanopropellers are shown in Figs. 3.20(a) and 3.20(b). The schematic formation process of the structure is illustrated in Fig. 3.20(c).

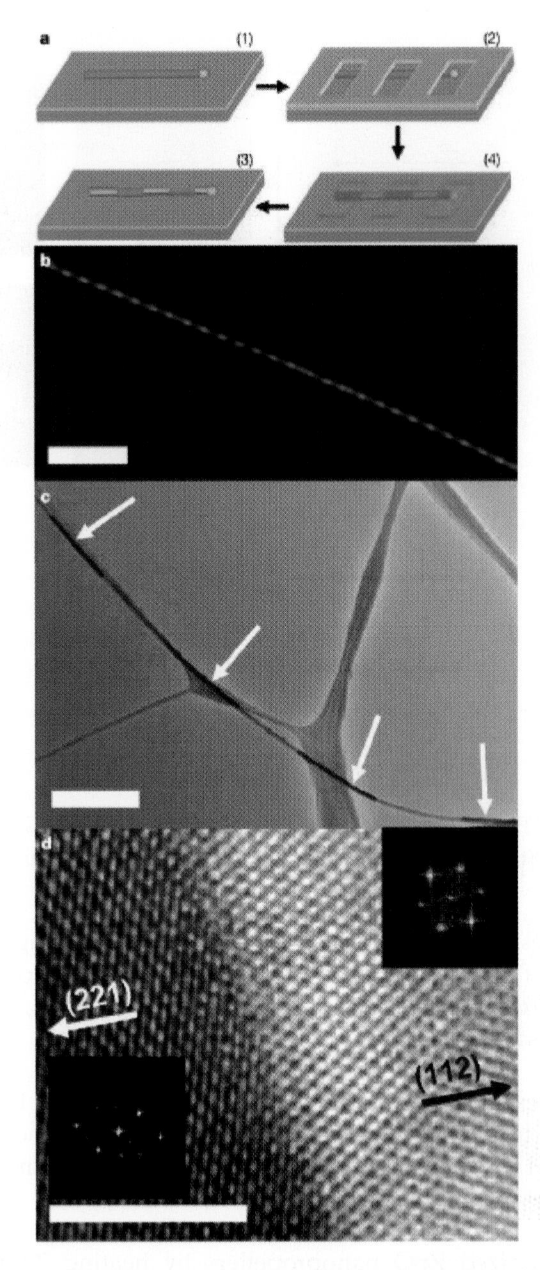

Fig. 3.16. (a) Schematic illustration of the formation process of the superlattice nanowires. (b) Dark field optical image. (c) TEM and (d) HRTEM images of the structure.[48]

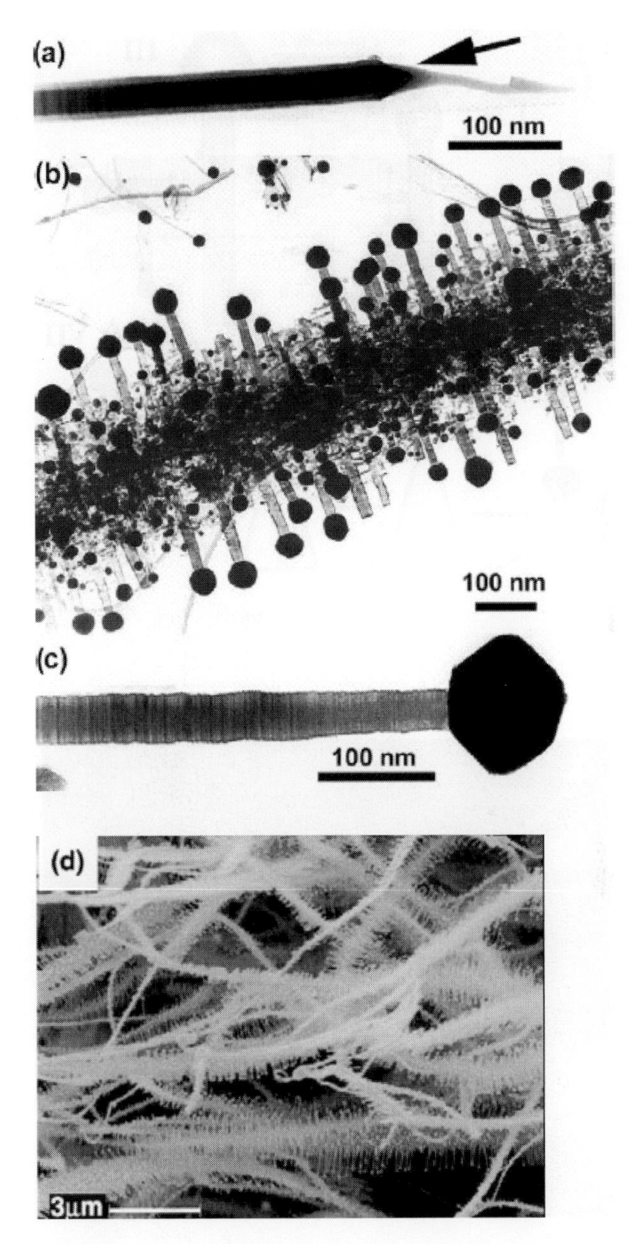

Fig. 3.17. (a)–(c) TEM and (d) SEM images of the fishbone structure. (e) The proposed growth process of this structure.[61,62]

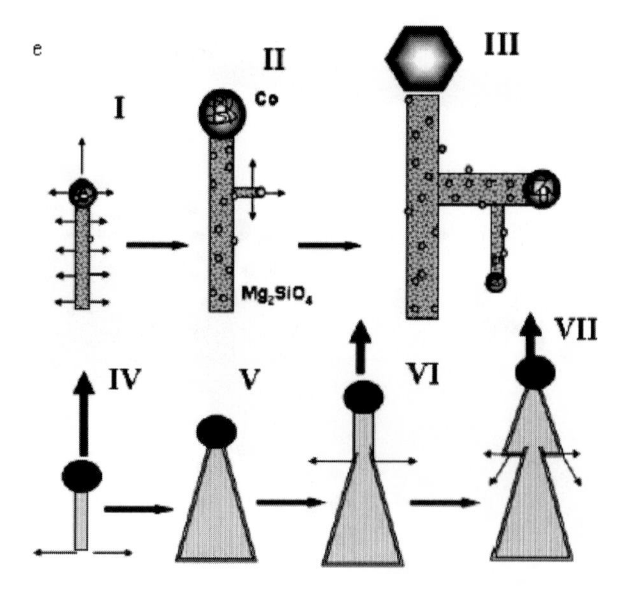

Fig. 3.17. (*Continued*)

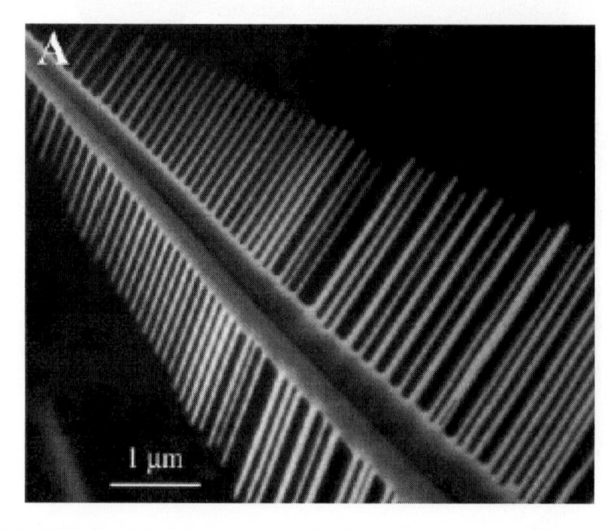

Fig. 3.18. (a) SEM (b) TEM images, (c) SAED pattern, (d) HRTEM images, and (e)–(h) images of the ZnO nanocomb structure.[63]

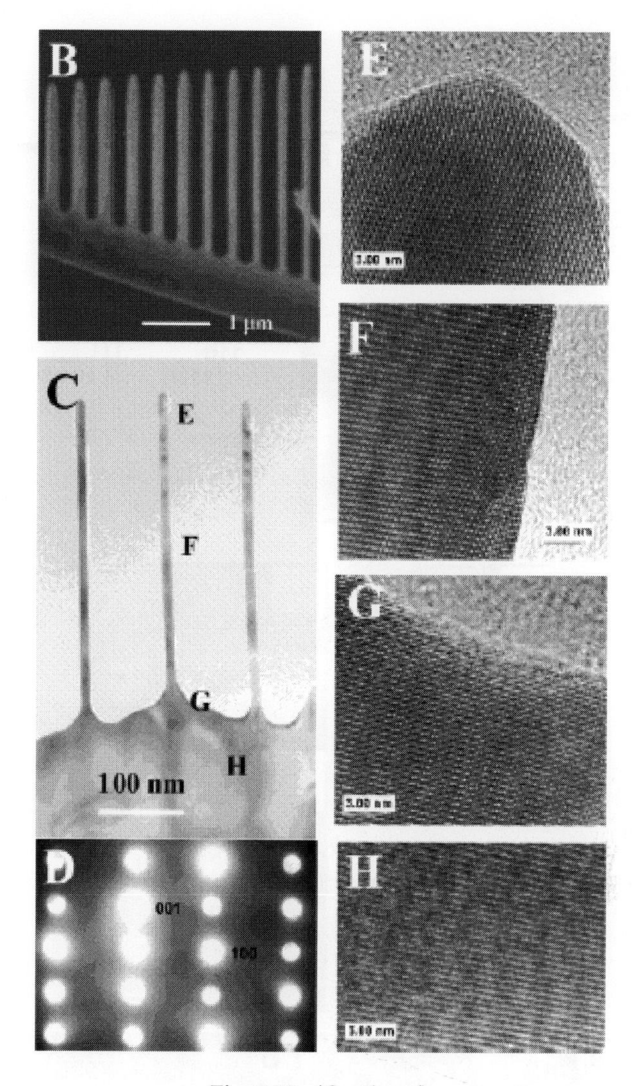

Fig. 3.18. (*Continued*)

Yan in our laboratory synthesized the MgO hierarchical structure by heating a mixture of Mg and Sn powders.[68] SEM images of the structure are shown in Fig. 3.21.

Ye in our laboratory synthesized Si/SiO_2 hierarchical structures by heating a mixture of SiO and SnS powders.[69] SEM images of the structure are shown in Fig. 3.22.

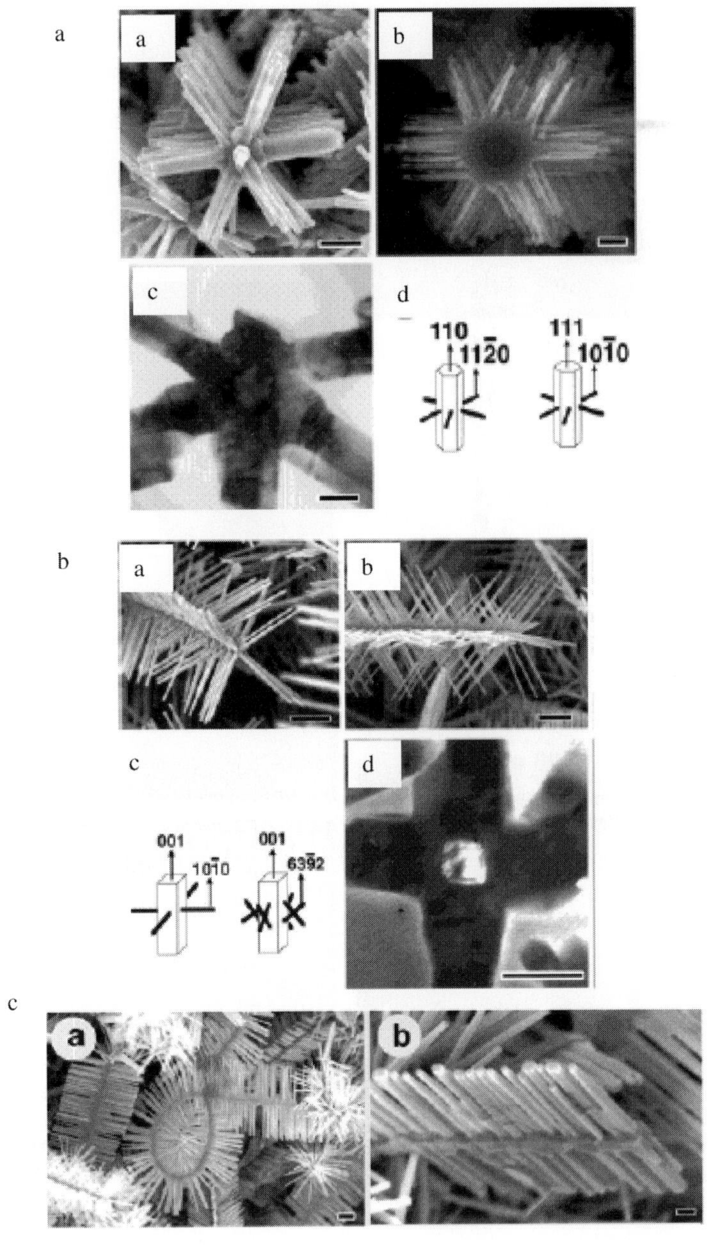

Fig. 3.19. SEM and TEM images of ZnO hierarchical nanostructures.[64]

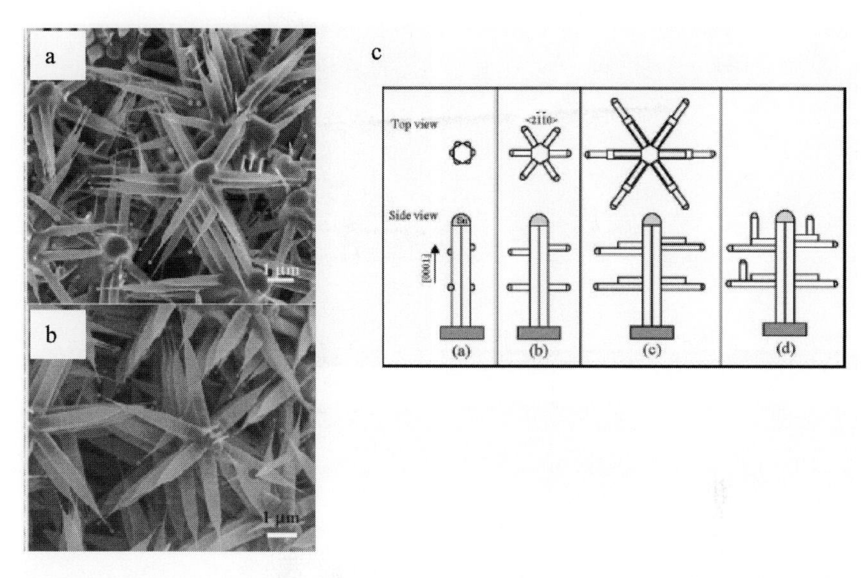

Fig. 3.20. (a) and (b) SEM images of the nanopropellers, and (c) the schematic formation process of the structure.[67]

Fig. 3.21. SEM images of MgO hierarchical nanostructures.[68]

Fig. 3.22. SEM image of Si/SiO_2 hierarchical structures.[69]

Following our work, Hu synthesized SiO_2/Si hierarchical structures by heating a mixture of SiO and SnO powders.[70] TEM images of the structure are shown in Fig. 3.23.

Wang have also synthesized SiO_2/Si hierarchical structures by heating a mixture of SiO and Sn powders.[71] SEM images of this structure are shown in Fig. 3.24(a), and the schematic illustration of the formation process of this structure is exhibited in Fig. 3.24(b).

Lan synthesized nanohomojunctions of GaN and nanoheterojunctions of InN on GaN by a two-step approach.[72] They first grew GaN nanowires by the VLS mechanism, then deposited a Au catalyst on the nanowires to direct the further growth of GaN or InN nanowires to form the nanohomojunctions or nanoheterojunctions. SEM and TEM images of the structures are shown in Fig. 3.25.

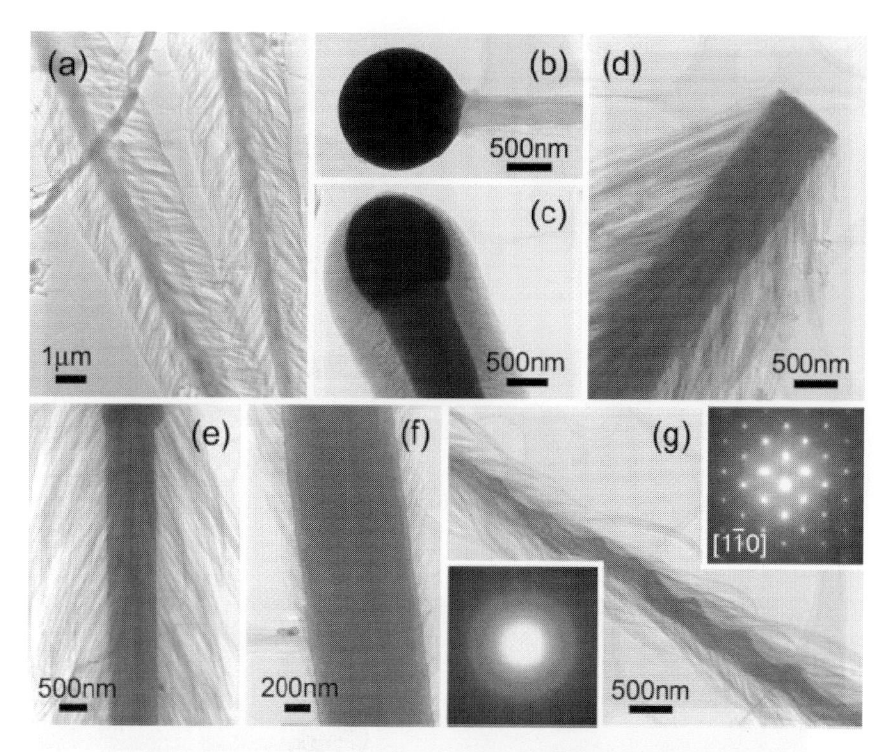

Fig. 3.23. TEM images of SiO_2/Si hierarchical structures.[70]

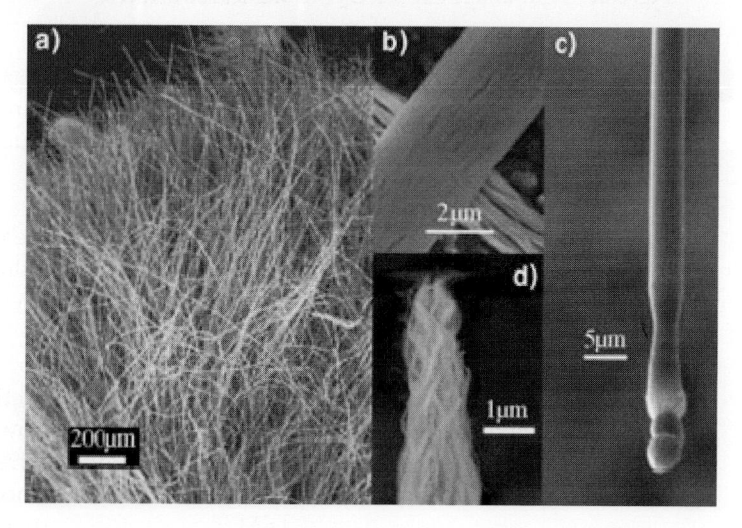

Fig. 3.24. (a) SEM images of SiO_2/Si hierarchical structures, and (b) the proposed formation process of this structure.[71]

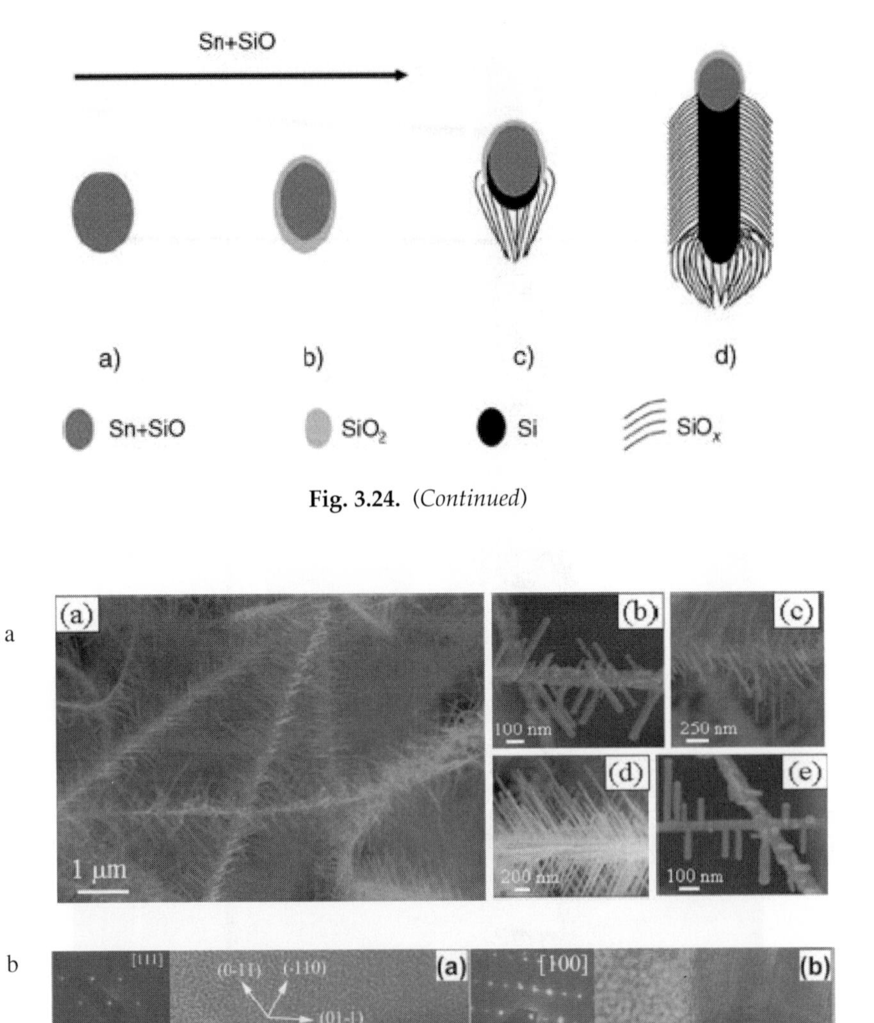

Fig. 3.24. (*Continued*)

Fig. 3.25. (a) SEM and (b) TEM images of GaN/GaN nanohomojunction and GaN/InN nanoheterojunction structures.[72]

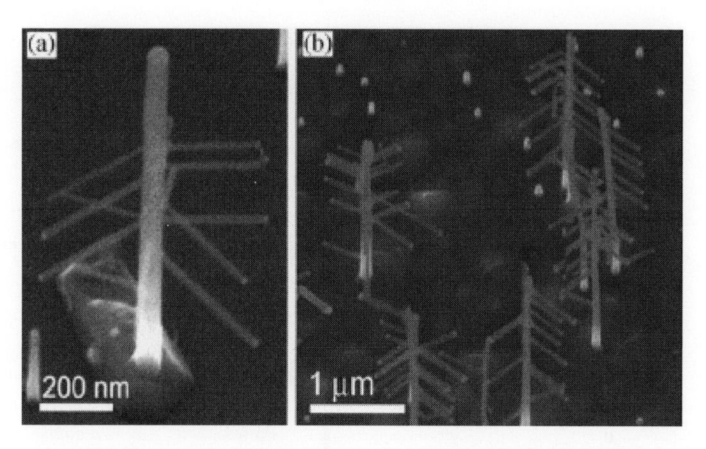

Fig. 3.26. SEM images of the GaP nanotree structure.[74]

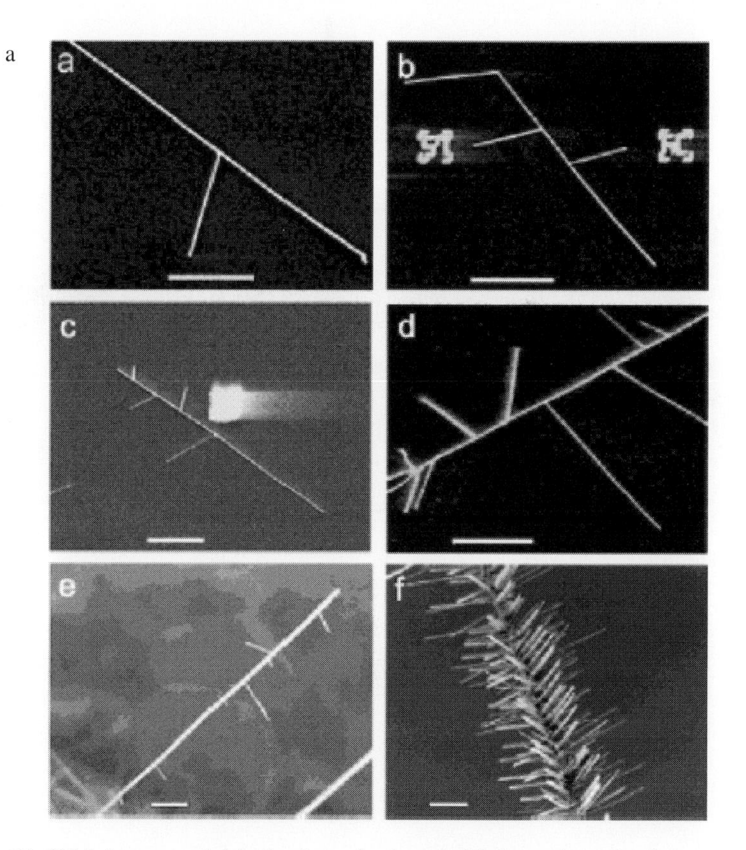

Fig. 3.27. SEM images of (a) the branched and (b) hyperbranched structures. (c) Schematic illustration of the growth model of the structures.[75]

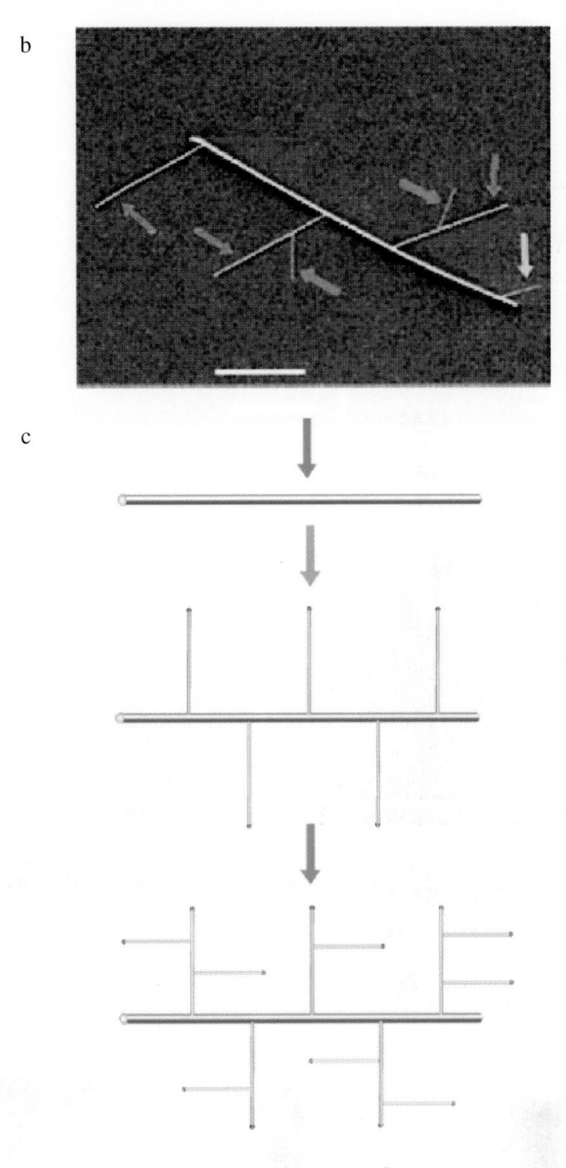

Fig. 3.27. (*Continued*)

Branched nanotree structures have been synthesized.[73–76] Dick synthesized GaP nanotrees by the approach similar to Lan's.[74] SEM images of the nanotrees are shown in Fig. 3.26.

Wang synthesized Si-branched and hyperbranched structures by a similar method.[75] SEM images of the branched and hyperbranched structures are shown in Figs. 3.27(a) and 3.27(b), respectively. The schematic illustration of the growth model of the structures is shown in Fig. 3.27(c).

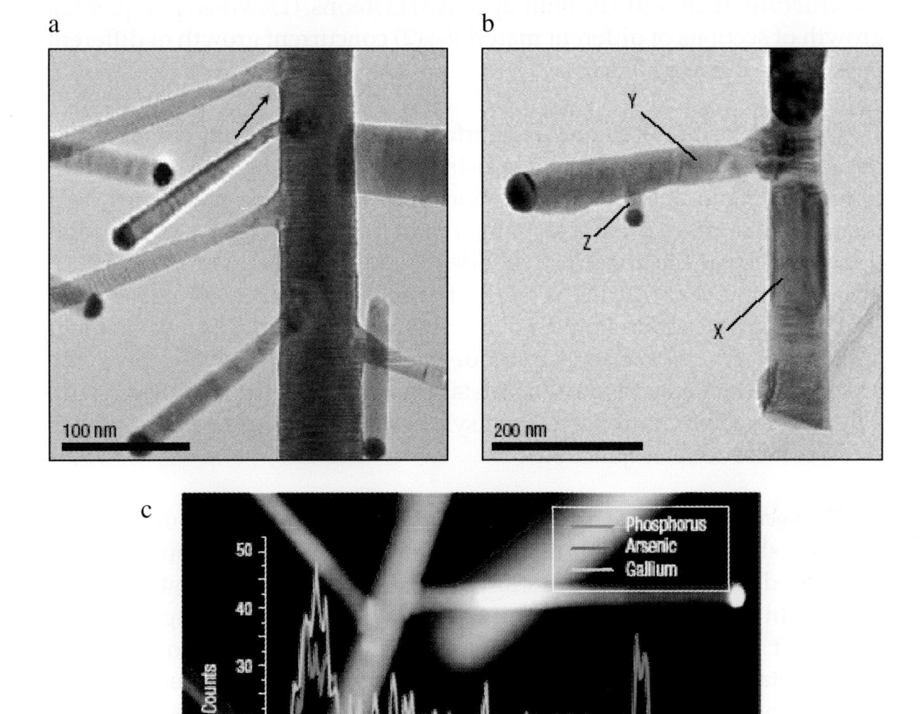

Fig. 3.28. TEM images of GaP nanotrees.[76]

Dick synthesized nanotrees with a more complex structure by a similar approach.[76] The nanotree is composed of a GaP main stem, branches and hyperbranches containing the GaP-GaAsP double heterostructure. TEM images and elemental mapping of the nanotrees are shown in Fig. 3.28.

3.3 Concluding remarks

The design and synthesis of heterostructures are dependent on the knowledge of the growth mechanism of one-dimensional building blocks. Heterostructures then could be built up in two fashions: (1) two-step sequential growth of sections of different materials; (2) concurrent growth of different parts of the building blocks.

The first fashion is more straightforward, and has been employed to design a wide variety of nanoheterostructures, including coaxial and multiple shell nanocables, heterojunction and superlattice nanowires, and branched nanotrees. In this fashion, a material section is grown first, and then metal catalysts are selectively deposited on the surface or at the tip of this section to direct the growth of another section in the subsequent step through the vapor–liquid–solid growth mechanism. The arrangement and size of the second section are determined by the metal catalyst. By controlling the deposition of the metal catalysts, complex nanotrees with a hyperbranch structure have been synthesized,[75,76] and even the branch could consist of heterojunctions.[76]

The second fashion is a more delicate case, and is not well-understood. In this fashion, the growth of hierarchical and branched nanostructures has been demonstrated. It generally involves low melting point metals at catalyst. In the growth process, one material with a lower melting point will grow first by the vapor–solid mechanism. Congruently, the low melting point metal deposits on the surface of the previously formed material section, and acts as the catalyst to direct the growth of the second section by the vapor–liquid–solid mechanism. Because the deposition of the metal catalyst on the surface of the preformed material section is concurrent with the growth of the materials, it does not need the post-deposition of the metal catalyst. The as-formed alloys are more uniform in size than the particles synthesized by the colloid chemistry method. Therefore, the branches are more homogeneous in size.

The design and fabrication of complex heterostructures with control on the assembly of the material sections, such as size and shape of the building blocks, are a fundamental issue in device miniaturization. The progress made in this field will definitely speed up this process.

Bibliography

1. A. Eychmuller, A. Mews, and H. Weller, *Chem. Phys. Lett.* **208**, 59 (1993).
2. A. Hasselbarth, A. Eychmuller, R. Eichberger, M. Giersig, A. Mews, and H. Weller, *J. Phys. Chem.* **97**, 5333 (1993).
3. J. W. Haus, H. S. Zhou, I. Honma, and H. Komiyama, *Phys. Rev.* **B47**, 1359 (1993).
4. A. T. Yeh, G. Cerullo, U. Banin, A. Mews, A. P. Alivisatos, and C. Shank, *Phys. Rev.* **B59**, 4973 (1999).
5. W. Jaskolski and G. W. Bryant, *Phys. Rev.* **B57**, R4237 (1998).
6. J. Hu, T. W. Odom, and C. M. Lieber, *Acc. Chem. Res.* **32**, 435 (1999).
7. Y. Xia, P. Yang, Y. Sun, Y. Wu, B. Mayers, B. Gates, Y. Yin, F. Kim, and H. Yan, *Adv. Mater.* **15**, 353 (2003).
8. M. Law, J. Goldberger, and P. Yang, *Annu. Rev. Mater. Res.* **34**, 83 (2004).
9. Z. L. Wang, *Annu. Rev. Phys. Chem.* **55**, 159 (2004).
10. L. J. Lauhon, M. S. Gudiksen, and C. M. Lieber, *Phil. Trans. R. Soc. Lond.* **A362**, 1247 (2004).
11. K. Suenaga, C. Colliex, N. Demoncy, A. Loiseau, H. Pascard, and F. Willaime, *Science* **278**, 654 (1997).
12. Y. Zhang, K. Suenaga, C. Colliex, and S. Iijima, *Science* **281**, 973 (1998).
13. G. W. Meng, L. D. Zhang, C. M. Mo, S. Y. Zhang, Y. Qian, S. P. Feng, and H. J. Li, *J. Mater. Res.* **13**, 2533 (1998).
14. L. J. Lauhon, M. S. Gudiksen, D. Wang, and C. M. Lieber, *Nature* **420**, 57 (2002).
15. H. M. Lin, Y. L. Chen, J. Yang, Y. C. Liu, K. M. Yin, J. J. Kai, F. R. Chen, L. C. Chen, Y. F. Chen, and C. C. Chen, *Nano Lett.* **3**, 537 (2003).
16. B. Geng, G. Meng, L. Zhang, G. Wang, and X. Peng, *Chem. Commun.* 2572 (2003).
17. Y. Li, C. Ye, X. Fang, L. Yang, Y. Xiao, and L. Zhang, *Nanotechnology* **16**, 501 (2005).

18. Q. Li and C. Wang, *J. Am. Chem. Soc.* **125**, 9892 (2003).

19. K. W. Chang and J. J. Wu, *Adv. Mater.* **17**, 241 (2005).

20. X. Fan, X. Meng, X. Zhang, S. K. Wu, and S. T. Lee, *Appl. Phys. Lett.* **86**, 173111 (2005).

21. S. Y. Bae, H. W. Seo, H. C. Choi, D. S. Han, and J. Park, *J. Phys. Chem.* **B109**, 8496 (2005).

22. J. Zhan, Y. Bando, J. Hu, Y. Li, and D. Golberg, *Chem. Mater.* **16**, 5158 (2004).

23. J. Hu, Y. Bando, J. Zhan, and D. Golberg, *Appl. Phys. Lett.* **85**, 2932 (2004).

24. B. Liu, Y. Bando, C. Tang, and F. Xu, *Appl. Phys.* **A80**, 1585 (2005).

25. F. Qian, Y. Li, S. Gradecak, D. Wang, C. J. Barrelet, and C. M. Lieber, *Nano Lett.* **4**, 1975 (2004).

26. L. Manna, E. C. Scher, L. Li, and A. P. Alivisatos, *J. Am. Chem. Soc.* **124**, 7136 (2002).

27. J. Zhang, L. Zhang, F. Jiang, and Z. Dai, *Chem. Phys. Lett.* **383**, 423 (2004).

28. W. Han and A. Zettl, *Appl. Phys. Lett.* **81**, 5051 (2002).

29. L. Dai, X. Chen, X. Zhang, T. Zhou, and B. Hu, *Appl. Phys.* **A78**, 557 (2004).

30. X. Kong, Y. Ding, and Z. L. Wang, *J. Phys. Chem.* **B108**, 570 (2004).

31. H. W. Seo, S. Y. Bae, J. Park, H. Yang, and B. Kim, *J. Phys. Chem.* **B108**, 6739 (2003).

32. J. Hu, Y. Bando, and Z. Liu, *Adv. Mater.* **15**, 1000 (2003).

33. R. Ma and Y. Bando, *Chem. Phys. Lett.* **367**, 219 (2003).

34. Q. Li and C. Wang, *Appl. Phys. Lett.* **82**, 1398 (2003).

35. X. Wang, P. Gao, J. Li, C. J. Summers, and Z. L. Wang, *Adv. Mater.* **14**, 1732 (2002).

36. K. W. Chang and J. J. Wu, *J. Phys. Chem.* **B109**, 13572 (2005).

37. Z. L. Wang, Z. Dai, P. Gao, Z. Bai, and J. L. Gole, *Appl. Phys. Lett.* **77**, 3349 (2000).

38. J. Hu, Y. Bando, Z. Liu, T. Sekiguchi, D. Golberg, and J. Zhan, *J. Am. Chem. Soc.* **125**, 11306 (2003).

39. J. Zhan, Y. Bando, J. Hu, T. Sekiguchi, and D. Golberg, *Adv. Mater.* **17**, 225 (2005).

40. J. Hu, M. Ouyang, P. Yang, and C. M. Lieber, *Nature* **399**, 48 (1999).

41. B. J. Ohlsson, M. T. Bjork, A. I. Persson, C. Thelander, L. R. Wallenberg, M. H. Magnusson, K. Deppert, and L. Samuelson, *Physica* **E13**, 1126 (2002).

42. W. I. Park, G. Yi, M. Kim, and S. J. Pennycook, *Adv. Mater.* **15**, 526 (2003).

43. C. P. T. Svensson, W. Seifert, M. W. Larsson, L. R. Wallenberg, J. Stangl, G. Bauer, and L. Samuelson, *Nanotechnology* **16**, 936 (2005).

44. J. Hu, Y. Bando, J. Zhan, and D. Golberg, *Adv. Mater.* **17**, 1964 (2005).

45. Y. Wu, R. Fan, and P. Yang, *Nano Lett.* **2**, 83 (2002).

46. M. T. Björk, B. J. Ohlsson, T. Sass, A. I. Persson, C. Thelander, M. H. Magnusson, K. Deppert, L. R. Wallenberg, and L. Samuelson, *Nano Lett.* **2**, 87 (2002).

47. M. Gudiksen, L. J. Lauhon, J. Wang, D. C. Smith, and C. M. Lieber, *Nature* **415**, 617 (2002).

48. Y. Wu, J. Xiang, C. Yang, W. Lu, and C. M. Lieber, *Nature* **430**, 61 (2004).

49. L. Manna, E. C. Scher, and A. P. Alivisatos, *J. Am. Chem. Soc.* **122**, 12700 (2000).

50. Y. W. Sun, S. M. Lee, N. J. Kang, and J. Cheon, *J. Am. Chem. Soc.* **123**, 5150 (2001).

51. F. Gao, Q. Lu, S. Xie, and D. Zhao, *Adv. Mater.* **14**, 1537 (2002).

52. Y. Dai, Y. Zhang, Q. Li, and C. Nan, *Chem. Phys. Lett.* **358**, 83 (2002).

53. X. Sun, X. Chen, and Y. Li, *J. Cryst. Growth* **244**, 218 (2002).

54. Y. Jun, Y. Jung, and J. Cheon, *J. Am. Chem. Soc.* **124**, 615 (2002).

55. L. Manna, D. J. Milliron, A. Meisel, E. C. Scher, and A. P. Alivisatos, *Nat. Mater.* **2**, 382 (2003).

56. C. Jiang, W. Zhang, G. Zou, W. Yu, and Y. Qian, *J. Phys. Chem.* **B109**, 1361 (2005).

57. X. Fang, C. Ye, L. Zhang, J. Zhang, J. Zhao, and P. Yan, *Small* **1**, 422 (2005).

58. B. Liu, Y. Bando, C. Tang, D. Golberg, R. Xie, and T. Sekiguchi, *Appl. Phys. Lett.* **86**, 083107 (2005).

59. G. W. Ho, A. S. W. Wong, D. J. Kang, and M. W. Welland, *Nanotechnology* **15**, 996 (2004).

60. Y. Zhu, W. K. Hsu, M. Terrones, N. Grobert, H. Terrones, J. P. Hare, H. W. Kroto, and D. M. R. Walton, *J. Mater. Chem.* **8**, 1859 (1998).

61. Y. Zhu, W. K. Hsu, W. Zhou, M. Terrones, H. W. Kroto, and D. R. M. Walton, *Chem. Phys. Lett.* **347**, 337 (2001).

62. S. Xie, W. Zhou, and Y. Zhu, *J. Phys. Chem.* **B108**, 11561 (2004).

63. H. Yan, R. He, J. Johnson, M. Law, R. J. Saykally, and P. Yang, *J. Am. Chem. Soc.* **125**, 4728 (2003).

64. J. Lao, J. Wen, and Z. Ren, *Nano Lett.* **2**, 1287 (2002).

65. J. Lao, J. Huang, D. Wang, and Z. Ren, *Nano Lett.* **3**, 235 (2003).

66. J. Lao, J. Huang, D. Wang, and Z. Ren, *J. Mater. Chem.* **14**, 770 (2004).

67. P. Gao and Z. L. Wang, *Appl. Phys. Lett.* **84**, 2883 (2004).

68. P. Yan, C. Ye, X. Fang, J. Zhao, Z. Wang, and L. Zhang, *Chem. Lett.* **34**, 384 (2005).

69. C. Ye, L. Zhang, X. Fang, Y. Wang, P. Yan, and J. Zhao, *Adv. Mater.* **16**, 1019 (2004).

70. J. Hu, Y. Bando, J. Zhan, X. Yuan, T. Sekiguchi, and D. Golberg, *Adv. Mater.* **17**, 971 (2005).

71. H. Wang, X. Zhang, X. Meng, S. Zhou, S. Wu, W. Shi, and S. T. Lee, *Angew. Chem. Int. Ed.* **44**, 6934 (2005).

72. Z. H. Lan, C. H. Liang, C. W. Hsu, C. T. Wu, H. M. Lin, S. Dhara, K. H. Chen, L. C. Chen, and C. C. Chen, *Adv. Funct. Mater.* **14**, 233 (2004).

73. Y. Zhu, H. Hu, W. K. Hsu, M. Terrones, N. Grobert, J. P. Hare, H. W. Kroto, D. R. M. Walton, and H. Terrones, *Chem. Phys. Lett.* **309**, 327 (1999).

74. K. A. Dick, K. Deppert, T. Martensson, W. Seifert, and L. Samuelson, *J. Cryst. Growth* **272**, 131 (2004).

75. D. Wang, F. Qiang, C. Yang, Z. Zhong, and C. M. Lieber, *Nano Lett.* **4**, 871 (2004).

76. K. A. Dick, K. Deppert, M. W. Larsson, T. Mårtensson, W. Seifert, L. R. Wallenberg, and L. Samuelson, *Nat. Mater.* **3**, 380 (2004).

Chapter 4
Quasi-Zero Dimensional Nanoarrays

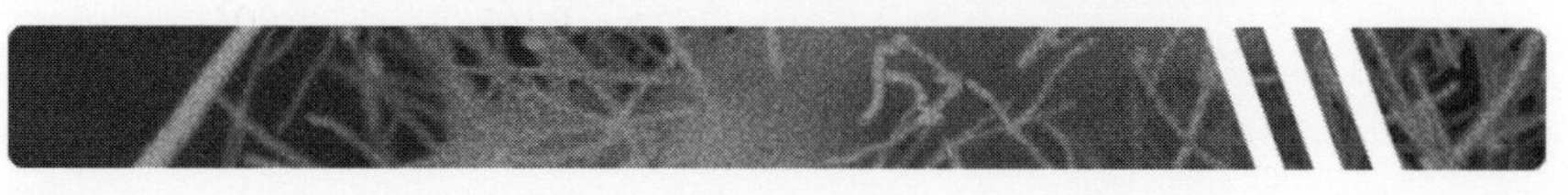

- Synthesis of two-dimensional colloid crystals
- Nanoarrays on two-dimensional colloidal crystal templates

Chapter 4

Quasi-Zero Dimensional Nanoarrays

4.1 Synthesis of two-dimensional colloid crystals

Fabricating smaller and smaller surface nanostructures is one of the important research directions in the nanomaterial field for their superior physical and chemical properties. Their preparation methods continue to increase. For example, traditional lithography techniques, including photolithography,[1,2] X-ray lithography,[3,4] electronic beam lithography,[5] AFM lithography,[6,7] and STM lithography,[8] usually have complex manipulation, high cost and low yield etc. Therefore, it is necessary to look for new methods, which can overcome the above-mentioned disadvantages. This is an important measure for promoting the production of nanomaterials on a large scale, and application of nanomaterials and nanodevices fabrication. The two-dimensional colloidal crystal method is one of the new methods.

The colloidal crystal is a special dispersed system with the particle diameters ranging from 50 nm to 100 μm. Generally, we name the dispersed particles the "colloidal particles". They represent random thermal movement in the liquid medium. The feature of these colloid particles is the uniformity of the size, shape, composition and surface properties of the particles. In the preparation process, under the drive of the minimum free energy of particle interfaces, during the nucleation and growth process these amorphous particles tend often towards forming the sphere shape. Now, commonly used monodispersive colloidal particles are prepared by the inorganic oxide deposition method[9] and the emulsion polymer method.[10,11] For example, the monodispersive SiO_2 colloidal spheres can be prepared by the hydrolysis of $Si(OC_2H_5)_4$, and the monodispersive colloidal spheres of polystyrene can be prepared by polymerization of styrene. Now, there are monodispersive colloidal goods. They may be used in scientific research and industry production.

In recent years, the study of the colloidal crystal has attracted much attention. In nature, there exist many colloidal crystals such as virus, butterfly wings, and opal etc. Opal is a kind of colloidal crystal, which is composed of SiO_2 colloidal spheres. The colloidal crystal is one typical ordered colloidal system and a long-range ordered structure formed from the self-organization of monodispersive colloidal particles (usually, they are polystyrene or SiO_2 colloidal spheres).[11] The colloidal crystal can be a kind of functional material, and can be used in optical gap materials, detectors, and optical lines etc. At the same time, the colloidal crystal can be used as the template for synthesizing other materials, such as two and three-dimensional ordered pore materials, ordered spheres, hollow sphere arrays, periodic nanoparticle arrays, and even ordered one-dimensional nanotubes, nanorods, nanopillar arrays etc. Therefore, the study of the colloidal crystal and the use of the colloidal crystal as a template is highlighted internationally.

In the following, we will emphatically introduce the preparation technique and structural characterization of the two-dimensional colloidal crystal.

The two-dimensional colloidal crystal is a long-range ordered monolayered or double layered structure of a self-organized formation of monodispersive colloidal spheres. The monodispersive colloidal spheres may form the two-dimensional colloidal crystal via self-organization on the solid substrate or the liquid surface.[12,13] The formation process of two-dimensional colloidal crystals may be divided into three types, in which the type of the colloidal crystal forming on the gas-solid interface is a riper formation process. Its applications are also very extensive.[14] All two-dimensional colloidal crystals obtained by the three formation processes are composed of many crystal domains. The number of colloidal spheres contained by the largest crystal domain may reach 10^6.[16] The spheres in the colloidal crystal are in contact with each other. The change of the lattice constant can be realized via modulating the colloidal sphere diameter. In the following, the three formation processes will be simply introduced.

(1) Formation on the liquid-gas interface.[15–18] On the liquid-gas interface, the two-dimensional colloidal crystal formed via self-organization, and then it was transformed onto the substrate by using the Langmuir–Blodgett method. Before using this method, the surface of colloidal spheres must be embellished, so that while being dispersed onto the liquid-gas interface, the particle volume of each colloidal sphere would be immersed into the liquid and hence the two-dimensional

colloidal would be formed via the colloidal spheres attracting each other. The crystal property of the two-dimensional colloidal crystal can be improved via altering the sphere sizes, concentration, the surface hydrophilic property, the surface charge density and the type of support liquid.

(2) Electrophoretic deposition.[19,20] Firstly, a drop of colloidal solution was limited between two conductive glass slides. Then, an electric field was applied to two glass slides. As a result, the random arrangement colloidal crystal spheres with negative charge adsorbed on the anode moved each other to reach an equilibrium state and thus formed the two-dimensional colloidal crystal. The main modulating parameter for this process is the voltage applied to the glass slides.

(3) Self-organized formation of the two-dimensional colloidal crystal on the solid-gas interface.[21] The edge of the colloidal sphere suspension has a larger curvature diameter and higher evaporation rate, and thus the colloidal crystal nucleates firstly at the edge. With further evaporation of the solvent, the solvent convection led the colloidal spheres to be transferred to the edge. The concave moony faces formed between neighboring spheres. Under the action of the surface tension, the colloidal spheres formed the two-dimensional colloidal crystal via self-organization. In the synthesis process, the following conditions are required. The solvent should have a lower evaporation rate. The substrate should have a clean and flat surface (the conduction should be much smaller than the sphere diameter), a uniform chemical composition and good hydrophilic property. The surfaces of the substrate and the colloidal sphere should carry the same kind of charges. Under action of static electric repulsion, the colloidal spheres can maintain a better flow ability in the liquid membrane, resulting in the growth of the colloidal crystal. The diameter deviation of the colloidal spheres should be very small, so that large-scale two-dimensional colloidal crystals can be obtained. The forming process of the two-dimensional colloidal crystal is shown in Fig. 4.1.

The synthesis methods for two-dimensional colloidal crystals include mainly drop coating, spin-coating and perpendicular withdrawing.

4.1.1. Drop coating[22,23]

The drop coating technique was first proposed by Micheletto.[23,24] A droplet of the colloidal sphere suspension was deposited onto a substrate. Then

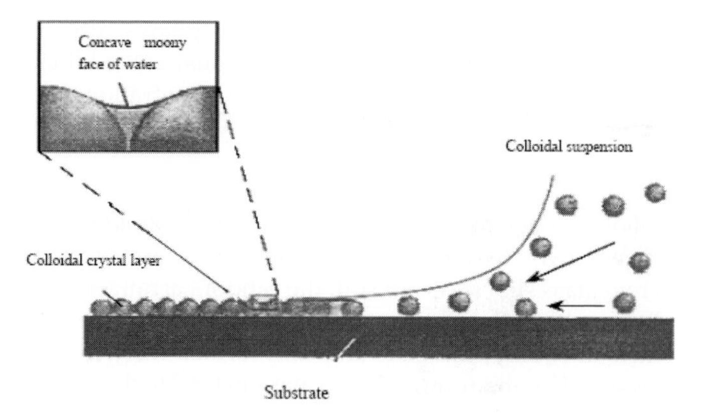

Fig. 4.1. The formation scheme for the two-dimensional ordered colloidal crystal.

the substrate with the colloidal sphere suspension was fixed onto a controlled temperature device, and placed into a closed box. The whole system was tilted about 3° to 15°. The temperature of this substrate was controlled to about 1°C higher than the surrounding temperature. After the solvent was completely evaporated, a monolayer colloidal crystal was formed on the surface of the substrate. The setup is shown in Fig. 4.2.

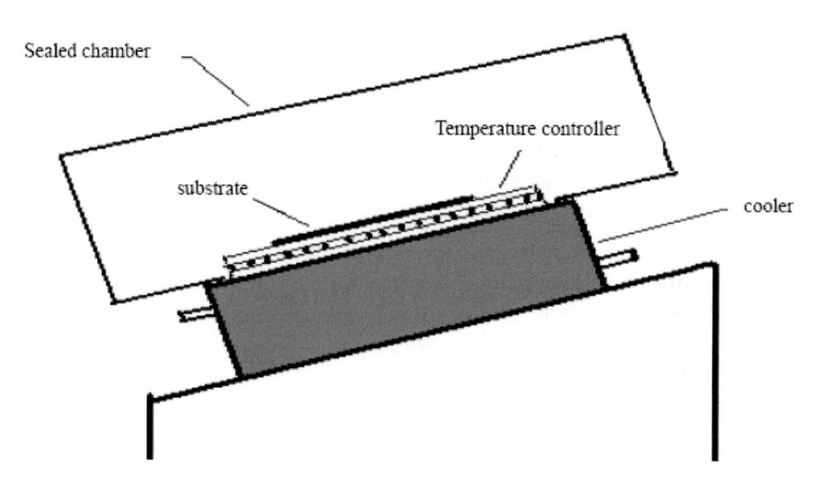

Fig. 4.2. The setup for synthesizing two-dimensional colloid crystals by using the droplet coating method.

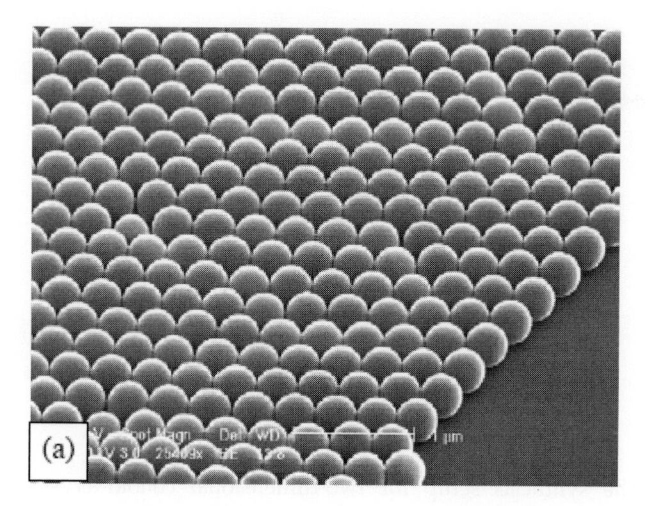

Fig. 4.3. The SEM image of the monolayered two-dimensional colloidal crystal section of ordered arrangement with a large area. The diameters of the colloid spheres are about 350 nm. The gaps between colloid spheres are the pore structure, and these large area gaps form the ordered pore structure.

Figure 4.3 shows the SEM morphology image of a single-layered two-dimensional colloid crystal. It can be seen that the two-dimensional crystal is composed of colloid spheres.

4.1.2. Spin-coating[19,24]

The spin-coating method is where a drop of colloid suspension with a certain colloidal volume is dropped onto a horizontal substrate and then this substrate is rotated at a certain speed. The drop of colloid suspension spreads rapidly on the substrate. When the solvent is evaporated, the colloidal spheres form a monolayered or double-layered colloidal crystal through self-organization. Generally, the smallest diameter of colloidal spheres is 50 nm. When the diameter of colloidal spheres is smaller than 50 nm, the dispersive diameters of spheres and higher charge density lead to the difficulty of effective self-organization for forming two-dimensional colloidal crystals. The suspension concentration and the rotating speed should be suitable, as a overly high concentration or overly low rotation speed can cause formation of double or many-layered structures. Inversely, some voids can form in the two-dimensional colloidal crystals. Usually, when the colloidal spheres are big, a smaller angle speed is used. For the inverse situation, the larger angle speed is used. When the sizes of the

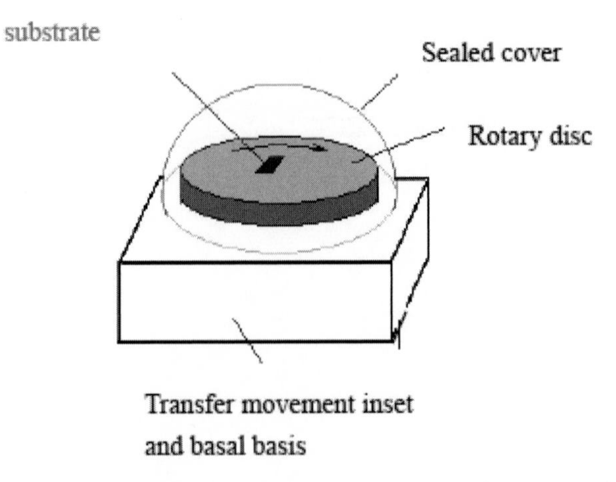

Fig. 4.4. The synthesis setup for two-dimensional colloid crystals by using spin-coating.

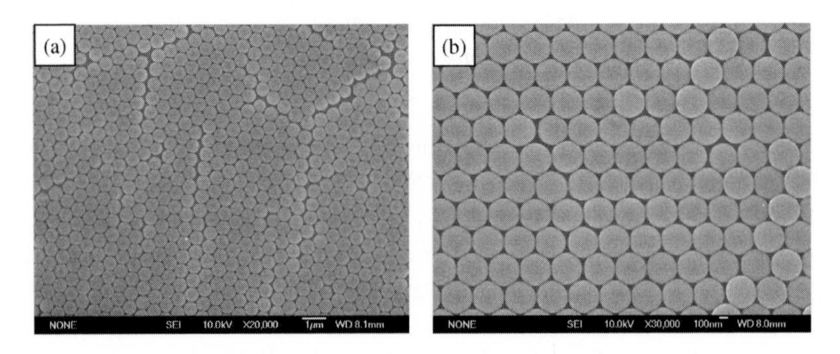

Fig. 4.5. SEM photographs of monolayer two-dimensional colloidal crystals. The diameters of the colloidal spheres are (a) 500 nm and (b) 350 nm, respectively.

spheres are larger, the concentration of the suspension may be suitably high. Figure 4.4 is a schematic drawing of the synthesis setup for two-dimensional colloidal crystals. Figure 4.5 shows the SEM photographs of two-dimensional colloidal crystals.

4.1.3. Perpendicular withdrawing[25]

The perpendicular withdrawing method is where the hydrophilic substrate is immersed in the colloidal suspension and then, the substrate is perpendicularly withdrawn at a certain speed (as shown in Fig. 4.6).

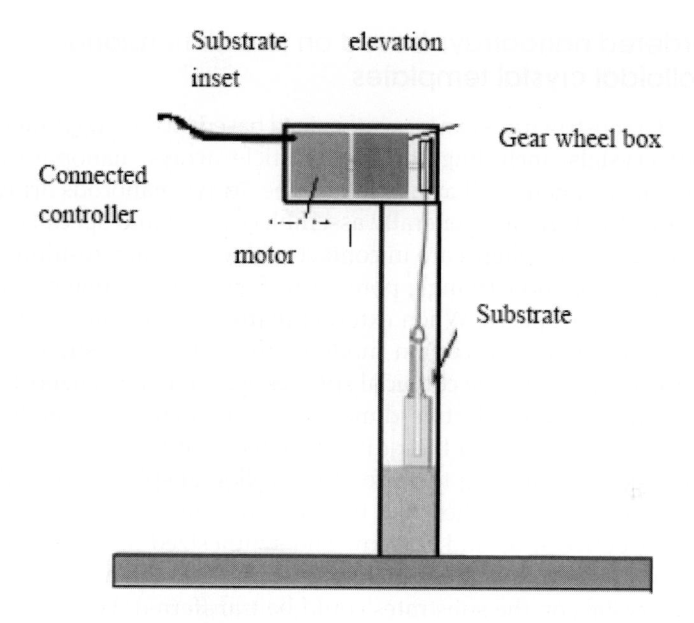

Fig. 4.6. The scheme for synthesizing two-dimensional colloidal crystals using perpendicular withdrawing.

The upper part of the substrate, which emerges from the surface of the colloidal suspension, spontaneously forms the colloidal crystal due to the solvent evaporation. With continuous elevation of the substrate, the colloidal crystal will grow continuously downwards. The withdrawing rate should be equal to the downward growth rate of the crystal along the substrate. The growth rate is associated with the concentration of the colloidal sphere suspension, the evaporation rate and the diameter of the colloidal spheres etc. The smaller the concentration of colloidal sphere suspension, the slower the evaporation rate of the solvent, and the larger the diameter of the colloidal spheres, the smaller will be the growth rate of the colloidal crystal and the slower will be the corresponding withdrawing rate of the substrate. When the concentration, temperature, humidity, and evaporation rate of the colloidal suspension are suitably controlled, the monolayered or double-layered colloidal crystal membranes form on the substrates via self-organization. W. P. Cai has synthesized the monolayered and double-layered two-dimensional colloidal crystals by using the perpendicular withdrawing method.

4.2 Ordered nanoarrays based on two-dimensional colloidal crystal templates

A series of nanostructures can be synthesized based on the two-dimensional colloidal crystals, including the nanoparticle arrays, nanopore arrays, nanoring arrays, nanowell arrays, nanotube arrays, nanorods arrays, etc. These nanostructures are generally assembled on flat and special cleaning substrates, and the spheres are in contact with each other, resulting in the formation of a periodic triangle pore array. Between the spheres and substrate are suitable spaces. When external matter or reaction ions fill these spaces and pores with a certain mode or the bottom substrate of pores are removed, and then, the colloidal spheres are removed, a nanostructure may be obtained. This is the two-dimensional colloidal crystal method. The size of these structures and their distribution density on the substrate can be controlled via changing the size of the colloidal sphere. Now, the colloidal sphere size for synthesizing two-dimensional colloidal crystals can be selected between 50 and 500 nm. The synthesized area can reach the square centimeter order.[22,23] Leiderer[26,27] found that the two-dimensional colloidal crystals on the substrates could be transferred. When water was used as the medium, the colloidal crystal could be transferred onto any substrate and even to the Cu net.

Based on two-dimensional colloidal crystals, the synthesis techniques for two-dimensional colloidal crystals can be mainly attributed to three aspects: (1) The two-dimensional colloidal crystal is used as the mask film to synthesize nanostructures. (2) The two-dimensional colloidal crystal is used as the substrate to synthesize the pore (or ring) structures. (3) Nanostructured arrays of particles or pore extending. So-called extending structures of particles or pores are that the new nanostructured arrays are synthesized based on nanostructured arrays synthesized directly by using the two-dimensional colloidal crystal method. Among the three aspects, nanoparticle arrays[24,28–30] and nanowell arrays[19,31–34] can be synthesized by using the two-dimensional colloidal crystals as the mask films. Here, the preparation process for nanoparticle arrays via nanosphere lithograph will be introduced simply.

The substrate coated with the two-dimensional colloidal crystal was placed in a vacuum plating film chamber. When in a high vacuum of $\leq 10^{-5}$ Pa, and the thermal evaporation, electronic beam evaporation and pulse laser deposition etc. were used, the atoms or molecules of the required material passed the interstices among the colloidal spheres, whose direction of movement was perpendicular to the substrate, resulting in depositing of these atoms or molecules onto the substrate and gradual growing. The deposition thickness was controlled by a quartz vibrator. After the deposition was finished, the colloidal spheres were dissolved by using dichloromethane, resulting in formation of the nanoparticle array.

Using this method, people obtained the metal nanoparticle arrays of Pt,[28] Au,[27,35] Ag,[24,29,30] Ni[36,37] etc., the metal oxide and sulphide nanoparticle arrays of TiO_2,[38] Y_2O_3,[39] and ZnS[39] etc., and the organic nanoparticle arrays of CoPt[24,29,30] etc.

The traditionary preparation methods for the two-dimensional structure include electronic beam lithography,[5] microcontact pressure print,[40] self-organization of polymer[41,42] etc. However, these methods have many disadvantages such as manipulation, expensive apparatus, the difficulty of obtaining large area two-dimensional structures, and the difficulty of controlling morphologies. The two-dimensional colloidal crystal template method overcomes these disadvantages. This makes synthesizing ordered pore films of different materials, morphologies and pore diameters etc. become possible. The nanopore array and nanoring array can be synthesized using electrochemical deposition[43,44] and solution dipping,[45] based on the two-dimensional colloidal crystal template.

In the following, we will introduce the synthesis and characterization of several typical nanoarrays and nanosheets based on two-dimensional colloidal crystal templates, including ordered pore arrays, nanostructured porous films, ordered polymer hollow-sphere, convex structure arrays and nanosheets reported by Cai's group.

4.2.1. Ordered pore arrays

Based on the two-dimensional (2D) colloidal crystal template, Cai's group successfully synthesized various kinds of large-scale 2D ordered pore arrays including metals, semiconductors, and compounds porous films by electrochemical deposition (ECD) or a solution-dipping technique.[43,45–49].

4.2.1.1. *ZnO-ordered pore arrays based on electro-deposition and colloidal monolayers*

Cai[46] synthesized a large-scale ZnO-ordered pore array based on a 2D colloidal monolayer template with controllable morphology using the one-step ECD technique. The pore morphology was controlled from hemispherical to a well-like structure by changing the deposition potential. The preparation process of ZnO-ordered pore arrays was as follows. First, 2D monolayer colloidal crystals were prepared. Then, the colloidal monolayer on the glass substrate was transferred onto the conducting substrates. Briefly, when the glass substrate coated with the PS monolayer was aslant immersed into distilled water in a cup, the integrated monolayer lifts off and floats on the surface of the distilled water. Then, an ITO-glass was used to pick them up. The conducting substrates with the monolayer were covered by an aluminum frame and insulating tape in the marginal parts of the monolayer, and was heated at 350 K for about 3 minutes to sinter the PS on the ITO before it was immersed in the electrolyte aqueous zinc nitrate solution (0.05 M) as the working electrode for ECD. A zinc sheet (99.99%

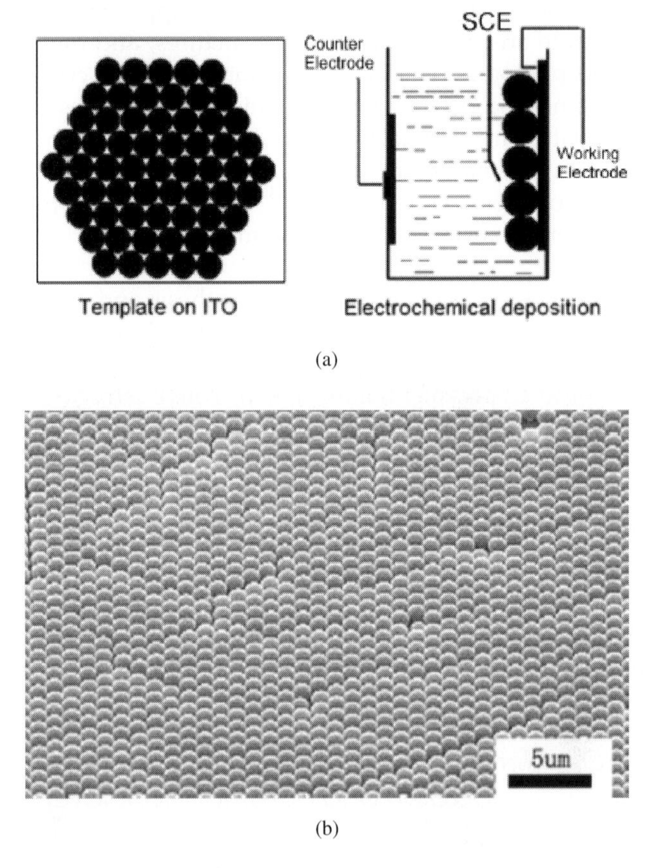

Fig. 4.7. (a) Illustration of electrodeposition. (b) A typical FE-SEM image of a 2D colloidal monolayer crystal.

purity) acted as the counter electrode. A saturated calomel electrode (SCE) was used as the reference electrode. The deposition device is schematically illustrated in Fig. 4.7(a). The deposition temperature was fixed at 355 K in a water bath. The deposition time was 2 h for all samples. Their experiment showed that electrodeposition is very sensitive to the potential and cannot occur below 0.8 V (relative to the SCE). When the potential is higher than 1.4 V, the deposition is too fast to be used. In this work, a deposition potential from 1.0 to 1.4 V was applied.

After deposition, the samples were washed with distilled water and then ultrasonically washed in methylene chloride (CH_2Cl_2) for about one minute to dissolve and remove the PSs, followed by washing with distilled water again and drying before further characterization.

Figure 4.7(b) is the field-emission scanning electronic microscope (FE-SEM) image of a typical 2D colloidal monolayer template, which has been transferred to an ITO-glass substrate and heated at 350 K for 3 minutes. After electrodeposition and the removal of PSs, we can obtain ordered pore arrays. Figure 4.8 shows this structure for the sample under a deposition potential of 1.0 V. The pores are hemispherical in shape and orderly arranged. Further experiments show that the array morphology depends on the deposition potential. With an increase of the potential, the pore skeleton evolves towards a thin wall. When the potential was increased

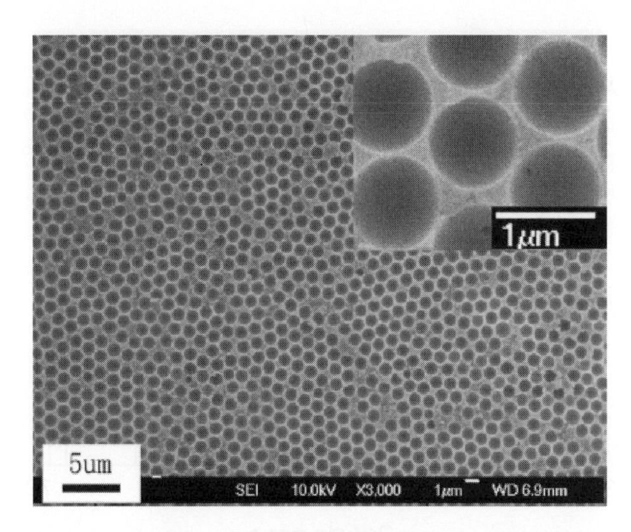

Fig. 4.8. FE-SEM image of pore arrays under the potential of 1.0 V. Inset: a magnified area.

up to 1.4 V, we found that the morphology of the porous film assumes a well-like structure, which is similar to that found in the carbon nanotubes and ZnO nanowire growth processes,[15,16] as illustrated in Fig. 4.9.

Figure 4.10 shows the XRD patterns for such ordered porous films prepared at different ECD potentials of 1.0 V and 1.4 V. The standard data of

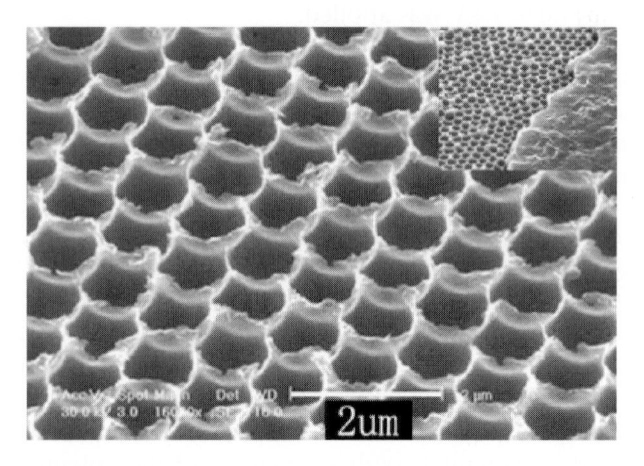

Fig. 4.9. FE-SEM image of pore arrays under the potential of 1.4 V. Inset: ultrasonic washing for a shorter time (see text).

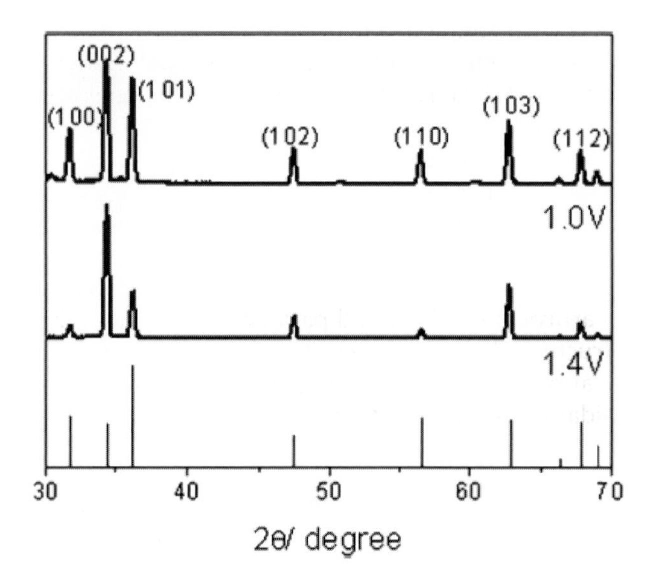

Fig. 4.10. XRD patterns for the samples at different potentials.

ZnO is also presented. All diffraction peaks can be identified as hexagonal wurtzite ZnO (JCPDS 36–1451, $a_0 = 3.253$ Å, $c_0 = 5.209$ Å) with the measured lattice constants a and c equal to 3.25 and 5.21 Å, respectively, ($c/a = 1.60$), which indicates the purity of the sample. In addition, ZnO crystal growth exhibits significant preferential orientation along the [001] crystal axis with an increase of the potential.

According to previous discussions,[50] it was suggested that the reduction of the nitrate (NO_3^- to NO_2^-) in a mild acid solution of Zn^{2+}, which results in increase of the pH value near the electrode surface, is crucial. With an increasing concentration of OH^-, $Zn(OH)_2$ will form and deposit on the cathode electrode. The deposited $Zn(OH)_2$ will subsequently decompose and form ZnO on the substrate at the temperature of the water bath. The morphology evolution of the ZnO-ordered pore array with deposition potential could be attributed to the different deposition rate or the growth rate of crystallites when the applied potential was changed. As we all know, the high potential should correspond to a high deposition rate. At a low electro-deposition potential, the crystallites on the substrate grew slowly and sufficiently filled the interstices among PSs in the monolayer, leading to hemi-spherical hollow arrays after 2 h of deposition and removal of PSs. On the contrary, at a higher potential, the crystallites grew rapidly and could not fully fill the interstices, resulting in a nanowall-like structure. Because of the high deposition rate, PSs were completely buried in the deposited film before the interstices were fully filled. Subsequent dissolution of PSs in CH_2Cl_2 solution and ultrasonic washing induced the upper part of the film to disconnect at the contact area between adjacent PSs. This has been confirmed in the inset of Fig. 4.9, which corresponds to the sample in Fig. 4.9 but with ultrasonic washing for a shorter time (only partially removing the upper film). We can see that the upper part of the film in some areas is still not disconnected. Also, the preferred growth of crystals at the higher potential may be attributed to the rapid deposition rate.

After the above results were reported, Cai[47] found that the obtained ZnO-ordered arrays are scalable and parallel but all randomly oriented in the crystal structure. Therefore, they prepared the highly oriented ZnO-ordered pore arrays by choosing a proper cathodic electrode (substrate). The 2D colloidal crystal preparation was the same as the one described above.

Au/Si substrate was prepared by thermal evaporation of gold in a vacuum of 1×10^{-5} Pa. The Au deposition rate was about 0.5 Å/min. This was controlled by a film thickness monitor (FTM-V, Shanghai). A thickness of about 100 nm of gold layer was deposited onto a single-crystal silicon

(100) substrate with low resistivity smaller than 10 Ω cm. Scanning electron microscopy (SEM) images of the as-prepared gold electrodes showed a smooth surface and X-ray diffraction (XRD) measurements showed a strong preferential orientation of Au (111) planes parallel to the substrate. Some commercial glass slides coated with a layer of 25 nm conductive indium-tin oxide (ITO-glass) were used as received for reference. The square resistance is about 150 Ω/m^2.

Finally, the colloidal monolayer on the glass substrate was transferred onto the conducting substrates. Briefly, when the glass substrate coated with the PS monolayer was aslant immersed into distilled water in a cup, the integrated monolayer lifts off and floats on the surface of the distilled water. Then, an ITO-glass or Au/Si substrate was used to pick it up. The conducting substrates with the monolayer were covered by an aluminum frame and insulating tape in the marginal parts of the monolayer, and heated at 350 K for a short time (about three minutes) to sinter the PSs on them. Figure 4.11 shows the microstructure of the PS colloidal monolayer on ITO-glass substrate after such heating. Then they were immersed in the electrolyte of zinc nitrate aqueous solution (0.05 M) as the working electrode for ECD and the electro-deposition area was about 1 cm^2. A zinc sheet (99.99% purity) acted as the counter electrode. Galvanostatic and

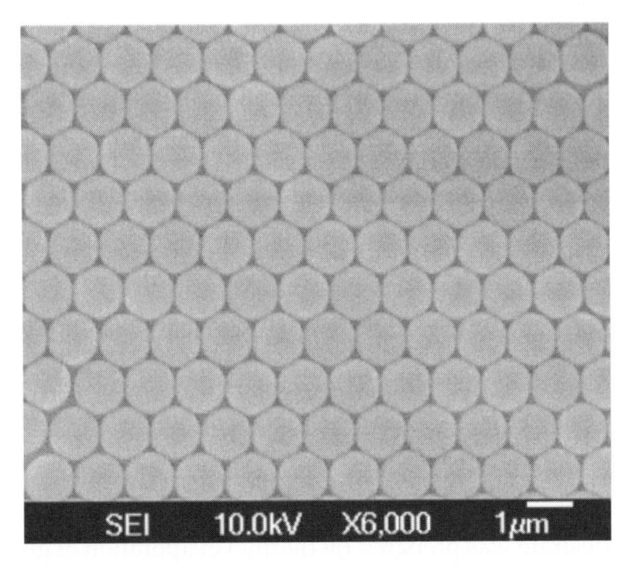

Fig. 4.11. FESEM of the PS colloidal monolayer on the ITO-glass substrate after heating at 350 K for 3 minutes.

potentiostatic cathodic deposition were employed on the Au/Si substrate and ITO-glass substrates, respectively. The deposition was performed at 355 K in a water bath and the deposition time was 120–150 minutes for different samples.

After deposition, the samples were washed with distilled water and then ultrasonically washed in methylene chloride (CH_2Cl_2) for about one minute to dissolve and remove the PSs, followed by washing with distilled water again and drying before further characterization. The morphology of the samples was examined by field emission scanning electron microscopy (FESEM) (JEOL-6700). XRD was measured on a Philips X'Pert diffractometer (40 KV, 40 mA) using Cu-K_α line (0.15419 nm).

Figure 4.12 shows the results for the samples on ITO-glass and Au/Si substrates, respectively, in which the standard powder XRD pattern of ZnO crystal is also presented. All diffraction peaks of the sample deposited on the ITO-glass substrate can be identified as of wurtzite ZnO and there exists a slightly preferred orientation of (002) in the polycrystalline ZnO film deposited in the potential range of 1.0–1.4 V. For the samples electrodeposited on the Au(111)/Si substrate, however, only two peaks at 34.4° and 38.2°, indexed by ZnO (002) and face-centered-cubic gold (111) peaks respectively, were observed, indicating that the film is highly oriented in the crystal structure and most of ZnO {001} and Au {111} planes are parallel

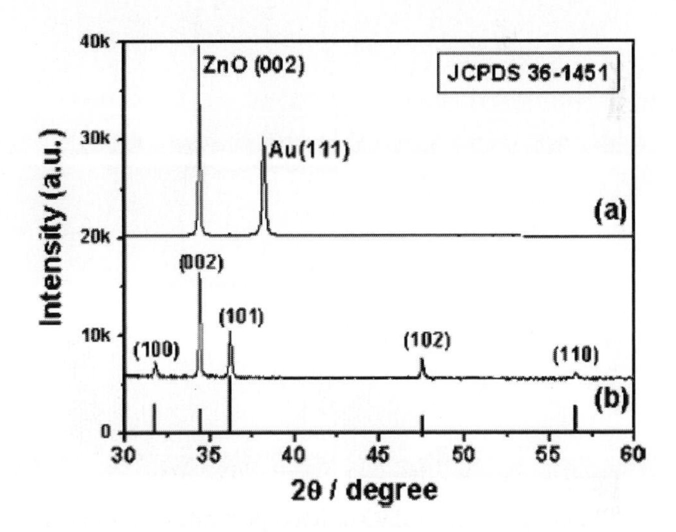

Fig. 4.12. XRD spectra of as-synthesized films electro-deposited on different substrates (a) Au/Si (0.9 mA, 120 minutes) and (b) ITO-glass (1.0 V, 120 minutes).

to the substrate surface [the peak of Si (400) is at 69°]. All samples electro-deposited on Au/Si substrates in the current range of 0.7–1.2 mA show the similar pattern of Fig. 4.12(a).

The morphologies for the samples electro-deposited on ITO-glass evolve from the truncated hollow spherical array to nanowall-like structured array with the deposition potentials changed from 1.0 to 1.4 V. Figure 4.13(a) shows the morphology corresponding to the deposition potential of 1.0 V, indicating the truncated spherical hollow array. For the samples on Au/Si substrates, however, the deposition currents have little effect on the film morphology, showing similar morphology in the current range of 0.7–1.2 mA, as shown in Fig. 4.13(b)–4.13(d), corresponding to the applied current of 0.9 mA. The pores are highly ordered in the film. This is the reverse replica of the PS colloidal monolayer template. From its magnified SEM image (Fig. 4.13(c)), we can see that the skeleton seems to be packed with the block-units ZnO nanosheets parallel to substrate surface, or that the morphology shows a step-structure from top view. However,

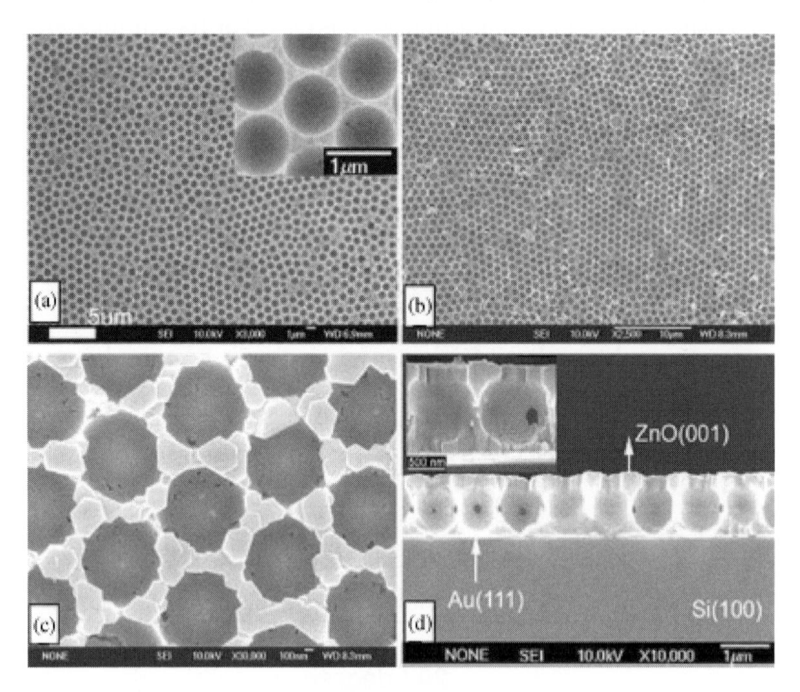

Fig. 4.13. FESEM morphology of ZnO-ordered pore arrays. (a) ZnO film on ITO-glass substrate (1.0 V, 120 minutes). (b) ZnO film on Au/Si substrate (0.9 mA, 120 minutes). (c) Magnified image of (b). (d) Cross-section image of (c).

the cross-section morphology (Fig. 4.13(d)) demonstrates that the skeleton is composed of densely aligned ZnO hexagonal nanocolumns with a small difference in heights. This is consistent with the ZnO wurtzite crystal structure and indicates that such films are of good crystal quality.

As for the ZnO film growth mechanism, a one-step cathodic electrochemical deposition reaction mechanism is widely accepted.[51,52] From the chemical reaction point of view, the reduction of NO_3^- results in the increase of OH^- ionic concentration in the solution. Then OH^- and Zn^{2+} will form $Zn(OH)_2$ (or $Zn(OH)_4^{2-}$) on the cathode electrode. The deposited $Zn(OH)_2$ will subsequently be decomposed and form ZnO on the substrate at the temperature of the water bath (~335 K). The electro-deposition of ordered pore arrays, such as gold and ZnO, on the ITO-glass substrate which is based on the colloidal monolayer template, has been well-illustrated in Cai's previous works.[43,45,46] They also found that the film morphology is sensitive to the electro-deposition parameters, such as applied potential and deposited time. Here, we mainly discuss the effect of the substrates. As known, the overvoltage for hydrogen evolution and therefore the surface pH could be different for two substrates. This would lead to different deposition rates and hence different surface morphologies of the films.[51] However, if ZnO nuclei were randomly oriented, the different overvoltage on the Au substrate would not induce the oriented film. The oriented growth of the porous ZnO film can be attributed to the oriented substrate, which leads to formation of oriented nuclei due to the lattice-match between the substrate and ZnO.

The electro-deposition conditions on the whole working electrode (substrate) in the electrolyte were homogeneous and, hence, the nucleation can occur at any site on the substrate that was not covered with the PS monolayer. Because the ITO-glass substrate is amorphous in structure and there is no influence of epitaxy during the initial nucleation period, the crystal nuclei of ZnO will be randomly oriented on the substrate, leading to the ZnO skeleton without obvious preferred orientation, as shown in curve (b) of Fig. 4.12. The slightly preferred orientation of (0002) can be attributed to the ZnO polar (0001) crystal plane, which has higher surface free energy compared with the other basal planes of $(01\bar{1}0)$ and $(2\bar{1}\bar{1}0)$. For the (111)-oriented Au/Si substrate, however, preferentially-oriented ZnO nuclei will be formed on the substrate so as to reach the lowest interface energy between the ZnO and the gold film. It is well-known that the interface energy is directly related with the lattice mismatch of interfaces.[53] To lower the interface energy, the orientation relationship between the ZnO nuclei and the Au (111)/Si substrate can be deduced as (1×1) ZnO (0001)

$[11\bar{2}0]$ / (1 × 1) Au (111) $[\bar{1}10]$. Figure 4.14 schematically shows such a relationship, which results in the smallest lattice mismatch between the ZnO film and the substrate. The lattice mismatch along ZnO $\langle 11\bar{2}0\rangle$ and Au $\langle \bar{1}10\rangle$ is about 12.7%. In addition, fast growth along the *c*-axis direction is energetically favorable due to the higher surface free energy of polar (0001) crystal planes. Thus, the (001)-oriented ZnO-ordered pore arrays with flat top surfaces and hexagonal nanocolumns were formed on the

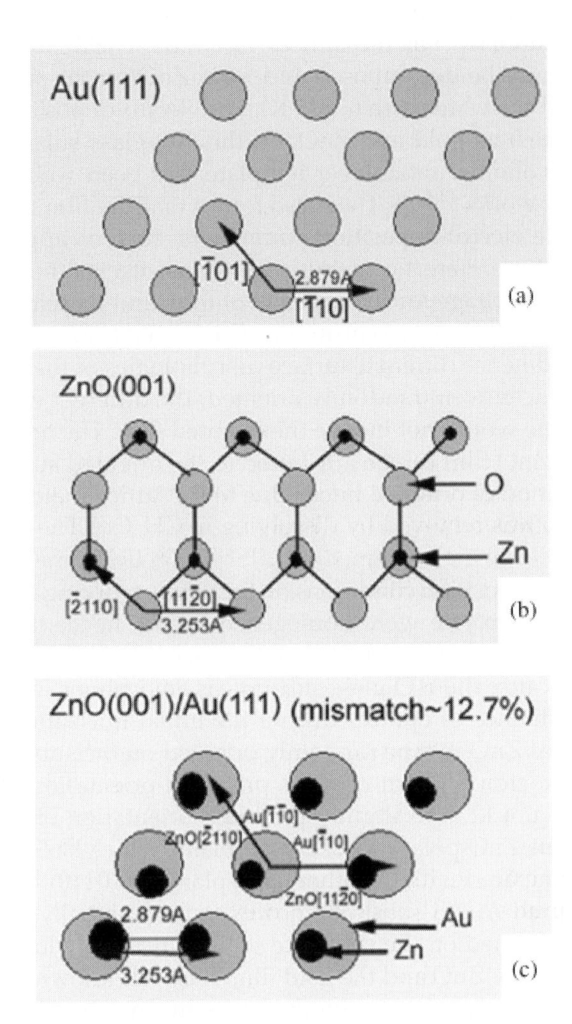

Fig. 4.14. Schematic drawings of crystal planes (a) Au (111), (b) ZnO (001), and (c) the epitaxial relationship of ZnO (001)/Au(111).

(111)-oriented Au/Si substrates by preferentially oriented nucleation and subsequent thermodynamically favored growth.

4.2.1.2. *Au-ordered through-pore arrays based on electro-deposition and colloidal monolayers*[43]

Based on electro-deposition and a colloidal monolayer, Cai have synthesized a series of ordered through-pore films, such as Cu, Zn, Ag, Ni, ZnO, Eu_2O_3 and Fe_2O_3. Section 4.2.1.1 described the synthesis and characterization of ZnO-ordered pore arrays. In this section, we chose gold as a good example to demonstrate the morphology-controlled fabrication of 2D ordered pore-arrays based on electro-deposition and a colloidal monolayer. The preparation process of the ITO-coated glass substrate with the PS monolayer was similar to that as described in section 4.2.1.1, except that this substrate-heating temperature was 383 K. This substrate was used as the working electrode electrolytic cell, with a graphite plate as the auxiliary electrode, and a saturated calomel electrode (SCE) as the reference electrode. A solution composed of $HAuCl_4$ (12 gL^{-1}), EDTA (5 gL^{-1}), Na_2SO_3 (160 gL^{-1}), and K_2HPO_4 (30 gL^{-1}) was used as the electrolyte. Its pH value was 5. A graphite plate and a saturated calomel electrode (SCE) were used as the auxiliary and reference electrode, respectively. The distance between the working electrode and the auxiliary electrode was about 4 cm. The electro-deposition was carried out at 45° and 0.7 V versus SCE. After deposition, the monolayer was removed by dissolving in CH_2Cl_2. The morphologies of all the samples were examined on a JSM-6700 field-emission scanning electronic microscope (FE-SEM).

Figure 4.15 shows the through-pore arrays after substrate-heating and subsequent electro-deposition for different times respectively. All samples were electro-deposited under a potential of 0.7 V versus SCE. Upon increasing the heating time of the monolayer-coated substrate before deposition (the PSs are 1000 nm in diameter), the openings at the film bottom evolved from an irregular shape (Figs. 4.15(a) and 4.15(b) to circles (Figs. 4.15(c) and 4.15(d)), and their sizes also increased due to heating-induced rise in contact area between PSs and the substrate. With a heating time of 40 minutes, the diameter of pores at the bottom is about 800 nm, while near that of the PSs, and the pores are nearly cylindrical, as shown in Fig. 4.15(d). The film thickness and openings at the film surface can be controlled by the electro-deposition time. Samples (a) and (b) in Fig. 4.15 were electro-deposited for 15 minutes. The pore diameters at the film surface are nearly equal to those of the PSs, and almost all the circles at the film

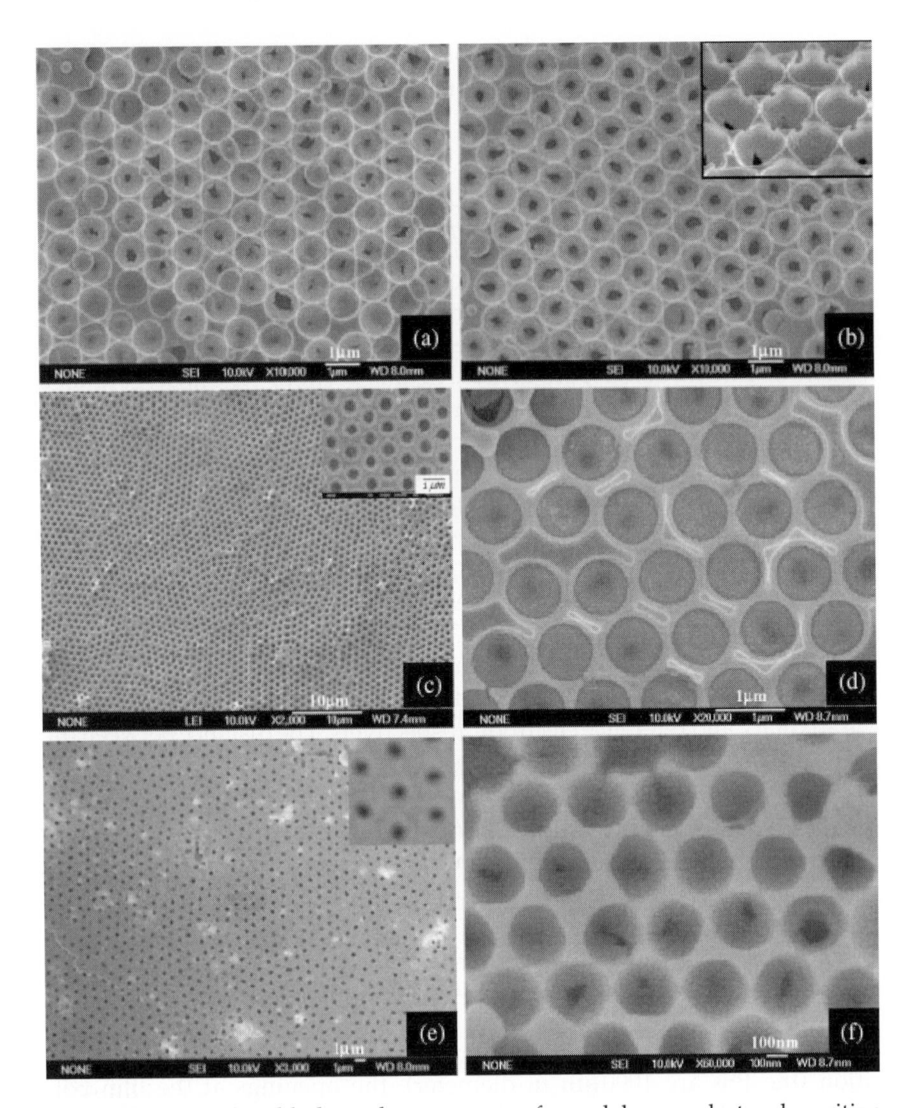

Fig. 4.15. Ordered gold through-pore arrays formed by an electro-deposition method based on a colloidal monolayer. The polystyrene sphere sizes in the monolayer are 1000 nm for (a)–(e) and 350 nm for (f). All the templates were heated to 110°C at an electro-deposition potential of 0.7 V versus SCE . The heating time and the deposition time are, respectively, (a) 2 minutes and 15 minutes; (b) 5 minutes and 15 minutes; (c) 16 minutes and 8 minutes; (d) 40 minutes and 8 minutes; (e) 16 minutes and 30 minutes; (f) 3 minutes and 5 minutes. The insets of (c) and (e) are the corresponding magnified images. The inset of (b) is a tilt view.

surface are in contact with each other. An approximate calculation according to a simple geometrical relationship between the spheres,[53] gives film thicknesses of 310 nm and 320 nm, respectively.

With a shorter electro-deposition time (8 minutes), the film thickness is thinner and the pore diameters at the film surface are smaller than those of PSs, therefore they become separated from each other due to the limitation of the template geometry, as shown in Fig. 4.15(c) (about 160 nm in thickness). However, if the electro-deposition is long enough, the film thickness will be greater than the distance between the center of the PSs and the substrate, and the pore diameters at the film surface will also be smaller than those of PSs, as illustrated in Fig. 4.15(e) (about 940 nm in thickness). Samples (c) and (e) were heated for the same time (16 minutes) before deposition, but the latter was electro-deposited for longer (30 minutes). Both show a similar morphology from the top view. However, there should be a tunnel between two adjacent pores for sample (e), which is induced by the contact area between two adjacent PSs due to heating before deposition.[54]

Similar results were obtained with monolayer templates composed of much smaller PSs. Figure 4.15(f) shows the ordered through-pore array fabricated in the 350 nm PS monolayer with heating and electro-deposition times of 3 and 5 minutes, respectively. It exhibits the same regularity as those formed in the 1000 nm PS monolayer, except for the pore sizes.

Interestingly, such an ordered porous film can be transferred integrally from one substrate to another by lifting it off on a water surface and picking it up with another substrate, as shown in Fig. 4.16. These substrates can be flat or curved, and the front and back surfaces of the films can be chosen according to the application requirements. Figures 4.17(a) and 4.17(b) show the ordered pore arrays on a mica substrate after transfer. These images correspond to the back surfaces of the deposited films shown in Figs. 4.15(b) and 4.15(c), respectively. We can thus see more clearly the morphologies of the openings at the pore bottom for the original through-pore films in Figs. 4.15(b) and 4.15(c). The sizes of the bottom openings are about 300 nm and 400 nm, respectively. Figures 4.17(c) and 4.17(d) show the ordered porous film on a type of porcelain tube used for a gas sensor. The film corresponds to that shown in Fig. 4.15(b). The inset in Fig. 4.17(c) is a photograph of the tube covered with the gold film and Fig. 4.17(d) is an enlarged image of the zone indicated in Fig. 4.17(c). We can see that the film covers the curved surface and that the microstructure has not changed, except that it looks like a tilt view of Fig. 4.15(b) due to the curved surface. This is of importance as it means that we can transfer the film to any sub-

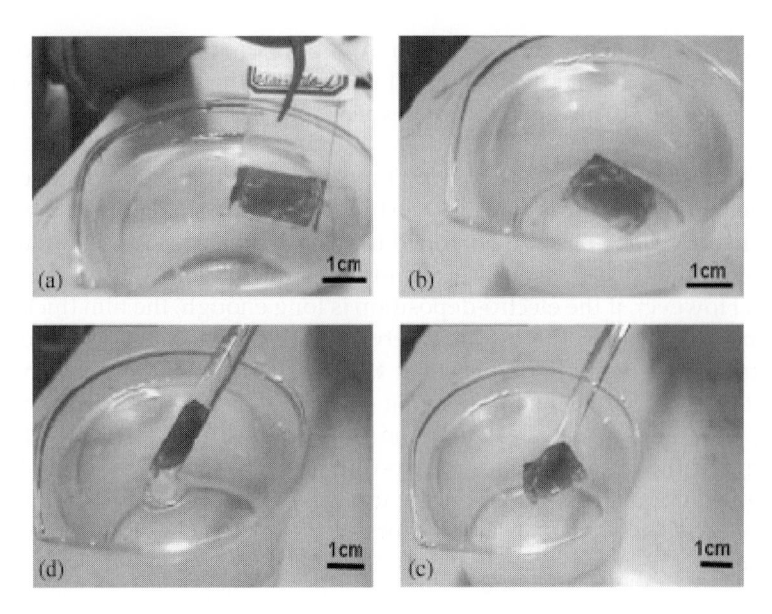

Fig. 4.16. Photographs showing the transfer of the ordered gold porous film on ITO-glass to a curved surface (glass rod). (a) The ITO-glass covered with gold film is dipped into water. (b) The gold film is floating on the water. (c) The gold film is picked up with a glass rod. (d) The gold film is transferred onto the curved surface of the glass rod.

strate, especially to a desired but insulating substrate, like mica or a curved surface, on which the film cannot be electro-deposited directly, thus overcoming the restriction of electro-deposition onto a conducting substrate.

Further experiments showed that such transferability is related to the electro-deposition rate (or deposition potential) and additive agents in the electrolyte, although it is mainly determined by the film-growth mechanism, which itself depends on the nature of the electrolyte. Here, the electrolyte contains $HAuCl_4$, Na_2SO_3, $C_{10}H_{14}N_2O_8Na_2 \cdot 2H_2O$ (EDTA) and K_2HPO_4. The $HAuCl_4$ is the source of the initial Au, some of which immediately forms Au(I).[55,56] Na_2SO_3 acts as a kind of complexing agent, EDTA improves the mechanical properties of the final films, and K_2HPO_4 mainly acts as an auxiliary complexing agent and buffering agent. When the ITO-glass covered with the PS monolayer is dipped into the electrolyte, HPO_4^{2-} is adsorbed onto the substrate before deposition. However, the substrate has already been structured with PSs. The PSs possess negative charge — the surfaces of the PSs were previously treated with SO_4^{2-} — and this will

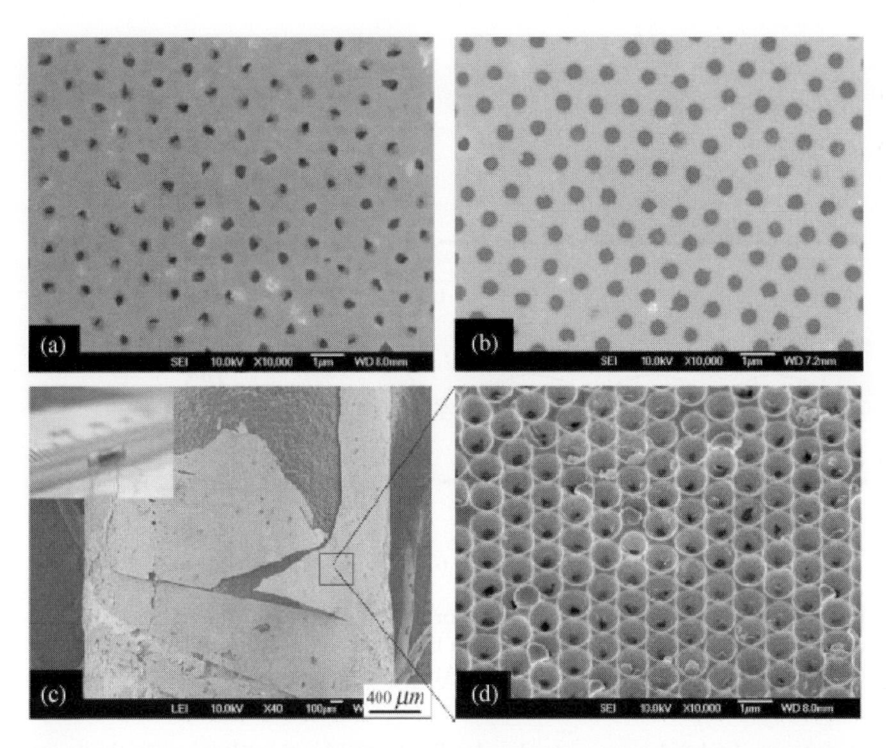

Fig. 4.17. Ordered gold porous films on a mica substrate and on the curved surface of a porcelain tube by lifting off the films from ITO-glass. (a) The back surface of the film shown in Fig. 4.15(b), on mica. (b) The back surface of the film shown in Fig. 4.15(c), on mica. (c) A low-power FE-SEM image of the gold film shown in Fig. 4.15(b), on the curved surface of a porcelain tube; the inset is a photo of the tube covered with gold film. (d) A magnified image of the zone shown in (c).

influence the properties of ITO around the spheres. As a result, the HPO_4^{2-} will be adsorbed onto the substrate away from the spheres, as illustrated in Fig. 4.18(a). (It is well-known that the adsorbates on the substrate impede the growth of a film.) When the electro-deposition begins, the Au crystal nuclei will naturally first form in the wedge-shaped regions (or corners) between the PSs and the substrate due to the low energy barrier of nucleation in this area (see Fig. 4.18(b)). Since the applied voltage (only 0.7 V versus SCE) during electro-deposition is not high enough to desorb the adsorbates from the substrate, the gold grows along the substrate and the PSs from the corners will also cover the adsorbed on the substrate. As a result, the thickness of the film at the edge of the pores (at the film surface) will always be thicker than that in the interstitial region between three

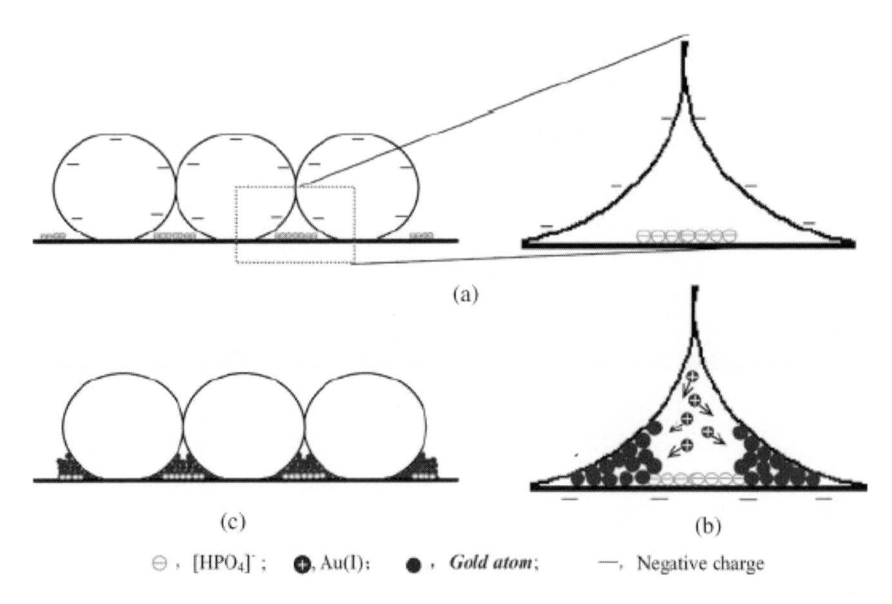

Fig. 4.18. Schematic illustration of the formation process of ordered gold porous films. (a) Monolayer-coated ITO-glass in the electrolyte before electro-deposition. (b) During initial electro-deposition. (c) After electro-deposition.

closely-packed spheres, as is clearly illustrated by the tilt view in the inset of Fig. 4.15(b).

In order to further understand the growth mechanism, we chose to examine an edge region of the sample shown in Fig. 4.15(d) (Fig 4.19). The pores are not ordered, but it is still possible to determine the details of the film growth. In the area without PSs, there are no deposited films; the film only grows around PSs. When a single PS is found on the substrate (zone (a) in Fig. 4.19), the gold grows around the sphere and forms a hemisphere-like shell of fixed thickness. However, when two spheres are arranged close to each other (zone (b)), the region between the two spheres and the substrate is obviously different due to the surface charge of the spheres. The amount of adsorbate is less, and as a result, the growth in such regions will be quicker. For the same reason, when three spheres are arranged close to each other (zone (c)), the growth in the region between the three spheres and the substrate is also quicker than that in other regions. A more obvious example is that when the spheres are arranged neither too close nor too far from each other (zone (d)). The pores (from the top view) are connected by a film of a certain thickness, and the thickness of the film between the pores is smaller than that surrounding the pores. On this basis, it is easy to understand the

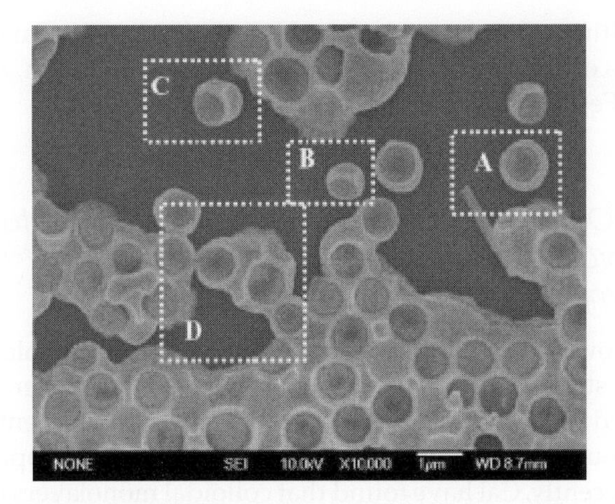

Fig. 4.19. Morphology in the edge region of the sample shown in Fig. 4.15(d). (a) The pore from one sphere. (b) Pores from two spheres in close contact. (c) Pores from spheres not in very close contact.

formation of the closely-packed pore array shown in Fig. 4.15 if all the PSs are closely packed in monolayers.

The existence of an adsorption layer in some areas between the Au film and the substrate (see Fig. 4.18(a)) weakens the adhesive force between the two, and means that the film can be stripped by the surface tension of water. This can be demonstrated more clearly by increasing the electro-deposition rate (or potential). When the potential is increased to 1.7 V versus SCE, the film does not float off in water. We believe that such a potential is high enough to remove the adsorbates quickly during the initial deposition, therefore the gold film has a much stronger adhesive force with the substrate,[57] and hence cannot be stripped. For the same reason, if we do not add K_2HPO_4 into the electrolyte, the film cannot be stripped at any potential, as there is no adsorption on the substrate. This situation has also been found for the electro-deposition of copper.

If the electro-deposition onto the monolayer-coated substrate is performed in two different electrolyte solutions separately, a bimetal (or bilayer) ordered porous film can be synthesized. Cai have obtained Au/Cu bilayer porous film (not shown here). Further experiments demonstrated that the strategy of electro-deposition based on a colloidal-monolayer template is universal and can be used for synthesis of other ordered through-pore arrays, including metals, semiconductors and metal oxides on any conducting substrate, or on an insulating substrate with a flat or curved surface by lifting off and transferring. Using this strategy, Cai have synthesized a series of ordered through-pore films, such as Cu, Zn, Ag, Ni, ZnO, Eu_2O_3, and Fe_2O_3.

4.2.1.3. *SnO_2 mono- and multi-layered nanostructured porous films based on solution-dipping templates*[48]

As is well-known, the polystyrene sphere (PS) colloidal monolayer on a substrate can be stripped off in water and transferred onto other substrates.[58] Cai have used this phenomenon in their previous experiments to prepare ordered porous films on a flat substrate by a solution-dipping template strategy.[59] Recently, Cai have found that colloidal monolayers on glass substrate can also be peeled off in precursor solutions other than water, while still retaining their integrity. Based on these findings, Cai present a simple but very effective strategy to directly fabricate porous films on curved (or non-flat) substrates by using a colloidal monolayer as a template. This procedure is schematically illustrated in Fig. 4.20. First, a flat ordinary glass substrate covered with a monolayer of PS sphere colloidal crystal (area of coverage more than 1 cm^2), is slowly immersed in a precursor solution. The colloidal monolayer can be stripped off from the substrate due to the surface tension of the solution and the difference in the wettability between the glass substrate and the colloidal crystal.[58,59] The monolayer then floats on the surface of the precursor solution (see photographs in Fig. 4.20). Due to capillarity effects, the interstitial space between the closely-packed spheres is filled with the solution. In the second step, the floating colloidal monolayer is picked up by a desired substrate with a flat or curved surface (a glass rod with a curved surface is shown here). The colloidal monolayer covers the curved surface of the substrate, and the interstitial spaces between the PS spheres and the space between the monolayer and the substrate is filled with solution. In the next step, the substrate, covered with the monolayer, is dried at a temperature slightly above the glass-transition temperature of the PS. As the solvent evaporates, the solute or hydrolyzate gradually deposits on the PS sphere's surfaces and on the substrate. Finally, the sample is calcined to remove the PS spheres, and an ordered porous film is thus formed on the curved surface of the substrate. Since this film grows directly on the surface during heat treatment, there is strong adherence between the film and the substrate. By using this method, people can fabricate the ordered porous films on any curved surface or flat surface. More importantly, the films can also be directly synthesized on relatively rough surfaces of commercially available ceramic tubes. An important application

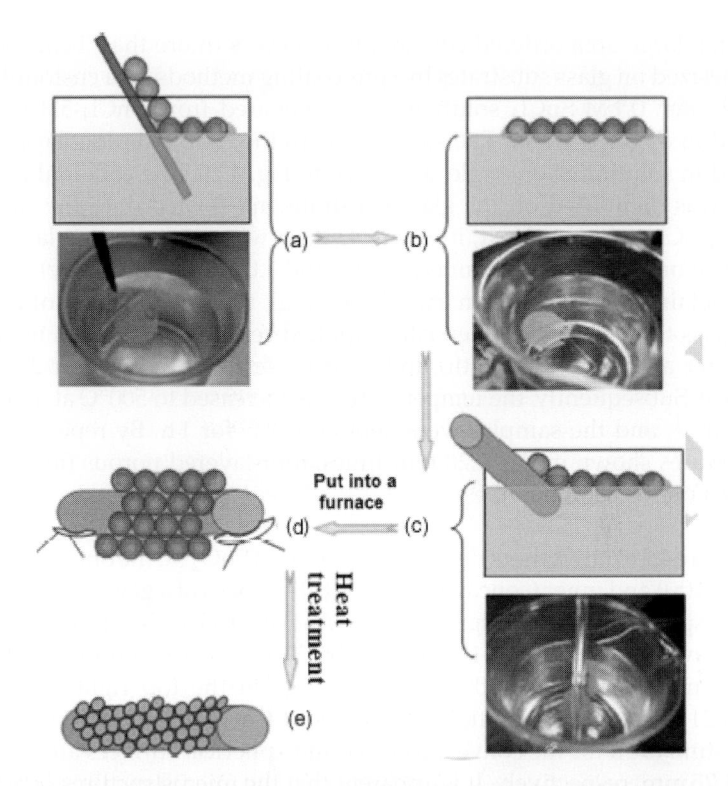

Fig. 4.20. Strategy for synthesizing ordered porous film on curved surface: (a) A flat glass substrate covered with a PS colloidal monolayer is dipped into the solution. (b) The colloidal monolayer floats onto the surface of the solution. (c) The monolayer is picked up using a (glass) rod with curved surface. (d) The rod with the monolayer and the solution is heated in a furnace. (e) An ordered porous film is formed on the curved surface after heat-treatment and removal of the PS spheres. The lower frames in (a)–(c) are photographs corresponding to manipulations described.

of this strategy is the direct construction of nanostructured gas sensors on ceramic tubes, which are used as substrates for conventional gas sensors.

Here, we have chosen SnO_2 as a model material to demonstrate the synthesis strategy described above. SnO_2 has been chosen as it is one of the most attractive materials for the semiconductor gas sensor due to its simple fabrication, high sensitivity, and ability to detect low levels of reducing gases. The SnO_2 mono- and multi-layered nanostructured porous films were synthesized as follows.

First, large-area ordered colloidal monolayers (more than 1 cm^2) were synthesized on glass substrates by spin-coating methods on a custom-built spin-coater. 0.1 M $SnCl_4$ solutions were prepared from $SnCl_4 \cdot 5H_2O$ dissolved in distilled water. The substrate with the colloidal monolayer was placed in solution at an angle, as shown in Fig. 4.20. The colloidal monolayer was then lifted off the glass substrate and floated onto the surface of the $SnCl_4$ solution. Next, it was picked up with an ordinary glass tube (with an outer and inner diameter of 1.2 and 1.0 mm, respectively), a steel ball (obtained from an ordinary ball-bearing), the inner surface of a broken glass tube, and a commercially supplied ceramic tube (2 mm in outer diameter and 5 mm in length), followed by drying at 120°C for 2 h in a furnace. Subsequently, the temperature was increased to 500°C at a rate of 6°Cmin^{-1}, and the samples were kept at 500°C for 1 h. By repeating the procedures shown in Fig. 4.20 four times, four-layered porous films were formed on the ceramic tubes.

Figure 4.21 shows the ordered porous films of SnO_2 on the outer (convex, Fig. 4.21(a)) and inner (concave, Fig. 4.21(b)) surfaces of a glass tube, as well as on a spherical surface (Fig. 4.21(c)), and a flat surface (Fig. 4.21(d)).These films have been fabricated using 0.1 M $SnCl_4$ precursor solutions and colloidal monolayers with 1000 nm PSs. Insets in the top right corner of Fig. 4.21 show low-magnification images of the corresponding samples. Curvature radii for the convex, concave and spherical surfaces are 0.6, 0.5, and 1.25 mm, respectively. It is apparent that the microstructures of all the ordered porous films are similar. The pore openings at the film surfaces are nearly circular, and their diameters are obviously smaller than the diameters of the PS spheres. A small triangular hole exists at the interstitial position between three closely-packed spheres (from top view), and a tunnel hole exits in the pore walls between two adjacent pores (see Fig. 4.21(a) and the inset of Fig. 4.21(b), bottom left). X-ray diffraction images (not displayed here) show that the films are composed of SnO_2 after drying at 120°C for 2 h. However, after removal of PS spheres by dissolution in CH_2Cl_2, such films are easily destroyed. After heat treatment at 500°C, these films adhere so strongly on the curved surfaces that they cannot be removed or damaged, even by ultrasonic washing.

Now, we shall further discuss the formation of porous films on curved surfaces. Formation of the ordered porous films can be described to occur in three steps, as depicted in an exaggerated way in the schematic in Fig. 4.22. After the transfer of the solution containing the colloidal monolayer onto the curved surfaces, the organization of PS spheres does not

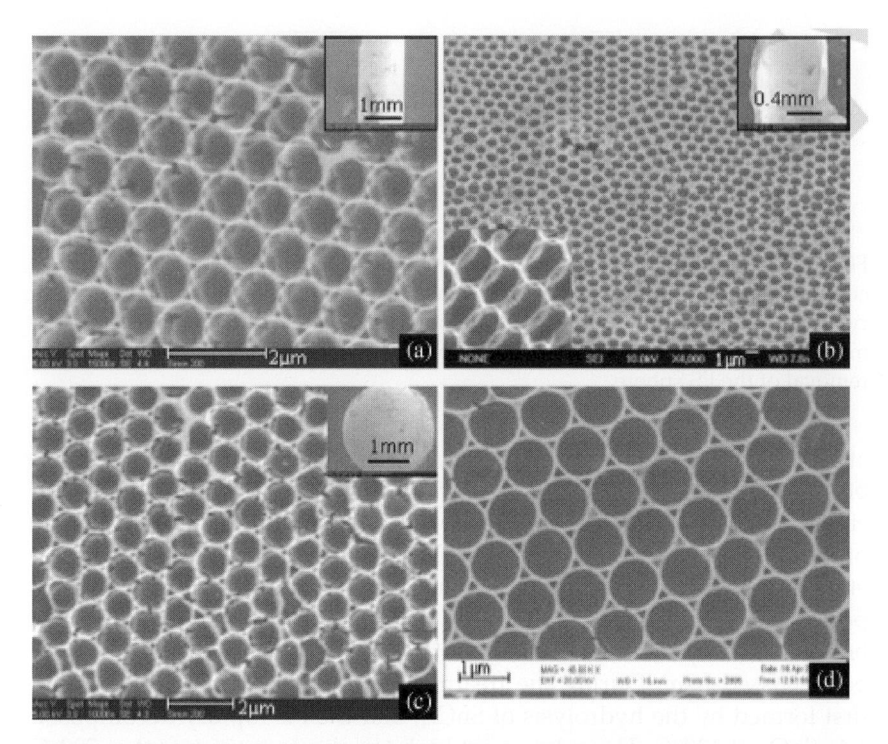

Fig. 4.21. FESEM images of SnO_2 ordered porous films on curved surfaces: (a) on the external surface of a glass tube (diameter: 1.2 mm); (b) on the internal surface of the glass tube (diameter: 1.0 mm); (c) on the surface of a steel sphere (diameter: 2.5 mm) and (d) on a flat surface. The insets are the corresponding low-magnification images. Diameters of the PS sphere are 1000 nm; the concentration of the precursor ($SnCl_4$) is 0.1 M.

change significantly as the curved substrates has a large curvature radius, and thus the PS spheres are still closely and orderly-arranged. The solution meniscus will be formed in the interstitial space between three adjacent PS spheres because of surface tension, as shown in Fig. 4.22(a). During drying at 120°C, near the glass-transition temperature of the PS spheres, the solvent in the template evaporates, and the PS spheres are deformed or sintered, inducing an increase in the contact area between two adjacent PS spheres and between the PS sphere and the substrate (from point contact to planar contact). When the solutions are saturated due to evaporation, further drying leads to the precipitation of the solute (hydrolyzate) onto the surfaces of the PS spheres and the substrate. Finally, the precipitated hydrolyzates form shell structures surrounding the PS spheres (Fig. 4.22(b)).Upon calcination

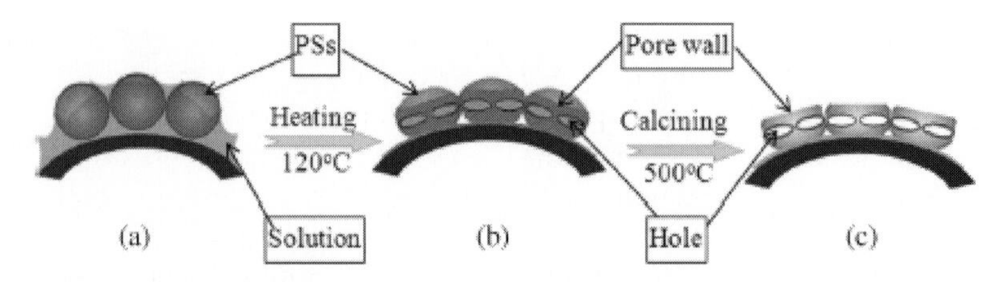

Fig. 4.22. Schematic demonstration of the growth of an ordered porous film on a curved surface, depicted in an exaggerated way. (a) A colloidal monolayer on a curved surface, with solution filling the interstices. (b) Complex of the colloidal monolayer and the SnO_2 skeleton. (c) The ordered porous film on the curved surface removal of the PS sphere.

at 500°C, the PS spheres are burnt out, and the final porous film composed of the shells, is left behind on the curved surface (Fig. 4.22(c)). Since precipitation begins when the solution level is above the tangent-point plane of the PS spheres,[59] the openings of the shells are smaller than the diameters of PS spheres and there is a triangular hole between three adjacent pores, when viewed from above. The film morphology and thickness can be controlled by the concentration of the precursor solutions and will not be discussed here. During the treatments mentioned above, $Sn(OH)_4$ is first formed by the hydrolysis of $SnCl_4$, and then is rapidly decomposed into SnO_2 at 120°C. The subsequent high-temperature treatment at 500°C induces strong adherence between the SnO_2 porous film and the curved surface.

Since the formation of the porous layer is based on the PS sphere colloidal monolayer template, the film can be made reproducible and has uniform thickness, microstructure, material composition and hence properties. The formation of the multi-layer film should be similar to the deposition of the first layer, as discussed above. The surface covered with the first porous layer acts as a substrate for the second layer, as shown in Fig. 4.20. The microbreaks in the first layer are remedied when they are placed in the precursor solution to grow the second layer. In addition, the first layer has the same thermal expansion coefficient as the second layer, which leads to fewer breaks in the second layer. Thus, the microstructure of every subsequent layer should be better than the previous layer.

Finally, it should be pointed out that the strategy presented here is universal and flexible. In addition to the SnO_2 porous films, it can also be used for synthesis of many other oxide-ordered porous films on any desired substrate with a flat or curved surface, for example, Fe_2O_3, TiO_2, ZnO, WO_3,

In_2O_3, etc. Doped porous films are also easily obtained by adding corresponding dopant ions into the precursor solutions to improve the sensing properties. In addition, mesoporous/macroporous gas-sensor films with improved sensitivity have been produced by combining the strategy in Fig. 4.20 with sol-gel technology (not shown). More importantly, based on the strategy in Fig. 4.20, we can fabricate hetero-multi-layered nanostructured porous films by alternately using two or more different precursor solutions. Such films could be used for testing multi-gases at the same time.

4.2.1.4. *Fe_2O_3-ordered pore arrays based on solution-dipping templates and colloidal monolayer*[45]

A 1 cm × 1 cm monolayer colloidal crystal was synthesized onto the substrate by spin-coating in a custom-built spin-coater.[60] This crystal can be transferred to any desired substrate, as illustrated in Fig. 4.23 (from a glass to a silicon substrate). $Fe(NO_3)_3$ solutions with different concentrations were prepared from $Fe(NO_3)_3{\cdot}9H_2O$ dissolved in distilled water. A drop of the solution was applied onto the substrate at the edge of the colloid monolayer with a quantitative pipette, whereupon the monolayer floated on the surface of the solution. The sample was then placed into an oven horizontally and dried at 80°C for 2 h, followed by heating at 400°C for 8 h to burn the latex spheres away and decompose the $Fe(NO_3)_3$ to Fe_2O_3. Finally, ultrasonic washing was performed for the calcined samples before sample characterization by scanning electron microscopy (STM), atomic force microscopy (AFM), and XRD.

Figure 4.24 shows some results for samples (on a glass substrate), with different precursor concentrations, dried in an oven at 80°C for 2 h and then heated at 400°C for 8 h, followed by ultrasonic cleanout (0.5 h). A series of pore arrays with different morphologies were obtained. X-ray diffraction (XRD) confirmed that the porous films consist of crystalline α-Fe_2O_3 (data not shown here). When the concentration is 0.02 M or higher, honeycomb structures are always formed, except for the pore shapes and diameters at the film surface, which depend on the precursor concentrations. A high concentration (0.8 M) gives rise to pores with nearly circular upper-end openings. The diameter of the openings at the film surface is obviously smaller than that of polystyrene sphere (PS) (Fig. 4.24(a)). There exists a small triangular hole at the interstitial position of the closely packed PSs and nanogaps on some pore walls (skeletons). The porous structure is more clearly revealed by tilted views. The porous film is actually a nearly

Fig. 4.23. Photos depicting the transfer of a 1 cm by 1 cm monolayer colloid crystal on glass substrate to a silicon substrate. (a) The monolayer on a glass substrate. (b) Lift-off in water. (c) Pick-up of the monolayer with an Si substrate.

spherical hollow array with truncated openings. The bottom of the spherical hollow is buried in the α-Fe_2O_3 film, as illustrated in Fig. 4.25(a), which was taken at the edge region of the film. In addition, there are holes in the pore walls between two adjacent pores (inset of Fig. 4.25(a)). As the precursor concentration decreases, the truncated shape of the pores at the film surface gradually becomes a regular hexagon (Figs. 4.24(b) and 4.24(c). The distance between the opposite sides of the hexagon is close to the diameter of the PSs. In addition, a clear triangular (for the 0.08 M solution) or spherical (for the 0.06 M solution) particle appears at interstitial positions of the closely-packed three spheres or nodes of the skeleton network, instead of a hole. No gaps were observed in the pore walls. When the precursor concentration is reduced to 0.02 M, a through-pore structured film (with

Fig. 4.24. FESEM topographic images for samples with different concentrations of precursors solutions, dried at 80°C (2 h) and subsequently calcinated at 400°C (8 h), followed by ultrasonic cleanout (0.5 h). (a) 0.8 M. (b) 0.08 M. (c) 0.06 M. (d) 0.02 M. (e) 0.002 M. (F) 0.8 M. (a)–(e) correspond to the template with 1000 nm PSs, and (f) corresponds to the template with 200 nm PSs.

two open-ended pores) is formed (Fig. 4.24(d)). This is more clearly in a tilt view, as shown in Fig. 4.25(b).[61]

There we can see that the pore diameters at the bottom (circular) and top (hexagonal) of the pores are different, at about 500 and 1000 nm respectively. The pores in the samples above are all ordered and closely-arranged,

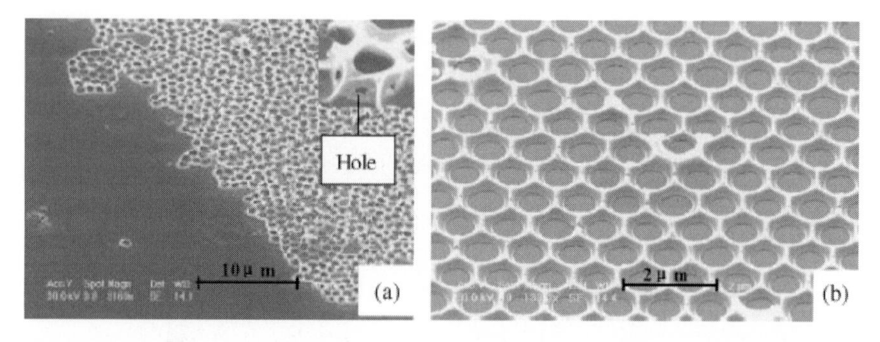

Fig. 4.25. Tilt views of ordered porous films. (a) The sample corresponding to Fig. 4.24(a) (in the edge region of the film). (b) The sample corresponding to Fig. 4.24(d).

bounded with the inter-connected net-like skeletons. However, as the concentration is reduced to a very low level (0.002 M), a ring array, rather than the closely-arranged pore array, is formed (Fig. 4.24(e)). Most of these rings are not perfectly circular. All the rings surround the positions where the latex spheres were originally located on the substrate and hence they were also arranged in hexagons. Further experiment revealed that the size of PSs does not significantly influence the morphology of the pore array. Figure 4.24(f) shows a typical result. When a precursor solution with a concentration of 0.8 M is applied onto the template with much smaller PSs (200 nm), the morphology of the porous film is similar to Fig. 4.24(a) except for the pore size. It should be mentioned that the center-to-center distance between adjacent pores or rings is always close to the diameter of the PSs, irrespective of the solution concentration.

Now let us briefly discuss the formation of the pore arrays with different morphologies. Changes in the shape of PSs during evaporation are essential to the morphologies. However, subsequent treatments are also important. When the colloidal monolayer template is dropped with the precursor solution, it floats on the top of the solution. The interstitial spaces of the closely packed PSs are then soaked with the solution due to capillary action, and a meniscus is formed on the solution surface, as illustrated schematically in Fig. 4.26(a). During subsequent drying at 80°C, the liquid surface and the colloidal monolayer gradually decreases due to evaporation of the solvent, thereby deforming the PSs. When the concentration of the solution reaches saturation point, further drying will lead to solute precipitation on the spheres' surface and the substrate. Finally, a nearly triangular hole is left between two adjacent

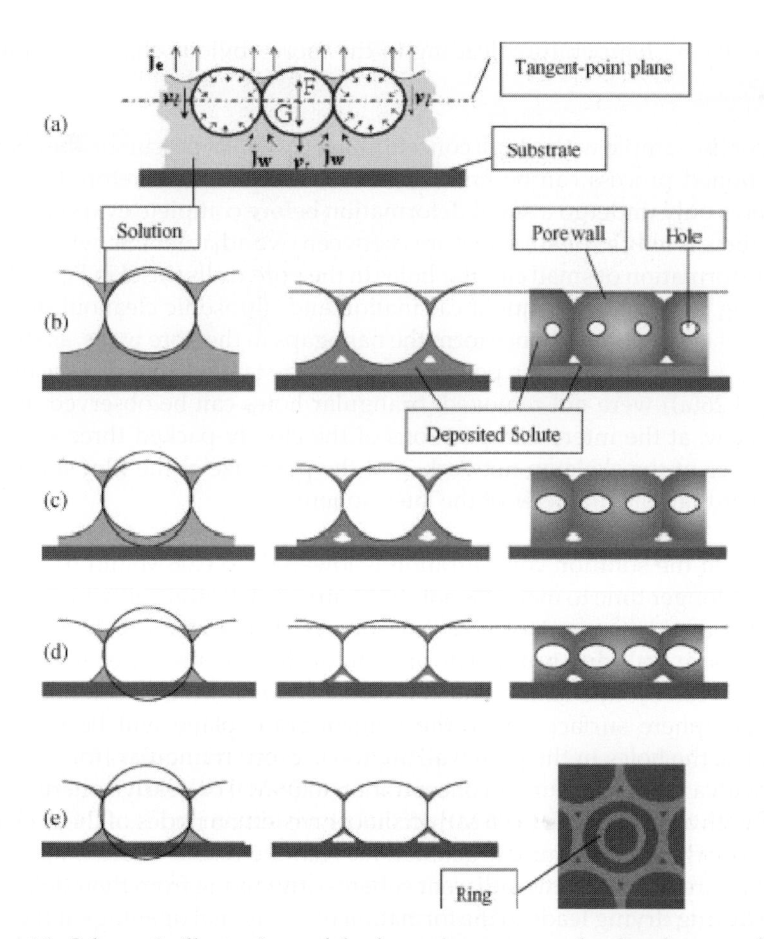

Fig. 4.26. Schematic illustrations of the formation process of ordered porous films for different concentrations of precursor solutions. (a) Latex spheres are floating on the solution. F and G denote the floatage and gravitation of the polystyrene spheres respectively. *Je* is the water evaporation flux, *Jw* the water influx, v_l and v_S are the descending rate of the solution surface and the colloid crystal, respectively. (b) High concentration (0.8 M). (c) 0.08 M and 0.06 M. (d) Low concentration (0.02 M). (e) Very low concentration (0.002 M). Left column: the solutions reach saturation. Middle column: complete evaporation of solvent. Right column: after removal of the template by dissolution. Figure (e) in this column is a top view.

spheres and the substrate due to lack of sufficient solution compensation (Figs. 4.26(b)–4.26(e)). This is similar to the pore-formation mechanism in a metal casting.[62] Obviously, the lower the solution concentration is, the longer the time to reach saturation induced by evaporation at the same drying temperature, leading to the more obvious change in sphere shape.

For the sample with a high concentration (0.8 M) of precursor, the above-mentioned process can be finished in a shorter time. Therefore the latex spheres only undergo a small deformation before complete evaporation,[63] and there is only a small contact area between two adjacent PSs, which leads to the formation of small circular holes in the pore walls (inset in Fig. 4.25(a) and Fig. 4.26(b)). Subsequent calcination and ultrasonic cleanout destroy some of the holes and hence form the nanogaps in the pore walls, as shown in Fig. 4.24(a). Because the pore walls above the tangent-point plane of PSs (Fig. 4.26(a)) were not removed, triangular holes can be observed, in the top view, at the interstitial positions of the closely-packed three spheres, or nodes of the skeleton network, and the pore size at the film surface is smaller than the diameter of the latex spheres.

When the solution concentration is lowered (to 0.08 M and 0.06 M), it takes a longer time to make the solution saturated, leading either to a larger contact area between two adjacent spheres (Fig. 4.26(c)) or larger holes in the pore wall due to the deformation of the spheres. After removal of the spheres by calcination and ultrasonic cleanout, the solute deposited on the sphere surface above the tangent-point plane will be removed because the holes in the pore wall are too big, but triangular [for the sample with a higher precursor concentration (0.08 M)] or nearly spherical [for that with a smaller one (0.06 M)] offshoots are left on nodes of the skeleton network due to the template geometry (Figs. 4.24(b) and 4.24(c)). The large contact area between two adjacent spheres originating from their deformation during drying leads to the formation of hexagonal openings at the film surface.

If the solution concentration decreases further (to 0.02 M), before solution saturation and solute precipitation, the latex spheres will come into contact with the substrate and this contact area will become large enough due to their deformation during drying (as illustrated in Fig. 4.26(d), pores with two-ended openings were formed (Fig. 4.24(d) and Fig. 4.25(b)).

For the same reason, when the concentration of the precursor is reduced to a very low level (0.002 M), only a thin solute deposition (shell) is preferentially formed on the free surface of the latex spheres above and below the tangent-point plane (Fig. 4.26(e)). Subsequent calcination at 400°C (8 h) and then ultrasonic vibration for a short time (a few minutes) removes the latex spheres but leaves the deposited shell, as shown in Fig. 4.27(a). There are two layers, and the top layer had slightly displaced

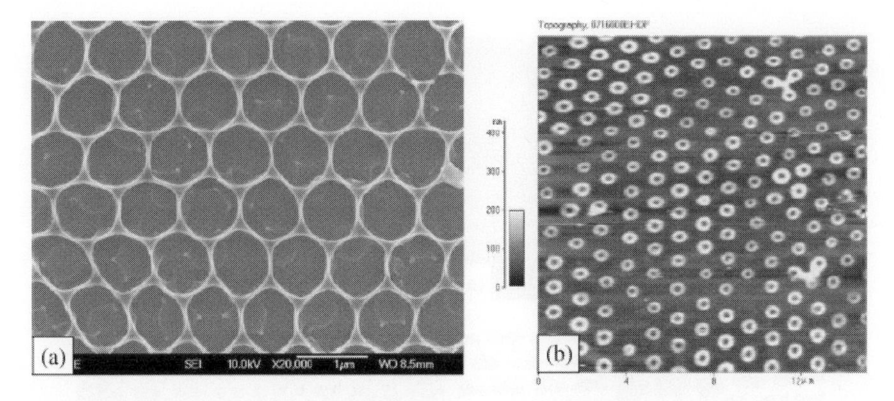

Fig. 4.27. The sample from the 0.002 M solution after drying at 80°C (2 h). (a) After subsequent heating at 400°C (8 h) and ultrasonic vibration for a few minutes (instead of 0.5 h). (b) After subsequent ultrasonic washing in dichloromethane solution for 0.5 h.

with respect to the bottom layer due to ultrasonic vibration. However, we can still see some irregular rings in the bottom layer. After ultrasonically washing for an additional half hour, the top layer is removed and the bottom layer, a ring array, can be seen (see Fig. 4.24(e)). Further experiments indicate that the non-circular shape of the rings results from calcination at 400°C after drying at 80°C. Ultrasonically washing the sample in dichloromethane solution after drying at 80°C, without calcination, gives an array of circularly rings (Fig. 4.27(b)).

Based on the discussions above, there should be some intermediate states during the formation of pores. Figure 4.28 shows such changes of the sample with a 0.02 M precursor solution. After drying the solution-dipped sample at 80°C for 2 h, the PSs have been deformed (Fig. 4.28(a)). The spheres have become concave on top and are nearly hexagonal, with greatly increased contact between neighboring spheres, which show that they had been sintered to a certain degree. If we put such a sample into dichloromethane to dissolve the spheres, instead of burning them away, and then perform an ultrasonic cleanout, a circular opening array (as seen from the top view) is left (Fig. 4.28(b)). The diameter of the openings are smaller than that of latex spheres, which is similar to the final structure of the sample from 0.8 M solutions (Fig. 4.24(a)). Further experiment revealed that there is a solute-deposition shell on the sphere surface above the tangent point plane in all samples, irrespective of the solution concentrations. Therefore, the final structure and morphology of the ordered pore array also depend on the subsequent

Fig. 4.28. Typical transitional states during porous film formation for the sample from the 0.02 M precursor solution. (a) After solution-dipping and drying at 80°C for 2 h. (b) After dissolution in dichloromethane for sample (a). (c) After calcinations at 400°C for 8 h for sample (a). (d) After ultrasonic vibration for a few minutes for sample (c).

template-removal method — burning, ultrasonic vibration or dissolution — in addition to the concentration of the solution. Upon calcinating the sample shown in Fig. 4.28(a), we can see a two-layer porous structure (Fig. 4.28(c)). The two layers are connected at the nodes of the skeleton network in the bottom layer. Subsequent ultrasonic washing for a few minutes removes the whole top layer almost completely, as shown in Fig. 4.28(d). Further ultrasonic vibration leads to the final structure in Fig. 4.25(b).

In summary, various orderly pore structured Fe_2O_3 films have been obtained using our solution-dipping template strategy. The porous film morphologies can be easily controlled by varying the concentration of the precursor solution and/or the treatment conditions. With a decrease of the concentration from a high to a very low level, the nanostructured complex (pore-hole, and pore-particle) arrays, through-pore arrays and even ring arrays can be attained. We can also control the pore size over a large range by changing the diameter of the template's latex spheres. The synthesis strategy presented here is universal and can be used for other metal or oxide ordered pore arrays. We have synthesized a series of other morphology-controlled ordered porous films, such as, zinc, ZnO, NiO, Co_2O_3, CuO, CeO_2, Eu_2O_3, Dy_2O_3, with different concentration precursor solutions, and also fabricated these structures on other substrates by transferring the colloidal monolayer from one substrate to a desired substrate before the precursor solution was dipped.

4.2.1.5. *In_2O_3-ordered pore arrays based on solution-dipping templates and colloidal monolayers*[49]

In this paragraph, we introduce the synthesis of In_2O_3 films structured with ordered pore arrays on glass and silicon substrates based on a colloidal monolayer template-directed solution dipping method. The experimental procedure is as follows.

The colloidal sphere monolayer with a large area ($>1\ cm^2$) was prepared on the clean glass substrate by spin coating on a custom-built spin coater. The monolayer is formed based on a capillary force-induced self-assembly process of PSs on the substrate with a smooth and hydrophilic surface. A 1000 nm PS suspension of 10 μl in volume was spin-coated at low speed ($\sim$100 rev./min.), while spin coating of 350 nm PSs of 4.0 μl was performed at a high speed ($\sim$850 rev./min.). Some of them were transferred onto a Si substrate by lifting off in water and picking up with the silicon wafer.

The precursor sols used in this work were prepared follows: $NH_3{\cdot}H_2O$ solution was added to the 0.5 M $InCl_3$ solution slowly until $In(OH)_3$ precipitated completely. The product was rinsed several times with millipore water until all Cl^- ions were removed (tested with $AgNO_3$ solution). A certain amount of the powder was dispersed in 0.25 M nitric acid and its pH value was controlled at 2.4–2.5 by gradually adding $NH_3{\cdot}H_2O$ solution.[64] Li[49] thus obtained stable, homogenous, and translucent precursor sols with indium concentrations of 0.2, 0.35, and 0.5 M, which had an obvious Dindal effect when a beam of laser (532 nm) passed through them.

In_2O_3-ordered pore arrays were fabricated as schematically illustrated in Fig. 4.29. A droplet of the preformed sol is first dropped onto the two-dimensional (2D) monolayer colloidal crystal with a quantitative

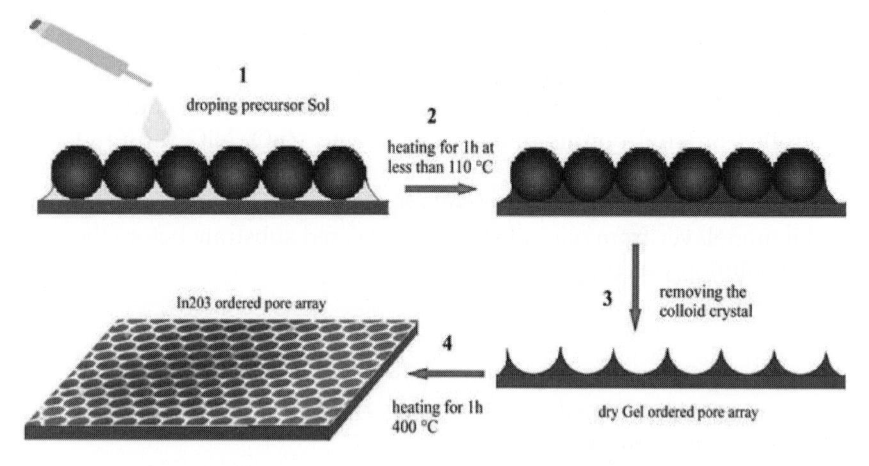

Fig. 4.29. Schematic illustration of the fabrication strategy of In_2O_3-ordered pore arrays.

pipette, which can infiltrate into the interstices between the substrate and the colloidal monolayer, followed by drying at 110°C and ultrasonically washing in methylene chloride (CH_2Cl_2) for 2 minutes to remove the template (the PS spheres can be dissolved quickly in the CH_2Cl_2 solution). Finally, the sample is annealed in air at 400°C for 1 h for the decomposition of $In(OH)_3$ into In_2O_3.

By a spin-coating method, a large-area colloidal monolayer (>1 cm^2) was successfully fabricated, as typically shown in Fig. 4.30, which was taken in the edge region of the PS film on a glass substrate. We can see that the PSs have crystallized and formed a monolayer with the closely-packed hexagonal lattice through a self-assembly process.

Figure 4.31 shows the pore arrays on glass substrates after removal of the templates and annealing at 400°C for the samples with different concentrations of the precursor sols. All exhibit a well-defined honeycomb shape from top view. With an increase of precursor concentration from 0.2 M to 0.5 M for the samples with a template of 1000 nm PSs, the depth of the pores in the array increases from less than the radius of the PSs (Fig. 4.31(a)) to larger than that of the PSs (Fig. 4.31(c)). But there is no obvious change of the pore size on the film surface (from the top view) with the concentration, since the pore depth is around the radius value at which the pore size at the surface from the top view is insensitive to the depth, as schematically illustrated in Fig. 4.32. Nevertheless, for the samples from templates with much smaller PSs (350 nm), the pore size at the film surface evolves significantly

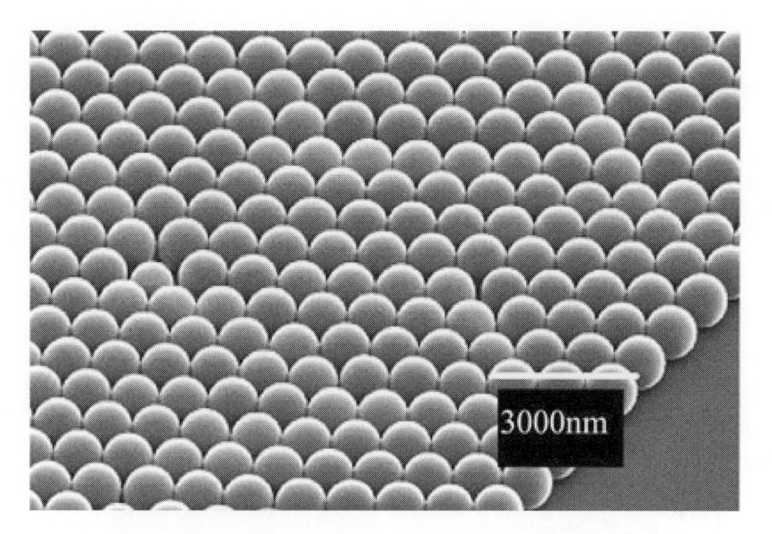

Fig. 4.30. FE-SEM image of 2D colloidal monolayer crystal on a glass substrate (in the edge region of the PS monolayer film).

with the concentration, as indicated in Figs. 4.31(d) and 4.31(e). The higher concentration leads to the deeper pores. In Fig. 4.31(e), the deeper pores can be seen clearly, and they seem to be hollow spheres with cut-off tops. The reason for the formation of pores with different depths is easily understood since higher-concentration precursors contain more solute per unit volume. For the samples with Si substrates, the morphologies are similar to those of the corresponding samples on glass substrates shown in Fig. 4.31. The morphologies of the ordered pore arrays are determined by precursor concentration and PS size, which are controllable. Figure 4.33 shows the XRD results for the ordered pore arrays before and after annealing. Before annealing (after removal of the monolayer), the skeleton is composed of $In(OH)_3$, as shown in Fig. 4.33(a), which is not well-crystallized. After annealing at 400°C in air, however, the skeleton has changed into the In_2O_3 crystal with body-centered cubic structure (see Fig. 4.33(b)). Its lattice constant is 10.1 Å, which is consistent with the standard value for the bulk (JCPDS 71-2195). The final ordered pore array was removed from the glass substrate by a scalpel and transferred onto a copper grid for TEM examination. It has been shown that the film is of an ordered pore structure, and the skeleton consists of In_2O_3 polycrystallites, as illustrated in Fig. 4.34. EDX analysis indicates the existence of only oxygen and indium in the skeleton, as shown in Fig. 4.35.

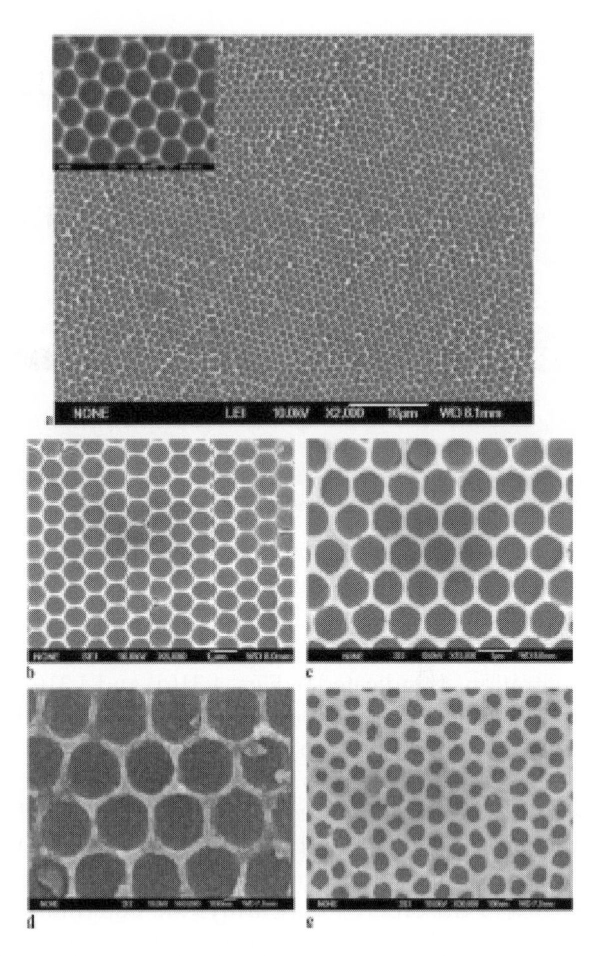

Fig. 4.31. FE-SEM images of ordered pore arrays on the substrate with different precursor concentrations after removal of PSs and annealing at 400°C. (a) and (d) 0.2 M, (b) 0.35 M, (c) and (e) 0.5 M. PSs' size is 1000 nm for samples (a), (b), and (c) and 350 nm for samples (d) and (e).

4.2.2. Two-dimensional ordered polymer hollow sphere and convex structure arrays based on monolayer pore films[65]

Li[65] reported that they have developed a two-step replication strategy to obtain 2D hollow polymer sphere arrays and convex structure arrays with ordered arrangement on a substrate based on the colloidal monolayer. Such hollow sphere arrays have the potential to be used for carriers of catalysts,

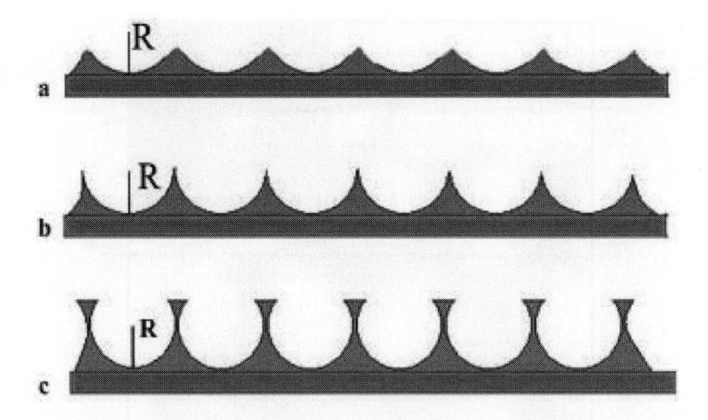

Fig. 4.32. A schematic illustration of the relationship between the depth of the pore and the precursor concentration. (a) lower concentration, (b) moderate concentration, and (c) higher concentration. *R*: the curvature radius of the pore.

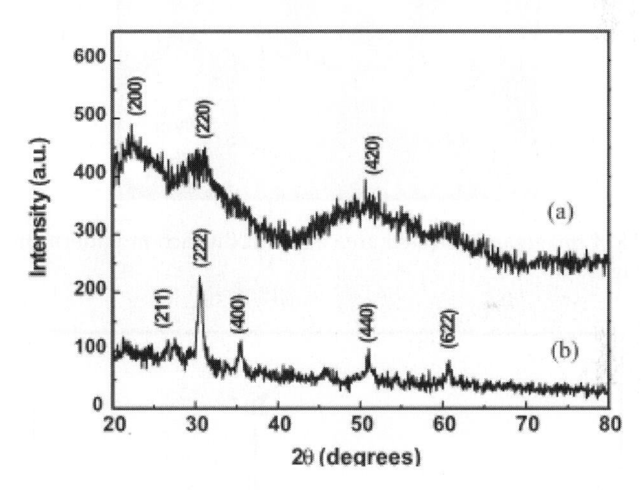

Fig. 4.33. XPD of the ordered pore arrays (a) before and (b) after annealing at 400°C corresponding to (a) $In(OH)_3$, and (b) In_2O_3 crystal, respectively.

enzymes, drugs delivery,[66–68] microreactor devices, microacetabula, and some other devices in medicine and biotechnology. The convex structure arrays could be useful in miniaturized optical components. The details are as follows.

Large-scale monolayer colloidal crystals (>1.0 cm in diameter) were prepared on the clean substrate by spin-coating[24] on a custom-built spin-coater.

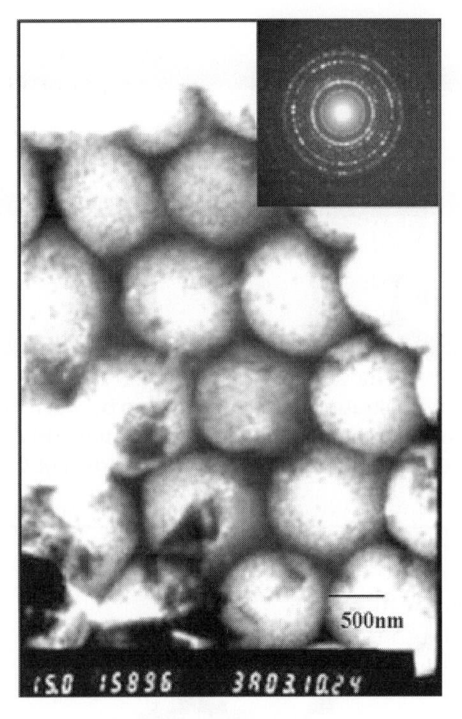

Fig. 4.34. TEM image and selected area electron diffraction pattern (inset) of the sample in Fig. 4.3(b).

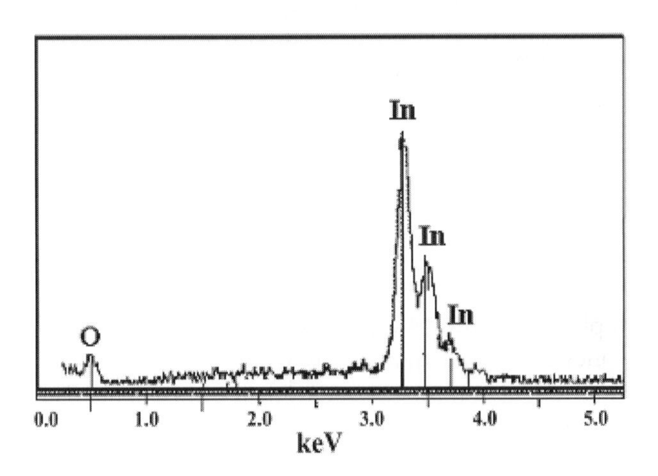

Fig. 4.35. EDX of In_2O_3-ordered pore arrays for the sample of Fig. 4.31(b).

A 2D FeO(OH)-ordered pore array (film) was first fabricated based on our solution dipping-template synthesis strategy, which was previously described in details,[45] as illustrated in steps 1–3 of Fig. 4.36(a). Subsequently, the polymers polyvinylalcohol (PVA, average molecular weight 71,000), polyvinyl pyrrolidone (PVP, average molecular weight 40,000), and polymethyl phenylsilane (PMPS, average molecular weight 41,000) were dissolved in distilled water, alcohol, and cyclohexane, respectively. The polymer solution (ranging from 20 to 50 g/l in this study) was dropped on the 2D FeO(OH)-ordered pore array until the pores were filled. Then the array full of the solution was capped with another glass substrate and reversed to locate the glass cover on the bottom. After sufficient drying at room temperature, the FeO(OH) pore array was removed by selective dissolution in 1 M oxalic acid solution. Finally, a 2D ordered polymer hollow sphere array was prepared by uncovering the top glass-substrate, as shown in steps 4–5 of Fig. 4.36(a). Similarly, if the pores of the FeO(OH) are truncated, depending on the $Fe(NO_3)_3$ concentration,[45] we can obtain truncated hollow sphere arrays. Furthermore, we can alternately drop the polymer precursor into the truncated pores of the FeO(OH) film and dry it until the pores are filled with the solid polymer. The filled arrays can be covered with another glass substrate and reversed. In this way, we can obtain a convex structure array, as displayed in Fig. 4.36(b).

By the spin-coating method, large scale monolayer colloidal crystals with hexagonal close-packed arrangement were fabricated on the glass substrate by a self-assembly process, as typically shown in Figs. 4.37(a) and 4.37(b). Based on this template, we can obtain centimeter-square-sized ordered pore arrays according to procedures 1 to 3 in Fig. 4.36. Figure 4.37(c) shows such an array prepared by the 1.0 M $Fe(NO_3)_3$ precursor solution. We can know that the pore depth is close to the diameter of PS. All the pores in the array are interconnected, and each pore has six small circular channels in its side wall induced by close-packed PS. XRD indicates that the skeleton of the pore array is composed of FeO(OH), as shown in Fig. 4.37(d). This inorganic pore array has many advantages as a second template for synthesis of the other functional materials with special ordered structures, such as simple preparation, regular pore morphology with spherical or truncated spherical shape, smooth pore wall, uneasy deformation, and low cost. It is more important is that it can be removed easily and quickly by selective dissolution in an oxalic acid solution.

Figure 4.38 shows the 2D arrays of PVA material based on the second template prepared by a 1.0 M $Fe(NO_3)_3$ solution with 1000 nm of PS (shown in Fig. 4.37(c)). Each unit in the arrays is a hollow sphere, which can be seen

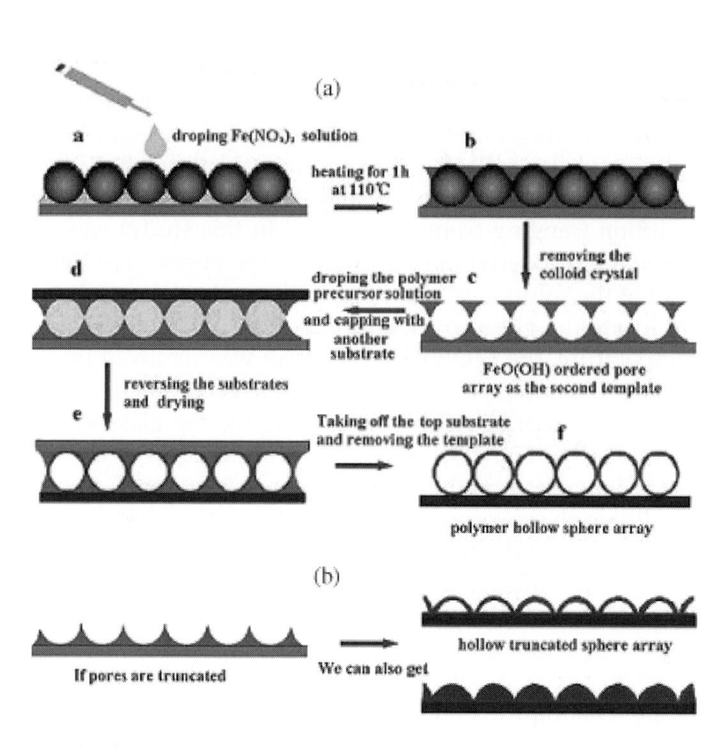

Fig. 4.36. The fabrication strategy of the 2D polymer (a) ordered hollow sphere array and (b) hollow and solid truncated sphere arrays.

more clearly in the TEM image shown in Fig. 4.38(e) and the FE-SEM image in the edge region of the sample shown in Fig. 4.38f). The hollow spheres are in contact with each other and arranged in order. When the concentration of PVA precursor solution is high enough (say 50 g/l), the hollow spheres in the array are integrated (Fig. 4.38(a)). When the concentration decreases to 30 g/l, a small hole appears on top of some hollow spheres (Fig. 4.38(b)), and if there is further decrease (to 20 g/l), the holes get bigger and there exist such holes on top of most hollow spheres in the array (Fig. 4.38(c)) typically larger than 100 nm in size. We know that each pore in the second template has six circular channels in its side wall (Fig. 4.37(c)). However, in the corresponding hollow sphere array, no such channels are found. This can be confirmed by two neighboring hollow spheres in a rectangular frame shown in Fig. 4.38(f) (see the side wall indicated by an arrow) and the TEM image in Fig. 4.38(e).

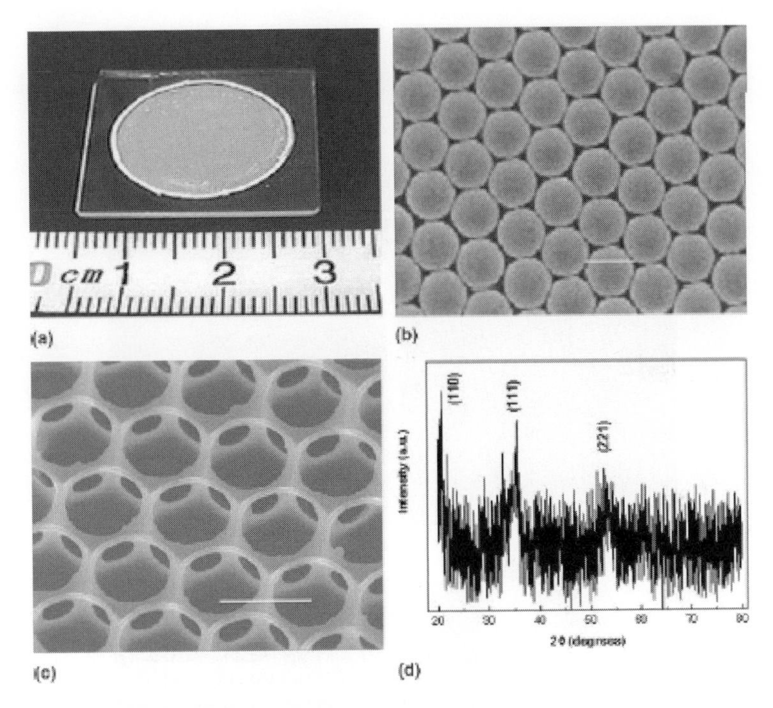

Fig. 4.37. Colloid crystal monolayer and its inverse opal made of FeO(OH). (a) and (b) are, respectively, photo and corresponding FE-SEM image for the PS monolayer colloidal crystal on glass substrate, (c) is the FE-SEM image of the inverse opal, prepared by 1.0 M $Fe(NO_3)_3$ precursor solution and dried at 110°C for 1 h, and (d) the corresponding XRD spectrum (indexes represent the crystal planes of FeO(OH)). The scale bar in (b) and (c) is 1 μm.

Obviously, in the precursor concentration range used in our experiments, when the precursor solutions of polymer evaporate and solidify, the polymer molecules will adhere to the pore wall of the second template and blockade channels in the side wall. Hence the final polymer spheres have the closed side wall (shell). Figure 4.38(d) shows the hollow sphere array fabricated by a lower PVA concentration (20 g/l) using the FeO(OH) template with smaller pore size (350 nm). Its morphology is similar to that with 1000 nm in pore diameter (Figs. 4.38(d) and 4.38(c)). There also exists a small hole on top of most hollow spheres with a typical size smaller than 50 nm. These small holes are controllable by precursor concentration.

The hollow spheres with small holes on the top could be useful in microreactor devices[69] that can endure acidic and alkaline conditions, selective permeability, and nutrient and drug delivery.[70] More interestingly,

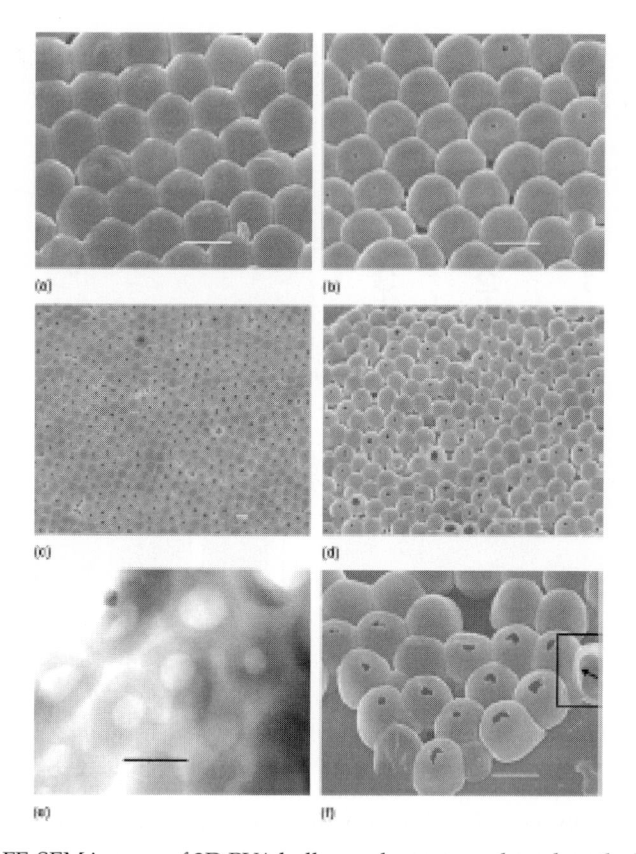

Fig. 4.38. FE-SEM images of 2D PVA hollow sphere arrays based on the FeO(OH)-ordered pore array obtained by 1.0 M $Fe(NO_3)_3$. Pore size is 1000 nm for (a)–(c) and 350 nm for (d). The concentrations of polymer precursor are (a) 50 g/l, (b) 30 g/l, (c) 20 g/l, (e) is the corresponding TEM image of sample (a) (by shaving the products from glass substrate), and (f) is the magnified SEM image in the edge region of sample (c). The scale bar is 1 μm.

such a hollow sphere with a small hole on its top could be used for microacetabula because of the elastic polymer. In addition, it could also be an ideal model for studying black body radiation in micro- or even nano-space because the small holes on top of the hollow spheres correspond to the nano-sized black body.

Obviously, under the same PVA concentration, the morphology of the hollow sphere arrays is mainly determined by that of the second template. In our previous work, the latter can be easily controlled by the concentration

Fig. 4.39. FE-SEM images of (a) the 2D FeO(OH) porous film prepared with 0.05 M precursor and (b) the corresponding 2D ordered PVA hollow sphere array fabricated by the 50 g/l polymer precursor. The scale bars are 1 μm.

of the precursor $Fe(NO_3)_3$.[45] Figure 4.39(a) illustrates the second template prepared with a much lower precursor concentration (0.05 M). It shows the pore array with regular hexagons from the top view. The corresponding hollow array is shown in Fig. 4.39(b) (the truncated hollow spheres can be confirmed by TEM observation, not shown here). It has a different morphology from that in Fig. 4.38(a). Generally, the hollow spheres in the final array will be a nearly spherical or truncated spherical shell, depending on the pore depth in the second template, as depicted in Fig. 4.36. If the pore depth is close to the diameter of PSs (as displayed in Fig. 4.37(c)), the final hollow sphere will be a nearly spherical shell (see Fig. 4.38(f)). Otherwise, it would be a nearly truncated (or hemi-) spherical shell (see Fig. 4.39(b)).

After five cycles of infiltration and drying with 30 g/l PVA precursor, the solidified PVA was full of pores in the second template, synthesized by

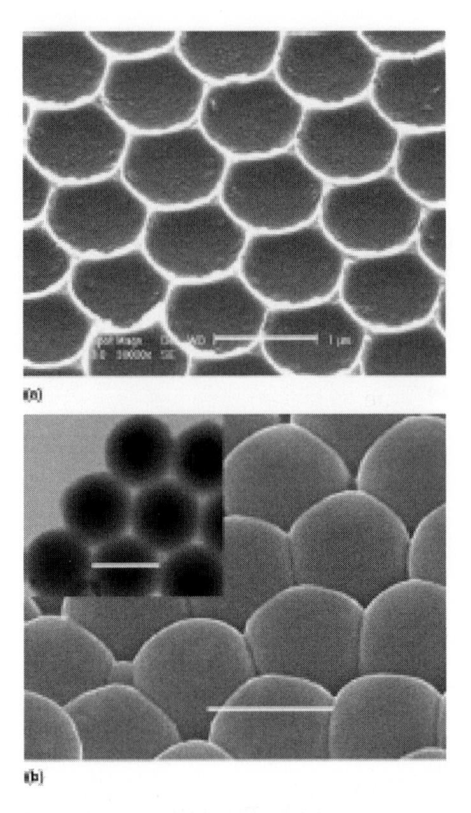

Fig. 4.40. FE-SEM images of (a) the 2D FeO(OH) pore array prepared with 0.08 M precursor and (b) corresponding PVA solid truncated sphere array fabricated by five cycles of infiltration with 30 g/l PVA precursor and drying. The inset in (b) is its corresponding TEM image. The scale bars are 1 μm.

1000 nm PS and 0.08 M $Fe(NO_3)_3$ solution (see Fig. 4.40(a)). After removal of the template, a hemisphere or convex structure array was fabricated, as shown in Fig. 4.40(b). TEM examination has demonstrated that the hemispheres are solid (see inset in Fig. 4.40(b)). Such arrays could be used for microlens arrays in the miniaturized optical devices.[71,72]

Similarly, we have also synthesized the PMPS and PVP 2D ordered hollow sphere arrays and convex structure arrays with similar morphology to that of PVA (not shown here). Further experiments indicate that many soluble polymer materials, that can solidify when being concentrated, can be used to fabricate the hollow sphere arrays by the strategy shown in Fig. 4.36.

The reason for hollow sphere formation can easily be understood. The solvents can evaporate through the rim cracks between the two glass substrates during drying. With solvents evaporating gradually, the polymer will concentrate further and adhere to the pore wall of the second template. After the FeO(OH) skeleton is removed, the sphere shell will be left. The holes on the top of the hollow spheres are associated with precursor concentration. From Fig. 4.38, we can know that it is not caused by removing the capping glass, since a high concentration of polymer leads to integrated hollow spheres without holes. The formation of the holes can be attributed to evaporation of the solvent. During drying, the level of the solution surface within pores of the second template decreases and the concentration increases. Due to the affinity between the polymer molecules and pore walls, some polymer molecules will attach on the pore wall. Obviously, the formed shell on top of the spherical or truncated spherical hollow is the thinnest. If the concentration of the polymer precursor solution is low enough, a hole will be formed on the top because of the overly-thin shell.

4.2.3. Au nanoparticle arrays[73]

Nanoparticles can be periodically arranged on solid supports by nanosphere lithography,[29] which is a general, simple, and low-cost method. The acquired two-dimensional nanoparticle arrays have been proposed for applications in, for example, sensors,[74,75] photonics,[76,77] magnets,[78] catalysts,[79–81] and data storage.[82] The intrinsic properties of such nanostructure arrays are determined by factors such as the size, shape, crystallinity, and composition of the nanoparticles as well as the geometry and interparticle spacing of the array.

Here, we discuss the laser morphology manipulation of gold nanoparticle arrays formed by nanosphere lithography[60,83] and describe their corresponding optical and structural evolution. This study demonstrates that laser irradiation is a good way of controlling the morphology of nanostructured materials and hence their properties, and also may prove to be a valuable new method for the fabrication of functional nanostructures.

Gold particle arrays were fabricated by nanosphere lithography, as described elsewhere.[29,83] Briefly, glass substrates were ultrasonically cleaned in acetone and then in ethanol for 1 h. Surfactant-free colloidal polystyrene spheres (PS) were purchased from Alfa Aesar. The spheres have a diameter of 1000 nm and size distribution of less than 5%, and are dispersed in water at 2.5 wt.%. Large-scale monolayer colloidal crystals

(>1.0 cm^2 in diameter) were prepared on the clean glass substrate by spin-coating in a custom-built spin-coater. The colloidal monolayer was then lifted off and floated by immersion in water. In this way, the monolayer could be transferred onto a variety of substrates such as quartz or ITO, as previously described.[27,43,45] Next, substrates coated with the 2D colloidal crystals were mounted on a sample holder and transferred into an ultra-high vacuum chamber. Gold was thermally evaporated under a base pressure of 10^{-6} Pa and deposited at a rate of ~0.2 nm/minute. A quartz crystal microbalance was used to monitor the thickness of the deposited film. After deposition to a 70 nm thickness, the samples were immersed in methylene chloride under sonication to remove the nanosphere mask, leaving a highly-ordered array of gold particles on the surface.

The gold nanoparticle array was then irradiated by laser pulses from a Nd:YAG laser operating at 1 Hz at the third harmonic wavelength of 355 nm with a nominal pulse width of 7 ns. The laser pulses were unfocused with an energy density of 15 mJ/cm .

Figure 4.41 shows an as-prepared Au nanoparticle array on an ITO substrate. As can be seen, the particles are hexagonally arranged with a P6 mm symmetry and have triangular cross-section with a height of about 70 nm. Figure 4.42 shows the morphological evolution versus the number of laser pulses. After about 40 laser pulses, the three sharp corners of each

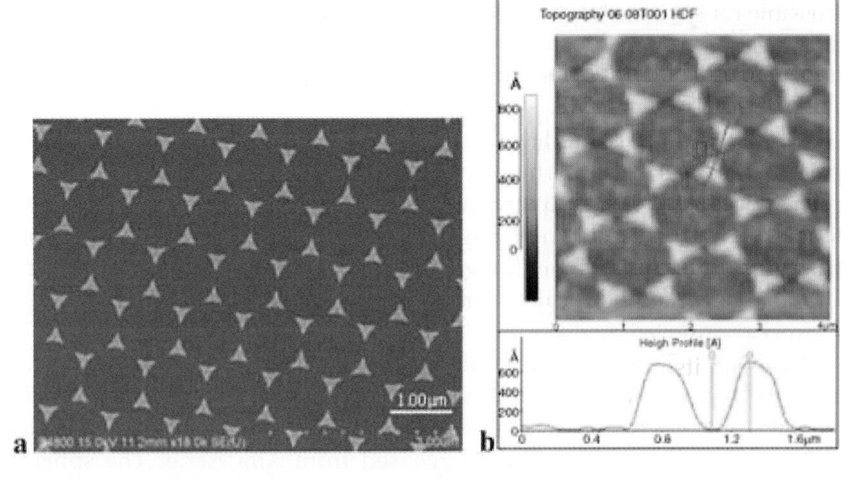

Fig. 4.41. (a) FE-SEM and (b) AFM images of a gold particle array fabricated with 1000 nm diameter polystyrene sphere colloidal monolayers before laser irradiation.

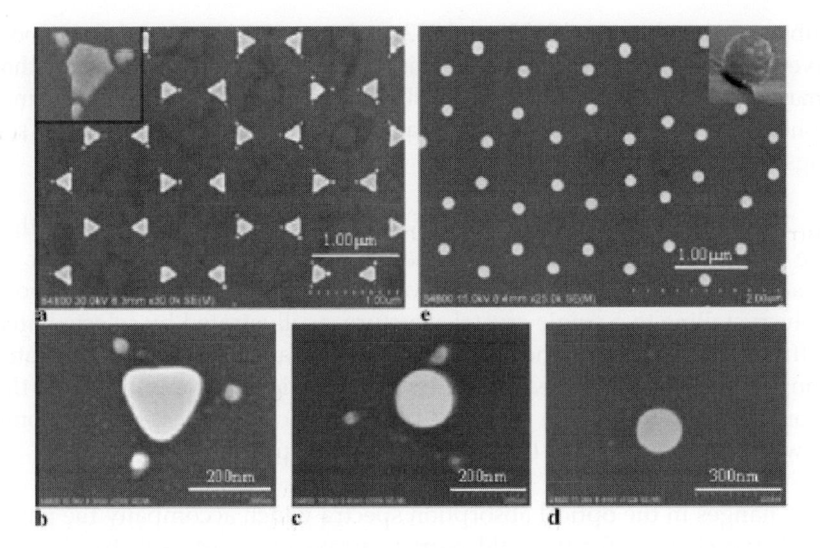

Fig. 4.42. Morphology of gold particle arrays on the ITO substrate after 355 nm laser irradiation (15 mJ/cm^2 per pulse) for different numbers of pulses. (a) 40 times; (b) 60 times; (c) 80 times; (d) 100 times; (e) 500 times.

particle become separated from the main body of the particle and three nanogaps of about 30 nm are formed in each particle, as demonstrated in Fig. 4.42(a). This morphology is particularly intriguing because it might be possible for such an array to be used as a substrate for molecular switching devices.[84] As the number of laser pulses is increased to 100, the nanoparticles at the corners become smaller and almost disappear, while the main body of the particle evolves from a polyhedron to a rounded and finally to a nearly fully circular shape, as illustrated in Figs. 4.42(b)–4.42(d) (from the top view). This demonstrates that the morphology of the nanostructured arrays can be manipulated by laser radiation through appropriate selection of the number of pulses. Applying more than 100 laser pulses did not induce any further changes but the complete disappearance of the nanoparticles at the corners and the edge sides of the original particles. Figure 4.42(e) shows a sample having been irradiated by more than 500 pulses, and its morphology is similar to that of the sample irradiated for about 100 pulses indicating that the particle has reached its equilibrium shape after 100 pulses. Furthermore, tilted observation has shown that the final particles are nearly spherically-shaped, as demonstrated in the inset of Fig. 4.42(e), which is also consistent with Kawasaki's reports.[85] Similar morphological evolutions were also observed for gold particles on quartz substrates irradiated with 355 and 532 nm laser wavelength pulses, respectively, suggesting a generality to this technique. Finally, it is noted that many of the particles seem to move slightly during irradiation, or the final spherical particles are not exactly located at the centers of the triangles (see Figs. 4.42(c) and 4.42(d)).

Transmission electron microscopy (TEM) examination has shown that the individual gold particles in the array are polycrystalline prior to irradiation. Laser irradiation, however, leads to the transformation from polycrystalline to a single crystal structure, as illustrated by selected area diffraction patterns of the single particles scraped from the substrates shown in Fig. 4.43. It means that laser irradiation not only modifies the morphology but also the structure of the particles, indicating that some new properties, sensitive to the structure, will appear.

Changes in the optical absorption spectra which accompany the particle shape changes for the gold particle arrays on quartz substrates were measured versus the number of laser pulses as shown in Fig. 4.44. The as-prepared samples show a very broad adsorbance peak centered around 680 nm, together with a shoulder extending well into the near infrared region, which indicates that the peaks are composed of at least two peaks. The peak at 680 nm decreases and disappears as the laser irradiation is increased up to 100 pulses. In addition, after irradiation by about 60 pulses, another peak emerges around 550 nm. After about 100 pulses, the peak

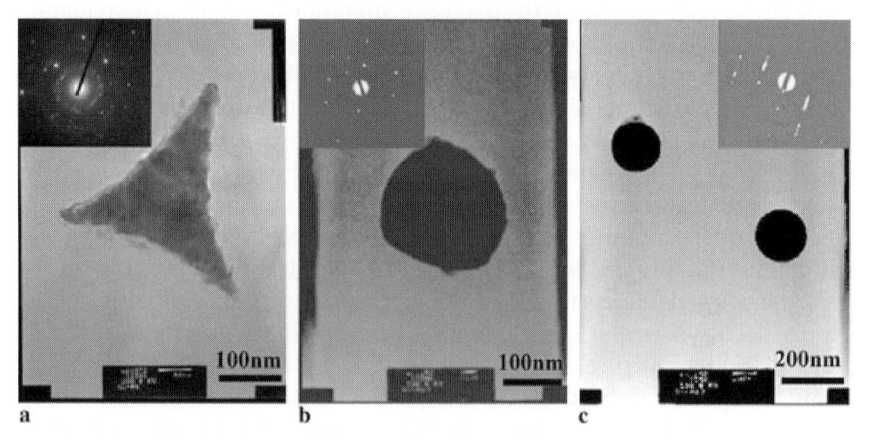

Fig. 4.43. TEM images of single gold particles. (a) As-prepared triangular gold particles. (b) and (c) after irradiation by 100 and 500 laser pulses, respectively. The insets show the corresponding selected electron diffraction patterns.

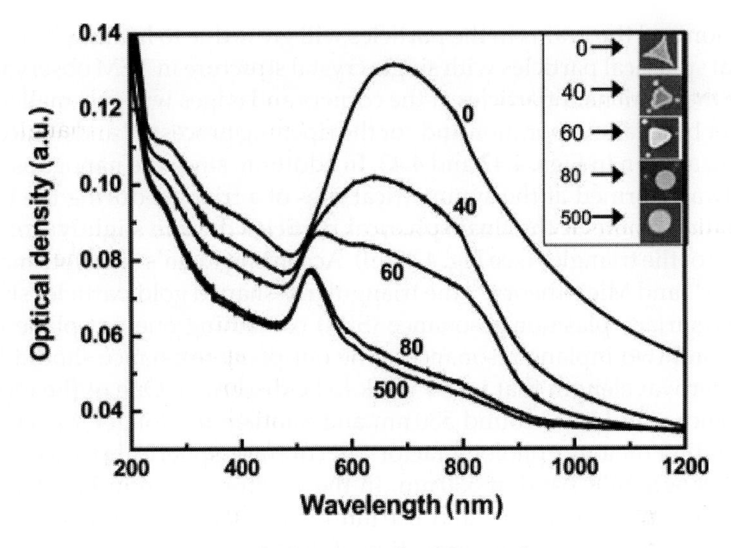

Fig. 4.44. Evolution of the optical spectra of gold particle arrays on quartz versus laser irradiation with the indicated number of laser pulse. Inset: the corresponding morphology of individual dots at each pulse number.

shifts to 530 nm. This peak shows very little dependence on the number of laser pulses beyond 100 laser pulses. This is in agreement with the lack of morphology change.

Now let us have a brief discussion. Although the complete understanding of the mechanism for the laser induced morphology evolution requires more and deeper work, here we can explain the results above based on the particle–laser interaction. The particle arrays in this study were prepared by physical vapor deposition on the PS monolayer. It is well-known that such vapor deposition usually leads to the formation of triangular non-compact particles (with a porosity) consisting of ultra-fine nanoparticles or grains, which exhibit polycrystalline electronic diffraction, as seen in Fig. 4.43(a). Obviously, laser irradiation will heat and sinter the triangular particles. After laser irradiation for a short time (e.g. 40 pulses), the main body of a triangular particle will become compact due to heating-induced sintering of ultrafine nanoparticles. Such compactness will result in the contraction of the triangular particles, and the contraction will induce formation of nanogaps and nanoparticles at the corners and the edge sides of the triangular particles, because of the edge effect and the interaction between the particles and the substrate (see Fig. 4.42). With an increase of laser irradiation time, the compact particle will be spheroidized by surface atomic diffusion and the grains in the particles will grow due to heating,[86] leading to final spherical particles with single crystal structure in TEM observation, while the small nanoparticles at the corners and edges will get smaller and smaller by local-evaporation and/or the ripening process,[86] and finally disappear, as seen in Figs. 4.42 and 4.43. In addition, since the nanogaps were not always formed at the symmetrical sites of a triangle during the initial irradiation, some of the final spherical particles deviate slightly from the centers of the triangles (see Fig. 4.42(e)). According to Jin's and Kott-mann's work[87,89] and Mie's theory,[90] the triangularly-shaped gold particles should exhibit surface plasmon resonance (SPR) containing one out-plane resonance and two inplane resonances. The out-plane resonance should be at a shorter wavelength that is too weak to be discerned. One of the inplane resonance should be around 530 nm and another at a longer wavelength (depending on a morphology factor). In contrast, spherical particles have only a single SPR band at 530 nm. In the spectra shown in Fig. 4.44, the broad absorption band around 680 nm for the sample before irradiation can be attributed to the superposition of the two bands of the triangularly shaped particles. Subsequent irradiation induces a spheroidization of the particles, leading to a decrease and eventual disappearance of the shape-dependent band around 680 nm, leaving to only the single SPR band at 530 nm, corresponding to a spherical particle. The variations of spectra can reflect the morphological changes of the particle arrays and, further, the related information of the laser. This shows the potential application of such a method in the fabrications of data storage devices.

In summary, the morphology manipulation of gold nanoparticle arrays formed by nanosphere lithography was performed by laser irradiation. Laser pulses induced amorphology evolution from triangularly to nearly spherically-shaped particles and a structural evolution from poly to single crystal. The corresponding optical absorption spectra were found to also change significantly with particle shape, and thus this material could be used for super-density optical data permanent storage. This study provides not only a good way to control the morphology of nanostructured materials and hence their properties, but also introduces a new tool for the fabrication of specific future nanodevices by area-selective treatment.

Bibliography

1. G. M. Wallraff and W. D. Hinsberg, *Chem. Rev.* **99**, 1801 (1999).
2. T. Ito and S. Okazaki, *Nature* **406**, 1027 (2000).

3. H. I. Smith and M. L. Schattenburg, *IBM . J. Res. Dev.* **37**, 319 (1993).
4. J. P. Silveman, *J. Vac. Sci. Technol.* **B15**, 2117 (1997).
5. T. W. Ebbesen, H. J. Lezec, and H. F. Ghaemi, *Nature* **391**, 667 (1998).
6. G. Y. Liu, S. Xu, and Y. Qian, *Acc. Chem. Res.* **33**, 457 (2000).
7. R. D. Piner, J. Zhu, C. A. Mirkin *et al.*, *Science* **283**, 661 (1999).
8. J. A. Stroscio and D. M. Eigler, *Science* **254**, 1319 (1991).
9. W. Stöber, A Fink, and E. Bohn, *J. Colloid. Interface Sci.* **26**, 62 (1968).
10. P. Pieranski, *Contemp. Phys.* **24**, 25 (1983).
11. I. Piirma, *Emulsion Polymerization*, Academic Press, New York, 1982.
12. C. A. Murray and D. H. V. Winkle, *Phys. Rev. Lett.* **58**, 1200 (1987).
13. A. T. Skjeltorp and P. Meakin, *Nature* **335**, 424 (1988).
14. B. Q. Cao, W. P. Cai, Y. Li *et al.*, *Phys.* **33**, 127 (2004).
15. A. J. Hurd and D.W. Schaefer, *Phy. Rev. Lett.* **54**, 1043 (1985).
16. H. H. Wickman and J. N. Korley, *Nature* **393**, 445 (1998).
17. M. Kondo, K. Shinozaki and L. Bergstrom, *Langmuir* **11**, 394 (1995).
18. M. Szekeres, O. Kamalin, P. G. Grobet *et al.*, *Colloids and Surfaces A: Physicochem. Eng. Aspects* **227**, 77 (2003).
19. H. W. Deckman and J. H. Dunsmuir, *Appl. Phys. Lett.* **41**, 337 (1982).
20. M. Trau, D. A. Saville, and I. A. Aksay, *Science* **272**, 706 (1996).
21. Y. Li, W. P. Cai, F. Q. Sun *et al.*, *Phys.* **32**, 153 (2003) (in Chinese).
22. R. Micheletto, H. Fukuda, and M. Ohtsu, *Langmuir* **11**, 3333 (1995).
23. N. D. Denkov, O. D. Velev, P. A. Kralchevsky *et al.*, *Langmuir* **8**, 3183 (1992).
24. J. C. Hulteen, D. A. Treichel, M. T. Smith, M. L. Duval, T. R. Jensen, and R. P. V. Duyne, *J. Phys. Chem.* **B103**, 3854 (1999).
25. A. S. Dimitrov and K. Nagayama, *Langmuir* **12**, 1303 (1996).
26. F. Burmeister, C. Schafle, P. Leiderer *et al.*, *Langmuir* **13**, 2983 (1997).
27. F. Burmeister, C. Schafle, B. Keilhofer, C. Bechinger, J. Boneberg, and P. Leiderer, *Adv. Mater.* **10**, 495 (1998).

28. U. C. Fischer and H. P. Zingsheim, *J. Vac. Sci. Technol.* **19**, 881 (1981).

29. J. C. Hulteen and R. P. van Duyne, *J. Vac. Sci. Technol.* **A13**, 1553 (1995).

30. C. L. Haynes and R. P. van Duyne, *J. Phys. Chem.* **B105**, 5599 (2001).

31. H. W. Deckman, J. H. Dunsmuir, S. Garoff *et al.*, *J. Vac. Sci. Technol*, **B136**, 333 (1988).

32. H. W. Deckman and J. H. Dunsmuir, *J. Vac. Sci. Technol.* **B1**, 1109 (1983).

33. H. W. Deckman and T. D. Moustakas, *J. Vac. Sci. Technol.* **B6**, 316 (1988).

34. A.-J. Haes, C. L. Hayhes, and R. P. van Duyne, *Mat. Res. Soc. Symp. Pro.* **636** (2001).

35. X. D. Wang, C. J. Summers, and Z. L. Wang, *Nano Lett.* **4**, 423 (2004).

36. M. Winzer, M. Kleiber, R. Wiesendanger *et al.*, *Appl. Phys.* **A63**, 617 (1996).

37. J. Rybezynski, V. Ebels, and M. Giersig, *Colloids and Surfaces A: Physicochem. Eng. Aspects* **219**, 1 (2003).

38. H. A. Bullen and S. J. Garrett, *Nano Lett.* **2**, 739 (2002).

39. F. Lenzmann, K. Li, A. H. Kitai *et al.*, *Chem. Mater.* **6**, 156 (1994).

40. Y. Xia, J. Rogers, K. E. Panl *et al.*, *Chem. Rev.* **99**, 1823 (1999).

41. M. H. Stenzel-Rosenboum, T. P. Davis, and A. G. Fanl, *Angew. Chem. Int. Edit.* **40**, 3428 (2001).

42. P. S. Shah, M. B. Sigman, and C. A. Stowell, *Adv. Mater.* **15**, 971 (2003).

43. F. Q. Sun, W. P. Cai, Y. Li, B. Q. Cao, Y. Lei, and L. D. Zhang, *Adv. Mater.* **16**, 1116 (2004).

44. P. N. Bartlett, J. J. Baumberg, S. Coyle *et al.*, *Farady Discuss.* **125**, 117 (2004).

45. F. Q. Sun, W. P. Cai, Y. Li, B. Q. Cao, Y. Lei, and L. D. Zhang, *Adv. Funct. Mater.* **14**, 283 (2004).

46. B. Q. Cao, W. P. Cai, F. Q. Sun, Y. Li, Y. Lei, and L. D. Zhang, *Chem. Comm.* **14**, 1604 (2004).

47. B. Q. Cao, F. Q. Sun, and W. P. Cai, *Electrochem. Solid-state Lett.* **8**, G237 (2005).

48. F. Q. Sun, W. P. Cai, Y. Li, L.C. Jia, and F. Lu, *Adv. Mater.* **17**, 2872 (2005).

49. Y. Li, W. Cai, G. Duan, F. Sun, B. Cao, F. Lu, Q. Fang, and I. W. Boyd, *Appl. Phys.* **A81**, 269 (2005).

50. M. Izaki and T. Omi, *J. Electrochem. Soc.* **143**, L53 (1996).

51. S. Penlon and D. Lincot, *J. Electrochem. Soc.* **145**, 864 (1998).

52. J. Y. Lee and Y. S. Tak, *Electrochem. Solid-state Lett.* **4**, C63 (2001).

53. P. N. Bartlett, J. J. Baumberg, S. Coyleb, and M. E. Abdelsalam, *Faratay Discuss. Chem. Soc.* **125**, 117 (2004).

54. P. Jiang, J. F. Bertone, and V. L. Colvin, *Science* **291**, 453 (2001).

55. A. He, B. Djurfors, S. Akhlaghi, and D. G. Ivey, *Plat. Surf. Finish.* **89**, 48 (2002).

56. W. Sun and D. G. Ivey, *Mater. Sci. Eng.* **B65**, 111 (1999).

57. R. Winand, *Electrochem. Acta.* **39**, 1091 (1994).

58. F. Burmeister, C. Schafle, and P. Leiderer, *Langmuir* **13**, 2983 (1997).

59. F. Q. Sun, W. P. Cai, Y. Li, B. Q. Cao, Y. Lei, and L. D. Zhang, *Materials Science and Technology* **21**, 500 (2005).

60. C. L. Haynes, A. D. Mcfarland, M. T. Smith, J. C. Hulteen, and R. P. van Duyne, *J. Phys. Chem.* **B106**, 1898 (2002).

61. For the 0.08 M and 0.06 M samples, the pores are oblate hemispherical hollows with nearly or regularly hexagonal openings at the film surface, but are only open at the upper end.

62. D. Verhoeven, *Fundamentals of Physical Metallurgy*, John Wiley Sons, New York, 1975.

63. After solute deposition, the deformation of the PSs will be limited during further drying treatment due to the restriction of the deposited solute shell.

64. R. B. H. Tahar, T. Ban, Y. Ohya, and Y. Takahashi, *J. Appl. Phys.* **82**, 865 (1997).

65. Y. Li, W. P. Cai, G. T. Duan, B. Q. Cao, and F. Q. Sun, *J. Mater. Res.* **20**, 338 (2005).

66. R. Castillo, B. Koch, P. Ruiz, and B. Delmon, Influence of preparation methods on the texture and structure of titania supported on silica, *J. Mater. Chem.* **4**, 903 (1994).

67. W. Meier, Polymer nanocapsules, *Chem. Soc. Rev.* **29**, 295 (2000).

68. H. Huang and E. E. Resen, Nanocages derived from shell cross-linked micelle templates, *J. Am. Chem. Soc.* **121**, 3805 (1999).

69. R. Djalali, J. Samson, and H. Matsui, Doughnut-shaped peptide nano-assemblies and their application as nanoreactors, *J. Am. Chem. Soc.* **126**, 7935 (2004).

70. A. D. Dinsmore, M. F. Hsu, M. G. Nikolaides, M. Marquez, A. R. Bausch, and D. A. Weitz, Colloidosomes: selectively permeable of colloidal particles, *Science* **298**, 1006 (2002).

71. Y. Lu, Y. Yin, and Y. Xia, A self-assembly approach to the fabrication of patterned, two-dimensional arrays of microlenses of organic polymers, *Adv. Mater.* **13**, 34 (2001).

72. E. Gu, H. W. Choi, C. Liu, C. Griffin, J. M. Girkin, I. M. Watson, M. D. Dawson, G. McConnell, and A. M. Gurney, Reflection/transmission confocal microscopy characterization of single-crystal diamond microlens arrays, *Appl. Phys. Lett.* **84**, 2754 (2004).

73. F. Sun, W. Cai, Y. Li, G. Duan, W. T. Nichols, C. Liang, N. Kkoshizaki, Q. Fang, and I. W. Boyd, *Appl. Phys.* **B81**, 765 (2005).

74. A. J. Haes, W. Paige Hall, L. Chang, W. L. Klein, and R. P. van Duyne, *Nano Lett.* **4**, 1039 (2004).

75. A. J. Haes, S. L. Zhou, G. C. Schata, and R. P. van Duyne, *J. Phys. Chem.* **B108**, 6961 (2004).

76. C. L. Hanes, A. D. Mcfarland, L. L. Zhao, R. P. van Duyne, G. C. Schatz, L. Gunnarsson, J. Prikulis, B. Kasemo, and M. Kall, *J. Phys. Chem.* **B107**, 7337 (2003).

77. C. L. Hanes and R. P. van Duyne, *Nano Lett.* **3**, 939 (2003).

78. R. M. Winze, M. Kleiber, N. Dix, and R. Wiesendanger, *Appl. Phys.* **A63**, 617 (1996).

79. Y. Tu, Z. P. Huang, D. Z. Wang, J. G. Wen, and I. F. Ren, *Appl. Phys. Lett.* **80**, 4018 (2002).

80. X. D. Wang, C. T. Summers, and Z. L. Wang, *Nano Lett.* **4**, 423 (2004).

81. K. Kempa, B. Kimball, J. Rybczynski, Z. P. Huang, P. F. Wu, D. Steeves, M. Sennett, M. Giersig, D. V. G. L. N. Rao, D. L. Carnahan, D. Z. Wang, J. Y. Lao, W. Z. Li, and Z. F. Ken, *Nano Lett.* **3**, 13 (2003).

82. J. Sort, H. Glaczynska, V. Ebels, B. Dieny, M. Giersig, and J. Rybcznski, *J. Appl. Phys.* **95**, 7516 (2004).

83. C. L. Haynes and R. P. van Duyne, *J. Phys. Chem.* **B105**, 13 (2001).

84. J. Lanhann, S. Mitragotri, T. N. Tran, H. Kaido, J. Sundaram, I. S. Choi, S. Hoffer, G. A. Somorjai, and R. Langer, *Science* **299**, 371 (2003).

85. M. Kawasaki and M. Hori, *J. Phys. Chem.* **B107**, 6760 (2003).

86. J. D. Verhoeven, *Fundamentals of Physical Metallurgy*, Wiley, New York, 1995.

87. R. C. Jin, Y. W. Cao, C. A. Mirkin, K. L. Kelly, G. C. Schatz, and J. G. Zheng, *Science* **294**, 1901 (2001).

88. J. P. Kottmann and O. J. F. Martin, *Appl. Phys.* **B73**, 299 (2001).

89. J. P. Kottmann and O. J. F. Martin, *Opt. Express* **6**, 213 (2000).

90. U. Kreibig and M. Vollmer, *Optical Properties of Metal Clusters*, Springer, Berlin, 1995.

Chapter 5
Nanoarray Synthesis and Characterization based on Alumina Templates

- Preparation of ordered channel AAM templates
- Synthesis and characterization of ordered nanoarrays

Chapter 5

Nanoarray Synthesis and Characterization based on Alumina Templates

5.1 Preparation techniques of ordered channel AAM (anodization alumina membrane) templates

As early as 1953, Keller,[1] who worked in the alumina lab of the American Alumina Company, reported preparation and structure features of ordered channel AAM templates obtained by the electrochemical method and electromicroscopy observations, respectively. However, due to the technology limitation, the channel distribution of alumina membranes synthesized by them was disordered. After 20 years, many researchers made a lot of studies into synthetic methods. It was not until the early 90s that Moskovits's group[2] synthesized alumina templates with the porous density of 10^{10}–10^{11}/cm^{-2}. Then, Martin[2–5] synthesized alumina templates with the pore diameters ranging from 12 to 250 nm.

In 1995, Masuda,[6] who worked in Kyoto University, first reported that they synthesized alumina membranes composed of highly-ordered and hexagonally close-packed arrays. After that, they improved the synthesis technique to obtain the alumina template, containing almost perfect and hexagonal channel arrays with long-range ordering (ordered channel AAM templates).[7]

In the following, preparation and structure features of ordered channel AAM templates will be systematically introduced, and the ordered channel array formation mechanism of alumina templates will be analyzed simply.

5.1.1. Preparation of ordered channel AAM templates

We simply named the ordered channel AAM template "AAM" in the following. The preparation process of AAM includes three main steps: pretreatment, anodization and post-treatment.

(1) Pretreatment

The pure alumina plate (purity: 99.999%, thickness: 0.2 to 0.3 mm) is first washed in turn in acetone and ethanol. After that, the alumina plate is annealed in a vacuum at 450°C, so that the internal mechanical stress of alumina plate is eliminated and at the same time, the grains in the alumina plate grow. Then, the electrochemical polishing is carried out in a mixture solution of perchloric acid and ethanol. Finally, the polished alumina plate is washed several times using deionized water, followed by drying in air. Therefore, an alumina plate with a clean and polished surface is achieved.

(2) Anodization

Anodization of the highly pure alumina plate is a typical self-organization process dependent generally on the conditions of anodization, including the type and concentration of electrolyte, the anode voltage, the temperature, the oxidation time, and subsequent etching treatment. Figure 5.1 shows the preparation process of ordered channel AAM template during using two-step anodization. As a typical example, the procedure of AAM template prepared in the oxalic acid solution is as follows. Firstly, a highly pure (99.999%) alumina plate was used as the starting material. Prior to anodization, the alumina plate is annealed in a vacuum at 450°C to remove the internal mechanical stress and for recrystallization. Then, a mirror surface was achieved by electropolishing at 23 V in a mixture solution of perchloric acid and ethanol. The first anodization was conducted at 40 V in a 0.3 M oxalic acid solution at 10°C for 4 h (Fig. 5.1(a)). After that, the alumina plate with a porous alumina film was immersed in a mixture solution of 6 wt.% H_3PO_4 and 1.8 wt.% H_2CrO_4 at 60°C for 7 h to remove the porous alumina film (Fig. 5.1(b)). Second anodization was carried out under the same conditions as those used in the first anodization, except that the anodization time was elongated to, for example, 10 h (Fig. 5.1(c)). In the second anodization process, with the increase of the anodization time, the porous diameter distribution becomes more uniform, and thus the highly ordered channel alumina membrane template was formed.

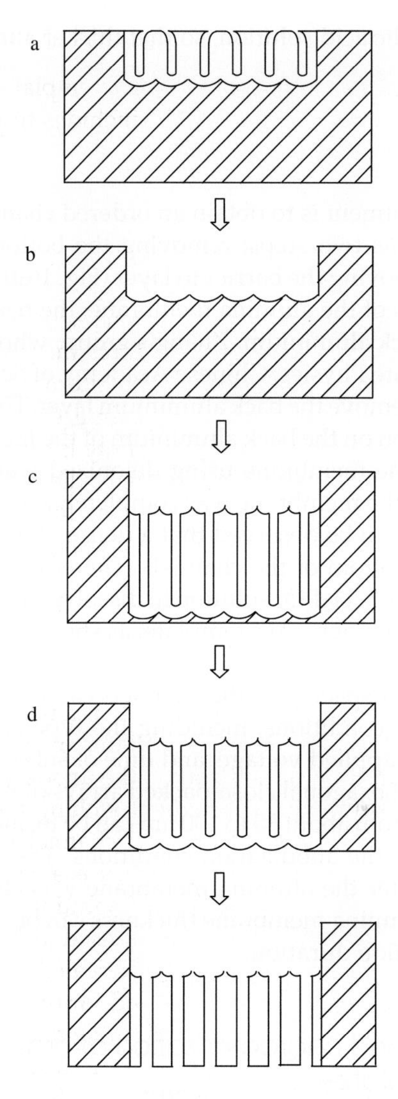

Fig. 5.1. The preparation procedure of ordered channel AAM templates.

In the sulphuric acid solution, alumina templates with smaller diameter pores can be also synthesized by using the similar synthesis technique. This is because under a certain condition, the alumina etching speed caused by sulphuric acid is slower than that caused by oxalic acid. Therefore, the second anodization time in the sulphuric acid solution is generally longer than that in the oxalic acid solution, so that thicker alumina templates are synthesized.

(3) Post-treatment

The aim of post-treatment is to obtain an ordered channel AAM template. This process contains two steps: removing the bottom back aluminium (Fig. 5.1(d)) and removing the barrier to layer (Fig. 1(e)). In order to remove the back aluminium of the alumina membrane, the researchers used toxic HgCl to etch the back aluminium. Zhang's group, who are working in the Institute of Solid State Physics, Chinese Academy of Science, first used the non-toxic $SnCl_4$ to remove the back aluminium layer. They dropped the saturated $SnCl_4$ solution on the back aluminium of the membrane (Fig. 5.1(c)) and then washed the membrane using deionized water. As a result, the back aluminium of the membrane was completely eliminated (Fig. 5.1(d)). From this figure, it can be observed that a dense aluminium barrier layer still existed at the bottom of the channels. Therefore, the membrane was immersed in the 5 wt.% H_3PO_4 solution to dissolve the barrier layer, resulting in the ordered channel AAM template, as shown in Fig. 5.1(c).

The diameter and spacing of the channels can be modulated by varying the anodization conditions, including the type and concentration of the electrolytes, the applied voltage, and time of subsequent etching treatment. For example, hexagonal close-packed arrays of AAM with selectable diameters ranging from about 35 to 100 nm can be formed in the oxalic acid solution by varying the anodization conditions. The channels diameters can be increased after the alumina membrane was etched in the H_3PO_4 solution, and the alumina membrane thickness can be changed by varying the second anodization duration.

5.1.2. Structure and characterization of ordered channel AAM templates

The structure scheme of the AAM template is shown in Fig. 5.2. It can be seen that the ordered channel AAM template is composed of many regular hexagonal cells. These structure cells present a hexagonal close-packed distribution. In other words, the AAM template is composed of hexagonal close-packed and highly-ordered channel arrays. The channel axis is perpendicular to the surface of the AAM template. For the AAM template without post-treatment, a barrier layer is located between the channel bottom and the alumina plate.

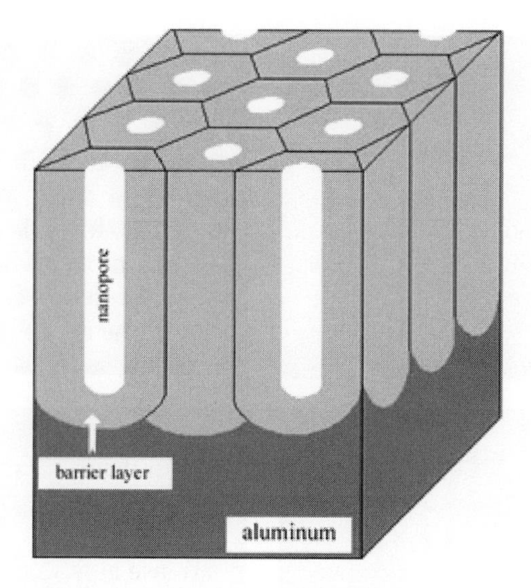

Fig. 5.2. The structure scheme of AAM templates.

Figure 5.3 is the SEM image of the AAM template with a highly-ordered channel arrays which was synthesized in a 0.3 mol/L oxalic acid solution at 40 V (DC). The surface morphology image of the ordered nanochannel array AAM template at low magnification is given in Fig. 5.3(a). The perfectly ordered region is about 1—4 μm^2, corresponding to the channel (or pore) density of about $10^{10}/cm^{-2}$. In a grain range, a highly-ordered hexagonal close-packed alumina template can be synthesized (Fig. 5.3(b)), and the channel diameters are 50 nm and spacing between neighboring channels is 100 nm. From the cross-section of the template, it can be seen that the channel of the template presents a uniform and parallel alignment state (Fig. 5.3(c)). Moreover, all the channels are perpendicular to the template surface (Fig. 5.3(d)).

5.1.3. Exploration of ordered channel formation mechanism

As an example, when the oxalic acid solution is used as the electrolyte, the change curve of the electrolytic current with the electrolytic time is shown in Fig. 5.4. Apparently, the electrolytic current decreases rapidly from the original peak value to the minimum, and then rises slowly to reach the highest value. This anodization process can be divided into four steps. In the first step, the electrochemical reaction made a uniform and dense oxidation film form on the highly pure aluminium surface. The conductivity

Fig. 5.3. SEM photographs of the alumina template. (a) The low magnification surface morphology; (b) the high magnification surface morphology; (c) the cross-section image; (d) the surface and cross-section image.

of aluminium changed from the high conductive state to the low conductive state. Therefore, the current decreases rapidly. In the second step, the pores nucleated randomly on the oxidation membrane. In the third step, the pores grew gradually and hence the pore position modulated spontaneously. Finally, the hexagonal close-packed structure was formed, so that the free energy of the system reached the minimum value and thus the structure was very stable. The fourth step was the stable process of the pore position modulating each other. With ordered pore growth and the pore length increased, the current gradually reached the stable value.

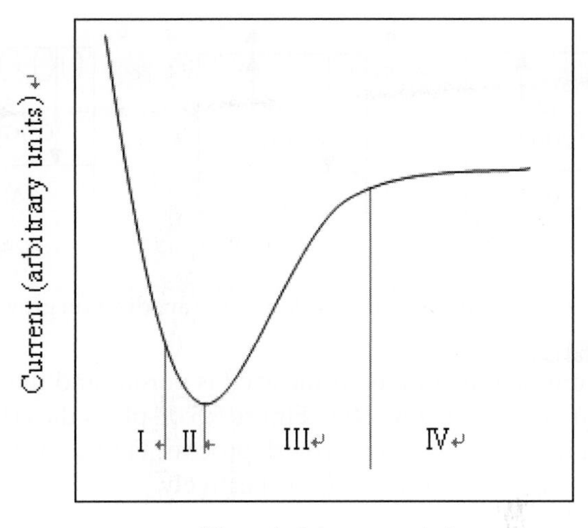

Fig. 5.4. The change of the current with time in the anodization process.

Now, there are a large number of research results on the formation mechanism of the porous alumina membrane. After Keller[1] proposed the structure model of the porous alumina membrane, people initially thought that under anodization conditions, the oxidation layer membrane formed on the aluminium surface consisted of crystallized and amorphous alumina, and in the acid solution, the amorphous component was dissolved and thus the crystallized alumina frame remained, which was the porous alumina membrane.[8] However, the experimental results obtained by Thompson[9] indicated that the main reason for the formation of anodic porous alumina was as follows. The microstructure of the alumina plate surface was not level and smooth, so that the current distribution of the surface was not uniform and thus the surface protuberances grew and formed the ridgy frame. The regions surrounded by the ridgy frame provided beneficial conditions for the formation of the porous structures of alumina membranes. Further, Thompson[10] developed this model and thought that in the anodization process, hydrolysis and precipitation of aluminium ions played a key role in the formation of nanopores. Xu[11,12] investigated in depth the relation between the anodization conditions and established the "critical current density" model. The main content of this model is as follows. For the anodization process of the electrolyte with certain density and temperature, there always exists a critical current density, J_c, when the current density of the circuit $J > J_c$, the alumina membrane synthesized is completely dense and the corresponding current efficiency is 100%, when

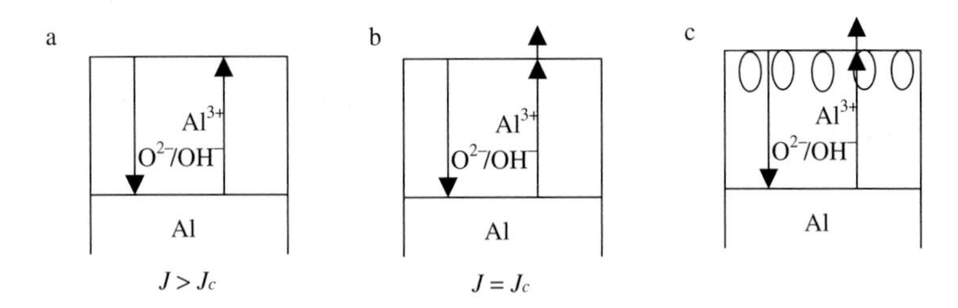

Fig. 5.5. The critical current density effect in the anodization growth process.

$J < J_c$, the alumina membrane synthesized is porous and the corresponding current efficiency is below 60%. Figure 5.5 displays the critical current density effect in the anodization growth process. Figures 5.5(a)–5.5(c) correspond to $J > J_c, J = J_c$ and $J < J_c$, respectively.

When $J > J_c$, under the electric field action, O^{2-}/OH^- ions penetrate the OE interface and move towards the MO interface. Similarly, the Al^{3+} ions move from the MO interface towards the OE interface and at the OE interface, Al^{3+} ions are directly combined with oxygen to form Al_2O_3. When $J = J_c$, parts of Al^{3+} ions penetrate the OE interface and enter the electrolyte and the other parts of Al^{3+} ions are combined with oxygen to form Al_2O_3 which become the component of the alumina membrane. When $J < J_c$, all Al^{3+} ions penetrate the OE interface and enter the electrolyte. The discovery of the critical current density effect makes people able to draw out the outline of the dynamic behavior law of positive and negative ions in the anodization process and reveal the internal liaison between the dense type oxidation membrane and the porous type oxidation membrane.

In 1998, Miiller's group[13,14] proposed a volume expansion stress model. This model indicates that in the stable oxidation process, the volume of Al_2O_3 formed after Al is oxidated become large. The volume expansion leads each pore to produce a stress action on its circumference. The uniform action of this stress makes the pores' self-organization arrangement form a hexagonal close-packed arrangement pattern with the lowest system energy.

5.2 Synthesis and characterization of ordered nanoarrays

One-dimensional nanoarrays are usually synthesized by means of a porous template. Main porous templates have two kinds: the track-etch polymer template[15] and the ordered channel AAM template (the alumina membrane templates). The preparation process of the track-etch polymer template is as follows. After the polyster or polycarbonate plates with the thickness of 6 to 20 m were irradiated by nuclei, many concave damage hotels were left on the plates. Then, these plates were treated by chemical etch, so that these concave holes become cylindric pores with uniform diameters, which were random distributed. For these polymer templates, the smallest pore diameter may reach 10 nm and the pore density is about $10^9/\text{cm}^{-2}$. The cylindric pores are not perpendicular to the surface of the plate. However, the alumina membrane template has the following features: the channel distribution is uniform, ordered and hexagonally close-packed, and the channels are parallel to each other and perpendicular to the surface. Therefore, the ordered nanoarrays of one-dimensional nanowires and nanotubes can be synthesized be means of the alumina templates.

In Sec. 5.2, we mainly introduce the synthesis and characterizations of ordered nanoarrays based on the AAM templates, and also introduce a few synthesis experiments of ordered nanoarrays based on other templates.

5.2.1. Ordered nanoarrays of elements

The elemental ordered nanoarrays can be synthesized by alumina templates using many synthesis methods. Usually, the ordered nanoarrays of metals and semi-metals may be prepared by using electrochemical deposition of corresponding materials inside the alumina membrane templates. The ordered nanoarrays of semiconductors can be prepared inside the alumina membrane templates by combining CVD (chemical vapor deposition) or electrochemical deposition with other techniques.

5.2.1.1. *Ordered nanoarrays of metal nanowires and nanotubes (Pb, Ag, Cu, Au)*

In recent years, a lot of nanowires and nanotubes of metal were made. People prepared the nanoarrays of metal nanowires and nanotubes using the alumina membrane templates, polymer templates and the other templates etc.[16–27]

In this paragraph, we will introduce the synthesis of nanoarrays of one-dimensioned Cu, Ag, and Pb nanowires/nanotubes by using the alumina membrane templates.

(1) Cu nanotube and nanowire array

All current methods in the synthesis of the nanotubes need to use the so-called "molecular anchor" to prepare the alumina membrane template. Wang[28] presented the synthesis of copper nanotubes in the alumina membrane template using a direct electrochemical deposition technique which does not need to employ the "molecular anchor" to treat the alumina membrane template. Alumina membrane templates were prepared by using the same method as that described in Sec. 5.1.1. The pore bottom of alumina membrane templates were opened by chemical etching in 5 wt.% phosphoric acid solution. A layer of Au was sputtered onto one side of the membrane served as an electrode. The electrochemical method was employed to deposit metal Cu into the pores of the alumina membrane.[12] The electrolyte contained 0.2 M $CuSO_4{\cdot}5H_2O$ and 0.1 M H_3BO_4. The electro-deposition was carried out at a constant current density (2.5 mA/cm^2) with carbonate serving as the counter electrode at room temperature for 2 h. The pH of the solution was controlled in the range 4.5 to 5.0 by adding 0.1 M H_2SO_4.

The sample with the nanotubes attached in the membrane was characterized by X-ray diffraction (XRD, MXP-18AHF) (see Fig. 5.6). From Fig. 5.6, we can see that the diffraction peaks in the range $40° < 2\theta < 78°$ can be indexed as (111), (200), and (220) planes of the face-centered cubic structure of the metallic copper. The diffraction peak intensity of the (220) plane is much higher than those of the other two peaks, which indicates that the Cu

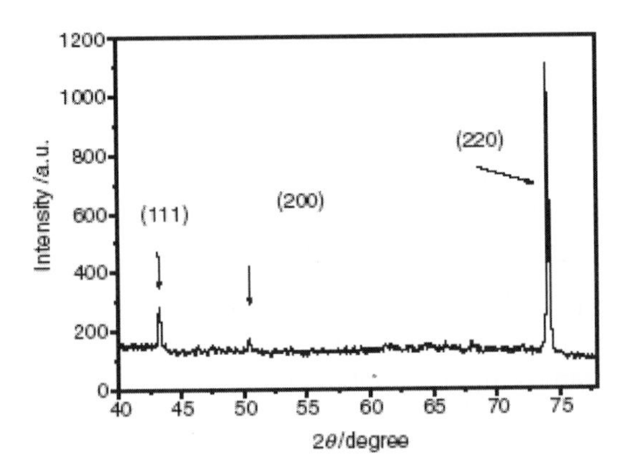

Fig. 5.6. X-ray diffraction of Cu nanotubes embedded in the alumina membrane. The diffraction peaks could be indexed as (111), (200), and (220) planes of face-centered cubic copper.

nanotubes in the alumina membrane are oriented along the [110] direction. Figure 5.7(a) shows the scanning electron microscopy (FESEM, JSM-6700F) image of the Cu nanotubes after removal of the alumina membrane. The results display a highly-ordered array of the Cu nanotubes. The diameter of the nanotubes is about 40 nm, which is in good agreement with the diameter of the alumina membranes. Figure 5.7(b) shows a top-view SEM image after removal of the top layer of the alumina membrane, denoting the open-ends of the Cu nanotubes.

The common approach to obtain the nanotubes in the alumina membrane is to use the "molecular anchor" to treat the pore's walls in the membrane chemically so that the deposited metal can stick onto the

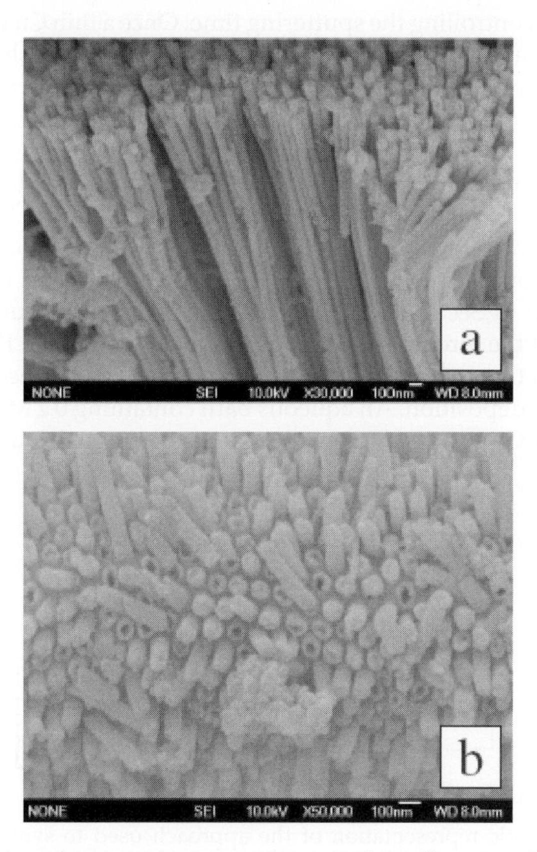

Fig. 5.7. Scanning electron microscopy images of the Cu nanotube arrays after dissolution of the alumina membrane. (a) a cross-section image; (b) a top-view image.

walls. For example, the alumina membranes for producing gold and nickel nanotubes were prepared by attaching the cyanosilane and methyl-γ-diethylenetriaminopropyldimethoxysilane to the pore walls inside the membranes as the "molecular anchor" respectively.[26,27] The selection of the "molecular anchor" usually plays an important role in the synthesis of the different metal nanotubes. However, their methods in producing the Cu nanotubes did not use any "molecular anchor". Instead, they controlled the thickness of the sputtered Au layer on the membrane carefully and found that it is a key factor for synthesizing the metal nanotubes. The Au layer should be thin enough not to cover the open ends of the pores in the membrane in order to produce the nanotubes. Otherwise the nanowires would be produced if the Au layer were so thick that it covered all the open ends of the pores. The thickness of the Au layer can be controlled accurately by controlling the sputtering time. Once a thin Cu layer attached on the pore walls in the alumina membrane, it created a shielding to the applied electrical field to prevent the further deposition in the nanotubes. The process of the nanotubes growth is illustrated in Fig. 5.8.

In the following, template synthesis of single-crystal Cu nanowire arrays by electro-deposition (EVD) will be reported. Gao[29] used the alumina membrane templates to prepare single-crystal Cu nanowire arrays. The preparation method of alumina membrane templates is the same as that described in Sec. 5.1.1. A gold layer (200 nm) was sputter-deposited on one side of the through-hole anodic alumina membrane PAA template, and served as the working electrode in a conventional three-electrode cell for Cu electro-deposition. An aqueous bath containing 0.2 M $CuSO_4 \cdot 5H_2O$ and 0.1 M H_3BO_3 was used to prepare Cu nanowires using a potentiostatic

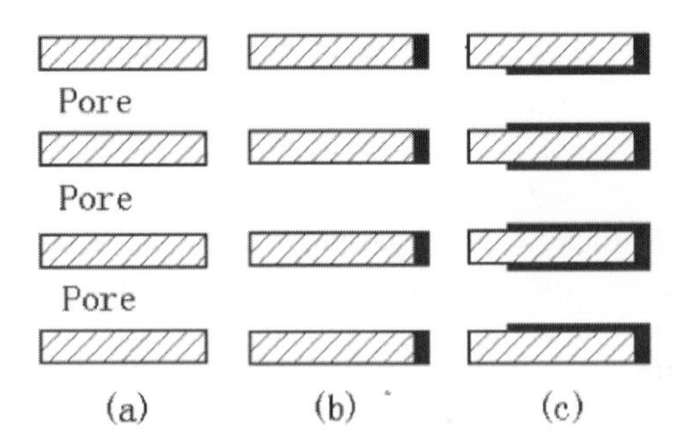

Fig. 5.8. Schematic representation of the approach used to synthesize Cu nanotube arrays in the alumina membrane. (a) Cross-section of the alumina membrane; (b) Au layer sputtered on one side of the alumina membrane; (c) Cu nanotubes grown along the pore walls.

ECD technique. The pH of the solution was controlled in the range 4.5 to 5.0 by adding 0.1 M H_2SO_4 solution. A glassy carbon plate and a saturated calomel electrode (SCE) were used as the counter electrode and reference electrode, respectively. Prior to the ECD experiment, the counter electrode and the reference electrode were cleaned by deionized water and dried in air. The ECD experiment was carried out at −0.15 V (SCE) for 10–20 minutes at room temperature to obtain the Cu nanowire arrays. After ECD, the sample surfaces were polished using a 50 nm SiC polishing sandpaper in order to get rid of the excess Cu particles sticking to the surface. The sample was then washed with ethanol and deionized water in turn, and dried in air before characterization.

SEM observation was performed to verify that Cu had been deposited in the form of wires within the nanochannels of the AAM template. Figures 5.9(a) and 5.9(b) reveal that large quantities of Cu nanowires with high packing density have been fabricated. Figure 5.9(b) shows that the lengths of the Cu nanowires are about 30 μm, corresponding well with the thickness of the AAM template used. It is important to point out that the AAM template had an array of densely-packed parallel nanochannels, which were arranged in hexagonal fashions, with channel diameters and lengths that could be controlled by changing the anodization parameters. Hence, the geometrical characteristics of the Cu nanowires could be controlled by choosing the proper type of AAM template.

The individual Cu nanowires were characterized by TEM after the PAA template had been thoroughly dissolved. High magnification TEM images (Figs. 5.10(a) and 5.10(b)) showed that the Cu nanowires were dense, continuous and uniform in diameter throughout the entire length of the wires. The diameter distribution of the wires was obtained using the statistical results of 20 wire diameters obtained from TEM images. These showed that the diameters of the Cu nanowires varied from 58 to 62 nm, with an average of 60 nm. This is in good agreement with the channel diameters of the AAM template used. However, Yi[30] have shown that the diameter of a single-crystal Pb nanowire prepared by pulse ECD within a track-etched polycarbonate template may change by more than 20% along its length, indicating the advantages of the PAA membrane in template synthesis of 1D nanostructures.

The crystalline structures of the individual Cu nanowires were investigated by SAED experiments. Lots of individual Cu nanowires were characterized, and we always observed a single set of diffraction spots in the SAED patterns (Fig. 5.10(c)), indicating that the Cu nanowires were single-crystal

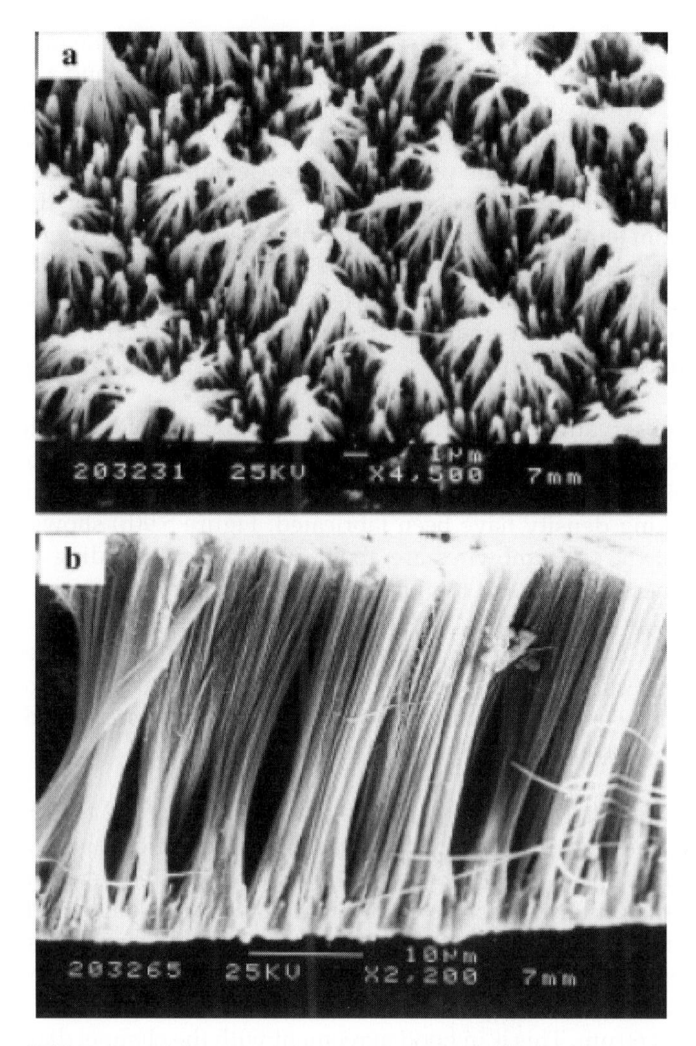

Fig. 5.9. SEM image of Cu nanowire arrays removing the AAM template: (a) top view and (b) cross-section view.

with a face-centered cubic (fcc) structure, which were further confirmed by our HRTEM and XRD investigations. A typical HRTEM image of individual Cu nanowires (Fig. 5.11) reveals that the {111} lattice fringes with lattice spacing around 0.208 nm are approximately vertical to the axis of the nanowire, and that the growth plane is one of the {111} planes. It is also found during the HRTEM investigation that no grain boundaries were observed along the Cu nanowires, and hence that the Cu nanowires are

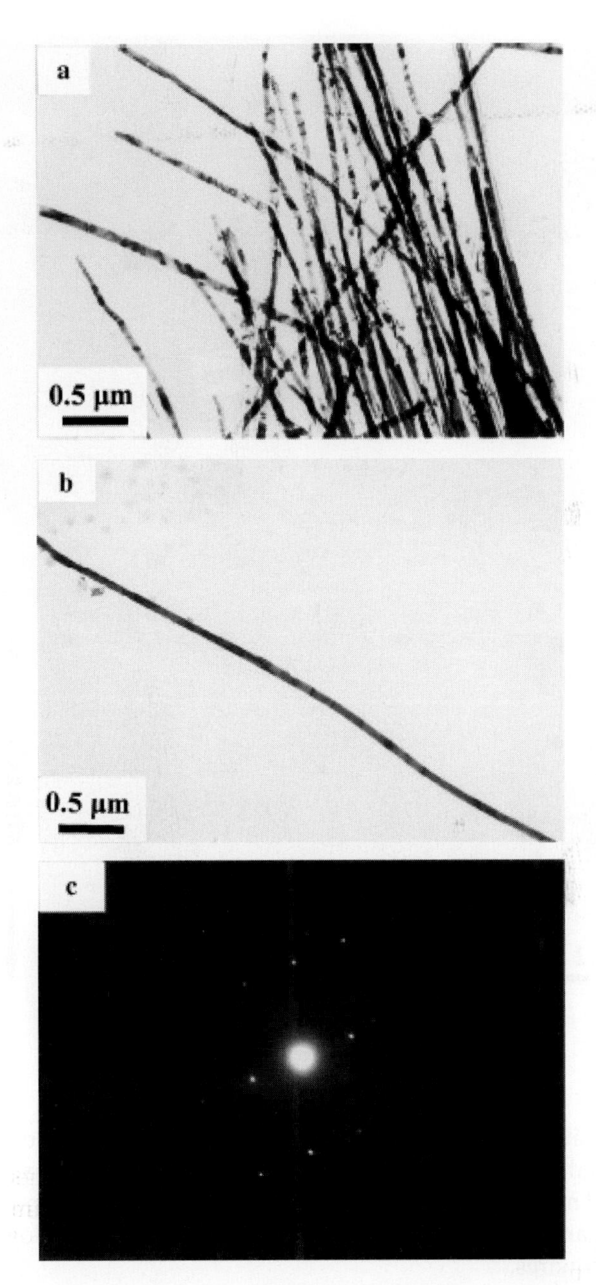

Fig. 5.10. TEM investigations of individual Cu nanowires: (a) high magnification TEM image of the Cu nanowires; (b) high magnification TEM of a single Cu nanowire; (c) corresponding SAED patterns of the Cu nanowire shown in (b).

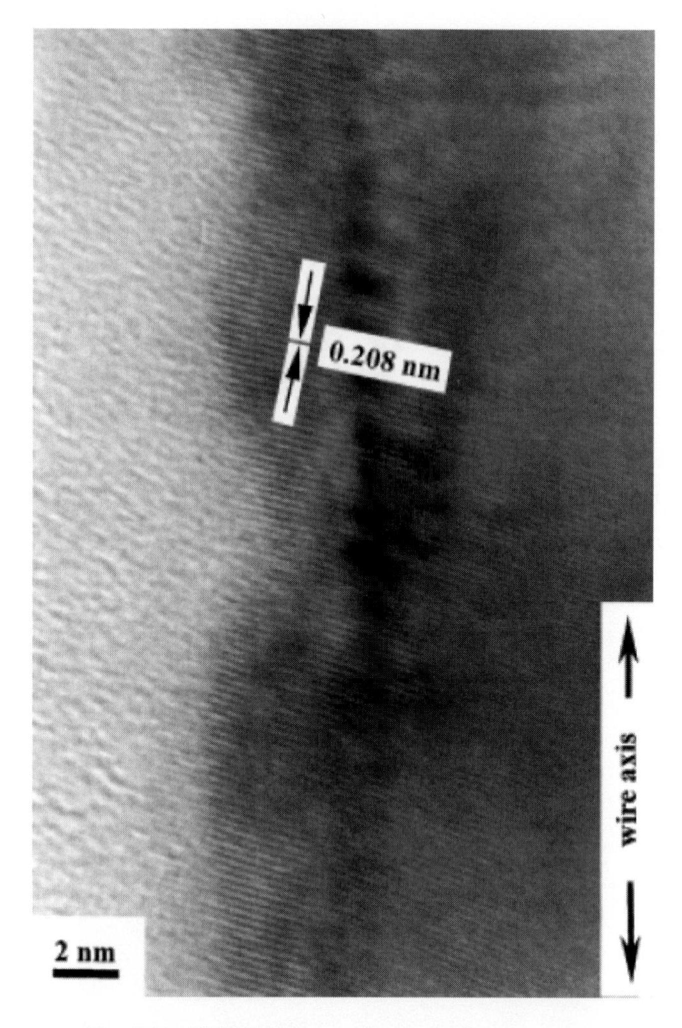

Fig. 5.11. HRTEM image of a single Cu nanowire.

essentially single-crystal. The uniformity of the wire diameters and the single-crystal nature are of importance for studying the transport properties of metal nanowires,[31] and the techniques recently developed for studying the 1D nanowires and/or nanotubes[32–35] can also be employed to study our Cu nanowires.

The XRD technique was used to investigate the phase structure and crystal orientation of the Cu nanowire arrays. From Fig. 5.12, it can be

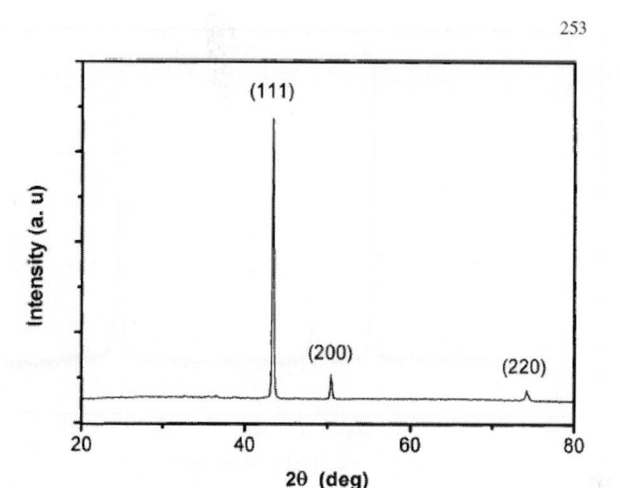

Fig. 5.12. XRD spectra of the as-prepared Cu nanowire arrays.

seen that there are three Cu peaks in the XRD spectra of the Cu nanowire arrays, with the highest at the (111) crystal plane and the other two lower peaks at (200) and (220). All the intense XRD peaks were compared with those at the same positions as the polycrystalline Cu standard, which indicated that the Cu nanowires with face-centered cubic crystal structures had been fabricated. There exists a preferred growth direction, the (111) crystal plane of the Cu nanowire arrays, which was also observed during HRTEM investigation. By normalizing the peak areas to the peak intensities of a polycrystalline bulk Cu standard,[30] it was found that more than 95% of the Cu nanowires in the arrays are orientated along a direction perpendicular to the (111) crystal plane. It could be deduced that every Cu nanowire has the same crystalline structure.

(2) Ag nanowire arrays[36]

The through-hole AAM with ordered nanochannels was prepared via a two-step anodization process as described in Sec. 5.1.1. A layer of gold was sputtered onto one side of the AAM to serve as the working electrode in a two-electrode electrochemical cell. The electrolyte contained a mixture of 300 gl^{-1} $AgNO_3$ and 45 gl^{-1} H_3BO_3 solutions and was buffered to pH = 2.5 with nitric acid. The electro-deposition was carried out for 10 h at a constant current density (2.5 $mAcm^{-2}$), with graphite serving as the counter electrode at room temperature.

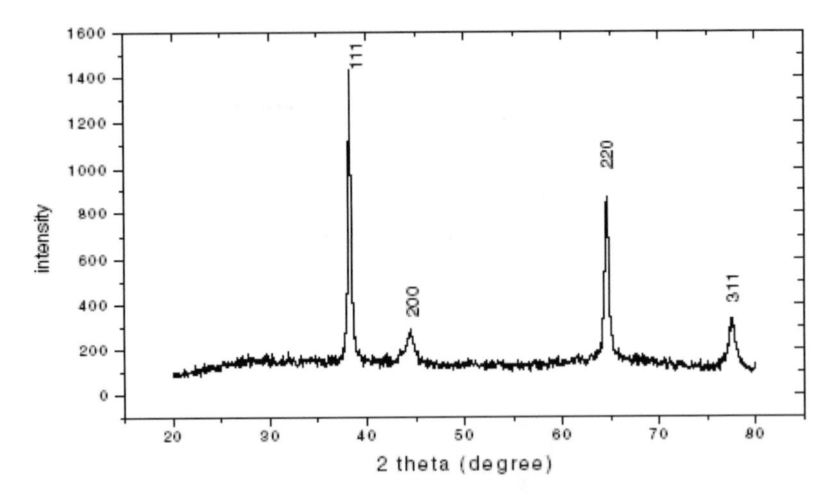

Fig. 5.13. XRD spectrum of the Ag nanowires arrays embedded in the AAM.

The X-ray diffraction pattern of the prepared Ag/AAM sample is shown in Fig. 5.13. The broad peak resulting from the AAM indicates that the AAM is amorphous. The four peaks are found to be very close to (111), (200), (220) and (311) of bulk Ag, indicating that the face-centered cubic (fcc) structure of bulk Ag is preserved in these wires. Figure 5.14 shows the SEM images of the Ag nanowire arrays prepared in the AAM template. It can be seen that large quantities of Ag nanowires with high packing density have been fabricated. The TEM image of the prepared sample (Fig. 5.15(a)) shows a number of nanowires with an average diameter of about 90 nm, which is nearly equal to the channel diameters of the AAM used. A TEM image of a single nanowire is shown in Fig. 5.15(b). The SAED pattern taken from this nanowire corresponds with that of Ag (Fig. 5.15(c)).

(3) Au nanowire arrays

Zhang[37] prepared the naonoarrays of Au nanowires using the AAM templates. An important advantage of this work is that the Au nanowires can be diameter-controllable and well-defined. The preparation process of the Au nanowire arrays is as follows. The through-hole AAM with ordered nanochannels was prepared via a two-step anodization process as described in Sec. 5.1.1. In order to prepare the nanoarrays of Au nanowires with different diameters, the AAM templates with selectable diameters ranging from 35 nm to 100 nm were be synthesized in oxalic acid solution by varying the anodizing conditions and the time of the pore widening

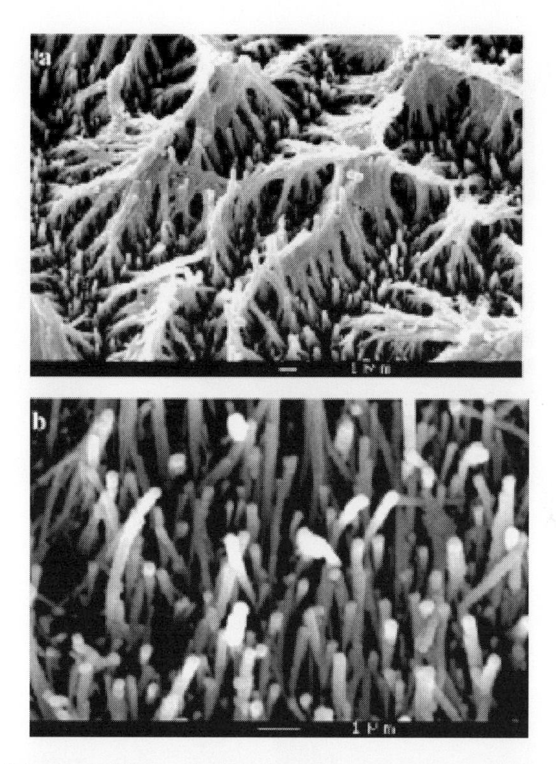

Fig. 5.14. SEM image of the ordered Ag nanowire arrays: (a) low-magnification image; (b) high magnification image.

treatment. (The AAM template channels were widened by chemical etching in aqueous phosphoric acid.) In order to fabricate the nanowire arrays, a layer of Au thin film was deposited as an electrode on one side of the anodic porous alumina membrane using a vacuum evaporation apparatus. Au nanowires were electro-deposited from the following solution: 12 gl^{-1} $HAuCl_4$, 160 gl^{-1} Na_2SO_3, 5 gl^{-1} EDTA, 30 gl^{-1} K_2HPO_4, 0.5 gl^{-1} $CoSO_4$; pH 9.0. The cell voltages were kept at 0.8 V.

Figure 5.16 shows the SEM images of hexagonal close-packed arrays of porous alumina. Figure 5.16(a) shows the SEM image of porous alumina, which was anodized in 0.3 M oxalic acid at 17°C at 35 V: the average pore diameter and inter-pore distance are 35 nm and 100 nm, respectively. Figure 5.16(b) shows the SEM image of porous alumina that was anodized in 0.3 M oxalic acid at 12°C at 40 V: the average pore diameter and inter-pore distance are 45 nm and 105 nm, respectively. Figure 5.16(c) shows the SEM image of porous alumina, which was anodized in 0.25 M oxalic acid at 5°C

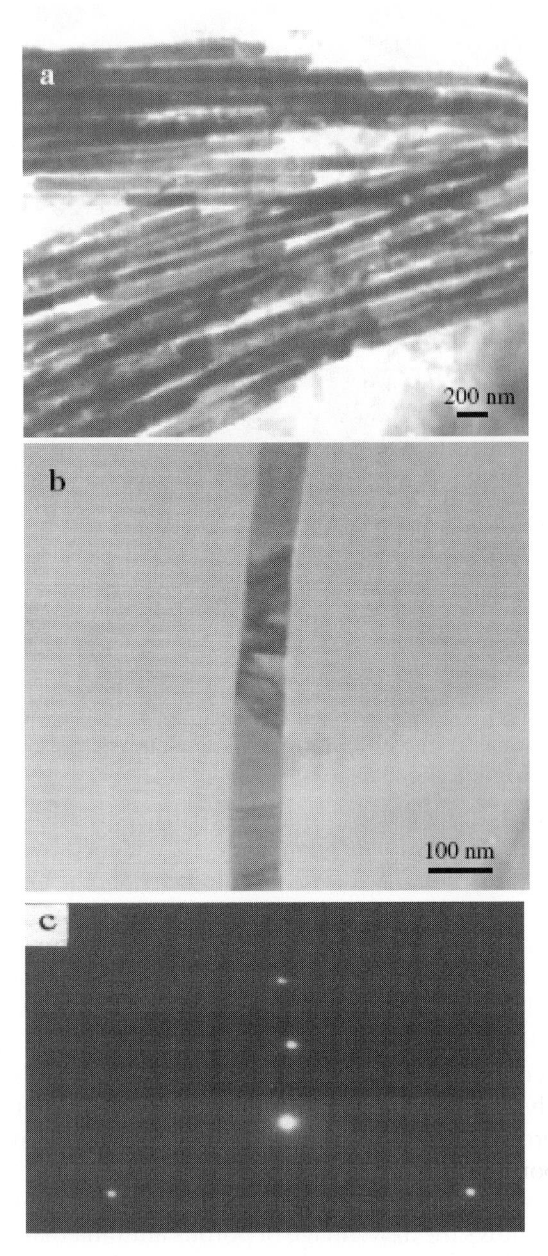

Fig. 5.15. (a) TEM image of Ag nanowires after removing the AAM. (b) and (c) TEM and SAED patterns from an individual nanowire.

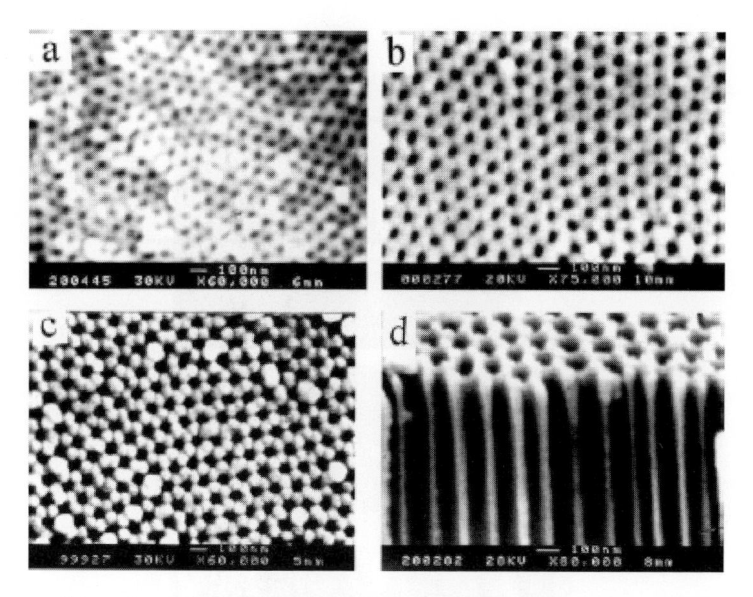

Fig. 5.16. Nanochannel alumina templates: (a) SEM image of the surface view of porous alumina anodized in 0.3 M oxalic acid at 17°C at 35 V; (b) SEM image of the surface view of porous alumina anodized in 0.3 M oxalic acid at 12°C at 40 V; (c) SEM image of the surface view of porous alumina anodized in 0.25 M oxalic acid at 5°C at 60 V; (d) oblique cross-section views of porous alumina anodized in 0.25 M oxalic acid at 3°C at 70 V.

at 60 V; the average pore diameter and inter-pore distance are 70 nm and 130 nm, respectively. Figure 5.16(d) shows an oblique cross-section view of porous alumina which was anodized in 0.25 M oxalic acid at 3°C at 70 V, in which the cross-sections of the parallel cylindrical pores can be clearly seen; the average pore diameter and inter-pore distance are 95 nm and 135 nm, respectively.

Figure 5.17 shows the SEM images of highly-ordered Au nanowire arrays prepared in the AAM templates with 95 and 70 nm diameter pores. It can be clearly seen that the Au nanowires are of an equal height and have a highly-ordered tip array. The exposed Au nanowires retain the size and shape of the pores in the template. Figure 5.18(a) shows a TEM image of one of the Au nanowires, which is prepared in the AAM template with 45 nm diameter pores. The Au nanowire is straight and has a uniform diameter of about 45 nm and length about 5 μm. A selected-area electron diffraction (SAED) pattern [Fig. 5.18(a), inset] of the nanowire could be indexed for the [110] zone axis of single crystal Au. The single-crystal structure was further

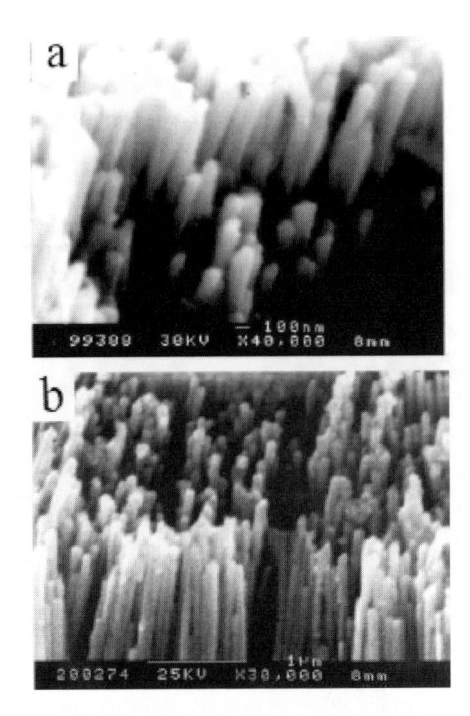

Fig. 5.17. SEM images of the highly-ordered array of Au nanowires prepared in the AAM templates with pore diameters about (a) 95 nm and (b) 70 nm.

confirmed by HRTEM images of the nanowires. The HRTEM image shown in Fig. 5.18(b) shows clearly the single-crystal structure of the nanowire and the (−111) and (002) lattice fringes with interplanar spacings of around 0.24 nm and 0.20 nm, respectively. It can be seen that only the edge of the nanowire is visible because the central part of the nanowire is too thick to be transmitted by the electron beam. Figure 5.18(c) shows a TEM image of an Au nanowire, which is prepared in the AAM template with 70 nm diameter pores. A selected-area electron diffraction (SAED) pattern (Fig. 5.18(c), inset) of the nanowire shows that the nanowire presents a polycrystalline structure. The results indicate that the template with a smaller pore diameter is favorable for the formation of single crystal nanowires, while the template with a larger pore diameter is favorable for the formation of polycrystalline nanowires. Template syntheses of single crystal metal and semiconductor nanowires have been reported by using electro-deposition[38,39] and other methods.[40] However, there are few reports on the growth mechanism because the situation becomes more complex by the presence of the template. An explanation of the nanowire growth mechanism within the

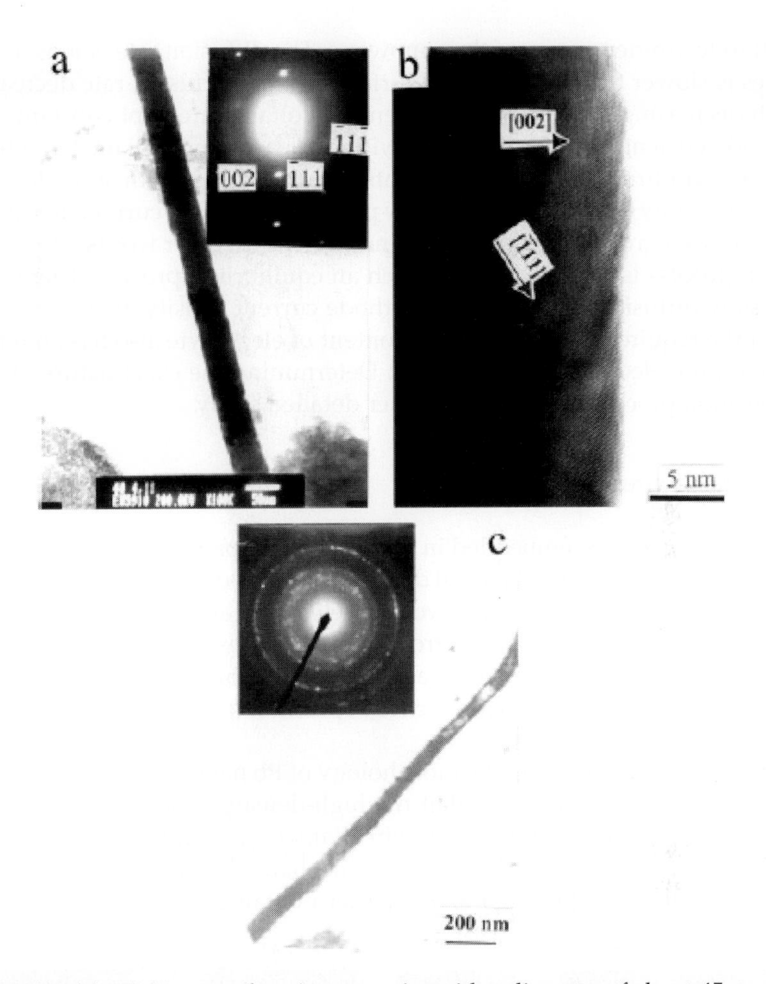

Fig. 5.18. (a) TEM image of an Au nanowire with a diameter of about 45 nm; the inset shows the SAED of the nanowire, which is consistent with the Au single crystal indexed as Au [110]. (b) HRTEM image of the single crystalline Au nanowire. (c) TEM image of an Au nanowire with a diameter of about 70 nm; the inset shows the SAED of the nanowire, which is consistent with polycrystalline Au.

AAM template must remain speculative. Since the templates have parallel arrays of pores of uniform diameter, we assume that the pores are one-dimensional and the surface of the Au substrate is smooth. Therefore, the Au-coated AAM template can be considered as a porous electrode, which is a combination of single pores. It has also been proved that the electrochemical processes could be diffusion-limited within the porous electrodes.[41] Therefore, it can be inferred that the diffusion rate should have an effect on the electro-deposition of the nanowires. The diffusion rate of ions in the pores is slower than on the plane surface and the diffusion rate decreases with decreasing pore diameter. When the voltages are kept constant, the cathode current densities decrease with decreasing pore diameter, which has been confirmed by our experiments. The oriented growth of single crystals could only be obtained under low polarization (small current densities) conditions to avoid the formation of new nuclei. In other words, the deposition process is expected to approach an equilibrium process. Therefore, the slow diffusion rate and small cathode current density are essential to meet the requirement. Besides, the content of electrolyte also has an influence on the electrochemical process. Determining the exact nature of the deposition process will require further detailed study.

(4) Pb nanowire arrays[42]

Pb nanowire arrays, embedded in AAM, were prepared in a two-electrode system using the electrochemical deposition method.[42] The AAM template with a Au-coated served as the working electrode and the count electrode is a graphite electrode. The electrolyte was composed of 30 g/l $Pb(NO_3)_2$ and 45 g/l H_3BO_3, and the pH was 2.5. The electrolyte deposition current density was 2.5 mA/cm^2. The electric deposition time was 6 to 8 h.

Figure 5.19(a) illustrate the morphology of Pb nanowire arrays without the template, which indicate that the high-density nanowires have been synthesized in the AAM. The nanowire diameter is about 40 nm (as shown in Fig. 5.19(b)), corresponding to the nanochannel diameter in the AAM. Moreover, the inset shows a SAED pattern obtained from the individual

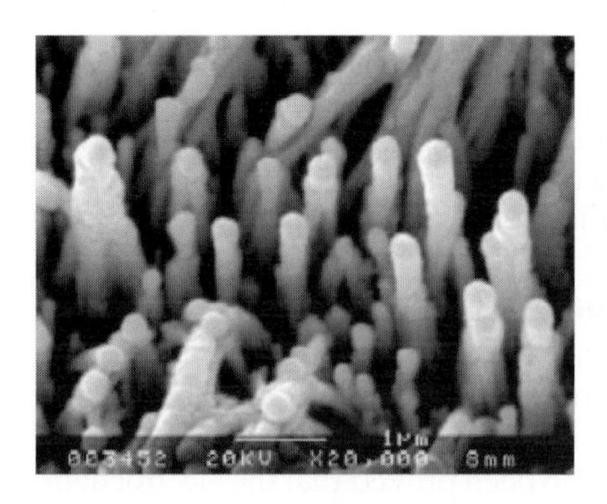

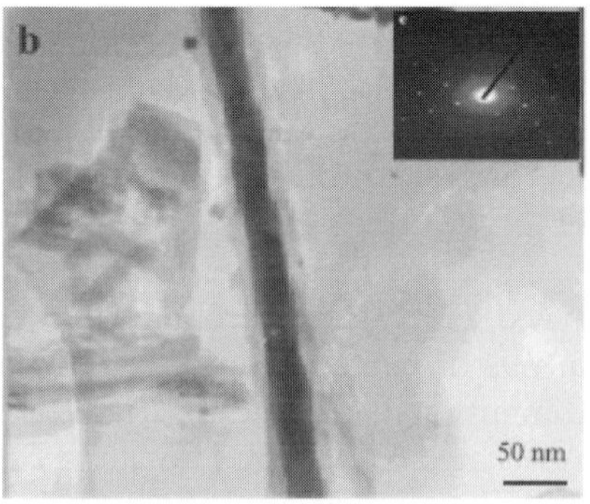

Fig. 5.19. (a) A SEM image of a lead nanowire array, (b) TEM image of a single nanowire.

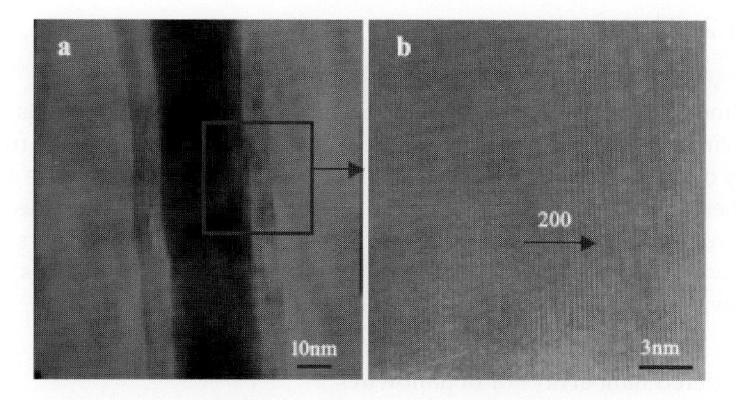

Fig. 5.20. HRTEM image of a single lead nanowire.

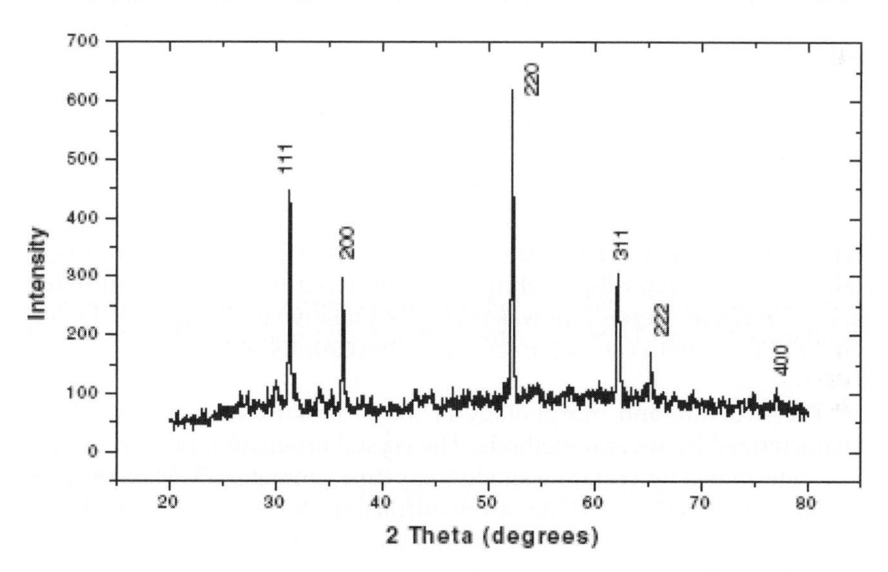

Fig. 5.21. XRD pattern of the lead nanowire arrays.

nanowire and the HRTEM image implies each nanowire is single crystalline (Fig. 5.20).

The XRD pattern of the Pb nanowire arrays is shown in Fig. 5.21, which indicates that the Pb nanowires have face-centered cubic crystal structure. The six peaks of the Pb nanowires are formed to be very close to the (100), (200), (220), (311), (222) and (400) peaks of bulk Pb.

5.2.1.2. *Ordered nanoarrays of semimetal nanowires and nanotubes*

Semimetal Sb and Bi are very important thermoelectric materials. Synthesizing ordered nanoarrays of these materials is beneficial to further study of the new kind of thermoelectric materials. The Bi nanowire arrays synthesized by using the AAM templates and polymer templates have been reported, but the reports on synthesis of ordered nanoarrays of Sb nanowires are very small. Zhang[43] prepared ordered nanoarrays of Sb nanowires by pulsed electro-deposition in the AAM templates.

(1) Ordered nanoarrays of Bi nanowires

Here, we will introduce the synthesis of ordered nanoarrays of single crystalline Bi nanowires,[45–48] Y-branched Bi nanowires[49] and Bi nanojunction nanowires.[50] Wang[45,46] synthesized single crystalline Bi nanowire arrays using DC electro-deposition in the AAM templates, which was fabricated by a two-step anodization process as described in Sec. 5.1.1. The Au layer was sputtered onto one side of the through-hole AAM template serving as the working electrode in a standard three-electrode electrochemical cell.[44]

The electro-deposition process of the Bi nanowires is similar to those reported previously.[51] The electrolyte contained 75 g/l bismuth nitrate pentahydrate. The deposition solution was buffered to pH = 0.9 with nitric acid. The electro-deposition was performed relative to the Ag/AgCl reference electrode with carbonate serving as the counter electrode.

The structure and morphology of the Bi nanowire arrays have been characterized by several methods. The crystal orientation of each sample was determined by a rotating anode X-ray diffractometer (D/Max-rA) with Cu-K_α radiation ($\lambda = 1.542$Å). X-ray diffraction (XRD) patterns are shown in Fig. 5.22. The broad peak results from the amorphous AAM. The peak of the Al (020) corresponds to the reflection of the surrounding aluminium of the AAM. For the sample that was synthesized from an AAM with a pore diameter of 50 nm, six peaks ((012), (104), (110), (202), (024) and (122)) were observed (Fig. 5.22(b)); but for the 20 nm sample, only four peaks ((012), (104), (110) and (122)) were found (Fig. 5.22(a)). The intensity ratio of (110) to (104) for 20 nm nanowires is much larger than that for the 50 nm ones. Compared with the peak positions of standard polycrystalline Bi (Fig. 5.22(c)), all observed peaks are found to be very close to the peak positions of bulk Bi. This indicates that the rhombohedral lattice structure of bulk Bi is preserved in these wires. Figure 5.23(a) shows the transmission electron

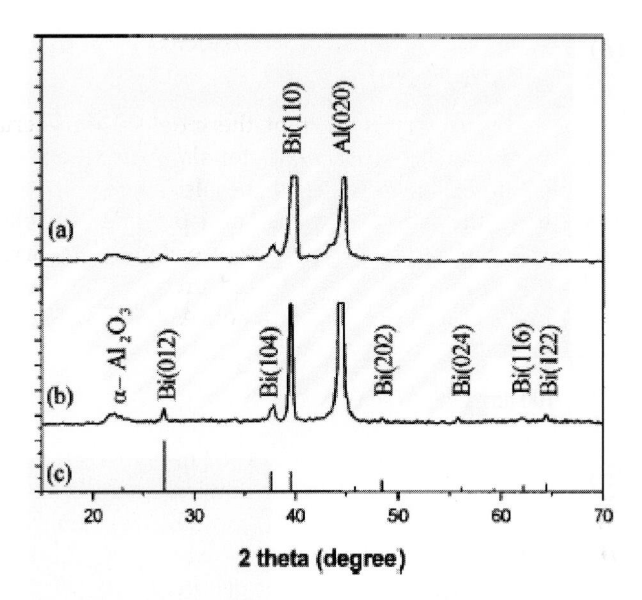

Fig. 5.22. XRD patterns of Bi nanowire arrays with average diameters of (a) 20 nm and (b) 50 nm, respectively. (c) The standard position and relative intensities of polycrystalline Bi.

microscopy (TEM) image of a row of nanowires which are 50 nm in diameter. The dark areas are Bi naowires. It can be seen that the nanowires in the nanochannels of AAM are highly-ordered. To confirm that the nanochannel arrays have been filled with continuous Bi nanowires, the AAM was dissolved in a solution of 5 wt.% sodium hydroxide, which would not attack the nanowires. Figure 5.23(b) shows the TEM image of a single 50 nm wire removed from the AAM, revealing that the nanowire is cylindrical in shape. The set in Fig. 5.23(b) shows selected-area electron diffraction of a single Bi nanowire. Along the nanowire, the diffraction patterns are essentially identical, revealing that the nanowire is single crystalline. We also found that the strongest peak of the Bi nanowires is (110), other than (012) for bulk Bi, and other peaks are very weak, indicating that most of the nanowires are oriented perpendicular to the (110) lattice plane. Therefore, it can be concluded that these nanowires are highly-oriented and essentially single crystalline, and that with the decrease of diameters, the orientation of the nanowires is better.

The one-dimensional growth is preferred and interesting. At the initial stage of the electro-deposition, the nanowires preferentially grow along

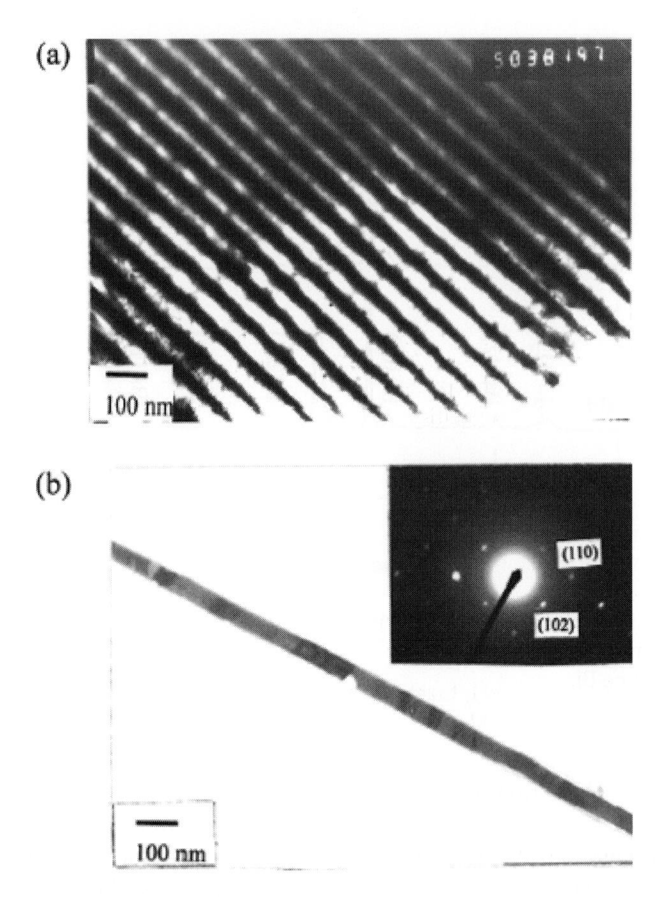

Fig. 5.23. (a) TEM image of a row of Bi nanowires which are 50 nm in diameter, and (b) TEM image of a single Bi nanowire which is 50 nm in diameter. The inset shows the electron diffraction pattern.

the (110) stacking face. Under constant current, the preferred growth is processed in the two-dimensional confinement of the channels of the AAM. Zhang[52] fabricated single crystalline Bi nanowires which are 13, 23, and 56 nm in diameter by a vacuum-melting and pressure-injection process.[52] It was found that 13 and 23 nm nanowires preferentially grow along (012), while 56 nm nanowires along (202). Therefore, it may be deduced that the nanowires preferred growth is closely related to the pore size of the AAM. Comparatively, our Bi nanowires with diameters of 20 and 50 nm all grow along (110). We suggest that the difference of preferred orientation may result from different fabrication methods.

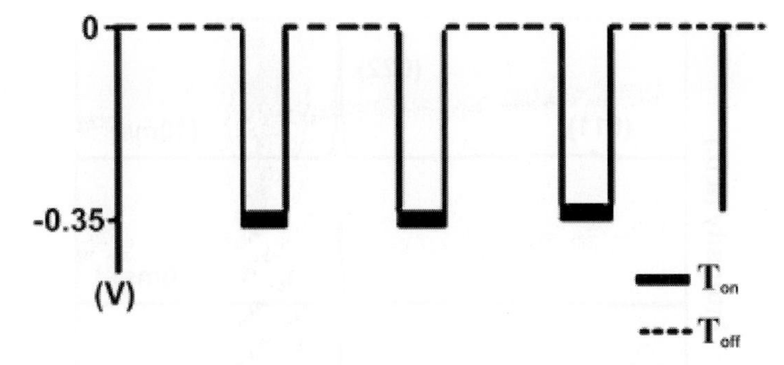

Fig. 5.24. Schematic of the pulse cycle for the pulsed electro-deposition. (T_{on} and T_{off} represent the deposition time and relaxation time of each pulse cycle, respectively.)

Li[47,48] reported that they synthesized Bi nanowire arrays using pulsed electro-deposition in the AAM templates. The synthesis process is as follows. A layer of Au which is about 60 nm in thickness was sputtered onto the bottom of the through-hole AAM template to serve as a conduction contact. The bismuth nanowire array was deposited from a plating solution consisting of $BiCl_3$ 40 g/l, tartaric acid 50 g/l, glycerol 100 g/l, NaCl 70 g/l and HCl 1 mol/l. The pH value of the electrolyte was adjusted to about 0.9 by adding appropriate amounts of aqueous ammonia 5 mol/l in order to avoid corrosive attack of the AAM.

The pulsed electro-deposition was carried out at −0.35 V, applied between the graphite anode and AAM cathode in a common two-electrode glass plating cell at 10°C with different pulse cycles, as shown schematically in Fig. 5.24. The relaxing time (T_{off}) between two successive pulsed cycles was varied to study the influence of the pulsed time on the structure of bismuth nanowires. The duty cycle pulsed deposition time (T_{on}) was kept at a constant, 10 ms.

Figure 5.25 shows the XRD patterns of bismuth nanowire arrays fabricated with different relaxation time together with the standard diffraction peaks of bismuth (JCPDS No. 5-0519). From the patterns, we can see that there is a very strong diffraction peak at $2\theta = 48.90°$ corresponding to the (022) plane of bismuth for samples fabricated with the relaxation time 30 and 50 ms. In addition, two peaks (011) and (022), the ratio of which is integral, are found for nanowire arrays fabricated with the relaxation time 10 ms. At the same time, all other peaks are very weak. The nanowires are a rhombohedral lattice structure as compared with the standard diffraction

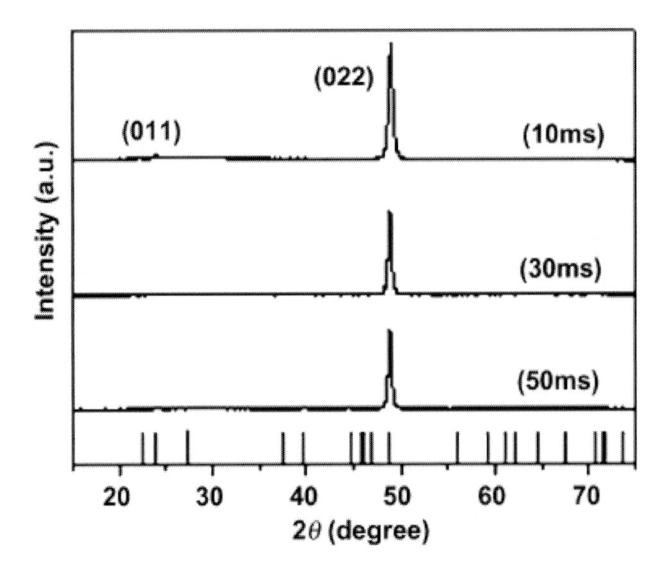

Fig. 5.25. XRD patterns of bismuth nanowire arrays with different relaxation time.

of bulk bismuth. It is worthy to note that the orientations of all the samples are the same in spite of the different relaxation time. This result indicates that bismuth nanowires grow preferentially along the [022] direction.

The typical SEM images of Bi nanowires with different diameters are shown in Fig. 5.26. Figure 5.26(a) shows the surface image of the Bi nanowire array after the AAM has been partly etched away, which

Fig. 5.26. SEM images of Bi nanowire arrays: (a) the top-view of Bi nanowire arrays with a diameter of 16 nm after partly eroding the AAM; (b) bundles of Bi nanowires with a diameter of 40 nm after completely eroding the AAM and; (c) surface-view of Bi nanowire array with a diameter of 73 nm.

demonstrates that the nanowires have the same height and are still standing on the substrate. Figure 5.26(b) shows the morphology of nanowires after the AAM has been completely etched away. All the nanowires have the same length with a very uniform diameter, implying that all the nanowires are simultaneously deposited into the AAM in a large area. The top view of nanowires after mechanical polishing is shown in Fig. 5.26(c). Apparently, all the pores are filled with nanowires in hexagonal closed-packed arrangement corresponding to the arrangement of pores of the AAM.

Figure 5.27 shows the TEM images of Bi nanowires fabricated with the relaxation time of 10 ms. A bundle of nanowires removed from the AAM can be clearly seen in Fig. 5.27(a), and the average diameter of the nanowires is about 20 nm, which is much smaller than the pore size of the template with the diameter of 80 nm, as clearly shown in Fig. 5.27(b), in which the nanowires are still within the pores of the AAM. A single nanowire and the corresponding SAED are shown in Fig. 5.27(c). One can see that the nanowire is relative uniform. The SAED shows that the nanowires are single crystalline with a rhombohedral lattice structure, which further confirms the XRD results. Figures 5.28 and 5.29 show TEM images of Bi nanowires fabricated with the relaxation time of 30 and 50 ms, respectively. The average diameter of Bi nanowires fabricated with the relaxation time of 30 ms is about 33 nm, and that of 50 ms is about 80 nm. The corresponding SAED patterns of a single nanowire also indicate that the nanowire is a single crystal. These results indicate that the diameters of Bi nanowires can be modulated by changing the relaxation time of pulsed electro-deposition. The length of the nanowires depends only on the thickness of AAM. From TEM images, one also can see that most nanowires are cylindrical in shape and identical in diameter, but there is some twist or rupture in the nanowires, which might have resulted from the mechanical force during the ultrasonication treatment of the TEM samples. The above results indicate that the diameters of Bi nanowires can be controlled from 20 to 80 nm (the pore size of the empty AAM) through modulating the relaxation time of pulse from 10 to 50 ms with the constant pulsed deposition time of 10 ms. It was also found that when the relaxation time was shorter than 10 ms, the crystallinity of nanowires gradually decreases, and in some cases even no nanowires can be fabricated. The selection of the optimal relaxation time is very important in the growth of single crystalline Bi nanowires.

Previous studies have shown that the degree of preferred orientation of Bi nanowire is imperfect and the peak positions changed randomly when the diameters of nanowires changed from 23 to 95 nm.[53,54] Comparatively,

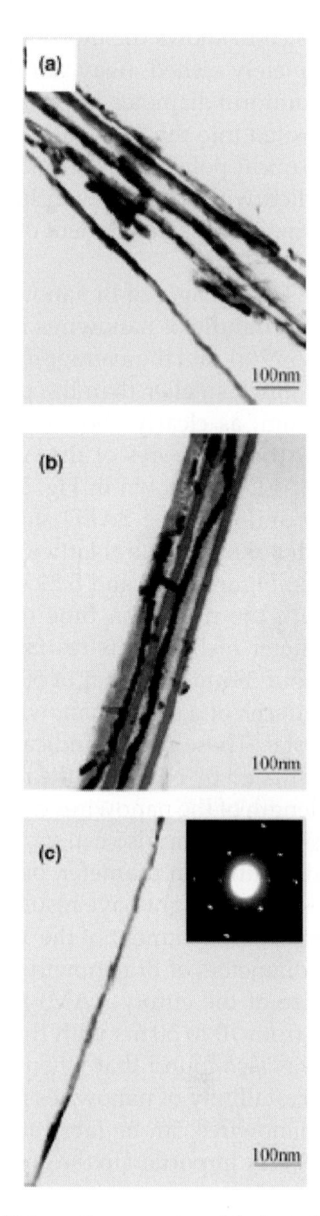

Fig. 5.27. TEM images of bismuth nanowires fabricated using the relaxation time 10 ms with the diameter of 20 nm: (a) free standing, (b) inside the template and (c) a single.

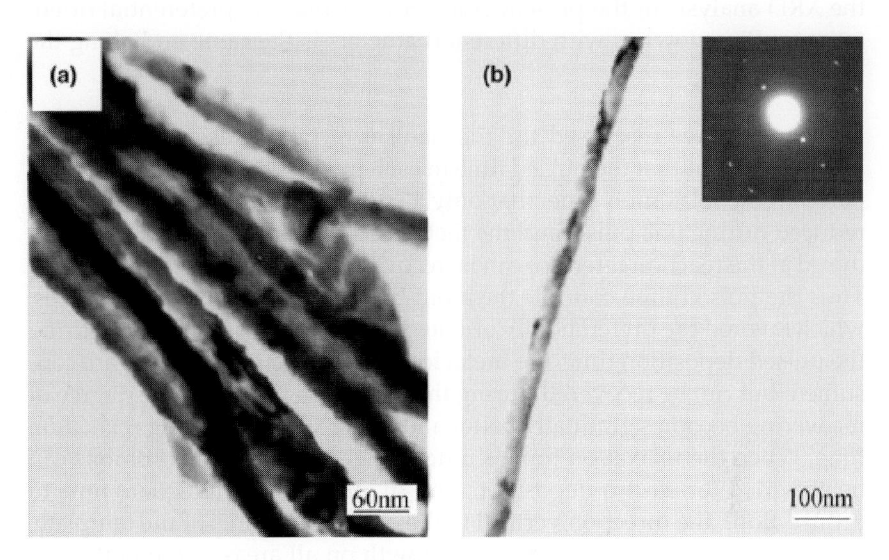

Fig. 5.28. TEM images of bismuth nanowires fabricated using the relaxation time 30 ms with the diameter of 33 nm: (a) free standing and (b) a single.

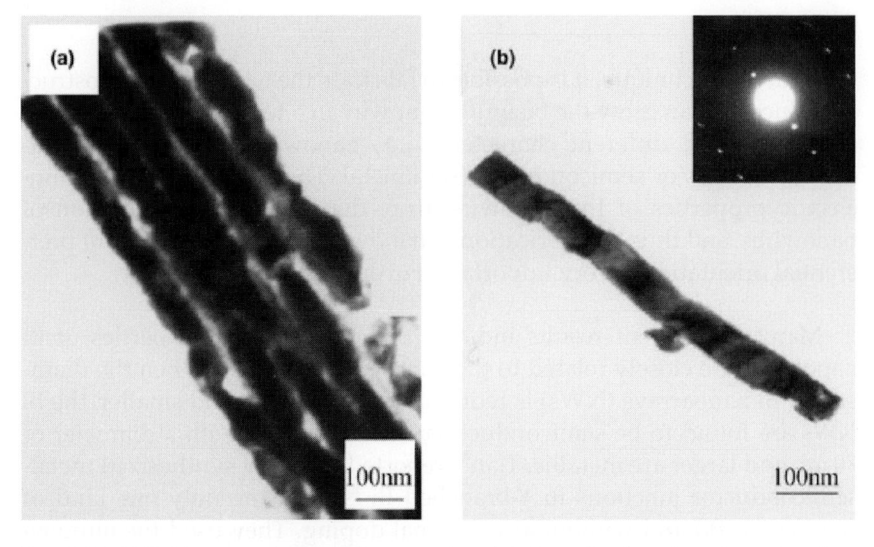

Fig. 5.29. TEM images of bismuth nanowires fabricated using the relaxation time 50 ms with the diameter of 80 nm: (a) free standing and (b) a single.

the XRD analysis in the present study reveals that the preferential orientation of Bi nanowires with different diameters is the same and along the [022] direction.

Zhang[43] have discussed the mechanism of pulsed electro-deposition and pointed out that the pulsed time in each pulse cycle was so short, compared to the relaxation time, that only a small number of metal ions were reduced during one pulse, and the metal ions' concentration gradient produced at the reaction interface can be recovered during the relaxation time. Thus the pulsed time controls the atom-by-atom growth of Bi nanowires, which favored the preferentially oriented growth of the nanowires. During the pulsed deposition time, the metal ions at the reaction interface are consumed, but can be recovered during the relaxation time, and the degree of recovering becomes gradually better and better with increasing relaxation time. When the relaxation time is much longer, enough metal Bi ions can be provided for electro-deposition, and Bi ions have an adequate time to diffuse along the direction vertical to the axes of the pores of the template, which provide initial atom by atom growth on all areas at the bottom of the pores, in which a continuous growth along the direction of pores takes place. But under a relative short relaxation time, the number of metal Bi ions provided and recovered is not enough, resulting in an initial growth only on part of the area at the bottom of the pores, which leads to thin nanowires.

With this technique, it is possible to fabricate the nanowire heterostructure since we can grow the bismuth nanowire in a template with different diameters (and different characters), i.e., nanowires with either semiconductor and/or semiconductor/semimetal. The transport and thermoelectric properties of the nanowire array depend on the orientation of nanowires, and thus the fabrication of nanowire arrays with different preferential orientation is very important. Further work is under way.

Many experiment results indicate that the transport proerties of Bi nanowires are closely related to the nanowire diameters. When the diameter of Bi nanoarrays (NWs) is reduced to about 50 nm and smaller, the Bi NWs are found to be semiconducting, while Bi NWs with a diameter of 70 nm and larger are metallic. Tian[49] reported that they synthesized metal-semiconductor junctions in Y-branched Bi NWs using only one kind of semimetal (Bi) and without any external doping. They used the alumina membrane with ordered Y-branched channel arrays as the templates to synthesize Y-branched Bi NWs arrays. The aim is to make the Y-branched Bi NWs in the AAM templates have large-diameter "stem" segments, which are metallic, and small-diameter "branched" segments, which are semiconducting. AAM templates with Y-branched nanochannels were fabricated by using a similar method reported previously.[56] High-purity Al foils were anodized in 0.3 M oxalic acid solution at 50 V_{DC} for 4 h. After chemically removing the original film in a mixed solution of phosphoric acid (6 wt.%) and chromic acid (1.8 wt.%) at 60°C for 6 h, a second anodization was performed under the same conditions for 4 h. The anodization voltage was then reduced to 35 V_{DC} (by a factor of $1/\sqrt{2}$), and another anodization was performed for 4 h. This will lead to nearly all large diameter channels (formed at 50 V) branching into two smaller diameter channels (formed at 35 V).[56] After the remaining Al was removed in a saturated $SnCl_4$ solution, the templates were immersed in a 5 wt.% phosphoric acid solution at 30°C for 90 minutes to remove the barrier layer, and to widen the nanochannels homogeneously over the entire pore length.[57] Figure 5.30 is a scanning electron microscope (SEM) image of the cross-section of the AAM template with Y-branched nanochannels. The inset of Fig. 5.30 shows the close-up of the region between stems and branches. The resulting AAM template consists of Y-branched nanochannels with stems and branches about 80 and 50 nm in diameter, respectively.

The electrochemical deposition of Bi into the Y-branched nanochannels of the AAM template was carried out using a similar method to the one previously reported.[24,25] A gold layer was sputtered onto the bottom side (with branched channels) of the AAM template serving as the working electrode in a three-electrode electrochemical cell. The electrolyte contained

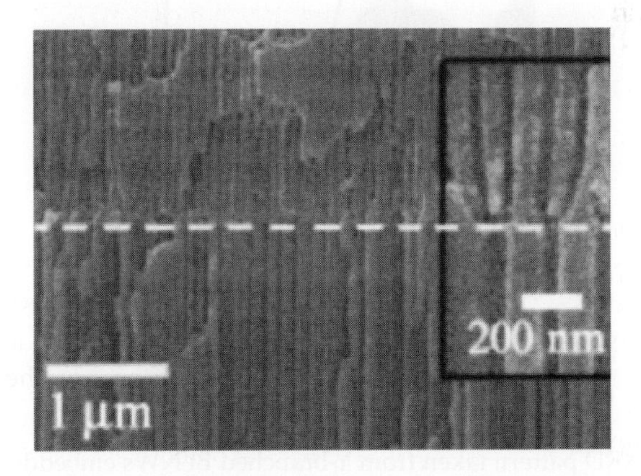

Fig. 5.30. A SEM image of the AAM template with Y-branched nanochannels. The inset shows the close-up of the Y-branched nanochannels. The dashed line shows where the Y-branches start to grow.

75 g/l $Bi(NO_3)_3{\cdot}5H_2O$, 65 g/l KOH, 50 g/l tartaric acid, and 125 g/l glycerol. The electrolyte was buffered to pH = 0.9 with HNO_3 solution. The electro-deposition was performed under −30 mV relative to the Ag^+/AgCl reference electrode. The X-ray diffraction (XRD) pattern (Fig. 5.31(a)) taken from the Y-branched Bi NWs embedded in the AAM template shows that all peaks are located very close to the peak positions of bulk Bi, revealing that the rhombohedral lattice structure of bulk Bi is reserved in the

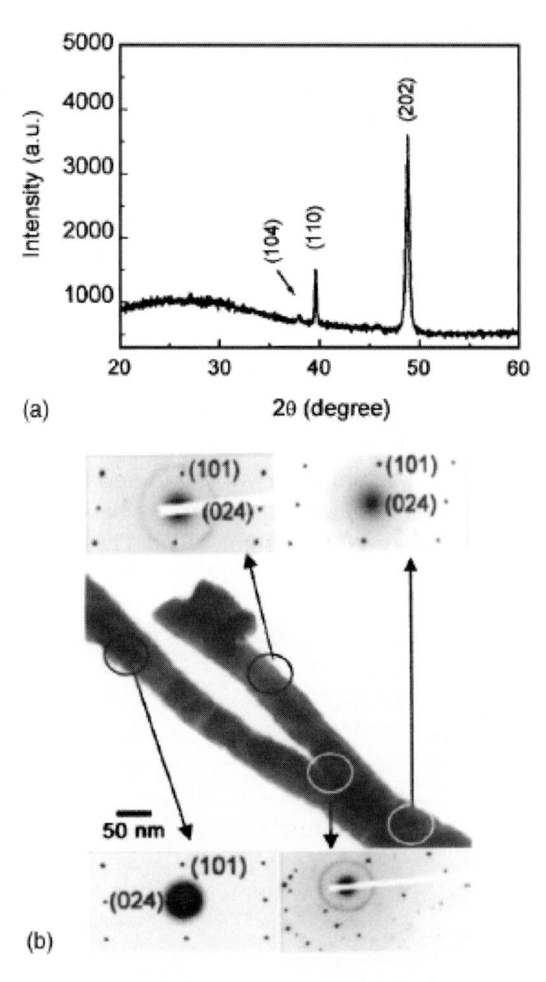

Fig. 5.31. (a) XRD pattern taken from Y-branched Bi NWs embedded in the AAM template. (b) TEM image of a single Y-branched Bi NW. The four SAED patterns were taken from the two branches, the stem, and the branch near the branching point, respectively.

Y-branched NWs. It should also be noted that the strongest peak of the Y-branched Bi NWs is (202), while other peaks are weak, indicating that most of the Bi NWs are oriented perpendicular to the (202) lattice plane. After dissolving the AAM template in a 5 wt.% NaOH solution, a transmission electron microscope (TEM) observation of a single Y-branched Bi NW (Fig. 5.31(b)) reveals that the stem of the NW is about 80 nm in dameter and the branches are about 50 nm in diameter. This is in agreement with those of the Y-branched nanochannels in the template. Selected-area electron diffraction (SAED) patterns taken from both the branches and the stem (Fig. 5.31(b)) reveal that the rhombohedral lattice structure is preserved in the Bi NWs. This is in agreement with the XRD results. Along the axes of both the stem and the branches, the diffraction patterns are identical, revealing that both the branches and the stem are essentially single crystalline, and the Bi NWs preferentially grow along the (101) stacking face. However, the diffraction pattern taken from the branch near the branching point (Fig. 5.31(b)) is not regular, indicating that this domain is polycrystalline, which may be caused by variations in the growth orientation of the branches.

In the following, we will introduce a new routine to the fabrication of Bi single crystalline tapering junction nanowire arrays reported by Li[50] According to experimental results, Li found that the crystalline nanowire can be obtained by modulating the pulse time in pulsed electro-deposition. Therefore, they synthesized Bi tapering junction nanowire arrays in the AAM templates by pulsed electro-deposition. The synthesis process is as follow. A layer of Au film (200 nm in thickness) was sputtered onto one side of the through-hole AAM template to serve as the working electrode in a two-electrode electrochemical cell, and a graphite plate was used as the count electrode. Bi junction nanowires were deposited from a solution containing $BiCl_3$ 10 g/l, tartaric acid 50 g/l, glycerol 95 g/l, and NaCl 50 g/l. The pH value of the electrolyte was adjusted to about 1.0 by adding appropriate amounts of 5 mol/l aqueous ammonia. The pulsed electrodeposition was carried out at −1.1 V with different pulse cycles as schematically shown in Fig. 5.32. In the first step, Bi nanowires with a diameter corresponding to the pore size of the AAM were deposited with a proper deposition time (the length of the nanowires increases with the deposition time). Then, Bi nanowires with a diameter smaller than the nanochannel size of the AAM were subsequently fabricated through decreasing T_{on} in the first step. Figure 5.33 shows the reaction scheme for fabricating Bi junction nanowire arrays and the dissolution of the AAM and the Au backing to obtain the free-standing junction nanowires.

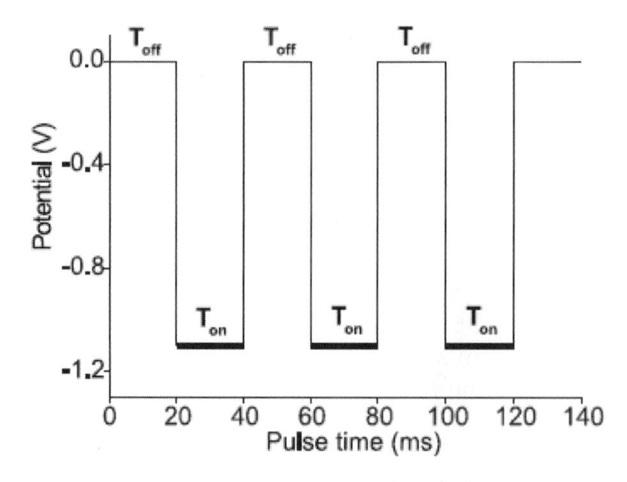

Fig. 5.32. Scheme of the pulse cycle for the pulsed electro-deposition. (T_{on} and T_{off} represent the pulsed deposition time and the relaxation time of each pulse cycle, respectively.)

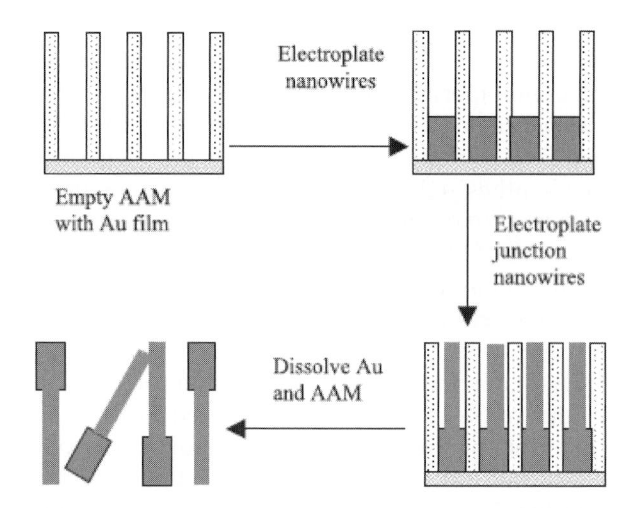

Fig. 5.33. Scheme of the fabrication of Bi tapering junction nanowires by pulsed electro-deposition in the AAM.

Figure 5.34 shows the TEM and SEM images of Bi junction nanowires with two segments of different diameters. Figure 5.34(a) shows Bi junction nanowires with the diameters of 45 nm and 25 nm fabricated in the AAM with the pore size of 45 nm using T_{on} of 35 ms and 26 ms respectively and keeping T_{off} at 40 ms, and Fig. 5.34(b) shows Bi junction nanowires with two

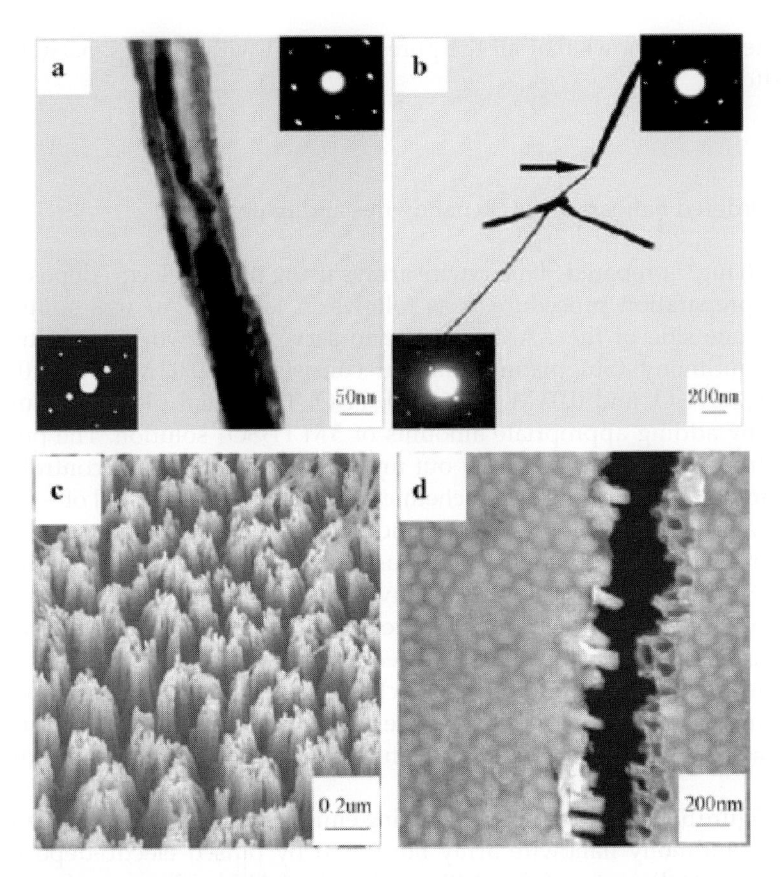

Fig. 5.34. TEM images of Bi junction nanowires: (a) S–S junction and (b) M–S junction. SEM images of (c) top and (d) bottom views of the M–S junction nanowire array after etching.

segments of 100 nm and 35 nm in diameter fabricated in the AAM with the pore size of 100 nm using T_{on} of 70 ms and 28 ms respectively and keeping T_{off} at 50 ms. From the electronic transport properties of Bi nanowires (see Ref. 58), we know that the semimetal–semiconductor transition will occur when the diameter of the Bi nanowires reduces to 60 nm, which suggests that the Bi junction nanowire shown in Fig. 5.34(a) is an S–S junction and that in Fig. 5.34(b) is a M–S junction. All the corresponding SAED images inserted in these TEM images indicate that the segments of the junction nanowires are all single crystals. The SEM image of the top view of the M–S junction nanowire array etched in NaOH solution reveals that the junction nanowires are large scale, and the SEM of the bottom view after etching the template shows that all the pores are filled with Bi (Figs. 5.34(c) and 5.34(d)).

(2) Ordered nanoarrays of Sb nanowires and nanotubes

Zhang[43] prepared Sb nanowire arrays using pulsed electro-deposition. The preparation procedure is as follows. A layer of Au was sputtered onto one side of the AAM template to serve as the working electrode. The antimony (Sb) plating solution consisted of 0.02 M $SbCl_3$, 0.1 M $C_6H_8O_7{\cdot}H_2O$, and 0.05 M $K_3C6H_5O_7{\cdot}H_2O$. The initial pH was adjusted to 2 by adding appropriate amounts of 5 M H_2SO_4 solution. The pulsed electro-deposition was carried out under modulated voltage control in a common two-electrode electrochemical cell at 12°C. A potential of −0.98 V was applied between the cathode and anode with a deposition current of about 12 mA/cm^2. The pulse time was 400 μs and the time between pulses was 800 μs. During the 400 μs negative potential pulse, the antimony was deposited on the pore ground. The relaxation time of 800 μs between two successive deposition pulses provided enough time for the concentration of the metal ions at the pore tips to achieve steady state through diffusion. During the relaxation time, the recovery of the ion concentration improved the homogeneity of the deposition and limited the hydrogen evolution.[59]

Figure 5.35(a) shows a field scanning emission microscopy (SEM) image of an antimony nanowire array fabricated by pulsed electro-deposition with a pore diameter of about 40 nm after the AAM has been partly etched away. Figure 5.35(b) is a SEM image of an antimony nanowire array grown on the AAM template with a prolonged deposition time, showing that almost all the pores are filled with antimony nanowires. A transmission electron microscopy (TEM) image of antimony nanowires liberated from AAM is shown in Fig. 5.35(c). These results demonstrate that the diameters of the antimony nanowires correspond satisfactorily to the pore diameter of the membrane used, and the antimony nanowires are uniform, smooth, straight and have the same height. The length of the nanowires is about 62 μm after deposition for 18 h, corresponding to a deposition rate of about 57.4 nm per minute. The X-ray diffraction pattern of an antimony nanowire array is shown in Fig. 5.35(d) together with the standard diffraction peaks of antimony (JCPDS No. 5-0562).[60] One can see that there is a very sharp diffraction peak at $2\theta = 41.98°$, and all other diffraction peaks are very weak. This demonstrate that all the nanowires have the same orientation. The sole sharp peak corresponds to $(11\bar{2}0)$ of hexagonal antimony. This

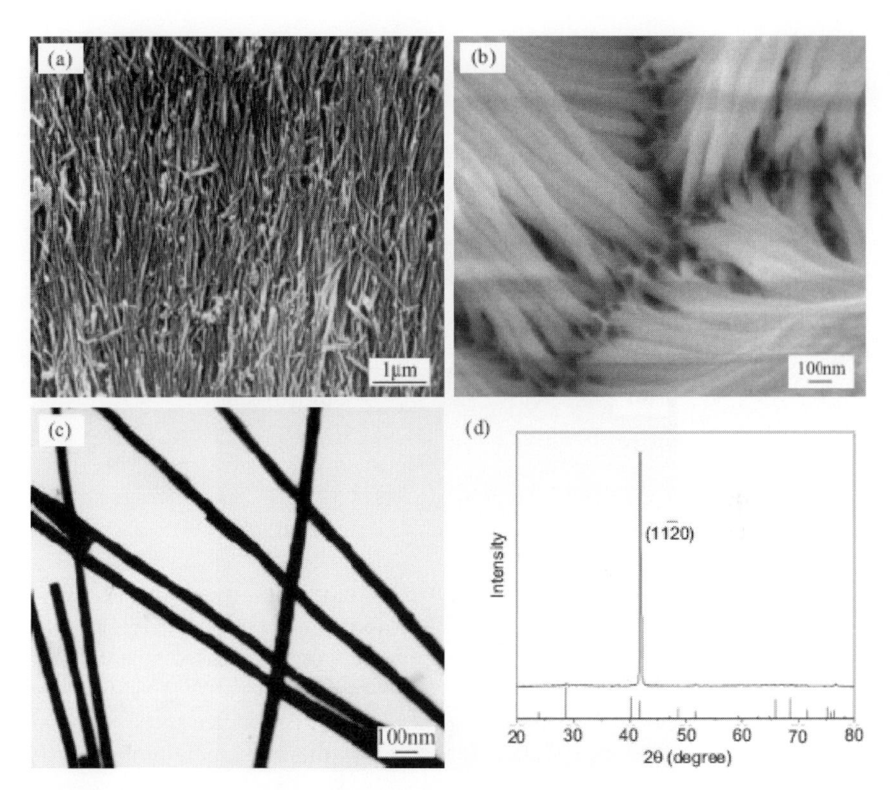

Fig. 5.35. (a) A typical SEM image showing the general morphology of the antimony nanowire array. (b) A typical FEM image showing the filling of the template and height variation of the nanowire. (c) A typical TEM image of antimony nanowires showing the morphology of individular nanowires. (d) XRD pattern of the antimony nanowire array; the sole diffraction peak indicates the same orientation of the nanowires.

result indicates that the antimony nanowires deposited in the AAM grow preferentially along the [11$\bar{2}$0] crystal direction.

A typical TEM image of a single antimony nanowire is shown in Fig. 5.36(a) together with the corresponding diffraction pattern (the inset in Fig. 5.36(a)). Due to the ultrasonic vibration used to prepare the TEM samples, the Sb nanowires are easily broken. It can be seen from the figure that the end of the nanowire is cut at an angle with respect to the nanowire direction. The diffraction spots of the (11$\bar{2}$0), (2$\bar{1}\bar{1}$0), and ($\bar{1}$2$\bar{1}$0) lattice planes correspond to the single-crystal hexagonal packed structure of antimony. Calculation indicates that the incident beam is parallel to the [0001]

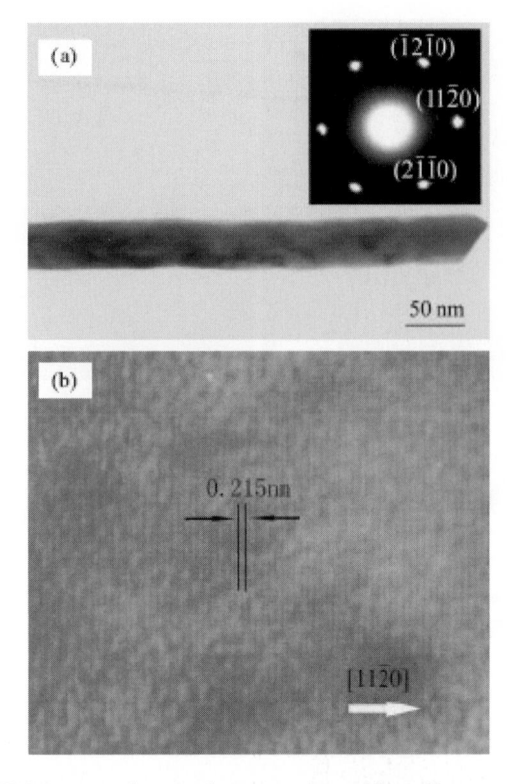

Fig. 5.36. (a) TEM image of a single nanowire with a diameter of 40 nm. (b) A HRTEM image of the same nanowire. The inset in (a) is the small-angle electron diffraction (SAED) pattern taken from the single nanowire.

direction. Figure 5.36(b) shows a high-resolution transmission electron microscopy (HRTEM) image of the same nanowire. It can be seen that the $(11\bar{2}0)$ lattice fringes with the lattice spacing around 0.215 nm are perpendicular to the axis of the nanowire. From this it can be deduced that the growth direction is along $[11\bar{2}0]$, which is consistent with the result of X-ray diffraction. An HRTEM image and the corresponding diffraction pattern of a single antimony nanowire are shown in Fig. 5.37, where the beam direction is along the $[10\bar{1}0]$ direction. Two-dimensional lattice planes can be clearly seen. The diffraction spots correspond to the (0003), $(01\bar{1}1)$, and $(0\bar{1}12)$ lattice planes with interplane spacing of about 0.37, 0.35 and 0.31 nm, respectively.

The use of pulsed electro-deposition leads to a uniform antimony deposition in the pores of porous alumina structures using the reported experimental conditions. It was found that on increasing the time per pulse cycle,

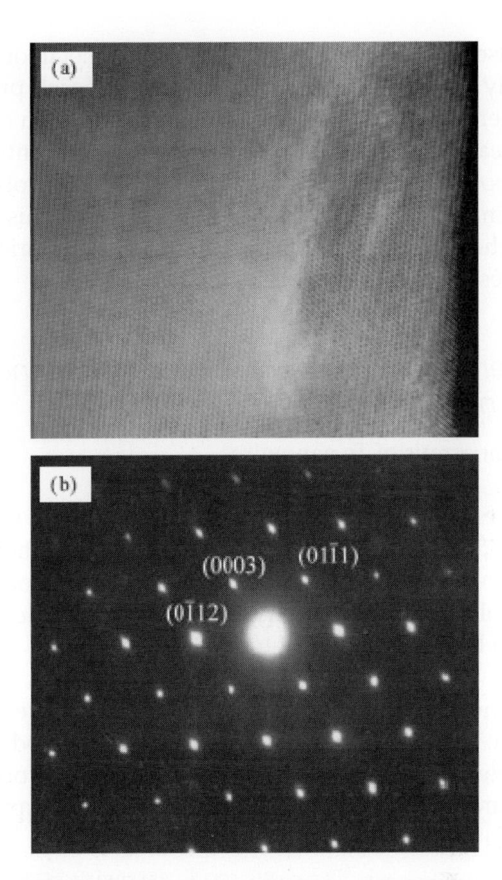

Fig. 5.37. (a) A HRTEM image of a single antimony nanowire with a diameter 40 nm; the (0003), (01$\bar{1}$1), and (0$\bar{1}$12) lattice planes are clearly resolved. (b) The corresponding SAED pattern recorded along the [10$\bar{1}$0] zone axis.

the crystallinity of the antimony nanowires decreased, and amorphous antimony nanowires were obtained under DC conditions using the same potential. During deposition, only a negative relaxation time was used. No additional pulse with positive potential was applied, as was used by Sauer in the fabrication of highly-ordered monocrystalline silver nanowire arrays.[61] The high reaction rate at the interface creates an increasing concentration gradient in the pores, which enhances the growth rate of the nanowires. As it is a semimetal, amorphous antimony is easily obtained at a high deposition rate of atoms. Thus, a relatively short pulse time was employed in this work. The pulse time in each pulse cycle was so short that only a small number of metal ions were reduced at the reaction interface during one pulse (less than one atomic layer). The metal ion concentration changes slightly near the reaction interface during each pulse. However, the time between pulses is sufficient for the concentration of metal ions to recover at the reaction interface. There is no evident concentration gradient during the deposition. Thus the pulse time, which controls the transport of atoms, dominates the growth rate of the nanowire. This atom-by-atom growth favors the perfect crystallinity and preferentially-orientated growth of the nanowires.

5.2.1.3. *Ordered nanoarrays of semiconductor nanowires and nanotubes*

(1) Ordered nanoarrays of Te nanowires

Now, the synthesis methods of nanowires and nanotubes for semiconductor Se and Te[62–66] mainly have the solution method or the hydrothermal method. However, the nanowires prepared by these methods disperse randomly in the solutions or on the Si substrates while ordered nanoarrays can be obtained by using the AAM templates.

Ordered nanoarrays of nanowires can be synthesized through two methods.[67] One is the electrochemical deposition method, for which the chemical reaction took place in the AAM templates to form directly the nanoarrays (sample A). The electrochemical reaction expressions are as follows:

$$HTeO_2^+ + 3H^+ + 4e^- \rightarrow Te(\downarrow) + 2H_2O, \tag{5.1}$$

$$HTeO_2^+ + N_2H_2 \rightarrow Te(\downarrow) + N_2(\uparrow) + 2H_2O + H^+. \tag{5.2}$$

The other synthesis method is the electrophoresis deposition in the AAM templates to prepare nanoarrays. The process of ordered Te nanowires arrays synthesized by using the electrophoresis deposition method is as follows. A small amount of hydrazine was added into the aqueous solution containing $HTeO_2^+$ and the solution was stirred. The chemical reaction took place (expression (5.2)) to form the colloidal particles. Then, in the three-electrode system, the colloidal particles under the electric field action nucleated and grew in the channels of the AAM template to form the ordered array of Te nanowires (sample B).

The XRD spectra of ordered arrays of Te nanowires are shown in Fig. 5.38. The peak position of the XRD of sample A and sample B are

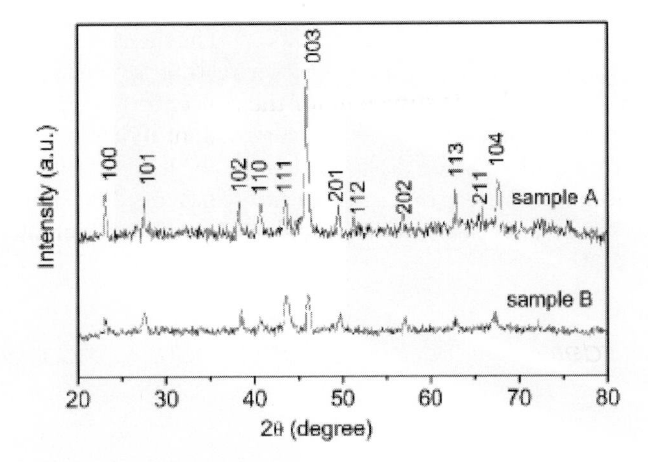

Fig. 5.38. XRD spectra of ordered arrays of Te nanowires.

similar and Fig. 5.38 also shows that the Te nanowires have a hexagonal structure. For sample A, the diffraction peak of (003) planes is stronger than those of other diffraction planes. This suggests that in the electrochemical process, Te nanowires in the AAM template grow preferentially along the normal direction of (001) phanes. But for sample B, the diffraction peak of (003) planes is not obviously stronger than those of other planes. This means that there is no preferential growth appearing during the growth of the Te nanowire. Also, the crystallization degree of the Te nanowires of sample B is poorer than that of sample A. This can be proved by TEM observations.

The morphologies and the microstructures of sample A and sample B are displayed in Fig. 5.39 and Fig. 5.40, respectively. From TEM photographs, it can be seen that for the two synthesis methods, the diameters of Te nanowires in the ordered arrays are uniform and consistent, and the surface of nanowires are smooth. From the HRTEM morphologies of a single Te nanowire and the corresponding SAED pattern of sample A, it can be observed that the Te nanowires in the array synthesized by the electrochemical deposition technique have a hexagonal single crystalline structure. But for sample B, the Te nanowires have a polycrystalline structure. This is mainly related to the dynamics of nucleation and growth of Te nanowires in the electrophoresis deposition process inside the AAM templates.

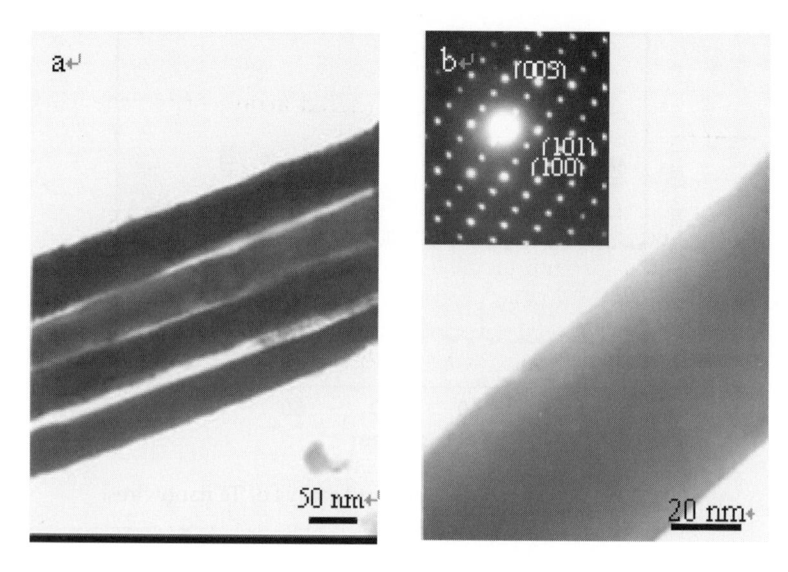

Fig. 5.39. (a) The TEM photograph and (b) the HRTEM morphology image and the corresponding SAED pattern of sample A.

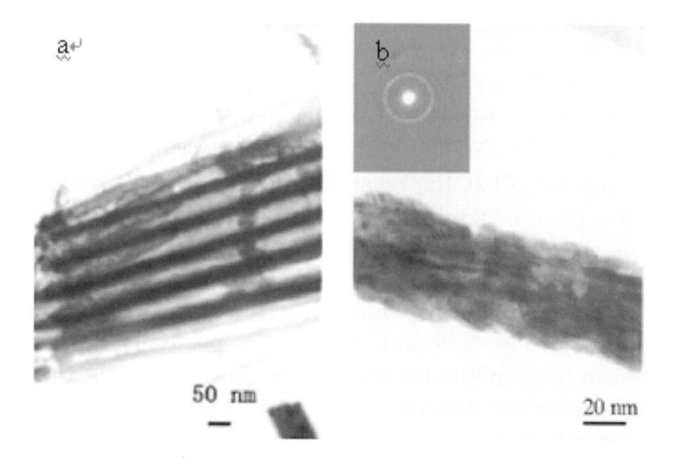

Fig. 5.40. (a) The TEM photograph and (b) the HRTEM morphology image and the corresponding SAED pattern of sample B.

(2) Ordered arrays of Si nanowires

Now, most Si nanowires prepared by many researchers present random distribution and entangled each other, so that they are difficult to be dispersed.

At the same time, these Si nanowires have more defects. Therefore, it is not beneficial to the applications and measurement of the properties of Si nanowires. Therefore, preparing the ordered arrays of Si nanowires has important significance for the proper study of Si nanowires and the development of the function devices associated with Si nanowires.

The ordered arrays of Si nanowires were prepared by combining CVD with the synthesis technique of the AAM templates.[68] On the basis of the AAM template, the synthesis process of ordered arrays of Si nanowires was divided into two steps. Firstly, the Au nanoparticles were introduced into the bottom of the pores in the AAM template as the catalysts by the AC (alternating current) electrochemical deposition method. Then, using the CVD (chemical vapor deposition) method, the Si nanowires on the surface of the nanocatalysts in the channels of the AAM template were made to grow due to the confinement action of the AAM template, resulting in the formation of the Si nanowire arrays. The diameter and length of the Si nanowires can be controlled by changing the pore diameter of the template and the reaction time. When Au nanocatalysts were introduced into the channels of the template, the alternating current working frequency was 60 Hz and the working voltage was 15 V, and electrolyte was the 2.5 g/l $AuCl_3$ solution with pH = 2.5. The preparation procedure of ordered arrays of Si nanowires was as follows. Firstly, the AAM template with the Au catalyst was placed in the Al_2O_3 crucible in the middle of the furnace, and the furnace was evacuated to 10^{-3} Torr. Then, N_2 was introduced into the furnace. After that, the furnace was evacuated again to 10^{-3} Torr. After the above process was repeated three times, H_2 was introduced into the furnace to make the flow ratio of N_2 with H_2 equal 10 : 1. Through the vacuum degree modulating, the total of the mixture gas of N_2 and H_2 was kept at 100 cm^3/minute. The reaction system was reduced at 500°C for 2 h. Finally, the furnace temperature was increased to 540°C and then the mixture gas of SiH_4 and N_2 was introduced into the furnace to grow Si nanowires in the channels of the AAM template. The working pressure during the reaction was 20 kPa. The volume ration of SiH_4 and N_2 was 1 : 10, and the flow speed was 120 cm^3/minute.

Figure 5.41 displays the SEM images of ordered Si nanowire arrays synthesized in the AAM template with the pore diameter of 38 nm. Figures 5.41(a) and 5.41(b) are the top view and the oblique view of Si-ordered arrays, respectively. It can be seen that the Si nanowires keep the ordering, the uniform diameters and the surface smooth of the AAM template. Figure 5.41(c) shows the sectional view. It can be observed that each nanowire filled each channel of the AAM template due to uniform

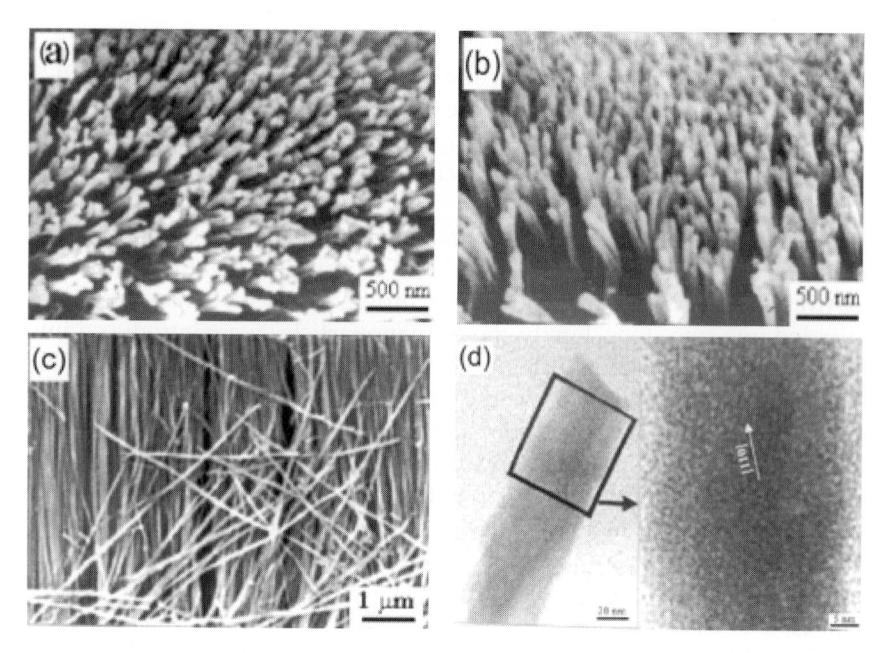

Fig. 5.41. The SEM morphologies and HRTEM lattice image of ordered Si nanowire arrays.

distributing of the catalysts in each channel. The length of the Si nanowire is about 10 m. The HRTEM observations of the Si nanowire structure reveal that the Si nanowires have a single crystalline structure.

The HRTEM lattice image of the Si nanowire (Fig. 5.41(d)) displays clearly the (110) plane. The lattice fringe is 0.384 nm, which is consistent with that of bulk Si. The growth mechanism of Si nanowires is the VLS mechanism. Namely, the Si atoms produced by thermal decomposition of SiH_4 diffuse into the channels of the AAM template, and dissolve in the Au nanoparticles. After the saturation is reached, the Si crystal nuclei begin to precipitate. Then, the Si nanowires grow epitaxially along the channel direction of the template.

(3) Ordered nanoarrays of carbon nanotubes

The CNTs have unusual structural, electronic, magnetic, thermal, and mechanical properties.[68,69] These peculiar properties have made CNTs particularly attractive as new materials for a variety of applications.[70–73] For example, these tiny elongated CNTs can be stronger than steel, lighter than aluminium, more conductive than copper.[68,69,74] Also, depending on different diameters and helicities, the CNT may exhibit the metallic property or the semiconductor property or the semimetallic property.[70] According to the above-described properties of CNTs, people developed the advanced scanning probe,[75–79] nanoelectronic devices[80–82] and the electronic source of field emission.[83–87]

The arc-discharge and laser evaporation are two main CNT preparation methods used to obtain highly quantitative CNTs. However, these methods possess some key problems, which need to be solved. Firstly, the synthesis of CNTs needs a high temperature of 3000°C to make the solid carbon become carbon atoms in vapor state. This limits the quantity of CNTs. Secondly, the CNTs prepared by the evaporation method present an entanglement state and mix with other types of carbon and metallic catalysts. Therefore, it is difficult to purify, manipulate and assemble CNTs and thus construct CNT devices. Developing controlled synthesis techniques of CNTs is beneficial for obtaining ordered structural systems of CNTs. It is very important to study the fundamental properties of CNTs and to explore the potential applications of CNTs as the molecular conducting wire base. The ultimate aim of synthesizing CNTs is to acheive atomic structure control of the positions and directions of CNT growth, helicities, diameters and morphology defects of CNTs etc. Recently, people used the chemical vapor deposition (CVD) method to acheive controlled growth of MWNTs on substrates. Moreover, people have already prepared longer and better oriented arrays of MWNTs on a large substrate.[88–91] Single-walled 0.4 nm CNT arrays have already been obtained using the pyrolysis of organic molecules in the channels of porous zeolite $AlPO_4$-5 (AFI) single crystals.[92] In recent years, the work on the controlled synthesis of CNTs has obviously made progress.

(a) Self-oriented regular arrays of MWNTs

In the CVD growth process, the early method of controlling the CNT's orientation is to allow the CNT to grow in a confined environment (for example, in the thin pores of mesoporous silicon or the channels of alumina). It was found that porous silicon is an ideal substrate for growing the organized nanotube. Fan[90] used the CVD method and porous silicon as substrates to synthesize the arrays of CNTs (Fig. 5.42). Porous silicon samples were obtained by electrochemical etching of p-doped n^+-type Si (100) wafers (diameter 2 inches, resistivity 0.008 to 0.018 ohm·cm). Etching was carried out in a Teflon cell using a Pt cathode for 5 minutes under

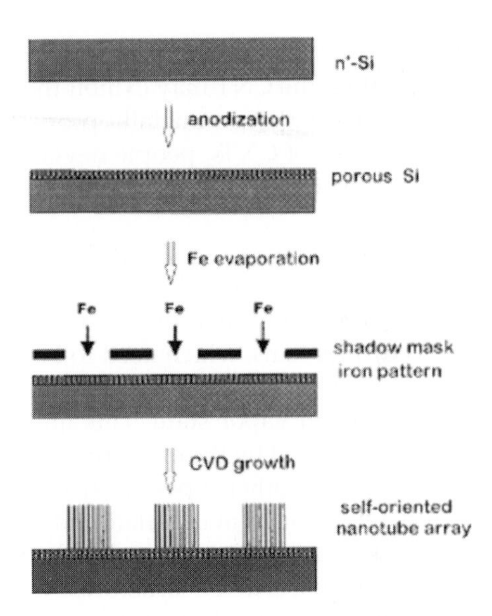

Fig. 5.42. Schematic process flow for the synthesis of regular arrays of oriented nanotubes on porous silicon by catalyst patterning and CVD.

backside illumination with a halogen lamp. The etching solution contained one part hydrogen fluoride (50% aqueous solution) and one part ethanol, and the anodization current density was kept constant at 10 mA/cm^2. The obtained porous silicon has a thin nanoporous layer (the pore diameters are about 3 nm) on top of a microporous layer (with micrometer pores). All substrates were patterned with Fe films (5 nm thick) by electron beam evaporation through shadow masks, containing squared openings with side lengths of 10 to 250 μm at pitch distances of 50 to 200 μm. Then, the substrates were annealed in air at 300°C and left overnight. In this process, the surfaces of the silicon and the iron were oxidized. In the case of porous silicon, the resulting thin oxide layer protected its porous structure from collapsing later in the high temperature process. The substrate was placed in a cylindrical quartz boat sealed at one end and then inserted into the center of a quartz tube reactor that was housed in a tube furnace. The furnace was heated to 700°C in flowing Ar. Ethylene was then allowed to flow at 1000 sccm for 15 to 60 minutes, after which the furnace was cooled to room temperature. The products were observed on a scanning electron microscope (SEM).

Figure 5.43 shows electron micrograghs of self-oriented nanotubes synthesized on n^+-type porous silicon substrates. It can be observed that

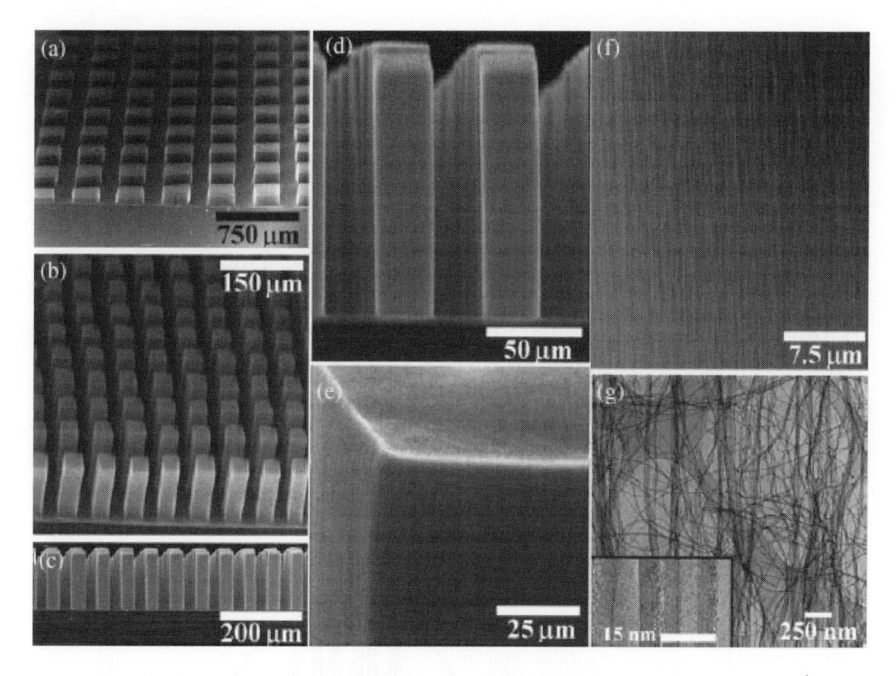

Fig. 5.43. Electron micrographs of self-oriented nanotubes synthesized on n^+-type porous silicon substrates. (a) SEM image of nanotube blocks synthesized on 250 μm by 250 μm catalyst patterns. The nanotubes are 80 μm long and oriented perpendicular to the substrate (see (f)). (b) SEM image of nanotube towers synthesized on 38 μm by 38 μm catalyst patterns. The nanotubes are 130 μm long. (c) Side view of the nanotube towers in (b). The nanotubes self-assemble such that the edges of the towers are perfectly perpendicular to the substrate. (d) Nanotube "twin towers", a zoom-in view of (c). (e) SEM image showing sharp edges and corners at the top of a nanotube tower. (f) SEM image showing that nanotubes in a block are well-aligned to the direction perpendicular to the substrate surface. (g) TEM image of pure multiwalled nanotubes in several nanotube blocks grown on a n^+-type porous silicon substrate. Even after ultrasonication for 15 minutes in 1,2-dichloroethane, the aligned and bundled configuration of the nanotubes is still evident. The inset is a high-resolution TEM image that shows two nanotubes bundling together. The well-ordered graphitic lattice fringes of both nanotubes are resolved.

three-dimensional regular arrays of nanotube blocks towers were formed on top of the patterned iron squares on the substrates (Figs. 5.43(a)–5.43(f)). Each nanotube block exhibits very sharp edges and corners in low-magnification SEM images, and no nanotubes are observed branching away from the blocks (Figs. 5.43(a)–5.43(e)). The width of the blocks is the same as that of the iron patterns. For CVD reaction times of 5 to 60 minutes, the height of the blocks is in the range of 30 to 240 μm. High-resolution

SEM (HRSEM) image (Fig. 5.43(f)) shows that the nanotubes within each block are well-aligned along the direction perpendicular to the surface. The aspect ratio of the blocks approaches about 5.

Figure 5.43(g) gives the transmission electron microscopy (TEM) image of nanotube blocks which were dispersed by the ultrasonic wave. Obviously, this ultrasonicated material still exhibits many aligned multiwalled bundles and some separated individual nanotubes. This clearly indicates that the nanotubes within each block are densely-packed and held together by van der Waals interactions. About 90% of these multiwalled nanotubes have diameters of 16 ± 2 nm.

Formation of these oriented regular arrays of carbon nanotubes belongs to the self-oriented nanotube growth mechanism (Fig. 5.44). During the initial stage of CVD, ethylene molecules are catalytically decomposed on the iron oxide nanoparticles. As supersaturation occurs, a nanotube grows

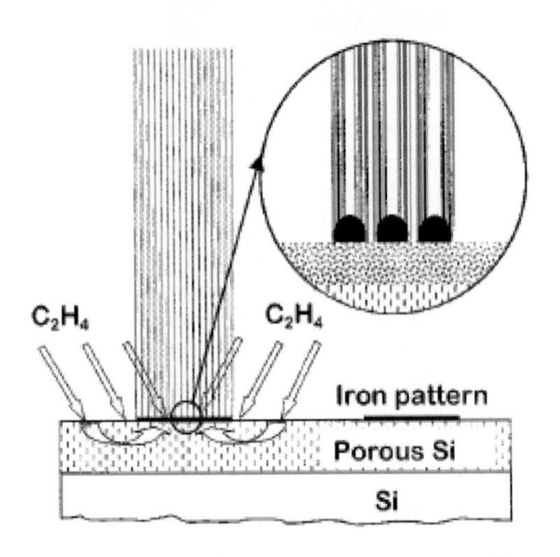

Fig. 5.44. Nanotube growth and self-orientation on porous silicon substrate. On the substrate, the iron oxide nanoparticles remain pinned down during CVD growth, and thus base growth operates. The ethylene gas molecules feed the growing nanotubes at the edge of the catalyst square by diffusing through the nanotubes and through the porous silicon layer. On a plain silicon substrate, the ethylene molecules can only feed the edge of the catalyst square and diffuse through the nanotubes. This results in a growth rate lower than would be observed on a porous silicon substrate.[28] The nanotubes within each block interact via van der Waals forces, which provide the overall rigidity for oriented growth. Some openings exist between the bundled nanotubes, allowing gas to diffuse.

off each of the densely-packed catalyst particles (average diameter 16 nm) and extends to the open space along the direction normal to the substrate. As the nanotubes lengthen, their outermost walls interact with those of neighboring nanotubes via van der Waals forces to form a large bundle with sufficient rigidity. This rigidity enables nanotubes to keep growing along the original direction. Even the outermost nanotubes are held by the inner nanotubes without branching away.

(b) The growth mechanisms of MWNTs[15]

The growth mechanism of CNTs has not been understood for a long time. In order to make CNTs applicable on a large scale in electronics etc., people must control the shapes, sizes, and even helicities etc. of CNTs during the synthesis of CNTs and thus, make the synthesized CNTs meet the application requirements. An indispensable prerequisite to such a control of nanotube structural homogeneity is an in-depth understanding of the nanotube growth mechanism. Since Iijima discovered CNTs in 1991, different growth models of CNTs have been theoretically proposed, for example, five-member rings–seven-member rings defect deposition growth,[93,94] lip-lip interaction growth,[95] step flow growth,[96] tip growth,[97,98] base growth[99] and extrusion mode growth[100] etc. Different growth models are applicable to different growth methods. However, direct experimental evidence to support any of the models is scarce. The *in situ* measurement of individual CNT growth process has not been realized. But, the growth of self-oriented MWNT arrays provides a chance to reveal experimentally the CVD growth mechanism of CNTs. Fan[100] firstly revealed experimentally the growth mechanism of MWCNTs using a ^{13}C isotope-labeling method. As we know, when CNTs are prepared by using the CVD method, the metallic catalysts are used. As a result, it is observed that one end of the resulting CNT has one catalyst particle existing, but the other end of the resulting CNT is hollow. This morphology can be formed by two mechanisms, which are tip growth and extrusion mode growth. According to the tip growth model, it is supposed that the catalyst particles play the nuclei of formation of CNTs in the CNT growth process, but once the formation is excited and the catalyst particles are capped by initial CNTs, the CNT growth begins to be carried out from the open ends of CNTs via continous deposition of carbon atoms on the dangling bonds of the open ends. When the temperature descends, the open ends of CNTs are closed and the growth stops. According to the extrusion growth model, it is supposed that, the metallic catalysts are the continuous growth positions of CNTs. The carbon atoms deposit on the catalyst particles, resulting in formation of the metal-carbon alloy particles. When the carbon atom concentration in the metal-alloy particles reaches the saturated concentration, the carbon atoms separate out from the alloy particles to form CNTs. The difference between the above-described two growth mechanisms is as follows.

In the former case, with growing of a CNT, the relative distance between the CNT growth end and the catalyst particle become larger and larger, but in the latter case, the relative distance between the CNT growth end and the catalyst particle is kept constant, as shown in Fig. 5.45.

In literature, the base growth model and top growth model are often reported. The base growth model is when the metallic catalyst particle adheres to the substrate and the top end of the CNT is closed, and does not contain the catalyst particle. The carbon source of the CNT growth arises from the interface position between the CNT and the catalyst particle (Fig. 5.46). The top growth model means that the catalyst particle located

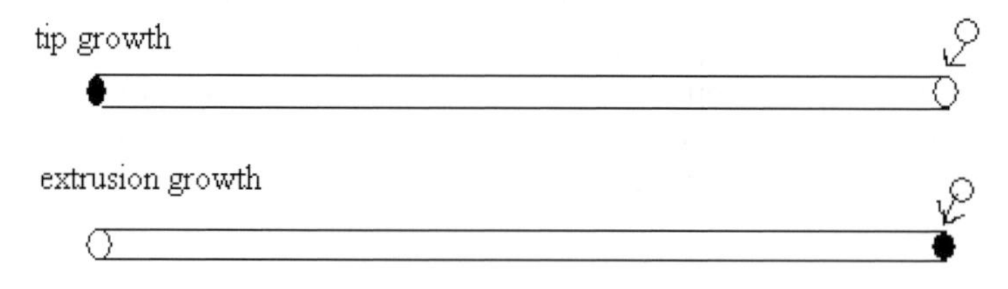

Fig. 5.45. Two possible CVD growth models of CNTs.

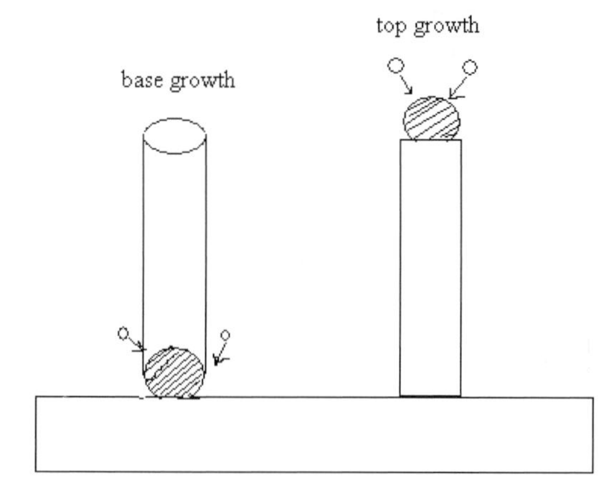

Fig. 5.46. The schematic of CNT CVD growth models. Right: top growth; left: base growth.

on the top of the CNT moves with the growth of the CNT. This moving particle provides the indispensable carbon source for the CNT growth. The two models do not have the intrinsic mechanism difference and both of them belong to the extrusion growth model. The only difference is that the catalyst particle stops on the substrate or is located on the tip of the CNT. This difference depends on the intensity of the adhesion between the catalyst particle and the substrate.

In order to reveal whether the CNT growth model belongs to the extrusion growth model or the tip growth model, Fan made the following isotope labeling experiment of carbon nanotubes. The experiment apparatus are shown in Fig. 5.47.

The porous silicon substrate was prepared by electrochemical etching.[90] A 5 nm thick iron film was electron-beam deposited onto the substrate and then annealed at 300°C overnight. Nanotube growth was carried out at 700°C in a 26 mm quartz tube housed in a tube furnace (Fig. 5.47). $^{12}C_2H_4$ or $^{13}C_2H_4$ flowed at 140 sccm, diluted by a co-flow of 260 sccm of argon under 1 atm pressure for the growth of MWNT arrays. The growth time was 1 minute, after which no purification or any other treatment step was applied to the sample. The ^{13}C-ethylene had a 99% ^{13}C purity. The ^{12}C-ethylene contained a negligible amount of ^{13}C isotope in its natural abundance of 1.1 atom%.

Two feeding sequences of isotope ethylene were used for the growth of ^{12}C-^{13}C nanotube junction arrays. In the first sequence, ^{12}C ethylene was introduced for 15 s and then switched to ^{13}C-ethylene for another 45 s. To avoid any possible disturbance to the growth process, no purge step was

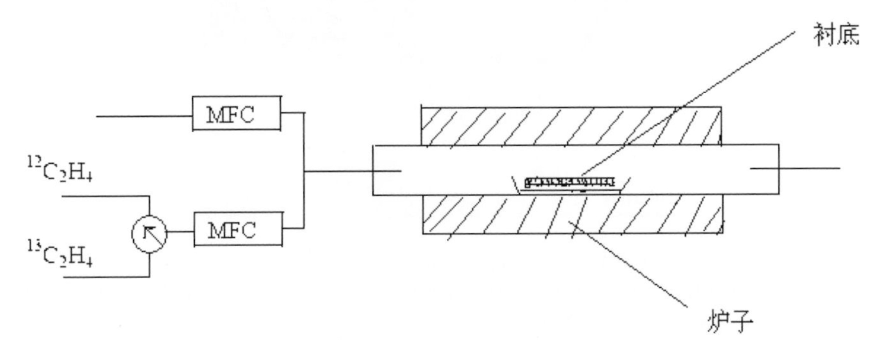

Fig. 5.47. The apparatus schema of CVD for ^{13}C isotope labeling. $^{12}C_2H_4$ and $^{13}C_2H_4$ are controlled by the change-over switch. MFC: the flowmeter.

taken to deplete ^{12}C-ethylene before introducing ^{13}C-ethylene. The second feeding sequence was inversed, in that ^{13}C-ethylene was first introduced for 15 s and then ^{12}C-ethylene for 45 s.

The as-grown MWNT arrays had uniform height of 10 μm, as shown in Fig. 5.48(a). For microscopy study, small bundles of MWNTs were pulled out from the array with an atomic force microscope (AFM) tip and carefully transferred onto a TEM grid, with their top and bottom ends recognized. TEM and X-ray fluorescence (EDX) were used to identify and locate the iron catalyst particles. It was observed that the iron particles exist only at the bottom ends of the nanotube, while the top ends have abundant capped empty nanotube tips (Fig. 5.48). This shows that the iron particles stayed on the substrate during the growth process due to the strong adhesion between the particles and the porous silicon surface. The TEM observation

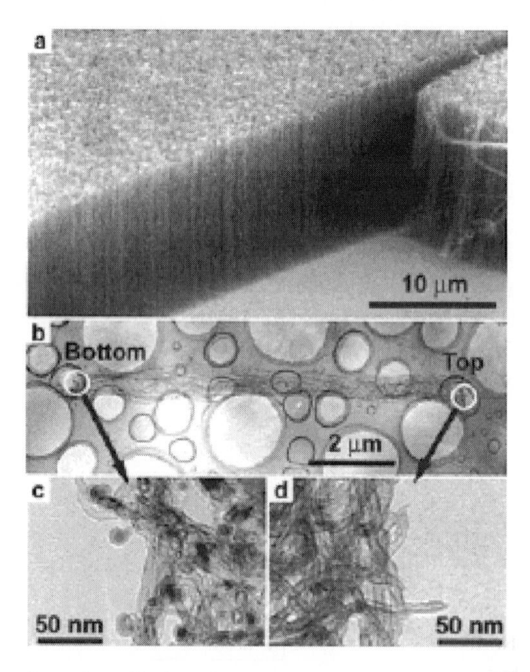

Fig. 5.48. Aligned multiwalled carbon nanotubes grown on porous silicon substrates. (a) Scanning electron microscopy picture of an as-grown MWNT array. (b) TEM image of a bundle of nanotubes; the left end is the bottom; the right end is the top. (c) A zoomed-in view of the bottom side of (b), where darker particles are catalytic iron particles. (d) A zoomed-in view of the top side of (b), where capped nanotube tips are visible.

of the individual nanotube shows that the nanotubes grew continuously through their lengths. The nanotubes had outer diameters of 10 nm and 6 to 8 inner graphic shells.

Figure 5.49(a) shows the Raman spectra of pure ^{12}C and ^{13}C nanotube arrays. It is clear that all Raman modes of ^{13}C nanotubes (ω_C^{13}) exhibited clear shifts to lower frequencies compared to those of ^{12}C nanotubes (ω_C^{12}), with a uniform ratio of ω_C^{13}: $\omega_C^{12} \approx 0.96$. For ^{12}C-^{13}C nanotube arrays obtained by the first feeding sequence, micro-Raman measurements revealed that at the top of the nanotube arrays, the nanotube consists of pure ^{12}C, while at the bottom, the nanotube consists of mostly ^{13}C (Fig. 5.49(b)).

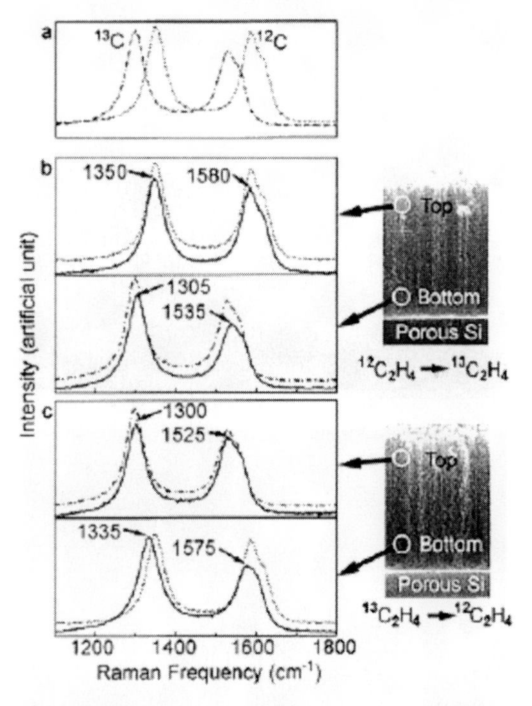

Fig. 5.49. Micro-Raman spectra of isotope-labeled MWNT arrays. (a) Pure ^{12}C nanotube array (dotted) and pure ^{13}C nanotube array (dash-dotted) spectra. (b) Nanotube arrays grown with ^{12}C ethylene first and then ^{13}C ethylene. (c) Nanotube arrays grown with ^{13}C ethylene first and then ^{12}C ethylene. The solid curves in the left plots are Raman spectra recorded at the locations circled in the right-hand side images. Pure ^{12}C or ^{13}C spectra are also plotted for comparison (upshifted for clarity).

Since ^{12}C was fed into the system earlier than ^{13}C, the result clearly reveals that the top segments of the nanotubes are formed chronically earlier in the growth process than the lower segments. These results indicate that the catalyst particles reside at the bottom of the nanotube array. This leads to a clear extrusion growth picture of the nanotubes as schematically shown in Fig. 5.50. The iron catalyst particle stays on the substrate throughout the growth process. Initially, a ^{12}C nanotube segment grows out from the catalyst to a certain length when ^{12}C ethylene is introduced into the system. As the gas source is switched to ^{13}C ethylene, a ^{13}C nanotube stem grows out from the catalyst, pushing the already grown ^{12}C segment upward

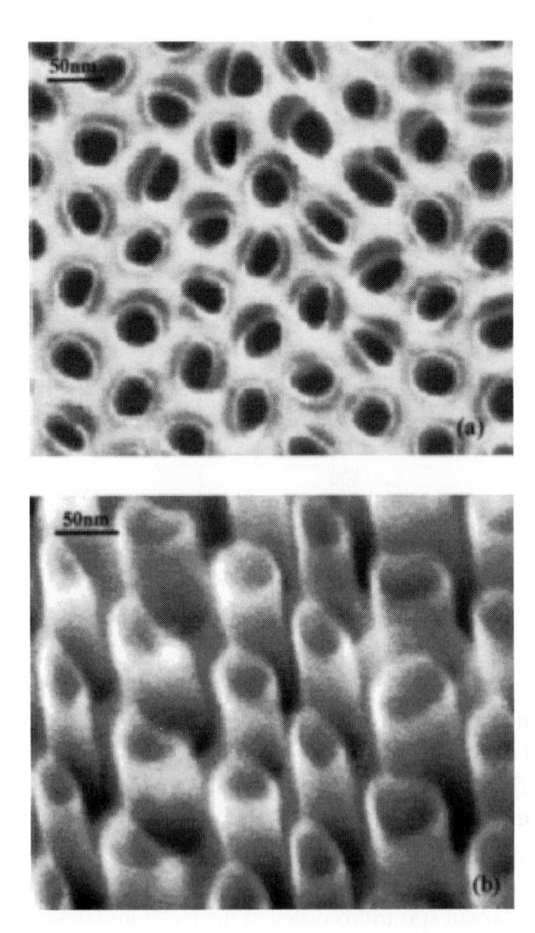

Fig. 5.50. (a) Ordered carbon nanotube arrays prepared by the AAM template method. (b) Parts of the AAM template being removed by the NaOH solution.

and forming a ^{12}C-^{13}C nanotube junction. Thus, the combined microscopy and microspectroscopy studies unambiguously reveal that the extrusion mechanism[72] operates in our CVD grown of MWNTs, and rules out any other possible growth mechanisms.

The second feeding sequence was inversed. In this case (Fig. 5.49(c)), the Raman spectra clearly revealed that the ^{13}C atoms were at the top segments of the nanotubes and the late introduced ^{12}C atoms were at the bottom segments of the nanotubes, which reinforces the nanotube growth mechanism described above.

From the Raman spectroscopy results shown in Fig. 5.49, one clearly sees that the top segments of the nanotubes consist of only the first introduced isotope carbon atoms. Moreover, TEM observations revealed that the nanotubes are composed of graphitic shells parallel to the tube axis and that the numbers of the shells are constant throughout the lengths of the nanotubes. Two conclusions can be drawn from these results. First, all the graphitic shells of the nanotube extrude from the catalyst particle; no direct deposition occurs on the grown tube stem. The absence of such over-coating is desired for the synthesis of clean nanotubes for device application. Second, no separate graphitic shell extrudes over other shells either in the outside or inside of the nanotube. Therefore, Fan conclude that synchronized extrusion occurs for all the shells of the multiwalled nanotubes from the catalyst, as suggested in the "Yarmulke" mechanism.[9]

(c) Single-walled carbon nanotube (SWNT) arrays[92,101]

Wang obtained single-walled 4 Å carbon nanotube arrays by the pyrolysis of tripropylamine molecules in the channels of porous zeolite $AlPO_4$-5 (AFI) single crystals.

AFI is a type of transparent microporous crystal containing one-dimensional channels packed in hexagonal arrays, with an inner diameter of 7.3 ± 0.1 Å, and the separation distance between two neighboring channels is 13.7 Å. The AFI host was synthesized according to the method described in Ref. 102. Tripropylamine (TPA) molecules were encapsulated in the channels during the crystal growth. The synthesis process of the SWNTs involves the pyrolysis of TPA molecules in the AFI channels in a vacuum of 10^{-4} torr at temperatures of 350–450°C and the formation of carbon nanotubes at 500–800°C. After the formation of SWNTs, the hexagonal arrays of SWNTs were achieved. This was shown in Ref. 92.

To observe the SWNTs in the AFI, the AFI framework was dissolved in 30% hydrochloric acid. The solution containing SWNT was then enriched and dispersed on a lacey carbon film for high-resolution TEM observations. From the SWNT images, the diameters of the SWNTs were determined, which were 0.42 ± 0.02 nm. These SWNTs were unstable under electron beam radiation, fading during observation in 10–15 seconds. For these 0.4 nm diameter SWNTs, there are three possible nanotube structures: these are the zigzag (5, 0) (diameter, $d = 0.393$ nm), the armchair (3, 3) ($d = 0.407$ nm) and the chiral (4, 2) ($d = 0.414$ nm), where (m, n) is a roll-up vector defining the tube symmetry. Of these, the zigzag (5, 0) tube can be exactly capped by half a C_{20} fullerence. Therefore, this structure is more likely to be formed. In these small diameter SWNTs, the strong bending of carbon bonds, plus the size confinement effect of the channels and the imperfection (point defects etc.) could make high energy regions be formed in these ultrasmall nanotubes. When these ultrasmall nanotubes are released from their channels, the defects cause instability. As a result, these nanotubes change into graphite stripes.

(d) Carbon nanotube arrays based on the AAM templates

Commonly, ordered carbon nanotube arrays are synthesized by combining the AAM (anodic alumina membrane) templates with catalysis growth. Firstly, metal particles (for example, Fe, Co, Ni and their alloys) were introduced into the channels of the AAM templates as the catalyst. Then, in the mixture of gas consisting of Ar or N_2 and hydrocarbon gas (for example, CH_4, C_2H_2 and C_2H_4 etc.), carbon nanotube arrays were prepared through the catalysis pyrolysis of hydrocarbon. Figure 5.50 shows carbon nanotube arrays prepared by this method.[103]

5.2.2. Ordered nanoarrays of binary compound nanowires

Binary compound nanowires contain mainly oxide nanowires, nitride nanowires, carbide nanowires, sulphide nanowires, and binary alloy nanowires etc. Here, we will introduce the synthesis of these compound nanoarrays.

5.2.2.1. *Ordered nanoarrays of alloy nanowires*

The rapid development of information science makes people study continually ultrahigh density information storage. Therefore, the study of the magnetic properties of ordered nanoarrays forms an upsurge. Ordered magnetic nanoalloy arrays may become one useful kind of perpendicular magnetic recording medium of ultrahigh density. Up to now, people have already synthesized ordered nanoarrays of ferromagnetic-ferromagnetic and ferromagnetic-nonalloy nanowires.[104–111] At the same time, many researchers have begun to investigate the thermoelectric properties of nanomaterials. Ordered nanoarrays of Bi-Sb semimetal alloy nanowires were synthesized.[112,113] Here, we will introduce the synthesis and characterization of ordered nanoarrays of Co-Cu, Fe-Ag, and Co-Ag nanowires.[114]

A layer of Au was sputtered onto one side of the AAM serving as the working electrode in a two-electrode electrochemical cell. The electrolytes and the electro-deposition conditions of Co-Cu, Fe-Ag, and Co-Ag alloy nanowire arrays were listed in Table 5.1. The electro-deposition was carried out at a constant current density, with carbonate serving as the counter electrode. After electro-deposition, these ordered alloy nanowire arrays embedded in the AAM were annealed at different temperatures T_A (300, 400, 500, and 600°C) for 1 h in a furnace under a flowing Ar atmosphere, respectively.

The typical top view (shown in Figs. 5.51(a)–5.51(c), respectively) and cross-section (shown in Figs. 5.52(a)–5.52(c), respectively) SEM images of Co-Cu, Fe-Ag, and Co-Ag nanowire arrays demonstrate that these

Table 5.1. Electrolytes and electro-deposition conditions of Co-Cu, Fe-Ag, and Co-Ag alloy nanowire arrays.

electrolytes (g/L)	Co-Cu	Fe-Ag	Co-Ag
$CoSO_4{\cdot}7H_2O$	50		40
$CuSO_4{\cdot}5H_2O$	0.5		
$AgNO_3{\cdot}7H_2O$		20	50
$FeSO_4{\cdot}5H_2O$		20	
CH_3COONH_4		10	40
$Na_2S_2O_3{\cdot}5H_2O$		100	
Na_2SO_3		40	
H_3BO_3	40		
NH_3H_2O			60
electro-deposition conditions	Co-Cu	Fe-Ag	Co-Ag
current density (mA/cm^2)	1.5	2.5	1.5
time (h)	6	6	6
temp (°C)	room temp	room temp	room temp

Fig. 5.51. Typical top-view SEM images showing the general morphology of the ferromagnetic-nonmagnetic alloy nanowire arrays: (a) Co-Cu; (b) Fe-Ag; (c) Co-Ag.

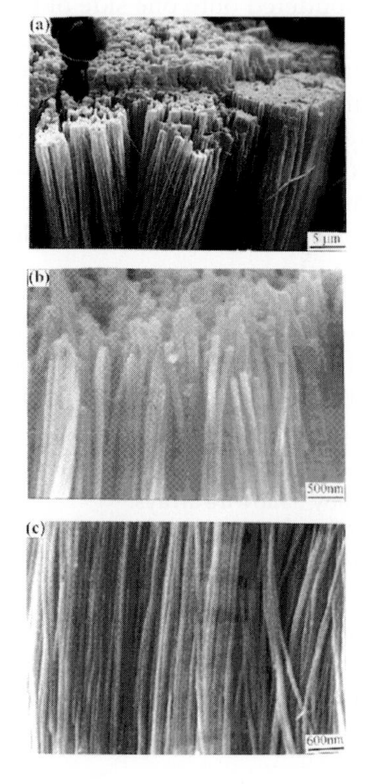

Fig. 5.52. Typical cross-section SEM images showing the general morphology of the ferromagnetic-nonmagnetic alloy nanowire arrays: (a) Co-Cu; (b) Fe-Ag; (c) Co-Ag.

deposited alloys, indeed, fill the nanochannels of the AAM uniformly and that the nanowires are apparently continuous, parallel, and ordered. The measured diameters of these alloy nanowires corresponded closely to the nanochannel diameters.

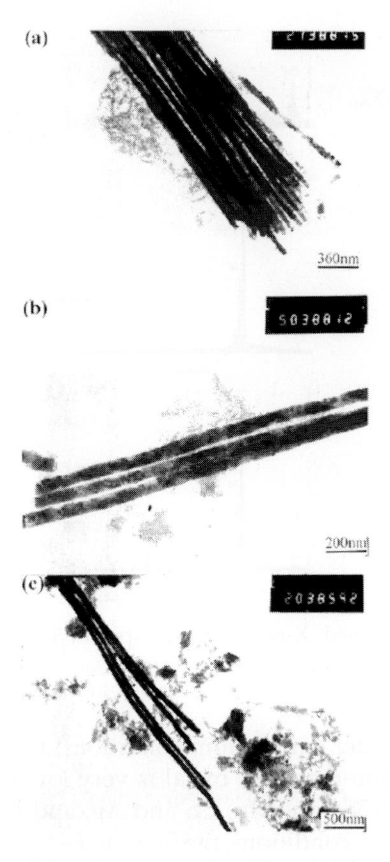

Fig. 5.53. TEM images of the alloy nanowires liberated from the AAM: (a) Co-Cu; (b) Fe-Ag; (c) Co-Ag.

The TEM images of Co-Cu, Fe-Ag, and Co-Ag nanowires liberated from the AAM are shown in Figs. 5.53(a)–5.53(c), respectively. It can be seen that these alloy nanowires are uninterrupted with a diameter of about 60 nm, which corresponds to the nanochannel diameter of the AAM used. Energy dispersed X-ray spectrometer (EDS) analysis attached on the SEM demonstrates that the atomic ratio of Co and Cu, Fe and Ag, and Co and Ag in the corresponding nanowires are close to 20 : 80, 20 : 80, and 14 : 86 (shown in Figs. 5.54(a)–5.54(c), respectively). The Co-Cu, Fe-Ag, and Co-Ag diffraction spectra of the as-deposited nanowire arrays embedded in the AAM (shown in Figs. 5.55(a)–5.55(c), the bottom curves) also indicate that these lines correspond to the face-centered-cubic (fcc) Cu, fcc-Ag, and fcc-Ag structure, respectively.

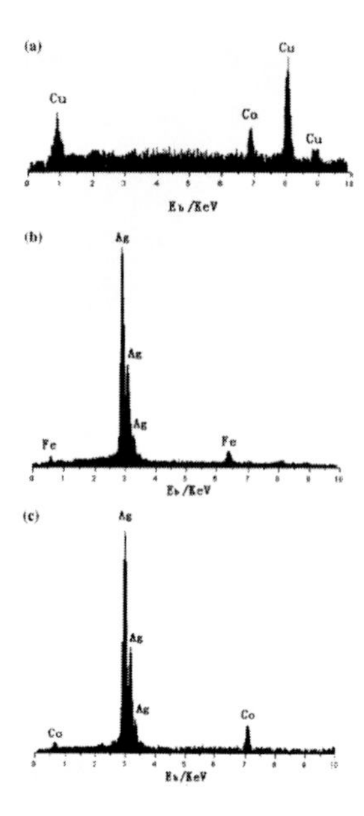

Fig. 5.54. Energy-dispersed X-ray analysis spectra of the alloy nanowires in Fig. 5.51: (a) Co-Cu; (b) Fe-Ag; (c) Co-Ag.

We know that, under equilibrium conditions, the mutual solubility of ferromagnetic and non-magnetic metal is very low in both the solid and liquid forms, such as Co and Cu, Co and Ag, and Fe and Ag. However, under non-equilibrium conditions, the Co/Cu, Fe/Ag, and Co/Ag systems could form metastable solid solutions. In this type of material, the annealing treatment results in the recrystallization of these metastable alloys into two separate phases, leading to the formation of small ferromagnetic metal particles that exhibit greatly enhanced magnetic properties because of their single-domain nature. These processes have already been proved by XRD and are shown in Fig. 5.55. It can be seen that the as-deposited samples are single-phase fcc-Cu, fcc-Ag, and fcc-Ag, respectively, but the Cu and Ag diffraction lines are shifted toward higher angles. Therefore, the as-deposited alloy nanowire arrays in this work are solid solutions. As shown in Fig. 5.55(a), the fcc-Cu structure is preserved up to $T_A = 400°C$. With further annealing at 500°C, however, the fcc-Co lines appear, indicating the recrystallization of the metastable Co-Cu alloy into fcc-Co and fcc-Cu.

As T_A increased over 500°C, the diffraction lines became progressively narrower due to the growth of particle size. Figure 5.55(b) indicates that the fcc-Ag structure is not changed up to $T_A = 300°C$. With further annealing at

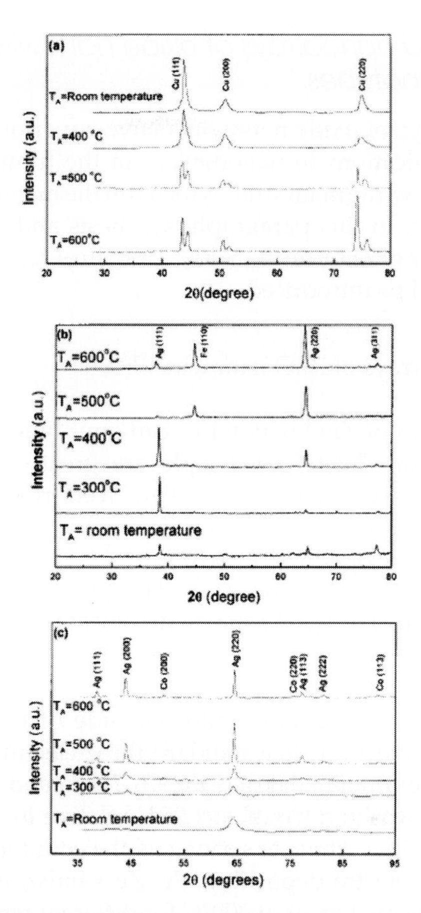

Fig. 5.55. X-ray diffraction spectra of the alloy nanowire arrays at room temperature, after annealing for 1 h at various temperatures T_A: (a) Co-Cu; (b) Fe-Ag; (c) Co-Ag.

400°C, however, the body-centered-cubic (bcc) α-Fe lines appear, indicating the recrystallization of the metastable Fe-Ag alloy into fcc-Ag and bcc-Fe. As T_A increased over 400°C, the diffraction lines became progressively narrower. The phase structural characterizations of the Co-Ag alloy nanowire arrays during the annealing process have been shown in Fig. 5.55(c). The fcc structure is preserved up to $T_A = 300$°C. With further annealing at 400°C, however, the fcc-Co lines appear, indicating the recrystallization of the metastable Co-Ag alloy into fcc-Co and fcc-Ag. As T_A increased over 400°C, the diffraction lines became progressively narrower due to the growth of particle size.

5.2.2.2. *Ordered nanoarrays of oxide nanowires and nanotubes*

Ordered nanoarrays of oxide nanowires have superior optical properties and potential applications in nanodevices in the future and thus attract much attention. Now, the main studies focus on the materials of ZnO, In_2O_3, SnO_2 and TiO_2 etc. In this paragraph, synthesis and characterization of ordered nanoarrays of ZnO, In_2O_3, SnO_2, TiO_2, Eu_2O_3 and Bi_2O_3 nanowires and nanotubes will be introduced.

(1) Ordered nanoarrays of ZnO nanowires[115]

Ordered nanoarrays of ZnO nanowires can emit blue-green fluorescence and ultraviolet light. Therefore, people are greatly interested in ZnO nanowire arrays. Many researchers spent much effort on the synthesis techniques of the ordered arrays of ZnO nanowires, including electrochemical deposition,[115,116] electrophoresis deposition,[117,118] sol-gel,[119] other solution methods,[120] and the CVD method.[121–123] Here, we will introduce synthesis and characterization of ordered nanoarrays of ZnO nanowires, which were made by Li[115]

A layer of Au was sputtered onto one side of the membrane serving as the working electrode in a standard three-electrode electrochemical cell. The electrolyte contained 80 g/l $ZnSO_4{\cdot}7H_2O$ and 20 g/l H_3BO_3. The electro-deposition was performed at 1.25 V relative to the Ag^+/AgCl reference electrode, with carbonate serving as the counter electrode at room temperature. After electro-depositing, the Zn nanowire arrays embedded in the AAM were heated in air at 300°C for different periods of time (from 0 to 35 h).

X-ray diffraction (XRD) spectra (Fig. 5.56) were taken from nanowire arrays with different heat treatment conditions. No peak associated with ZnO was found when the spectrum was taken immediately after Zn was electro-deposited into the nanochannels of the AAM (Fig. 5.56(a)). Even after being exposed to air at room temperature for one week, there is still no ZnO peak in the XRD spectrum. It can be seen clearly that with the increase of heat treatment time, the intensities of the Zn peaks become weaker and those of ZnO peaks become stronger (Figs. 5.56(b)–5.56(d)). After heat treatment at 300°C for 35 h, all of the Zn peaks disappeared (Fig. 5.56(e)), indicating that Zn deposited in the channels of the AAM had been oxidized completely. In addition, the peak positions and their relative intensities (Fig. 5.56(e)) are consistent with the standard powder

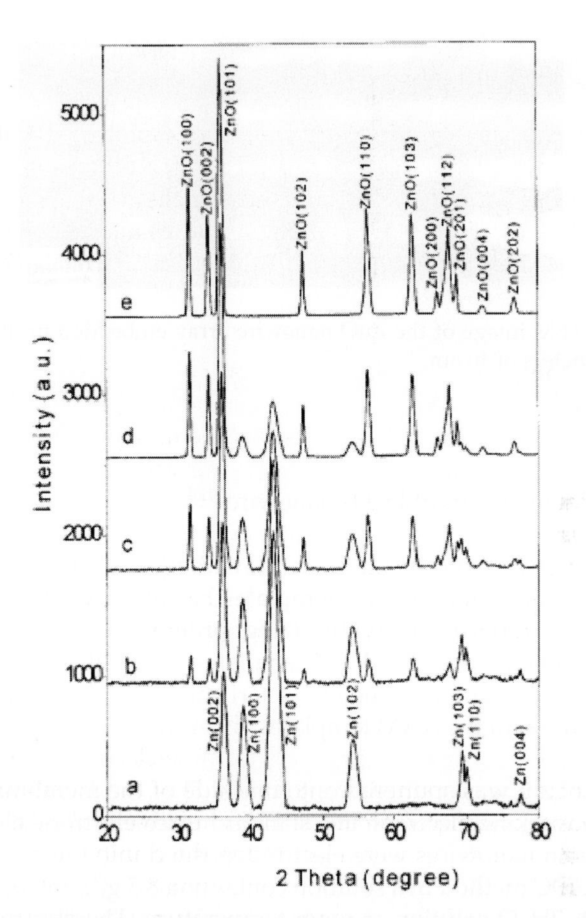

Fig. 5.56. XRD patterns of the Zn nanowire arrays embedded in alumina templates with diameters of 40 nm. (a) No heat treatment, (b) heated at 300°C for 8 h, (c) for 15 h, (d) for 25 h, (e) for 35 h, respectively.

diffraction pattern of ZnO, indicating that there is no preferred orientation and that the ZnO nanowires are polycrystalline. The broadening of ZnO peaks is due to the small particle size. Figure 5.57 shows a TEM (JEM-200 CX) image of the AAM with ZnO nanowires in its channels. Compared with the blank AAM, the dark and bright areas correspond to ZnO nanowires and Al_2O_3 supporting frame, respectively. The ZnO nanowires with diameters equal to those of the nanochannels are distributed in the AAM periodically at a constant interval and form a parallel aligned array.

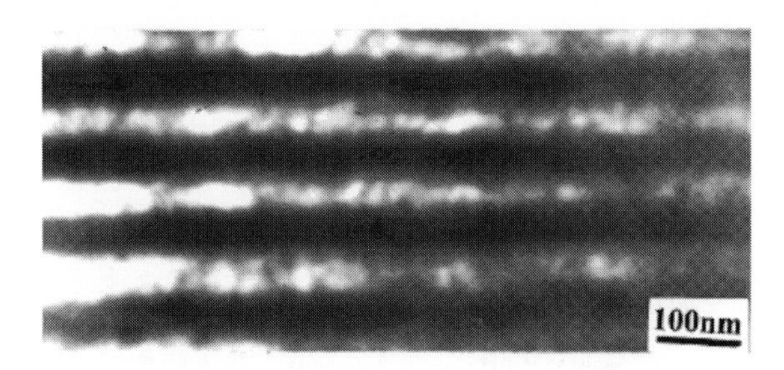

Fig. 5.57. A TEM image of the ZnO nanowire array embedded in the AAM with channel diameters of 40 nm.

(2) Ordered nanoarrays of In_2O_3 nanowires[124]

Using the one-step nanocasting route to synthesize In_2O_3 nanowire arrays embedded in mesoporous SiO_2 templates has already been reported.[125] However, the reports on synthesizing ordered nanoarrays of In_2O_3 nanowires embedded in the AAM templates are small. In the following, we will introduce synthesis and characterization of ordered nanoarrays of In_2O_3 nanowires in the AAM templates.[124]

A layer of Au was sputtered onto one side of the membrane that was used as the working electrode in a standard three-electrode electrochemical cell. The In nanowires were electro-deposited into the nanoholes by a three-probe DC method in a solution containing 8.5 g/L $InCl_3$, and 25 g/L $Na_3C_6H_5O_7{\cdot}2H_2O$ solution at room temperature. The electro-deposition was performed at -1 V (versus Ag/AgCl), with carbonate serving as the counterelectrode. After electro-deposition, the assembly systems were annealed in air at different temperatures to form ordered In_2O_3 nanowire arrays.

Figure 5.58 shows the X-ray diffraction (XRD) spectra of the In_2O_3/AAM assembly system oxidized at different temperatures for 12 h. XRD results indicate that the as-prepared nanowire arrays consist only of In, and after annealing at 923 K for 12 h, In partially transforms into In_2O_3. When the annealing temperature is increased to 973 K, the In diffraction peaks disappear and complete In_2O_3 nanowire arrays are obtained. The diffraction peaks of (222), (400), (440), and (622) correspond to the cubic In_2O_3 phase. This indicates that In_2O_3 nanowire arrays embedded in the AAM are cubic polycrystalline.

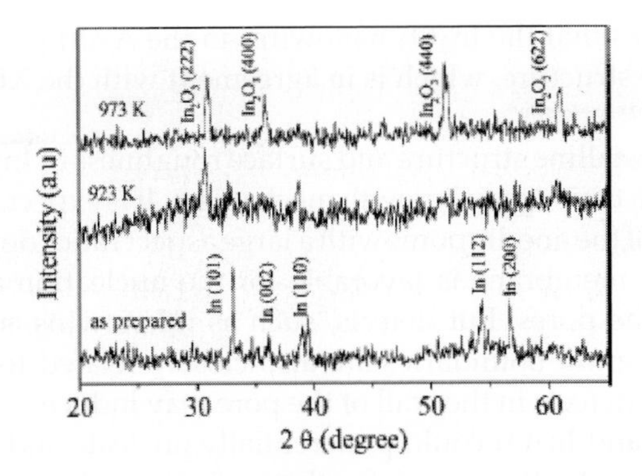

Fig. 5.58. XRD spectra of In_2O_3 nanowire arrays embedded in anodic alumina membranes (AAMs) annealed at different temperatures for 12 h.

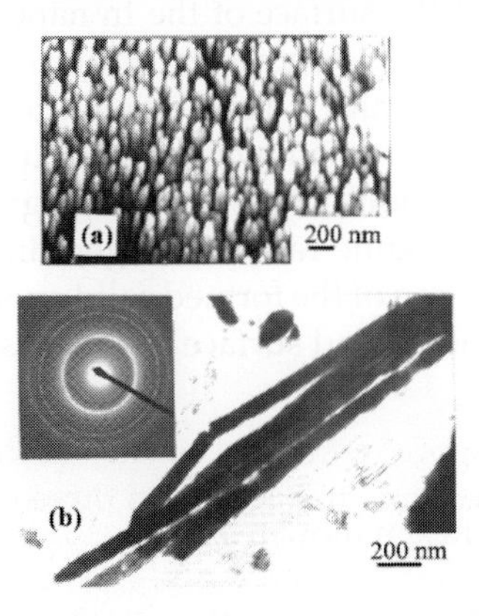

Fig. 5.59. (a) SEM image of In_2O_3 nanowire arrays embedded in the AAM. (b) TEM micrograph of In_2O_3 nanowires.

Figure 5.59(a) shows the scanning electron microscopy (SEM) (JOEL JSM-6300) top image of In_2O_3 nanowire arrays grown in a template after cleaning in an ultrasonic bath for 3 minutes, followed by a brief etching of the alumina in aqueous NaOH. It can be clearly seen that the In_2O_3 nanowires are basically of equal height and are uniformly embedded in the AAM. Figure 5.59(b) shows the transmission electron microscopy (TEM) (JEM-200CX) of In_2O_3 nanowires after the AAM was removed from the In_2O_3/AAM assembly system by dissolving the AAM in 0.5 M NaOH solution and washing it for several times with distilled water. It can be seen that the surface of the nanowire is not smooth and the diameter of the individual In_2O_3 nanowire is also not uniform. The corresponding TEM diffraction pattern indicates that the In_2O_3 nanowires in the AAM pores are of cubic polycrystalline structure, which is in agreement with the XRD results.

The polycrystalline structure and surface roughness of In_2O_3 nanowires may result from their special growth mechanism. It is expected that the capillary effect[126] of the anodic pores with a large aspect ratio (depth/diameter) in the alumina membrane is favorable for the nucleation and growth of nanowires in the pores. But defects, such as microtwins, stacking faults, and low-angle grain boundaries are also closely related to the nanowire growth.[127] The defects in the wall of the pore may induce a different orientation growth, and In_2O_3 could preferentially nucleate and grow on these defects. This may be the reason for the surface roughness and polycrystalline structure of the In_2O_3 nanowires. On the other hand, the oxidation process of In nanowires to In_2O_3 may be complicated. It can be deduced that oxidation begins at the surface of the In nanowires. At temperatures higher than 429.76 K, the In nanowires in the AAM are liquid. Consequently, when the temperature is high enough to lead oxygen to diffuse into the In nanowires, the In nanowires begin to be oxidized in random orientations (nuclei formation). These nuclei will grow further as oxidation continues, but not at the same rate, as oxygen diffusion depends on the crystallographic orientation of the formed solid crystallites, resulting in the formation of polycrystalline and surface roughness of In_2O_3 nanowires.

(3) Ordered arrays of SnO_2 nanowires[128]

It has been reported that nanocrystalline SnO_2 possesses excellent characteristics such as a high gas sensitivity and a short response time.[129,130] Up to now, studies on tin oxide have mainly focused on the preparations and physical properties of SnO_2 films and nanoparticles. Comparatively, very little work has been conducted on the one-dimensional nanostructural SnO_2 system. Here Zheng[128] first reported the preparation of a large-scale uniform SnO_2 nanowire array based on highly-ordered nanoporous alumina membranes (AAM) by electrochemical deposition and thermal oxidization. The microstructure properties and growth mechanism of SnO_2 nanowires embedded in the AAM are discussed.

The fabrication process involves three steps: (1) electrochemical generation of an alumina template with highly-ordered hexagonal arrays of nanochannels; (2) electro-deposition of pure metal Sn in the alumina template; (3) oxidation of Sn nanowire array embedded in the AAM in the air.

The alumina membrane was formed by a two-step anodization process as described in Sec. 5.1.1.

A layer of Au was sputtered onto one side of the membrane and used as the working electrode in a standard three-electrode electrochemical cell. Before mounting into the electrochemical cell, the AAM was immersed in deionized water under ultrasonic agitation for 2 minutes to clean the pollutant on the surface of the membrane and to expel the air bubbles from the pores. This step is very important for obtaining good nanowires and a homogeneous growth in the whole growing area. If there are gas bubbles in some pores, the growth of the nanowires cannot grow continuously and uniformly because of the embolism effect induced by the gas bubbles in the pores. The Sn nanowires were electro-deposited into the nanoholes by a three-electrode cell direct-current method in a solution containing 7 g/L $SnCl_2{\cdot}H_2O$, and 25 g/L $Na_3C_6H_5O_7{\cdot}2H_2O$ solution at room temperature. The electro-deposition was performed at −0.8 V (versus saturated Ag/AgCl), with graphite serving as the counter electrode. After the Sn nanowire array was electro-deposited into the AAM, the assembly system was annealed in the air at different temperatures to form an ordered SnO_2 nanowire array.

Figure 5.60 shows XRD patterns of the SnO_2/AAM assembly system annealed at different temperatures for 10 h. It can be seen that the

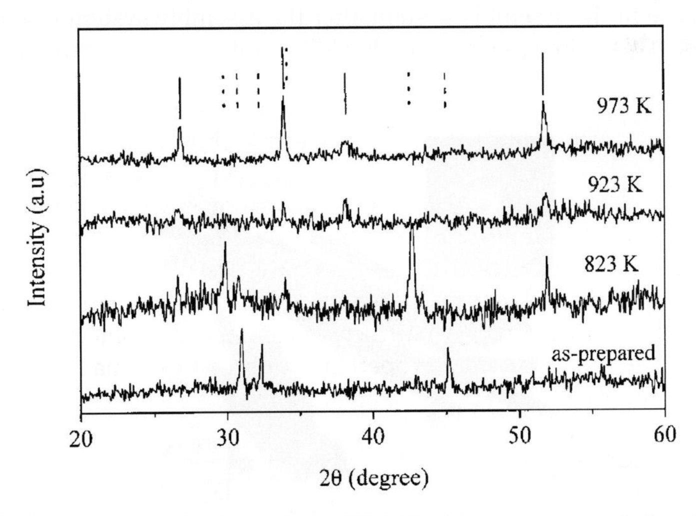

Fig. 5.60. XRD patterns of the SnO_2/AAM assembly system annealed at different temperatures for 10 h. -, SnO_2; …, SnO; - - -, Sn.

as-deposited nanowire array consists only of Sn, and after annealing at 823 K for 10 h, Sn transforms partially into SnO and SnO_2. The Sn and SnO diffraction peaks disappear when the annealing temperature increases to 923 K. This indicates that a complete SnO_2 nanowire array embedded in the AAM has formed and it is a polycrystalline cassiterite structure.

Figure 5.61 shows TEM micrograph of SnO_2 nanowires after the AAM was removed from the SnO_2/AAM assembly system. This figure shows that the diameter of SnO_2 nanowires is about 70 nm, which is basically equal to that of the pores of the AAM used. The length of the nanowires is in the range of hundreds of nanometers to several microns. Longer nanowires might be obtained by using AAM with deep holes, and thermal oxidization at higher temperatures and longer times. Figure 5.61 also shows that the surface of the nanowires is rough and the diameter of the individual SnO_2 nanowire is not uniform. The TEM diffraction pattern indicates that the SnO_2 nanowires in the AAM pores are a polycrystalline structure. This is in agreement with XRD results. These results may imply that the SnO_2 nanowires growth is achieved by piling up small particles, not atom-by-atom growth of crystallites during thermal oxidization.

Figure 5.62 shows the Raman spectrum of the SnO_2/AAM assembly system annealed at 923 K for 10 h. In this figure, only two broad Raman peaks denoted as A_{1g} and B_{2g} are present, which correspond to the Raman modes of polycrystalline cassiterite SnO_2.[129] This indicates that there is no SnO phase in the assembly system after the assembly system is annealed at 923 K. This is in agreement with XRD results. Since the broadening of

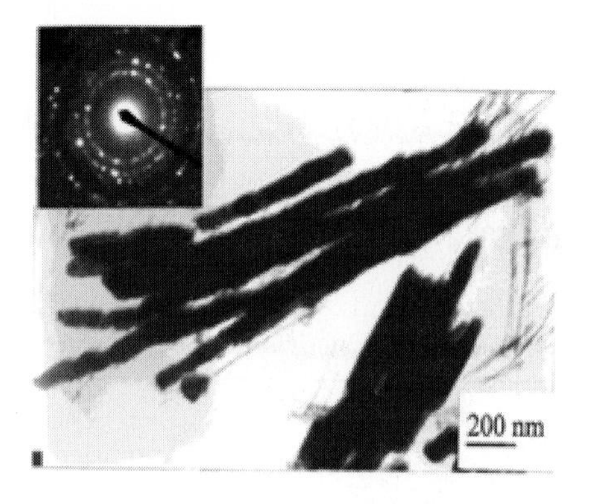

Fig. 5.61. TEM micrograph of SnO_2 nanowires and the corresponding electron diffraction pattern.

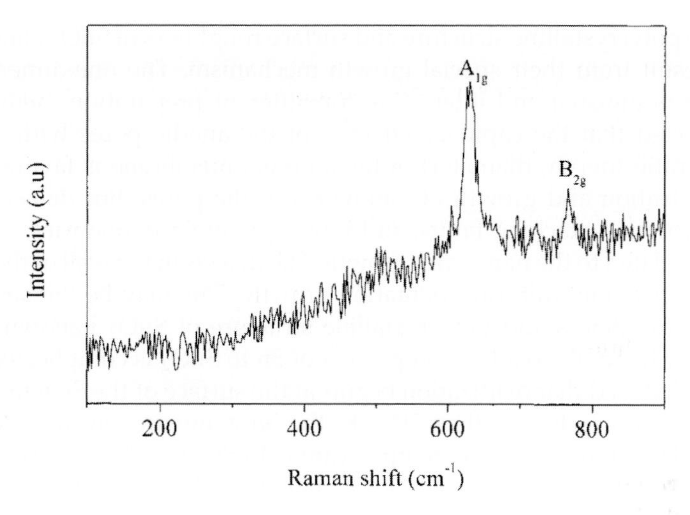

Fig. 5.62. Raman spectrum of the SnO_2/AAM assembly system annealed at 923 K for 10 h.

Raman peaks is the signature of nanocrystals, this Raman spectrum further indicates that SnO_2 nanowires are made up of nanoparticles.

Figure 5.63 shows an SEM top image of a SnO_2 nanowire array grown in a template. The SnO_2 nanowires are basically of equal height and are uniformly embedded in the AAM. The nanowires observed in Fig. 5.62 are oriented in different directions, which could result from the mechanical forces during dissolution and drying.

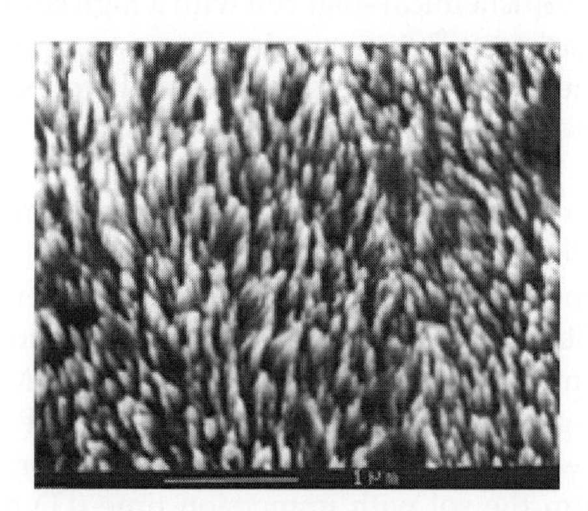

Fig. 5.63. SEM top image of a SnO_2 nanowire array embedded in the AAM.

The polycrystalline structure and surface roughness of SnO_2 nanowires may result from their special growth mechanism. The one-dimensional growth is unusual and interesting for different preparation methods. It is expected that the capillary effect[126] of the anodic pores with a large aspect ratio (depth/diameter) in the alumina membrane is favorable for the nucleation and growth of nanowires in the pores. But defects, such as microtwins, stacking faults, and low-angle grain boundaries, are also closely related to the nanowires growth. The defects in the wall of the pores may induce non-uniform orientation growth. This may be the source of surface roughness and polycrystalline structure of SnO_2 nanowires. On the other hand, the oxidization process of Sn to SnO_2 is complicated.[131] It can be deduced that oxidization begins at the surface of the Sn nanowires. At temperatures higher than 505.6 K, the Sn nanowires in the AAM are liquid. Therefore, when the temperature is high enough to allow oxygen to diffuse into the Sn nanowires, Sn nanowires begin to be oxidized in random orientations (nuclei formation). As oxidization continues, these nuclei grow further at different rates, because oxygen diffusion depends on the crystallographic orientation of the formed solid crystallites. The XRD results have indicated that oxidization does not occur directly from Sn to SnO_2, and the primary reactions taking place in the AAM may be the conversion of Sn to SnO, Sn to SnO_2, and SnO to SnO_2. These reactions may be another source of the formation of the polycrystalinity and surface roughness of SnO_2 nanowires.

(4) Ordered arrays of TiO_2 nanowires[132]

The recent interest in anatase TiO_2 was motivated by its key role in the injection process of a photochemical solar cell with a high conversion efficiency. Single crystalline anatase TiO_2 nanowire arrays were prepared in anodic alumina membrane (AAM) templates by a sol-gel process. The preparation process is as follows.

First, 8.5 ml of tetrabutyl titanate was added to 25 ml of ethanol. Then 4.5 ml of CH_3COOH was mixed with 1.8 ml of water and 12.5 ml of ethanol. The latter solution was added dropwise to the former with continuous stirring. The resultant alkoxide solution was kept standing at room temperature for 30 minutes, resulting in a TiO_2 sol. Five AAMs were immersed in the sol. The sol-gel transition took place gradually at 30°C for 10 h. The resultant wet gel was then dried in an oven at 60°C for 14 h. Four AAMs were removed from the sol with immersion time (IT) of 0.5 (A), 1.5 (B), 6 (C), and 10 h (D), respectively. The last AAM (sample E) was removed from the wet gel with IT of 24 h, 10 h in sol, and 14 h in wet gel. All samples were heat-treated immediately after removal from the sol or wet gel under the same conditions. The heat treatment was performed in a tube furnace in air. The temperature was first ramped to 400°C, then ramped to 650°C and held at this temperature for 2 h. Finally, it was ramped back down to room temperature. The surface TiO_2 films of all samples were removed by polishing the samples with grit sandpaper.

A highly-ordered TiO_2 nanowire (TN) array is shown in Fig. 5.64(a). It can be seen that the TNs are uniform and nearly parallel. The diameters of the TNs are about 6 nm, which is in good agreement with the pore diameters of the AAMs. Figure 5.64(b) shows the TEM image of two individual TNs and the selected area electron diffraction (SAED) pattern from one TN of sample E. The SAED pattern is consistent with the TiO_2 single crystal indexed as anatase TiO_2.[101] The X-ray diffraction pattern of sample E (Fig. 5.64(c)) shows that the TNs are anatase type with highly-oriented crystals in the (101) plane. In addition, a background halo pattern characteristic of an amorphous α-Al_2O_3 phase is also presented in Fig. 5.64(c). Figure 5.65 shows the TEM images of other samples. It can be seen that the TN in sample A (Fig. 5.65(a)) is located at the center of the pore and has a diameter of about 10 nm. When the IT increases (0.5, 1.5, 6, 10 h), the TNs become thicker (10, 20, 40, 60 nm) up to a limit of 60 nm in diameter. It seems that the TNs first form in the center of the pore and then extend to the pore wall. As we know, at the pH value used here the sol particles are positively charged, although weakly.[133,134] Meanwhile, the pore walls of

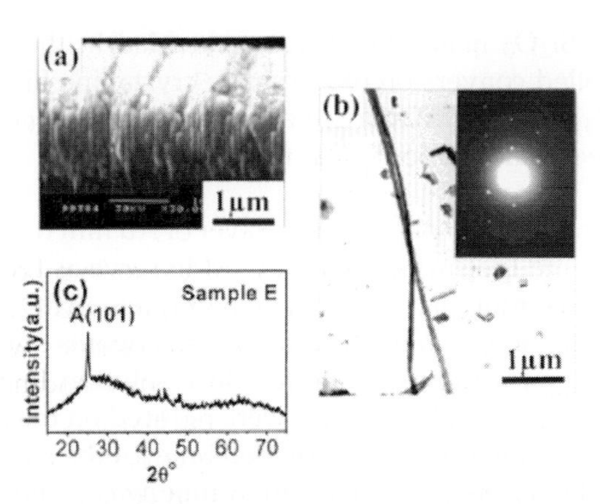

Fig. 5.64. (a) SEM image of TN arrays, (b) TEM image of TNs, and (c) XRD pattern of sample E. IT of sample E is 24 h. The inset in (b) is the SAED pattern[101] of one TN.

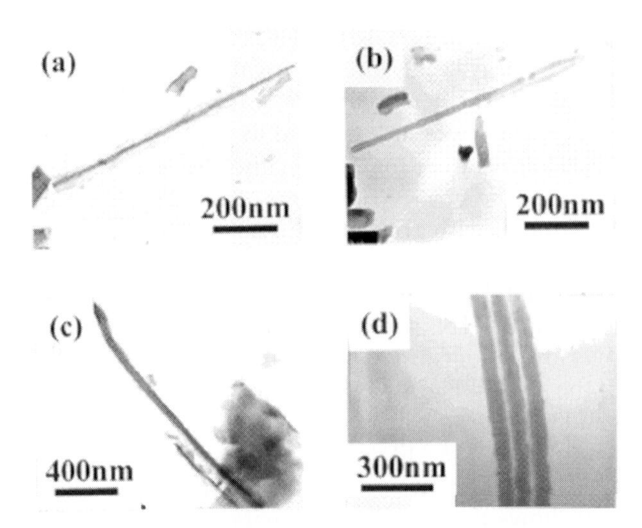

Fig. 5.65. TEM images of TNs in (a) sample A, (b) sample B, (c) sample C and (d) sample D. The corresponding IT is 0.5, 1.5, 6 and 10 h, respectively.

the AAM are also positively charged, so the density of sol particles is larger in the center area of the pores. Thus, it is not surprising that the TNs first form in the center area of the pore and then extend to the wall gradually.

(5) Ordered arrays of Bi_2O_3 nanotubes[135]

Li[135] prepared Bi_2O_3 nanotube arrays embedded in the AAM template through controlled conversion of the single crystalline Bi nanowire array to the Bi-Bi_2O_3 core-shell nanowire array to the single crystalline Bi_2O_3 nanoarray. The detailed process is as follows.

Before electrochemical deposition, a layer of Au film (thickness 200 nm) was sputtered onto the surface of the AAM to serve as the working electrode in a two-electrode plating cell, and a graphite plate was used as the counter electrode. Bi was deposited into the nanopores of the AAM using pulsed electrochemical deposition from a $BiCl_3$ solution. During the pulsed deposition time, T_{on}, metal Bi species were reduced on the pore ground of the AAM. The relaxing time, T_{off}, provided time for the recovery of the ion concentration. Here, the pulse deposition time was 35 ms and the relaxing time was 60 ms. As a result, single crystalline Bi nanowire arrays were obtained.

As the single crystalline Bi nanowire arrays embedded in the AAM were obtained, they were put into a furnace for high-temperature oxidation in air. Firstly, in the slow-oxidation mode, the Bi nanowires were subjected to a sequence of temperature steps (100°C/step) from room temperature to 500°C with a heating rate of 100°C/h, and the sample was held at each temperature step for 4 h. After the slow-oxidation process, the sample was cooled to room temperature with a slow cooling rate of 100°C/h. The slow-oxidation process, as schematically shown in Fig. 5.66(a), resulted in the formation of Bi-Bi_2O_3 nanowire arrays. In the next step, the top surface of the sample was carefully polished using Al_2O_3 nano-powders and rinsed with deionized water several times. Finally, in the fast-oxidization mode (Fig. 5.66(b)), the slowly oxidized sample was quickly heated from room temperature up to 750°C with a quick heating rate of 25°C/min. The sample was held at 750°C for 10 h and then cooled to room temperature with a slow cooling rate of 100°C/h. In the fast-oxidization mode, the Bi-Bi_2O_3 nanowire array was further converted to the single-crystalline Bi_2O_3 nanotube array.

Figure 5.67(a) shows the XRD pattern of the heat-treated Bi nanowire array in the slow-oxidization mode (Fig. 5.66(a)). All the diffraction peaks can be attributed to α-, β- and δ-Bi_2O_3 (JCPDF No. 76-1730, 78-1793 and 77-2008) and the remaining Bi in nanowires. Although it is known that the equilibrium phase of Bi_2O_3 at temperatures below 729°C is α-Bi_2O_3, δ-Bi_2O_3 was formed due to the oxidation of nanosized bismuth. In the experiments, they found that it is difficult to obtain pure phase Bi_2O_3 because the δ-Bi_2O_3 will transform into metastable β- and γ-Bi_2O_3 on cooling below 729°C, and usually, these phases will transform into α-Bi_2O_3 on further

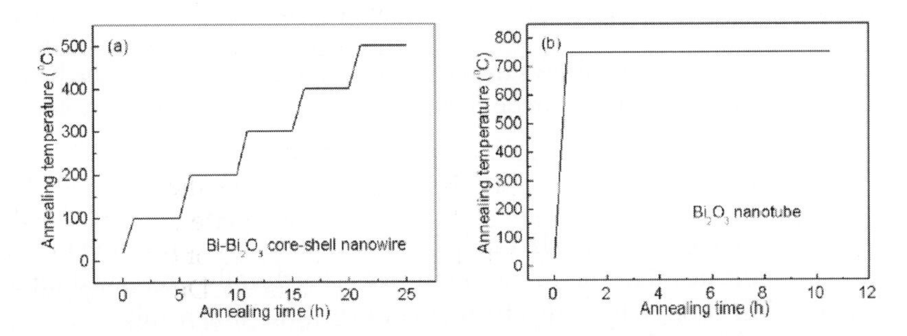

Fig. 5.66. The heating treatment process for (a) converting the Bi nanowire array to the Bi-Bi_2O_3 core-shell nanowire array in a slow-oxidization mode, and (b) further converting the core-shell nanowire array to the Bi_2O_3 nanotube array in a fast-oxidization mode.

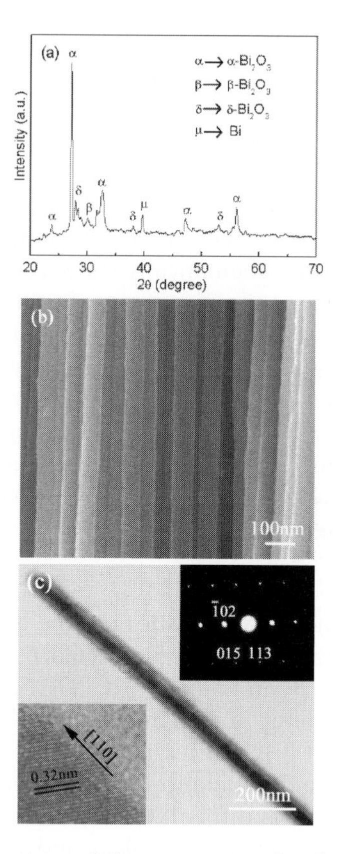

Fig. 5.67. (a) The XRD pattern of the nanoarray after heat-treating in the slow-oxidization mode, (b) the SEM image of the cross-section of the array, and (c) TEM and HRTEM images of nanowires.

cooling.[136] Figure 5.67(b) shows the cross-sectional SEM image of the heat-treated Bi nanowire array. It can be clearly seen that the uniform diameter of nanowires was perfectly preserved after heating treatment. From the TEM and HRTEM images shown in Fig. 5.67(c), a very uniform core-shell structure was formed after oxidization and an obvious interface was observed. The [110] crystal direction of HRTEM with SAED pattern of the center of the nanowire indicates that the Bi nanowire was structurally preserved after thermal oxidization, which also further confirms the XRD result.

Figure 5.68(a) shows the XRD pattern of the sample after thermal oxidization in the fast mode (Fig. 5.66(b)). The diffraction peaks can be ascribed to the δ-Bi_2O_3 and the AAM system phase. Figure 5.68(b) shows the SEM

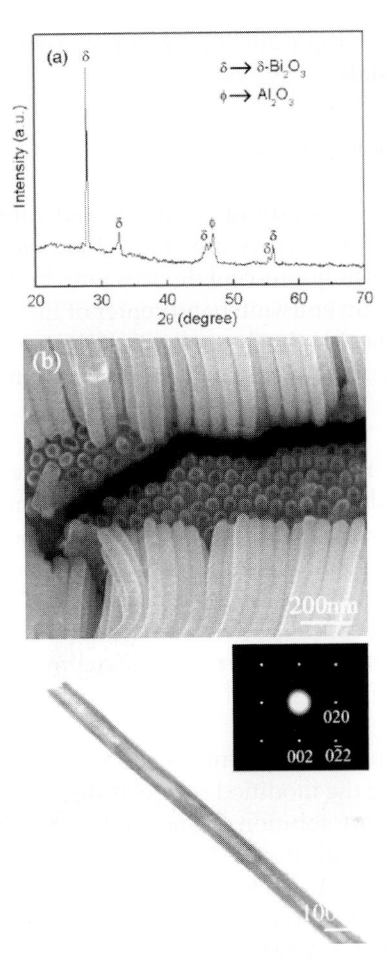

Fig. 5.68. (a) The XRD pattern of the nanoarray after heat-treating in the slow-oxidization mode and then in the fast-oxidation mode, (b) the SEM image of the top surface of the nanoarray, and (c) the TEM image of the single-crystalline Bi_2O_3 nanotube array.

image of the nanotube array after the AAM was partly etched in NaOH solution, which indicates that the large-area and ordered nanotube array was formed in the AAM after the two-step oxidization. From the top image of the nanotubes, it can be seen that the diameter of the nanotubes is about 75 nm, which is larger than the diameter (60 nm) of Bi nanowires. The TEM image (Fig. 5.68(c)) of a randomly selected single Bi_2O_3 nanotube also indicates that the diameter of nanotubes is uniform and about 75 nm. The SAED pattern inserted in the top-right corner of Fig. 5.68(c) indicates the nanotubes are single-crystals.

(6) Ordered arrays of Eu_2O_3 nanotubes[137]

Now, the reports on the study of synthesis and properties of rare earth compound nanostructure arrays are small. Rare earth compounds have potential applications in fluorescent devices with high properties. In many rare earth ions, the main emission band center of Eu^{3+} is located at 612 nm. Therefore, it can be used as the fluorescent catalyst of many materials. This attracts the attention of many researchers. Eu_2O_3 is an important phosphorescent material.

The nanotube arrays of inorganic compounds are usually synthesized by using electroless deposition, sol-gel etc. in the templates. Among these methods, the traditional sol-gel method is when the template is directly immersed in the corresponding solution to synthesize the arrays. The drive force of this synthesis process is the capillary role. This synthesis method requires the lower sol concentration due to the limit of the small channels of the templates. Therefore, the arrays of some materials are synthesized with difficulty by using the sol-gel route in the small channel templates.

Recently, Wu[137] successfully synthesized Eu_2O_3 nanotube arrays in the AAM templates using the modified sol-gel route. Firstly, the template was immersed in the mixture solution of urea and Eu nitrate for a certain time; and then the solution with the AAM template was kept at 80°C for 72 h. At this time, the Eu $[(OH)_x]$ (H_2O) sol was formed in both the solution and the template. Because the sol particles carried negative charge and the channel walls of the AAM template carried a little positive charge, the great numbers of sol particles adhered to the channel walls of the AAM template. After that, the AAM template was heat-treated in the tube furnace. During this period, gelation and crystallization took place to form nanotubes in the channels of the template. With increasing assembling time, the wall thickness of the nanotubes became large. The external diameter of the nanotubes can be controlled by changing the channel diameter of the template. Figure 5.69(a) shows the TEM photograph of Eu_2O_3 nanotubes after removing the AAM template. Figure 5.69(b) shows the TEM photograph of a single nanotube and the corresponding SAED pattern. From these figures, it can be seen that the external diameter of Eu_2O_3 nanotubes is about 70 nm and these tubes have a polycrystalline structure.

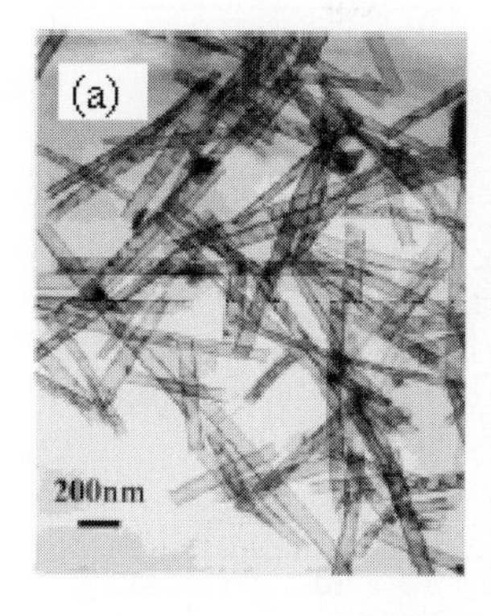

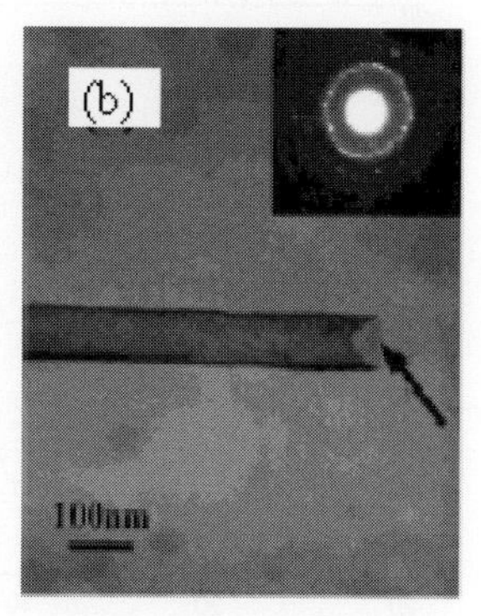

Fig. 5.69. (a) The TEM photograph of Eu_2O_3 nanotubes, and (b) the TEM photograph of a single nanotube and the corresponding SAED pattern.

5.2.2.3. *Ordered nanoarrays of sulphide, selenide, telluride and ionide nanowires*

By combining the AAM template technique with the electrochemical deposition technique, people may synthesize ordered nanoarrays such as CdS,[138,139] Ag_2S,[140] Bi_2S_3,[141] CdSe,[142–144] PbSe,[145] CaTe,[146,147] Bi_2Te_3,[148,149] and other binary compounds[150] nanowires. Here, we will introduce synthesis and characterization of ordered nanoarrays of Bi_2S_3, Ag_2S, CdSe, CdTe, GaN and AgI nanowires.

(1) Ordered nanoarrays of Bi_2S_3[141] and Ag_2S[140] nanowires

Firstly, we introduce the synthesis and characterization of ordered arrays of Bi_2S_3 nanowires. The electrochemical deposition of ordered arrays of Bi_2S_3 nanowires in the AAM template was carried out in a three-electrode system. The electrolyte contained $BiCl_3$, elementals and non-aqueous DMSO with Bi^{3+} may form complex ions. After electro-deposition, the sample was annealed in a highly pure Ar protection furnace, resulting in the obtainment of a better crystallization Bi_2S_3 nanowire array.

From the XRD spectrum (Fig. 5.70), it can be observed that there are no peaks of S and Bi, and the Bi_2S_3 nanowires have a monoclinic structure with the lattice parameters $a = 1.149$ nm, $b = 1.1304$ nm and $c = 0.3981$ nm.

Figure 5.71(a) shows the SEM image of the ordered array of Bi_2S_3 nanowires after removing parts of the AAM template. The TEM photograph

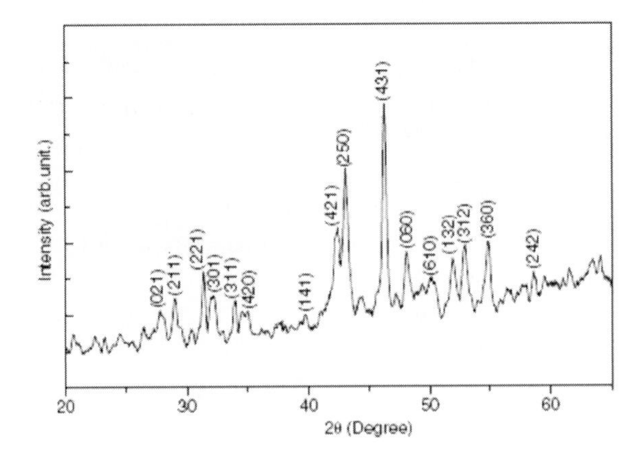

Fig. 5.70. The XRD pattern of the ordered array of Bi_2S_3 nanowires.

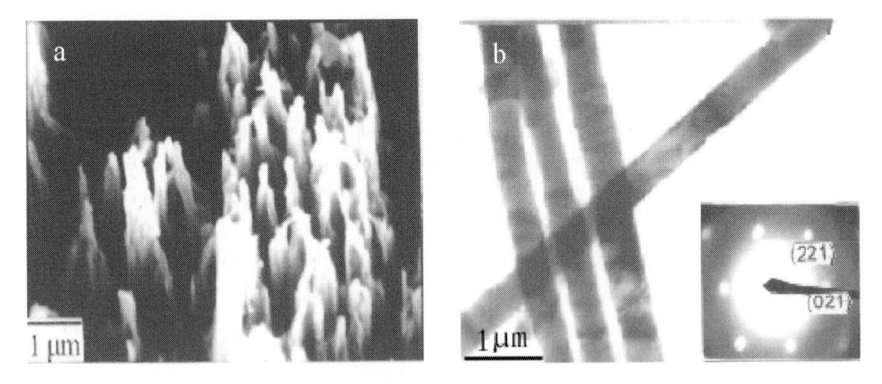

Fig. 5.71. (a) The SEM image of ordered array of Bi_2S_3 nanowires, and (b) the TEM image of ordered array of Bi_2S_3 nanowires and the corresponding SAEM pattern.

of Bi_2S_3 nanowires reveals that the diameters of these nanowires are basically consistent and about 40 nm. The SAED pattern shows that Bi_2S_3 nanowires have better crystallization and a monoclinic structure. This result is consistent with that of XRD analysis.

Electrochemical fabrication of ordered Ag_2S nanowire arrays will be introduced as follows.[140] Ordered three-dimensional arrays of Ag_2S nanocrystallite have been fabricated by using the reverse micelles method.[152] Here, Peng[140] demonstrated the generation of ordered Ag_2S nanowire arrays by DC electro-deposition into the nanochannels of the AAM template from a non-aqueous DMSO solution containing $AgNO_3$ and elementals. The detailed preparation procedure is as follows.

Electro-deposition of Ag_2S nanowire arrays was carried out in a solution containing 0.055 mol l^{-1} $AgNO_3$ and 0.19 mol l^{-1} elemental sulfur dissolved in DMSO similar to the data[151] at 120 ± 0.5°C and -0.8 V relative to the Ag/Ag^+ reference electrode for 60 minutes. The AAM with gold painting film served as the working electrode and a graphite electrode served as the counter electrode. After being removed from the electrolyte, the AAM with Ag_2S nanowires was immediately immersed in hot DMSO of about 80°C for 20 minutes, and rinsed with acetone, followed by hot distilled water. Finally, it was annealed in argon atmosphere at 350°C for 3 h to improve the crystallinity, and subjected to further analysis.

The X-ray diffraction (Fig. 5.72) of the annealed Ag_2S nanowire arrays embedded in the AAM shows the diffraction peaks $(-1\,0\,1)$, $(-1\,1\,1)$, $(1\,1\,1)$, $(-1\,1\,2)$, $(1\,2\,0)$, $(-1\,2\,1)$, $(1\,2\,1)$, $(-1\,0\,3)$, $(0\,3\,1)$, $(2\,0\,0)$, $(1\,0\,3)$, $(1\,3\,1)$, $(-1\,2\,3)$, $(-2\,1\,2)$, $(0\ 1\ 4)$, $(-2\,1\,3)$, $(-2\,2\,3)$ and $(-2\,0\,4)$ assigned to the monoclinic Ag_2S (β-Ag_2S) structure with constants of $a = 0.423$ nm, $b = 0.693$ nm, $c = 0.786$ nm and $\beta = 99.61°$. No unreacted Ag or S was detected. Furthermore, no obvious peaks for the amorphous AAM were observed.

A typical top-view and cross-sectional view of the Ag_2S nanowire arrays are shown in Figs. 5.73(a) and 5.73(b), respectively. From Fig. 5.73(a), it can be obviously seen that large scale Ag_2S nanowires have been formed. Figure 5.73(b) shows that these nanowires are ordered and uniform with

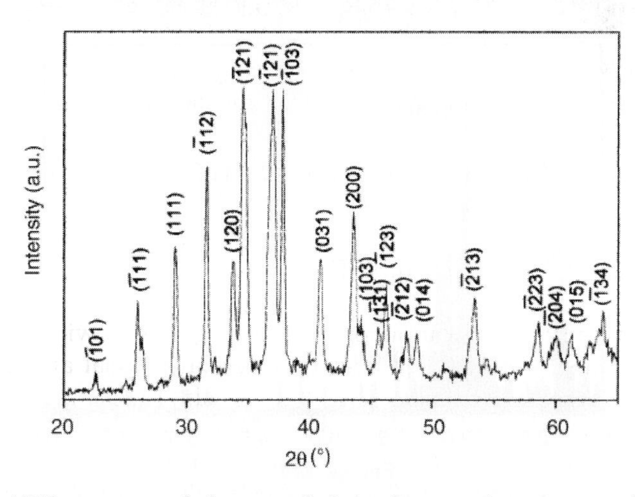

Fig. 5.72. XRD patterns of the annealed Ag_2S nanowire arrays embedded in the AAM.

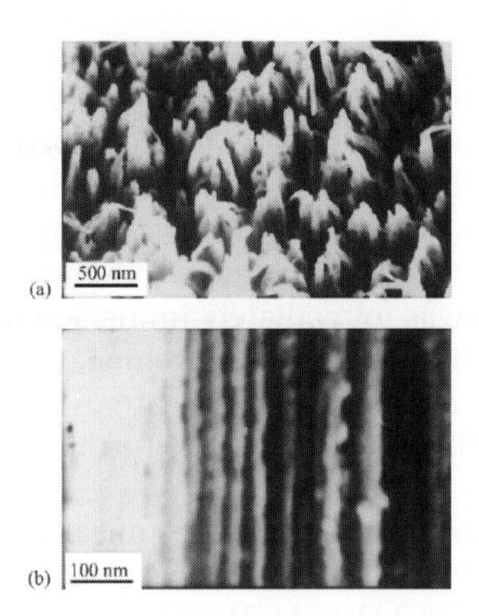

Fig. 5.73. SEM image of the Ag_2S ordered nanowires with diameter of about 40 nm and lengths of about 5 μm: (a) top-view of the nanowire arrays; (b) cross-sectional view of the nanowire arrays.

diameters of about 40 nm and lengths of about 5 μm, corresponding closely to the pore sizes of the AAM used. Furthermore, it can be seen that the nanowires can be filled uniformly and continuously into the pores of the AAM by DC electro-deposition. The chemical composition of the Ag_2S nanowires was determined by EDAX. The EDAX spectrum (Fig. 5.74) of the Ag_2S nanowires shows the peaks of Ag (La) and S (Ka). Quantitative

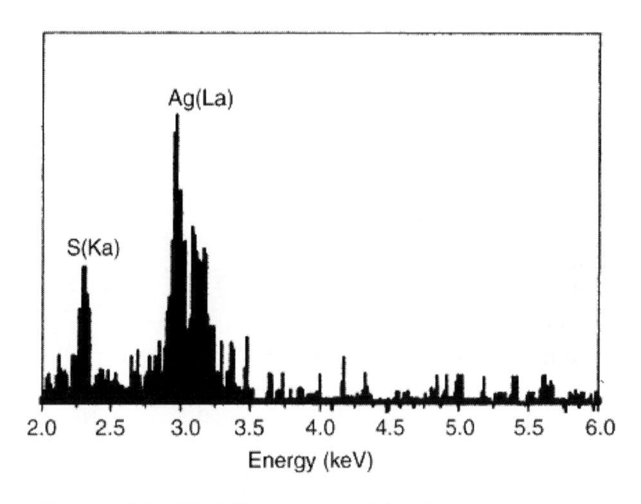

Fig. 5.74. The EDAX spectrum of Ag_2S nanowire arrays.

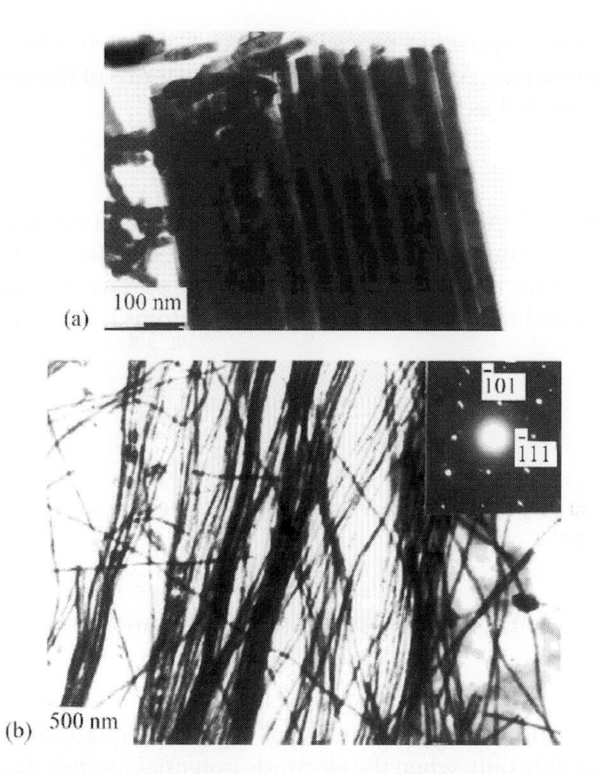

Fig. 5.75. TEM images of the Ag_2S nanowires with diameter of about 40 nm: (a) embedded in the nanochannels of the AAM; (b) several Ag_2S nanowires detached from the AAM. The inset is the SAED pattern taken from one of these Ag_2S nanowires.

analysis reveals that the atomic ratio of Ag to S is very close to a 2 : 1 stoichiometry.

Figure 5.75(a) shows the TEM image of a row of Ag_2S nanowires with diameters of about 40 nm embedded in the AAM. It can be seen that the nanowires in the nanochannels of AAM are highly-ordered and arranged tightly, corresponding to the high porosity of the AAM used, and the pores of the AAM are fully filled with Ag_2S nanowires. In Fig. 5.75(b), a typical TEM image of Ag_2S nanowires detached from the AAM shows that these nanowires are uniform with diameters of about 40 nm. The SAED pattern (inset in Fig. 5.75(b)) taken from one of these nanowires indicates that the Ag_2S nanowires are crystalline and have good agreement with the XRD results.

The electro-deposition procedure of Ag_2S nanowires from the DMSO solution containing $AgNO_3$ and elemental S in terms of the general scheme may be presented as follows:

$$2Ag^+ + S + 2e = Ag_2S. \tag{5.3}$$

In principle, the Ag_2S synthesis may occur via two different routes. The first one mainly involves two steps. Firstly, the elemental sulfur in the solution diffuses and absorbs on the electrode surface. Secondly, the electrochemical reduction of the absorbed sulfur atoms is followed by reacting with Ag^+:

$$S_{abs} + 2e = S^{2-}, \tag{5.4}$$

$$2Ag^+ + S^{2-} = Ag_2S. \tag{5.5}$$

The second pathway can be expressed in the following manner:

$$Ag^+ + e = Ag, \tag{5.6}$$

followed by a chemical reaction between sulfur and silver:

$$S + 2Ag = Ag_2S. \tag{5.7}$$

The formation of Ag_2S according to the above equations (Eqs. (5.4) and (5.5)) is possible only when the electrode potential reaches the potential for S^{2-} formation as well as for Eqs. (5.6) and (5.7) which reaches the reduction potential of Ag^+ to Ag. At 120°C (Ag^+ 0.055 M, and S 0.19 M, sweep rate 50 mV s^{-1}, cyclic voltammogram). Figure 5.76 shows that there is only one reduction peak at −0.2 V versus the Ag/Ag^+ reference electrode, which corresponds to the potential for silver not sulfur reduction because it has been investigated that S was not reduced to any significant extent up to −1.2 V versus Ag/Ag^+ in DMSO.[15,16] In present work, the deposition potential is −0.8 V versus Ag/Ag^+. Under such conditions, it can be seen that the reduction of Ag^+ on the surface of the electrode is more favorable than the reduction of S to S^{2-}. Therefore, it can be concluded that, by the second pathway, the Ag_2S nanowires were prepared more efficiently.

(2) Ordered nanoarrays of CdSe nanowires[139]

The synthesis of ordered nanoarrays of CdSe nanowires is similar to that of ordered nanoarrays of Bi_2S_3 nanowires. The electrolyte containing $CdCl_2$ and SeO_2, and a small amount of $NH_3{\cdot}H_2O$ was added into the electrolyte solution to modulate the pH value of the electrolyte solution. After the

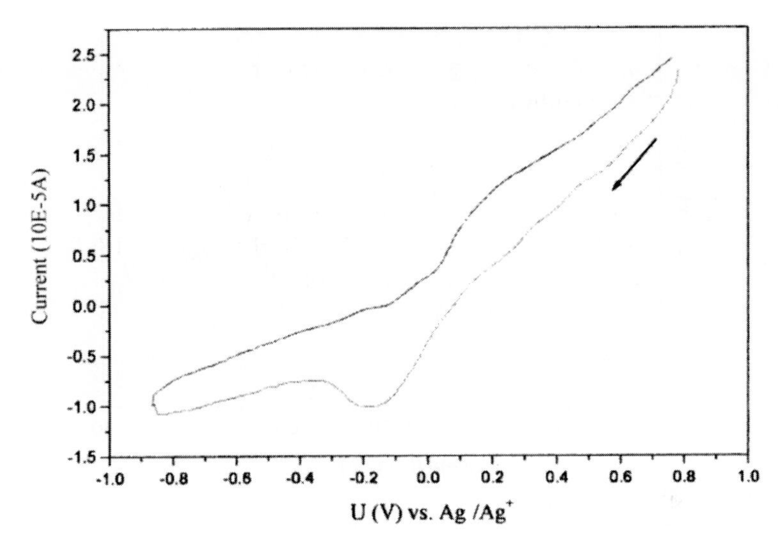

Fig. 5.76. The cyclic voltammogram for the reduction procedure of 0.055 M $AgNO_3$ and 0.19 M S in DMSO at the AAM/Au electrode. Sweep rate: 50 mVs^{-1}; area of the electrode: 0.5 cm^2.

electro-deposition was finished, the sample was annealed in a highly pure Ar protection furnace to cause the sample to be crystallized.

Figure 5.77 shows the XRD spectrum of ordered nanoarrays of CdSe nanowires. From analysis of this XRD spectrum, it can be obtained that CdSe nanowires possess a hexagonal wurtzite structure. Figures 5.78(a) and 5.78(b) demonstrate the SEM morphologies of ordered nanoarrays of CdSe nanowires, displaying the surface morphology and sectional morphology of ordered arrays, respectively.

(3) Ordered nanoarrays of CdTe nanowires[147]

In this paragraph, we will introduce an efficient route to fabricate CdTe nanowire arrays by potentiostatic electrochemical deposition in the nanochannels of the AAM template in aqueous solution containing $CdSO_4$ and $HTeO_2^+$.

Electrochemical deposition of CdTe nanowire arrays was carried out in a typical three-electrode electrochemical cell at room temperature. The counter-electrode was a pure graphite plate, a saturated calomel electrode (SCE) served as the reference electrode, and the PPA template with an Au

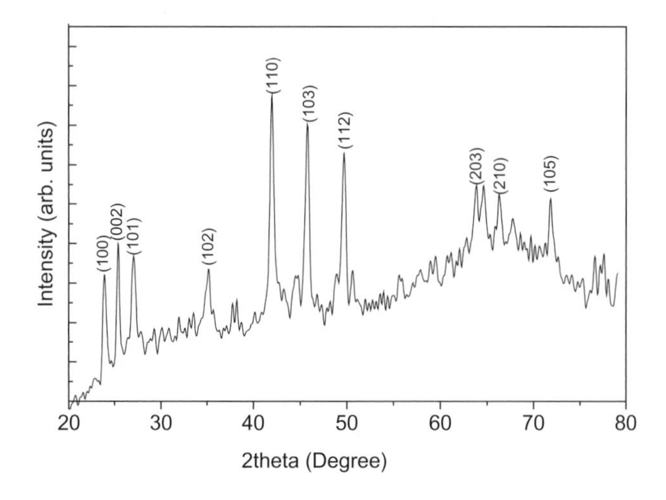

Fig. 5.77. The XRD spectrum of ordered arrays of CdSe nanowires.

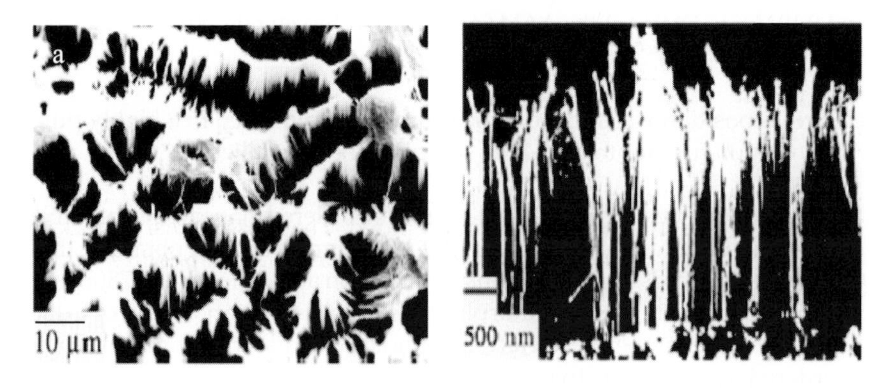

Fig. 5.78. The SEM photographs of ordered nanoarrays of CdSe nanowires. (a) the surface morphology; (b) the sectional morphology.

layer was used as the working electrode. The electrolyte solution contained a mixture of 1 M $CdSO_4$ and 3×10^{-4} M $HTeO_2^+$ that was added in the form of TeO_2. The deposition potential value was −580 mV versus SCE. The pH of the solution was adjusted to 2 by using 1 M H_2SO_4. After electrochemical deposition for 8 h, the obtained sample was rinsed several times with deionized water and dried in air at room temperature. Then, the sample was loaded into a quartz tube that was inserted into a furnace for annealing at 300°C for 1 h. Prior to annealing, the quartz tube was flushed with argon

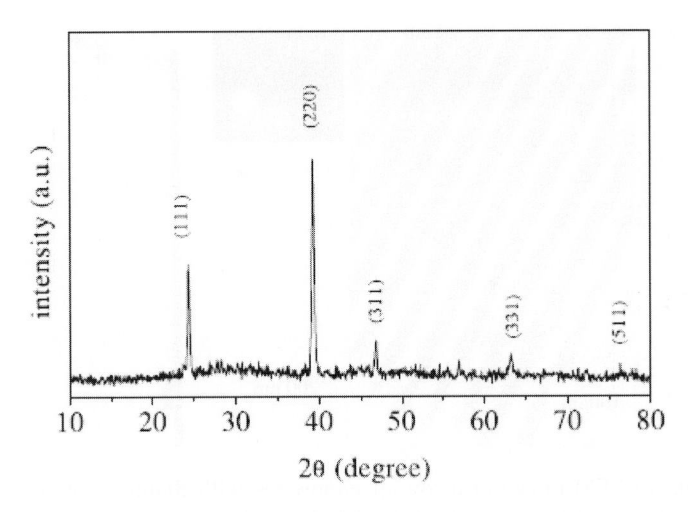

Fig. 5.79. XRD pattern of the CdTe nanowire arrays.

for 3 minutes, and then the argon flow was maintained at about 100 sccm at a pressure of 0.1 mbar.

The X-ray diffraction spectrum of the CdTe nanowire arrays is shown in Fig. 5.79. The intensive diffraction peaks well match the cubic phase structure of CdTe (JCPDS 15-77) with the lattice constant of $a = 6.48$ Å, indicating that cubic CdTe has been prepared. The most intensive peak in the XRD pattern indexed as (220) suggests that the CdTe nanowires would have a predominant growth direction, which will be testified to by HRTEM investigation in the following.

A typical TEM image of the fabricated CdTe nanowire arrays (Fig. 5.80(a)) shows that the nanowires embedded in the nanochannels of the AAM template are uniform, continuous, and highly-ordered. The dark areas are nanowires with a diameter of about 60 nm, corresponding to the channels in the AAM template. Figure 5.80(b) shows a single CdTe nanowire separated from the AAM template. It can be seen that the CdTe nanowire is straight and uniform with a smooth surface and has a diameter of about 60 nm. The selected-area electron diffraction (SAED) pattern taken from the single nanowire (inset of Fig. 5.80(b)) indicates that the CdTe nanowire is single crystalline with cubic phase structure.

The CdTe nanowire structure was further characterized by HRTEM. The lattice-resolved image (Fig. 5.81) reveals the (220) and (1 − 11) atomic

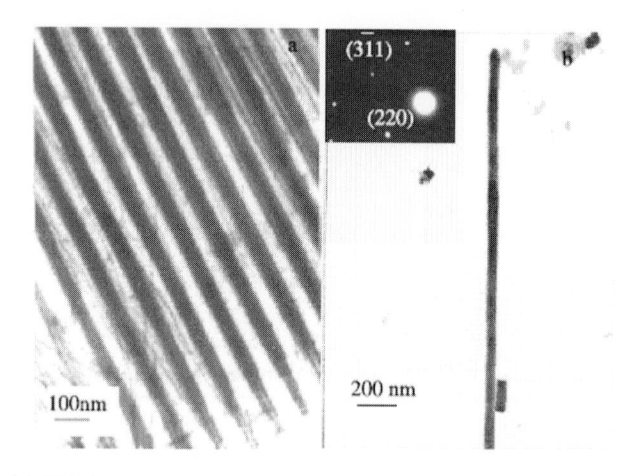

Fig. 5.80. (a) TEM image of a row of nanowires with diameter of about 60 nm embedded in the nanochannels of the AAM template. (b) TEM image of a single nanowire liberated from the AAM template with a diameter of about 60 nm and a length of about 15 μm. The inset at the top-right is an electron-diffraction pattern taken from this nanowire.

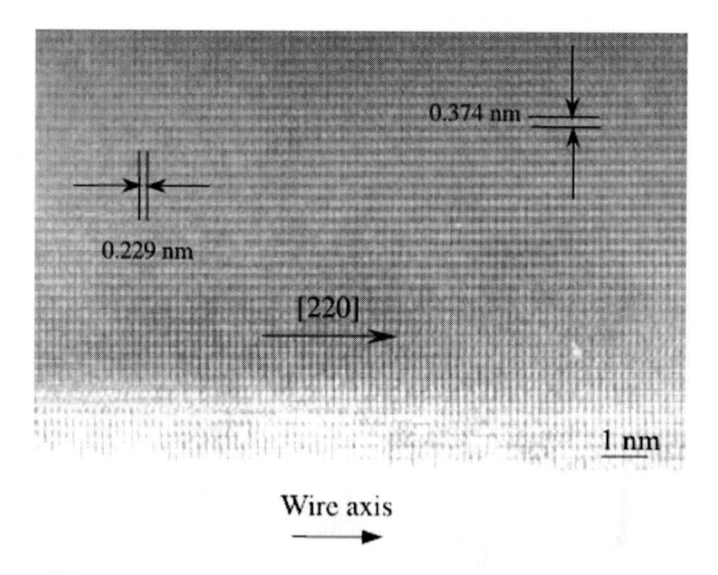

Fig. 5.81. HRTEM image of a single CdTe nanowire with a 60 nm diameter. The (220) and (111) lattice planes are clearly resolved.

planes with lattice spacings around 0.229 nm and 0.374 nm, respectively. The (220) planes are vertical to the long axis of the CdTe nanowire, revealing that the CdTe nanowire has a preferential [220] growth direction along the nanowire axis. It can also be confirmed by the XRD results (Fig. 5.79).

The chemical composition of the CdTe nanowires was determined using X-ray energy-dispersion analysis (EDAX) and XPS. The EDAX spectrum (Fig. 5.82) indicates that the sample is composed of Cd and Te elements with small amounts of Al and O. The Al and O peaks in the pattern come from the remaining AAM template. The atomic ratio of Cd and Te is about 1 : 1. The compositions of the CdTe nanowires embedded in the template were further confirmed by using XPS. The binding energy peak position of Te $3d_{5/2}$ is observed at 572.8 eV and the kinetic energy peak position of Te $3d_{5/2}$ is observed at 490.8 eV. These are in agreement with the data reported previously[21] and indicate that stoichiometric CdTe was formed.

The effect of the $HTeO_2^+$ concentration on the atomic ratio of Te to Cd in the as-deposited products was investigated in our experiment. According to the EDAX analysis results of the CdTe nanowire arrays, we obtained the ratio of Te to Cd in various concentrations of $HTeO_2^+$ (shown in Fig. 5.83) for otherwise constant electro-deposition conditions (1 M $CdSO_4$,

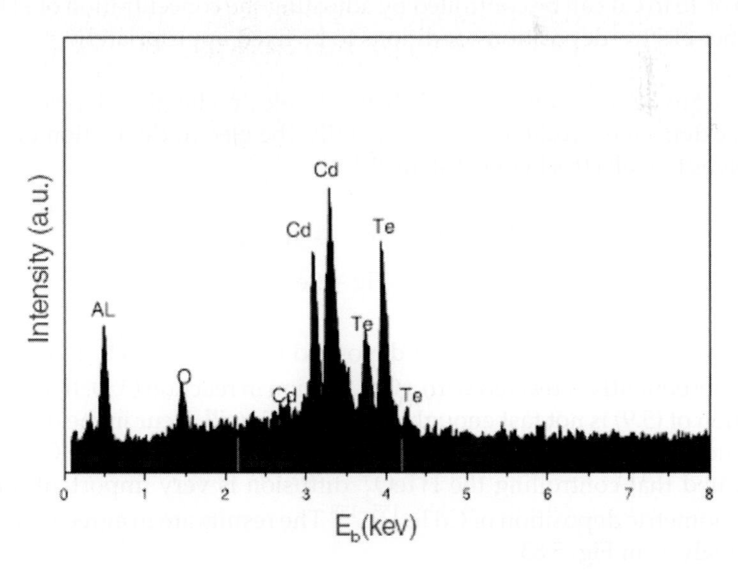

Fig. 5.82. EDAX spectrum of a CdTe/AAM sample.

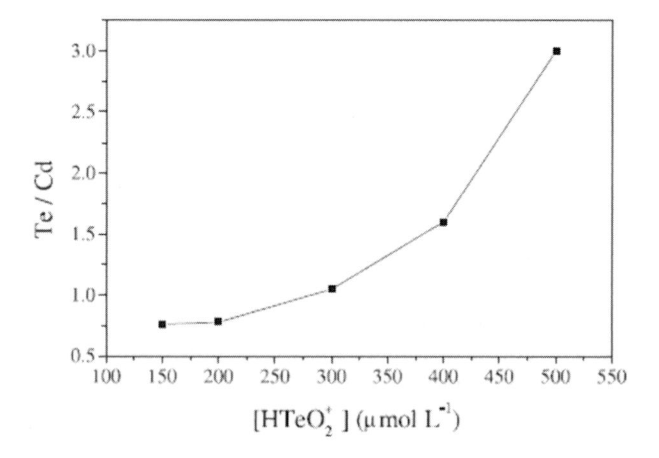

Fig. 5.83. Influence of $HTeO_2^+$ concentration on the ratio of Te to Cd.

the deposition potential −580 versus SCE, room temperature, pH 2). It indicates that the ratio of Te to Cd is influenced greatly by the concentration of $HTeO_2^+$ ions, and the necessity of using low $HTeO_2^+$ concentration in the electrolyte is evident. The optimum concentration of $HTeO_2^+$ to form stoichiometric CdTe nanowires is about 3×10^{-4} M. It also implies that the ratio of Te to Cd can be controlled by adjusting the concentration of $HTeO_2^+$ at other electro-deposition conditions to be fixed appropriately.

The growth mechanism of a CdTe film by electrochemical deposition has been extensively studied.[154–158] Generally, the electro-deposition of CdTe involves two electrochemical steps:[16,17]

$$HTeO_2^+ + 3H^+ + 4e^- \rightarrow Te + 2H_2O, \quad (5.8)$$

$$Cd^{2+} + Te + 2e^- \rightarrow CdTe. \quad (5.9)$$

In the reaction (5.8), Te is first deposited from the reduction of $HTeO_2^+$ and subsequently is reacted with a Cd^{2+} cation in reaction (5.9). If the rate of reaction of (5.9) is not fast enough, the excess Te will occur in the deposited product. Previous studies of the electrochemical deposition of a CdTe film indicated that controlling the $HTeO_2^+$ diffusion is very important for the stoichiometric deposition of CdTe.[155–158] The results are in agreement with our analysis in Fig. 5.83.

(4) Ordered nanoarrays of GaN nanowires[159]

GaN-based materials have been the subject of intensive research for blue and ultraviolet light emission[160,161] and high temperature/high power electronic devices.[162,163] Much effort has been made in synthesizing crystalline GaN membranes,[164,165] and nanoparticles.[166] However, to our knowledge, few reports about one-dimensional structures of GaN were found. Recently, Fan and co-workers[167] reported an interesting preparation method by which crystalline GaN nanorods were fabricated through a gas reaction in the presence of carbon nanotubes that functioned as a nanometer confined reactor. Here, Chen demonstrated an efficient approach for synthesis of single crystalline GaN nanowires on a large scale in the AAM template instead of in the carbon nanotubes.

The gas reaction of Ga_2O vapor[167,168] with a constant flow ammonia atmosphere was carried out in a horizontal quartz tube at a temperature as high as 1273 K. The AAM template was first put on top of mixtures of Ga and Ga_2O_3 with a mole ratio 4 : 1 inside an alumina crucible. The alumina crucible was placed in the hot zone inside the quartz tube and held in a constant flow ammonia atmosphere (300 standard cubic centimeters per minute) at 1273 K for 2 h.

After the reaction and cooling to room temperature, the membrane was characterized by a rotating anode X-ray diffractometer (D/Max-rA) with CuK_a radiation ($\lambda = 1.542$ Å). In the X-ray diffraction spectrum (XRD) (Fig. 5.84), the reflection peaks of (100), (002), (101), (102), (110), (103), (112), (201) correspond to the hexagonal wurtzite GaN phase with the lattice parameters $a_0 = 3.186$ Å, $c_0 = 5.178$ Å; $A(311)$, $A(222)$, $A(400)$, $A(440)$ correspond to the cubic χ-Al_2O_3 phase with the lattice parameter $a_0 = 7.95$ Å. The XRD results show that the hexagonal wurtzite GaN is in nanochannels of the anodic alumina membrane.

An Ar^+ laser with 514.5 nm incident wavelength and 200 milliwatts output power was chosen as the incident light source of a laser Raman scattering spectometer (Spex1403). In the Raman backscattering spectrum at room temperature (Fig. 5.85), the first peak at 420 cm^{-1} corresponds to the phonon vibration frequency of the $A_1(g)$ mode of Al_2O_3, and the peaks at 531, 560, 569, 733, and 745 cm^{-1} agree with the phonon vibration frequencies of A_1(TO), E_1(TO), E_2, A_1(LO), and E_1(LO) modes of the crystalline wurtzite GaN. Within experimental errors, our results agreed well with those of Azuhata.[169] They observed that the corresponding phonon

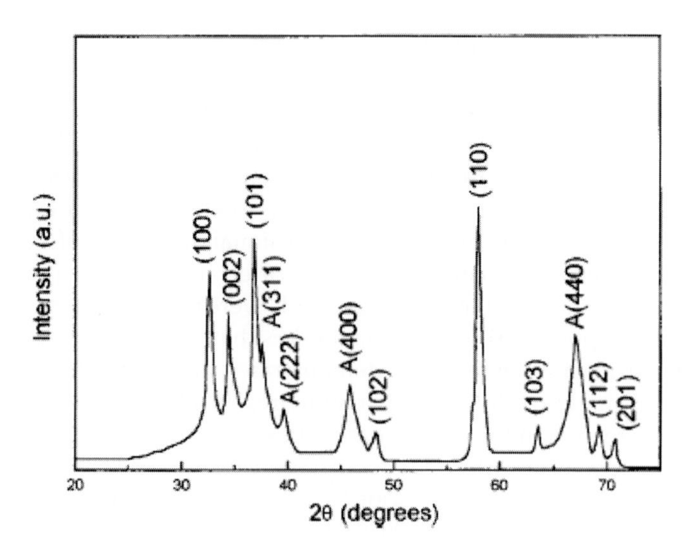

Fig. 5.84. X-ray diffraction spectrum of anodic membrane.

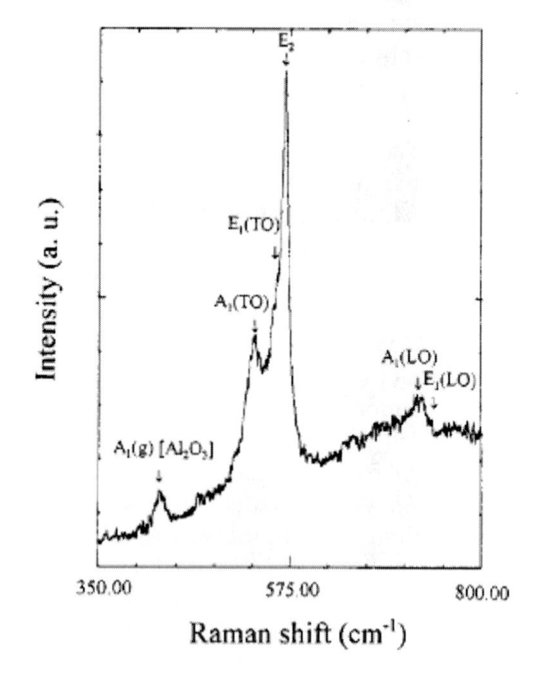

Fig. 5.85. Raman backscattering spectrum of anodic membrane.

frequencies in the wurtzite GaN epitaxial layer grown on sapphire are 418, 533, 561, 569, 735, and 743 cm^{-1}, respectively. Due to the small-size effect of the crystalline GaN formed in the nanochannels of the AAM template, the widening and intensity of each peak is greater compared with that in the GaN epitaxial layer.[169] The results of the Raman backscattering spectrum at room temperature are consistent with that of XRD.

The scanning electron microscopy (SEM) (JOEL JSM-6300) image of the membrane surface (Fig. 5.86(a)) indicates that many GaN nanowires with length up to hundreds of micrometers cross each other and are randomly distributed on the surface of the alumina membrane. The high magnification SEM image of the membrane's cross-section (Fig. 5.86(b)) shows that the GaN nanowires grow out of the AAM template and are twisted on the surface. Before the SEM observation, metallic Pt was evaporated on the surface of the specimens with the evaporation current of 7 mA for 45 minutes

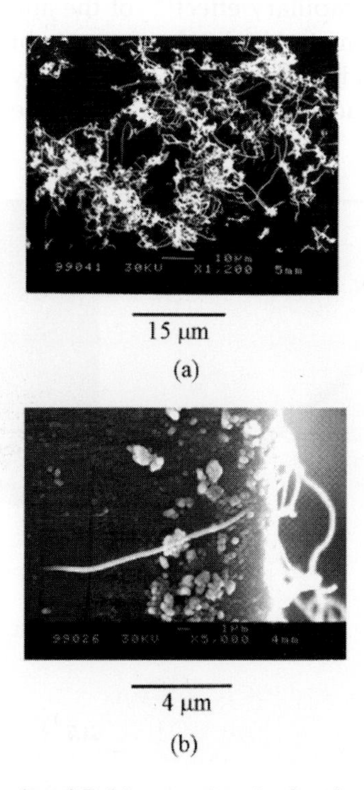

Fig. 5.86. SEM photographs of GaN nanowires in the alumina membrane. (a) Low magnification SEM image of the surface of anodic alumina membrane. (b) High magnification SEM image of the cross-section of the anodic membrane.

to form a conducting film to avoid electrostatic charging. The Pt resulted in an increase in diameter of the nanowires in the SEM investigation. From Figs. 5.86(a) and 5.86(b), it is clear that large scale GaN nanowires grow from the nanochannels of the anodic membrane pores and wind-like tenuous cotton fibers on the surface of the anodic membrane.

For transmission electron microscopy (TEM) observation, an alumina membrane was pulverized in ethyl alcohol, and then dispersed onto a carbon film on a copper grid. The high magnification TEM image (Fig. 5.87) shows that the average diameter of the GaN nanowires is approximately 14 nm. Many TEM images demonstrate that the diameter distribution of the GaN nanowires sharply peaked at 14 nm. The selected-area electron diffraction pattern (SAED) is consistent with the GaN single crystal indexed as wurtzite GaN[331] (the inset in Fig. 5.87).

The one-dimensional growth is preferable, unusual, and interesting. It is expected that the capillary effect[167] of the anodic pores with a large aspect ratio (depth/diameter) in the alumina membrane is favorable for the nucleation and growth in the pores for GaN nanowire formation. Defects, such as microtwins, stacking faults, and low-angle grain boundaries, are

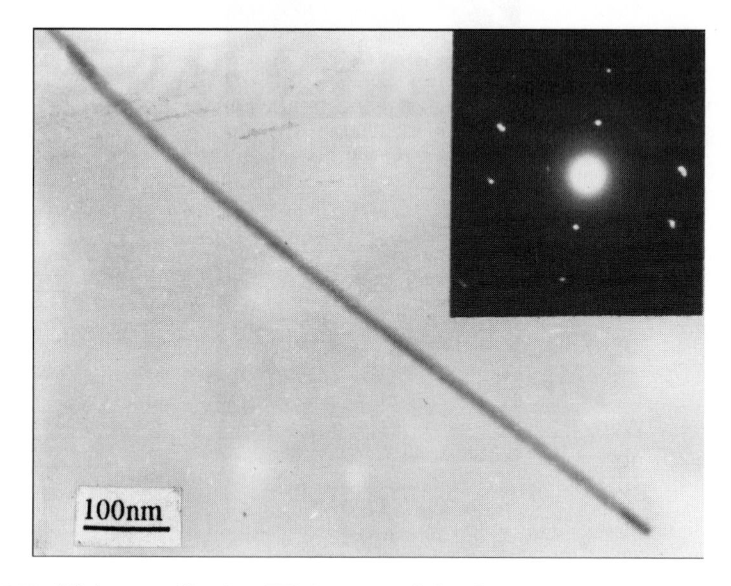

Fig. 5.87. High magnification TEM image of the GaN nanowire in the alumina membrane. The inset is the selected-area electron diffraction pattern[331] of an individual GaN nanowire.

closely related to the preferable GaN nanowire growth.[170] It is believed that the multiplication of screw dislocations,[171] which are generally found in the crystalline materials with the hexagonal close-packed structure, may play an important role in the unidirectional growth and formation of the single crystalline GaN nanowire. In the process of the nanowire growth, the gas reaction can be expressed as

$$4Ga(l) + Ga_2O_3(s) \rightarrow 3Ga_2O(g), \tag{5.10}$$

$$Ga_2O(g) + 2NH_3(g) \rightarrow 2GaN(s) + H_2O(g) + 2H_2(g). \tag{5.11}$$

The pressure of Ga_2O vapor generated from the reaction (5.11) at 1000°C is about 7 Torr.[172] Therefore, a large number of defects in the nanochannel walls of the anodic alumina membranes seem to serve as the centers of nucleation in the saturated reaction atmosphere, and the multiplication of screw dislocations results in unidirectional growth of single crystalline GaN nanowires.

(5) Ordered nanoarrays of AgI nanowires[173]

Silver iodide (AgI) is an important material mainly for two applications, in solid-state batteries on the basis of superionic conductivity, and as a photographic film material based on the photochemical reactions occurring in it. AgI has a rich phase diagram with several different solid phases existing,[174] and it may exist in two phases β-AgI and γ-AgI, at room temperature and ambient atmosphere.[175] In β-AgI, the iodine ions are arranged in hexagonal close-packed (hcp) lattice with the silver ions being tetrahedrally coordinated to each of the iodides. Thus, in β-AgI, the system is in a wurtzite structure. In γ-AgI, the iodine ions are arranged in a face-centered-cubic (fcc) lattice with the silver ions tetrahedrally coordinated to the iodine ions. At 420 K, β-AgI undergoes a first-order phase transition into the superionic phase, in which iodine ions form a body-centered-cubic (bcc) lattice. However, all of these works were performed on bulk AgI. It is still a challenge to synthesize aligned and well-distributed nanowire arrays of AgI.

Recently, Wang[173] had prepared the nano-AgI arrays in ordered porous alumina membrane[176] in which nano-AgI is composed of a mixed phase of β-AgI and γ-AgI. In this letter, highly oriented β-AgI nanowire arrays were prepared in alumina membrane, and *in situ* X-ray diffraction (XRD) was employed to study the AgI nanowire arrays. The negative thermal expansion behavior and the elevated phase transition temperature (higher than the usual β to α phase transition temperature of 420 K) from β-AgI to α-AgI have been found.

The fabrication of AgI nanowire arrays was described in a previous publication.[177] In brief, the resulting porous alumina membrane was held on a hole (with diameter $\sim$ 1 cm) located between two electrolytic cells. $AgNO_3$ and KI aqueous solution were poured into two electrolytic cells, respectively. After a direct voltage was applied between the two electrolytic cells, Ag^+ and I^- ions moved in the electric field and reacted in the channels of the alumina membrane. When the color of the alumina membrane changed into yellow, it indicated that the AgI was deposited in the channels.

Figure 5.88 shows a transmission electron microscopy (TEM, JEM200CM) image, from which it can be clearly seen that AgI nanowires were deposited in the channels of the alumina membrane. The diameter of the AgI nanowires is about 40 nm, in good agreement with the diameter of the channel.

In order to investigate the phase transition behavior of the AgI nanowire arrays, *in situ* high temperature X-ray diffraction fraction (XRD, MXP 18AHF) was performed in the range from 293 to 473 K under the ambient atmosphere. Temperatures were kept constant ($\pm 1°C$) for 30 minutes before each measurement, and scans were carried out for $20°C < 2 < 60°C$. Figure 5.89(a) shows one XRD pattern of the AgI nanowire arrays at 293 K. Only one diffraction peak was observed, which indexed to the (002) plane of β-AgI and the (111) plane of γ-AgI. In accordance with the commonly accepted conclusion,[178] one always gets γ-AgI in the presence of excess

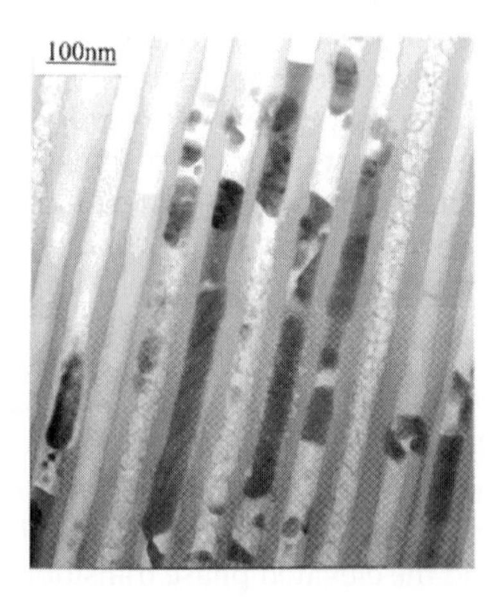

Fig. 5.88. TEM image of the AgI nanowires located in the channels of the ordered porous alumina membrane.

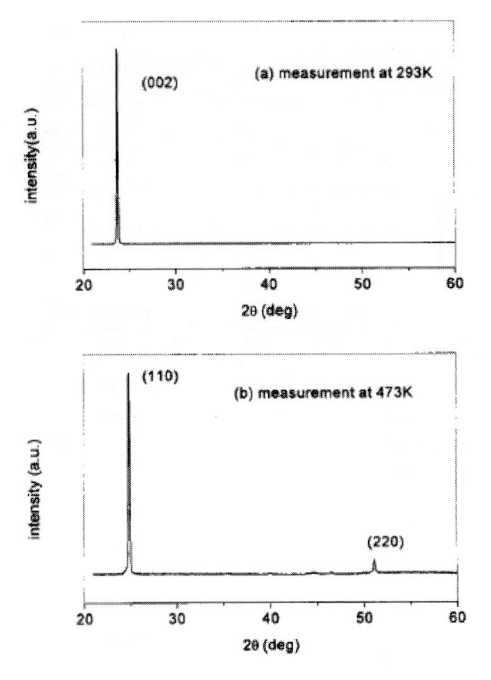

Fig. 5.89. XRD patterns of the AgI nanowire arrays, oriented along the (002) plane.

silver ions, whereas a large excess of iodine concentration produces β-AgI. In the present experiment, the iodine concentration was bound to be in the excess because the AgI nanowires were formed by diffusion of Ag^+ from $AgNO_3$ to KI solution.[179] Thus, we come to the conclusion that highly oriented β-AgI nanowire arrays were prepared in the channels of the alumina membrane. From 293 to 433 K, the X-ray diffraction patterns, ascribed to the (002) plane of β-AgI, show no significant difference. At 473 K, only α-AgI was found (see Fig. 5.89(b)), orienting along the (110) plane, which indicates the AgI nanowires transit from the β to α phase. Furthermore, with increase of temperature, the peak positions of the (002) plane of the β phase shift toward higher angle direction (see Fig. 5.90(a)), suggesting a negative expansion property. The lattice parameters of β-AgI as a function of the temperature are shown in Fig. 5.90(b), which clearly shows the variation of the lattice constants with temperature.

As previously presented, AgI undergoes a phase transition from the β to α phase at 420 K.[180,181] However, in our experiment the phase transition

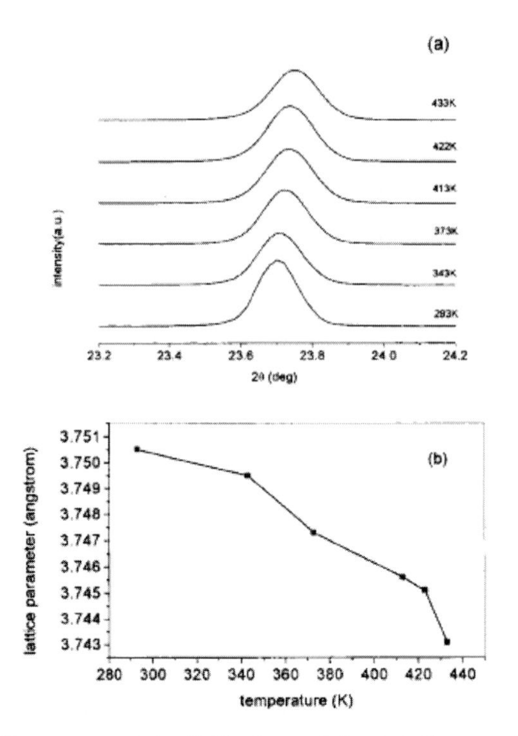

Fig. 5.90. (a) XRD patterns of the (002) plane of the β-AgI measuring from 293 to 433 K. (b) Temperature dependence of the lattice parameters of AgI nanowires from 293 to 433 K.

occurs at above 433 K at least, which clearly indicates that the AgI nanowire arrays in the alumina membrane have a higher phase transition temperature. At this temperature, the β phase orienting along the (002) plane of hexagonal close-packed structure transformed to the α phase orienting along the (110) plane of body-centered-cubic structure.

With the elevation of temperature, most materials usually show a positive thermal expansion property. This behavior can be understood by considering the effects of the anharmonic potential on the equilibrium lattice separations and is usually characterized by the Gruneisen parameter.[182] The negative thermal expansion, which represents lattice contraction with the elevation of temperature, was also observed among anisotropic systems,[183,184] where contraction along one crystallographic direction was usually accompanied by expansion along the others.[185]

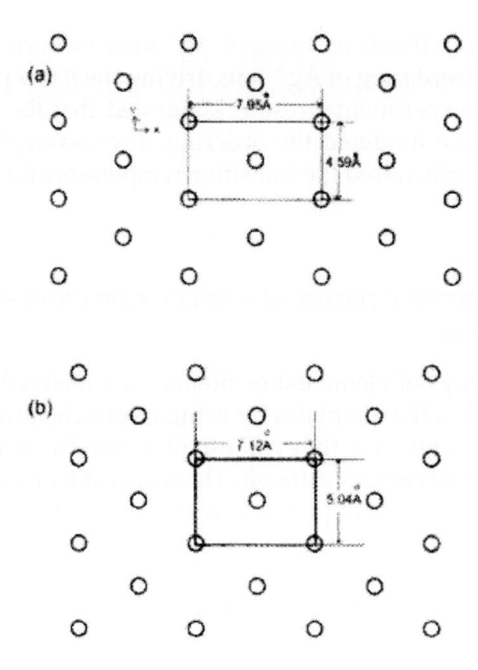

Fig. 5.91. (a) Iodine ions arrangement of the (002) plane of β phase and (b) the (110) plane of α phase.

The close-packed structure transforms into the body-centered-cubic structure in the AgI nanowires, which can be interpreted according to the Burgers mechanism.[186] In Fig. 5.91, for the transition from the (002) plane of hexagonal close-packed structure to the (110) plane of body-centered-cubic structure, the iodine ions will have a glide, implying that it will expand along the Y direction, resulting in the distance between the iodine ions changing from 4.59 to 5.04 Å, and contracting along the X direction (see Fig. 5.91(a)), resulting in the distance between the iodine ions changing from 7.95 to 7.12 Å. Meanwhile, the plane spacing contracted from 3.75 Å of the (002) plane of the β phase to 3.53 Å of the (110) plane of the α phase. Due to the fact that AgI nanowires are located in the channels of alumina membrane, the expansion in the radial direction will be hindered, resulting in a rising phase transition temperature. This indicated that the channels could enhance the stability of β-AgI. So far, it has not reached a consensus for what is responsible for the β to α phase transition. Madden and his co-workers[187] argued that the disordering tendency of the Ag^+ ions in the b phase could be taken as the driving force for the β to α phase transition. However, Seok and Oxtoby[188,189] argued that the structure change of the I^- lattice and the disordering of Ag^+ ions are correlated processes, rather than the disordering of Ag^+ ions driving the β to α phase transition. In any case, our experimental results suggested that the channels of the alumina membrane hindered the ordering–disordering transition of the Ag^+ ions, henceforth, raised the transition temperature for the β to α phase transition.

5.2.3. Ordered nanoarrays of ternary compound nanowires

Ordered nanoarrays of elemental or binary nanomaterials were successfully synthesized in the templates by using electrochemical deposition or CVD methods etc. However, the synthesis of ordered nanoarrays of ternary or polynery materials is very difficult. The main difficulty is how to realize co-precipitation of ternary or polynery components under the action of the external electric field. People synthesized nanowire and nanotube arrays using the electroless deposition method[190–193] and also many synthesize ternary nanoarrays.[194–196] We will introduce the synthesis and characterization of Co-Ni-P and Ni-W-P alloy nanowire arrays.

5.2.3.1. *Co-Ni-P alloy nanoarrays*[195]

Although there have been reports of fabrication of single element and binary alloy magnetic nanowire arrays by electro-deposition in anodic alumina membrane (AAM),[197,200] ternary alloy magnetic nanowire arrays have not been reported so far. Because the redox evolution potential and thermodynamic stability of these elements are different, it is very difficult to electro-deposit large-scale and uniform ternary alloy nanowire arrays in an aqueous solution, especially in a solution containing such element as phosphor (P). In addition, there are also some defects in the electro-deposition method, just as Prieto described.[201] Several factors affect the degree of pores filling in AAM. Firstly, pore-to-pore variations in nucleation rate arise from the presence of heterogeneities at the cathode interface, such as grain boundaries or adsorbed impurities. Secondly, even if wires are nucleated in all the pores, the rate of growth must be uniform so that all wires grow to the length of the pore simultaneously. Thirdly, if there are very small cracks in the AAM, even slightly larger than the pore diameter, deposition will occur predominantly in the cracks owing to point discharge and uneven current distribution in the pores of AAM. Here, Yuan[195] have succeeded in fabricating uniform size and shape of Co–Ni–P nanowire arrays on a large-scale by autocatalytic redox reaction in the AAM. Compared with the electrodeposition method, the autocatalytic redox method needs neither a supply of power nor a sprinkling of gold (Au, as a conductive layer) on one side of the AAM before the deposition, which is carried out via the redox reaction of an oxidizer and a reductant in an electrolyte solution. It is an autocatalytic self-assembly process which is promising for the production of uniform nanowire arrays on a large-scale, and more importantly it will make it possible to finely control the aspect ratio of the nanowire using pores of different diameters and AAM thicknesses. Control of the uniform size and shape of nanowire arrays on a large-scale is recognized as a very important issue in the fabrication of nanostructure and turned out to be a challenging problem.[15,18] In addition, the autocatalytic redox method can be applied to many other materials[1] and open up significant opportunities in the nanoscale fabrication of magnetic materials for ultra-high-density magnetic recoding.

The preparation process of Co-Ni-P alloy nanoarrays is as follows. The through-hole AAM template was first immersed in an aqueous solution of $SnCl_2$ ($10\,g\,l^{-1}$) for 1 minute and washed with distilled water 2 or 3 times, and then, the AAM template was further kept in a solution of $PdCl_2$ ($1\,g\,l^{-1}$) for 30 s and rinsed with distilled water several times again. Subsequently, Co-Ni-P nanowire arrays were deposited in the pores of the AAM from a solution of $15\,g\,l^{-1}$ $CoSO_4{\cdot}7H_2O$, $8\,g\,l^{-1}$ $NiSO_4{\cdot}6H_2O$, $22\,g\,l^{-1}$ $NaH_2PO_2{\cdot}H_2O$ and $60\,g\,l^{-1}$ Rochelle salt at 80–85°C.

Figure 5.91 shows a SEM image of Co-Ni-P alloy nanowire arrays prepared by autocatalytic redox reaction in an AAM. Figure 5.92(a) is a top-view image, Fig. 5.92(b) is a side-view image and Fig.5.92(c) is a SEM image of Co-Ni-P deposited in the pore wall (see bright dots) of the AAM for 20 s. We know from Fig. 5.92 that Co-Ni-P nanowires are grown out from the pores of the AAM. Therefore, the Co-Ni-P nanowire is about 70 nm in diameter, which is consistent with the pore diameter of the AAM. The autocatalytic redox reactions of Co-Ni-P nanowire arrays are as follows:

$$SnCl_2 + PbCl_2 = SnCl_4 + Pb, \tag{5.12}$$

$$Ni^{2+} + 2e = Ni, \quad E^\circ = -0.25V \tag{5.13}$$

$$Co^{2+} + 2e = Co, \quad E^\circ = -0.28V \tag{5.14}$$

$$H_2PO_2^- + 2H_2O = H_2PO_3^- + 2H^+ + 2e, \quad E^\circ = -0.50V \tag{5.15}$$

$$H_2PO_2^- + 2H^+ + e = P + 2H_2O. \quad E^\circ = -0.25V \tag{5.16}$$

The E° represents the standard single electrode potential. If the E° value of a reductant is lower than that of an oxidizer, it will be possible for the reaction between an oxidizer and a reductant to take place in the view of thermodynamics. The larger the electrode potential difference between an oxidizer and a reductant is, the higher the possibility of a redox reaction is. According to the E° values, Co, Ni and P could be reduced by NaH_2PO_2 in the electrolyte solution. However, these reactions among (5.13), (5.14), (5.15) and (5.16) cannot take place without some catalyst, which is controlled by kinetics. Pd atoms act as a catalyst in the reactions. The growth mechanism of Co-Ni-P nanowire arrays may be described as follows. Firstly, the $SnCl_2$ solution which is adhered to the pore wall of the AAM hydrolyzes to form $Sn(OH)_2$ in an aqueous solution, then, $PdCl_2$ is reduced to Pd atoms by $Sn(OH)_2$. Consequentially, these Pd atoms trigger the redox reactions. Once the redox reactions are triggered, the above reactions of (5.13), (5.14), (5.15) and (5.16) can be autocatalyzed. In fact, as a catalyst, the concentration of Pd atoms is very low so that no trace of it is seen on the EDS spectrum in Fig. 5.94. Since the autocatalytic redox reaction needs neither a supply of power nor a sprinkling of Au on one side of the AAM in the fabrication process of Co-Ni-P nanowire arrays, and since these redox reactions are autocatalytic self-assembly processes and the electrolyte solution concentration is uniform, Co-Ni-P nanowire arrays are formed where there is the electrolyte solution. The Co-Ni-P nanowire arrays are definitely going to be uniform in size and shape on a large-scale (see Fig. 5.92 SEM images), and all wires grow to the same length in the pore simultaneously, as before some electro-deposition defects can be overcome. The TEM and SAED images of a single Co-Ni-P nanowire are shown in Fig. 5.93. Figure 5.93(a) indicates that the Co-Ni-P nanowire is about 70 nm in diameter and 3 mm in length. Figure 5.93(b) shows that the Co-Ni-P nanowire is amorphous in structure, which is confirmed by the XRD spectrum as

Fig. 5.92. SEM images of Co-Ni-P nanowire arrays: (a) top-view image, (b) side-view image, (c) Co-Ni-P deposited in the pore wall (see bright dots) of the AAM for 20 s.

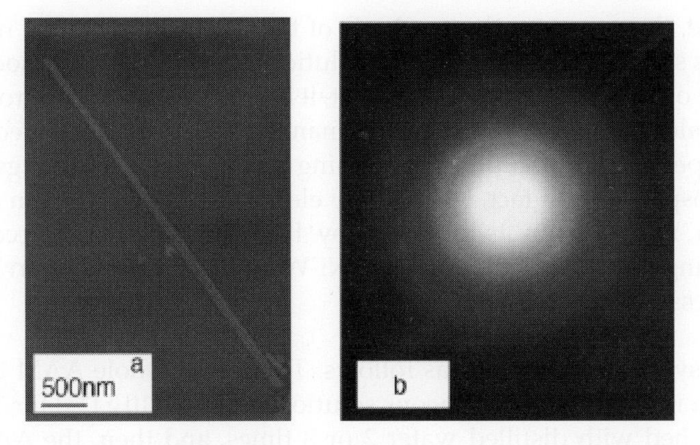

Fig. 5.93. (a) TEM image of the Co-Ni-P alloy nanowire with diameter of about 70 nm. (b) SAED of the Co-Ni-P alloy nanowire.

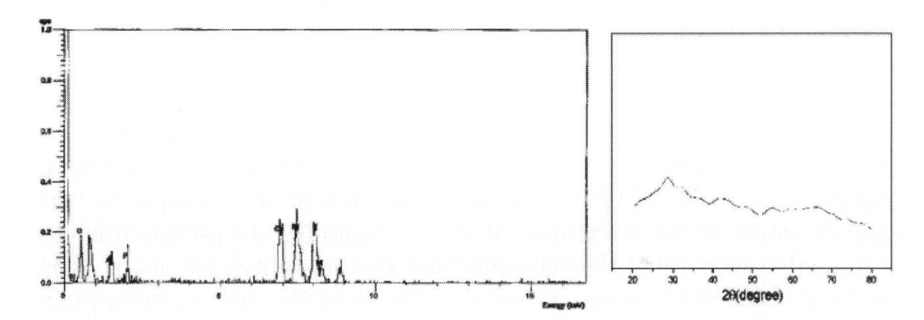

Fig. 5.94. The EDS (left) and XRD (right) of the Co-Ni-P alloy nanowire. The EDS show that the atomic ratio and the mass ratio of the Co-Ni-P nanowire are 42.82 : 40.48 : 16.70, and 46.58 : 43.87 : 9.55, respectively. The unlabeled peaks are associated with copper coming from the TEM grid.

shown in Fig. 5.94. This amorphous structure of Co-Ni-P nanowire arrays will meet the need of perpendicular anisotropy in magnetic properties. The EDS spectrum reveals that the nanowires are composed of Co, Ni and P, the atomic ratio is 42.82 : 40.48 : 16.70 and the mass ratio is 46.58 : 43.87 : 9.55.

5.2.3.2. *Ni-W-P alloy nanowire arrays*[196]

Although there have been reports of fabrication of single element and binary alloy magnetic nanowire arrays by electro-deposition in the AAM template, the report on the synthesis of ternary alloy magnetic nanowire arrays is small. Because the redox evolution potential and thermodynamic stability of these elements are different, it is very difficult to electro-deposit large-scale and uniform ternary alloy nanowire arrays in an aqueous solution, especially in a solution containing such elements as tungsten (W) and phosphor (P). In fact, W is hardly electro-deposited from an aqueous solution, much less in its ternary alloy. Here, Yuan[196] have succeeded in fabricating large-scale and uniform Ni-W-P alloy nnanowire arrays with electroless deposition in the AAM.

The synthesis process is as follows. The through-hole AAM template was first immersed in an aqueous solution of $SnCl_2$ ($10\,g\,l^{-1}$) for 1 minute and washed with distilled water 2 or 3 times, and then, the AAM template was further kept in a solution of $PdCl_2$ ($1\,g\,l^{-1}$) for 30 s and rinsed with distilled water several times again. Subsequently, Ni-W-P nanowire arrays were deposited in the pores of the AAM from a solution of $15\,g\,l^{-1}$ $NiSO_4{\cdot}6H_2O$, $10\,g\,l^{-1}$ $Na_2WO_4{\cdot}2H_2O$, $22\,g\,l^{-1}$ $NaH_2PO_2{\cdot}H_2O$, and $40\,g\,l^{-1}$ sodium citrate at 80–85°C.

Figure 5.95 shows an SEM image of Ni-W-P alloy nanowire arrays prepared by electroless deposition in an AAM. Figure 5.95(a) shows the view from the top and Fig. 5.95(b) shows the view from the side. Figure 5.95(a) indicates that the Ni-W-P alloy nanowire arrays are uniform in size and all pores of the AAM are filled up. Figure 5.95(b) reveals that the nanowires are parallel to each other and the nanowires are about 3 μm in length. The SEM images reveal that the Ni-W-P nanowire arrays fabricated by electroless deposition are large-scale and uniform in size and shape. The reactions in the fabrication of Ni-W-P nanowire arrays are as follows:

$$SnCl_2 + PdCl_2 = SnCl_4 + Pd, \tag{5.17}$$

$$Ni^{2+} + 2e = Ni, \quad E^\circ = -0.25\,V \tag{5.18}$$

$$WO_4^{-2} + 6e + 4H_2O = W + 8OH^-, \quad E^\circ = -1.05\,V \tag{5.19}$$

$$H_2PO_2^- + 3OH^- = HPO_3^{2-} + 2H_2O + 2e, \quad E^\circ = -1.57\,V \tag{5.20}$$

$$H_2PO_2^- + 2H^+ + e = P + 2H_2O. \quad E^\circ = -0.25\,V \tag{5.21}$$

The E° represents the standard single electrode potential. According to the E° value, Ni, W and P could be reduced by NaH_2PO_2 in the electrolyte solution. The Ni-W-P nanowire arrays are definitely going to be large-scale and uniform (see Fig. 5.95, SEM images). The TEM and SAED

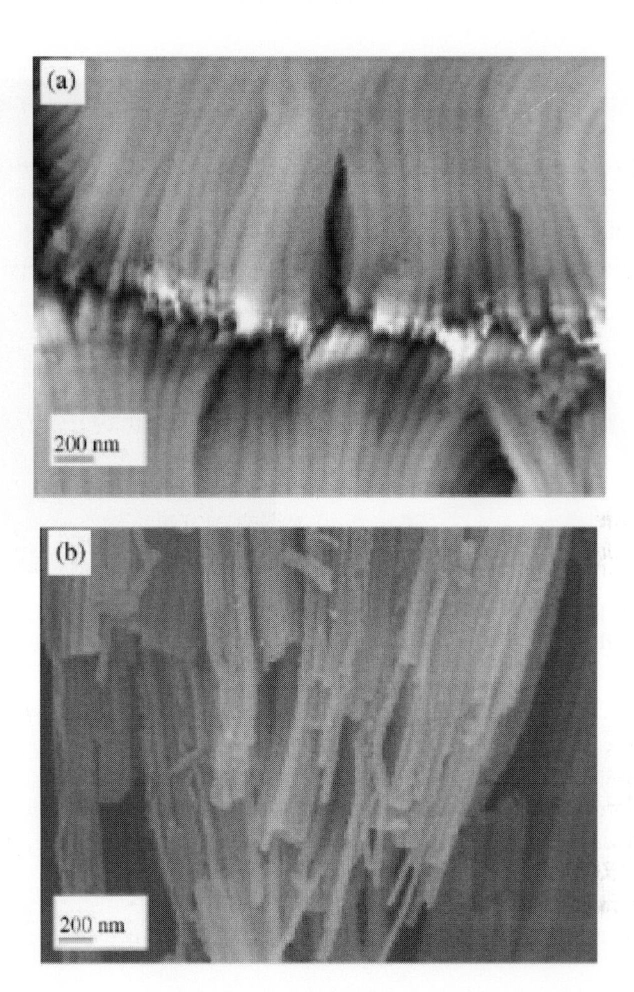

Fig. 5.95. SEM image of Ni-W-P nanowire arrays: (a) view from the top; (b) view from the side.

images of a single Ni-W-P nanowire are shown in Fig. 5.96. Figure 5.96(a) shows that the Ni-W-P nanowire is about 60 nm in diameter and 3 m in length. Figure 5.96(b) illustrates that the Ni-W-P nanowire is amorphous in structure, which is confirmed by the XRD spectrum shown in Fig. 5.97. This amorphous structure of Ni-W-P nanowire arrays will meet the need of perpendicular anisotropy in magnetic properties. The EDS shown in Fig. 5.98 reveals that the nanowires are composed of Ni, W and P, the atomic ratio is 65.79 : 6.51 : 27.70, and the mass ratio is 65.28 : 20.22 : 14.50.

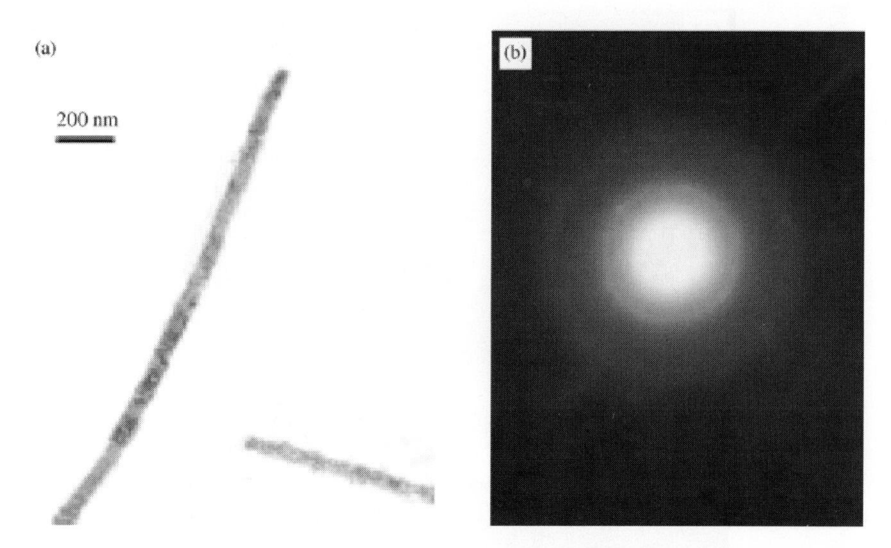

Fig. 5.96. (a) TEM image of Ni-W-P alloy nanowires with a diameter of about 60 nm. (b) SAED of the Ni-W-P alloy nanowire.

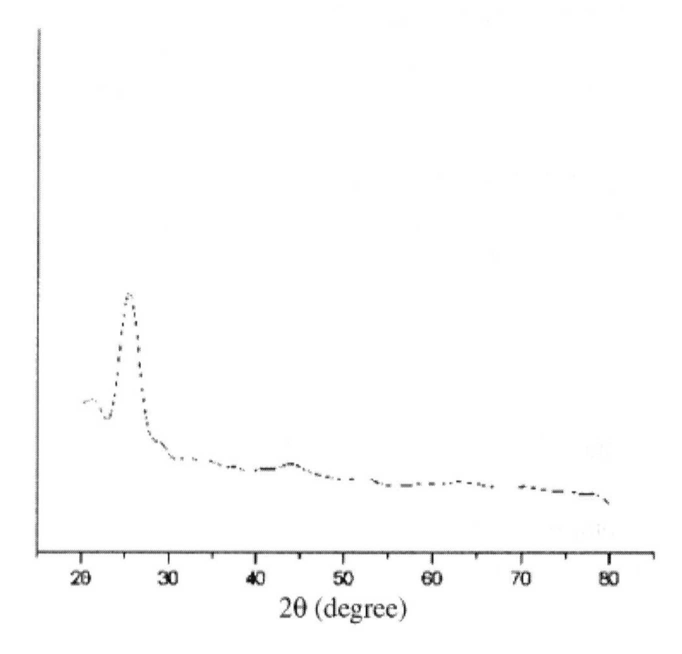

Fig. 5.97. The X-ray diffraction pattern for Ni-W-P nanowires.

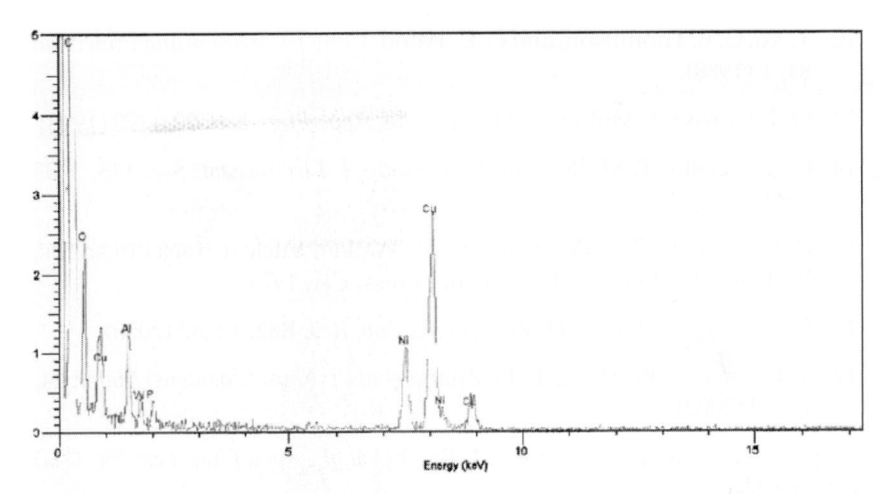

Fig. 5.98. The EDS of a Ni-W-P alloy nanowire. The atomic ratio of Ni : W : P is 65.79 : 6.51 : 27.70, and the mass percentage of Ni : W : P is 65.28 : 20.22 : 14.50.

Bibliography

1. F. Keller, M. S. Hunter, and D. L. Robinson, *J. Electrochem. Soc.* **100**, 411 (1953).
2. D. Aimawlawi, N. Coombs, and M. Moskovits, *J. Appl. Phys.* **70**, 4421 (1991).
3. C. R. Martin, *Science* **266**, 1961 (1994).
4. C. A. Foss, Jr., C. L. Hornyak, J. A. Stocked, and C. R. Martin, *J. Phys. Chem.* **96**, 1491 (1992).
5. M. J. Tierney and C. R. Martin, *J. Phys. Chem.* **93**, 2878 (1989).
6. H. Masuda and K. Fukuda, *Science* **268**, 1466 (1995).
7. M. Nakao and T. Tamamura, *Appl. Phys. Lett.* **71**, 2770 (1997).
8. J. W. Diggle, T. C. Downie, and C. W. Goulding, *Chem. Rev.* **69**, 365 (1969).
9. G. E. Thompson and G. C. Wood, *Nature* **290**, 230 (1981).
10. G. E. Thompson, R. C. Furneaan, G. C. Wood *et al.*, *Nature* **272**, 433 (1978).
11. Y. Xu, G. E. Thompson, and G. C. Wood, *Trams. Inst. Met. Finish.* **63**, 98 (1985).

12. Y. Xu, G. E. Thompson, and G. C. Wood, *Chin. J. Corrosion and Protection* **81**, 1 (1988).
13. O. Tessensky, F. Müller, and U. Gösede, *Appl. Phys. Lett.* **72**, 1173 (1998).
14. O. Tessensky, F. Müller, and U. Gösede, *J. Electrochem. Soc.* **145**, 3735 (1998).
15. R. L. Fleisber, P. B. Price, and R. M. Walker, Vuclear tracks in solids, Berkeley: University of California Press, CA, 1975.
16. M. Zheng, L. Menon, H. Zeng *et al.*, *Phy. Rev.* **B62**, 12282 (2000).
17. Y. T. Pang, G. W. Meng, L. D. Zhang *et al.*, *J. Phys.: Comdens. Mater.* **14**, 11729 (2002).
18. K. Nielsch, R. B. Wehrspohn, T. Barthel *et al.*, *Appl. Phys. Lett.* **79**, 1360 (2001).
19. D. J. Sellmyer, M. Zheng, and R. Skomski, *J. Phys.: Comdens. Mater.* **13**, 433 (2001).
20. H. Zheng, R. Skomski, L. Menon *et al.*, *Phy. Rev.* **B65**, 134426 (2002).
21. T. Gao, G. Meng, L. Zhang *et al.*, *Appl. Phys.* **A74**, 403 (2002).
22. T. M. Whitney, J. S. Jiang, P. C. Searson, *et al.*, *Science* **261**, 1316 (1993).
23. L. Vila, P. Vincent, P. L. Dauginet-De *et al.*, *Nano Lett.* **4**, 521 (2004).
24. C. A. Huber, T. E. Huber, M. Sadogi *et al.*, *Science* **263**, 800 (1994).
25. C. J. Brumlik, V. P. Menon, and C. R. Martin, *Mater. Res.* **9**, 1174 (1994).
26. C. J. Brumlik and C. R. Martin, *J. Am. Chem. Soc.* **113**, 3174 (1991).
27. J. C. Bovo, C. Y. Tie, Z. Xu *et al.*, *Adv. Mater.* **13**, 1631 (2001).
28. Y. H. Wang, C. H. Ye, X. S. Fang, and L. D. Zhang, *Chem. Lett.* **33**, 166 (2004).
29. T. Gao, G. W. Meng, J. Zhang, Y. W. Wang, C. H. Liang, J. C. Fan, and L. D. Zhang, *Appl. Phys.* **A73**, 251 (2001).
30. G. Yi and W. Schwarzacher, *Appl. Phys. Lett.* **74**, 1746 (1999).
31. Z. Zhang, D. Gekhtman, M. S. Dresselhaus, and J. Y. Ying, *Chem. Mater.* **11**, 1659 (1999).
32. X. Duan, Y. Huang, Y. Cui, J. Wang, and C. M. Lieber, *Nature* **409**, 66 (2001).

33. Z. I. Wang, *Adv. Mater.* **12**, 1259 (2000).

34. H. Ruda and A. Skik, *J. Appl. Phys.* **86**, 5103 (1999).

35. U. Landman, R. N. Barnett, A. G. Scherbakov, and P. Avouris, *Phy. Rev. Lett.* **85**, 1958 (2000).

36. Y. T. Pang, G. W. Meng, Q. Fang, and L. D. Zhang, *Nanotechnology* **14**, 20 (2003).

37. X. Y. Zhang, L. D. Zhang, Y. Lei, L. X. Zhao, and Y. Q. Mao, *J. Mater. Chem.* **11**, 1732 (2001).

38. D. Routkevitch, T. Bigioni, M. Moskovits, and T. M. Xu, *J. Phys. Chem.* **100**, 14037 (1996).

39. D. Xu, Y. Xu, D. Chen, G. Guo, L. Gui, and Y. Tang, *Adv. Mater.* **12**, 520 (2000).

40. B. B. Lakshmi, C. T. Partrissi, and C. R. Martin, *Chem. Mater.* **9**, 2544 (2000).

41. J. S. Dunning, D. N. Bennion, and J. Newman, *J. Electrochem. Soc.* **68**, 236 (1971).

42. Y. T. Pang, G. W. Meng, L. D. Zhang *et al.*, *Adv. Funct. Mater.* **12**, 719 (2002).

43. Y. Zhang, G. H. Li, Y. C. Wu, B. Zhang, W. H. Song, and L. D. Zhang, *Adv. Mater.* **14**, 1227 (2002).

44. T. M. Whitney, J. S. Jiang, P. C. Searson, and C. L. Chien, *Science* **261**, 16 (1993).

45. X. F. Wang, L. D. Zhang, J. Zhang, H. Z. Shi, X. S. Peng, M. J. Meng, J. Fang, J. L. Chen, and B. J. Gao, *J. Phys. D: Appl. Phys.* **34**, 418 (2001).

46. X. F. Wang, J. Zhang, H. Z. Shi, Y. W. Wang, G. W. Meng, X. S. Peng, and L. D. Zhang, *J. Appl. Phys.* **89**, 3847 (2004).

47. L. Li, Y. Zhang, G. H. Li, and L. D. Zhang, *Chem. Phys. Lett.* **378**, 244 (2003).

48. L. Li, Y. Zhang, G. H. Li, X. W. Wang, and L. D. Zhang, *Mater. Lett.* **59**, 1223 (2005).

49. Y. T. Tian, G. W. Meng, S. K. Biswas, D. M. Ajsyan, S. H. Sun, and L. D. Zhang, *Appl. Phys. Lett.* **85**, 967 (2004).

50. L. Li, Y. Zhang, G. H. Li, W. H. Song, and L. D. Zhang, *Appl. Phys.* **A80**, 1053 (2005).

51. K. Lui, C. L. Chien, P. C. Searson, and K. Y. Zhang, *Appl. Phys. Lett.* **73**, 143 (1998).

52. Z. Zhang, J. Y. Ying, and M. S. Dresselhaus, *J. Mater. Res.* **13**, 1754 (1998).

53. F. Ebrahimi, G. R. Bouren, M. S. Kell, and T. E. Matthews, *Nanostruct. Mater.* **11**, 343 (1999).

54. Y. M. Lin, S. B. Cronin, J. Y. Ying, and M. S. Dresselhaus, *Appl. Phys. Lett.* **76**, 3944 (2000).

55. Z. B. Zhang, D. Gekhtman, M. S. Dresselhaus, and J. Y. Ying, *Chem. Lett.* **11**, 1659 (1999).

56. J. Li, C. Papadopoulos, and J. Xu, *Nature (London)*, **402**, 253 (1999).

57. D. Li, Y. Wu, R. Fan, P. Yang, and A. Majumdar, *Appl. Phys. Lett.* **83**, 3186 (2003).

58. Z. B. Zhang, Y. Z. Sun, M. S. Dresselhaus, and J. Y. Ying, *Appl. Phys. Lett.* **73**, 1589 (1998).

59. K. Nielsch, E. Müller, A. P. Li, and U. Gösele, *Adv. Mater.* **12**, 582 (2000).

60. International Centre for Diffraction Data (ICDD), Newtown Square, PA, Powder Diffraction File, formerly JCPDS.

61. G. Sauer, G. Brehm, S. Schneider, K. Nielsch, R. B. Wehrsphohn, J. Choi, H. Hofmeist, and U. Gösele, *J. Appl. Phys.* **91**, 3243 (2002).

62. B. Gates, B. Mayers, Y. N. Xia *et al.*, *Adv. Funct. Mater.* **12**, 219 (2002).

63. C. H. An, K. B. Tang, Y. T. Qian *et al.*, *Eur. J. Inorg. Chem.* **17**, 3250 (2003).

64. H. Zhang, D. R. Yang, Y. J. Ti *et al.*, *J. Phys. Chem.* **B108**, 1179 (2004).

65. M. Mo, J. Zeng, Y. Qian *et al.*, *Adv. Mater.* **14**, 1658 (2002).

66. B. Mayers and Y. N. Xia, *Adv. Mater.* **14**, 279 (2002).

67. A. W. Zhao, C. H. Ye, L. D. Zhang *et al.*, *J. Mater. Res.* **18**, 2318 (2003).

68. P. I. F. Harris, Carbon Nanotubes and Related Structures — New Materials for the Twenty-First Century, Cambridge University Press, Cambridge, 2001.

69. R. Sato and G. Dresselhaus, Physical Properties of Carbon nanotubes, Impertial College Press, London, 1998.

70. M. S. Dressselhaus, G. Dresselhaus, and P. C. Eklund, Science of Fullerenes and Carbon Nanotubes, Academic Press, San Diego, 1996.

71. T. W. Ebbesen, *Phys. Today* **49**, 26 (1996).

72. B. I. Yakobson and R. E. Smalley, *Am. Sci.* **85**, 324 (1997).

73. C. Dekker, *Phys. Today* **52**, 22 (1999).

74. L. Pai, Intelligent Macromolecules for Smart Devices: From Material Synthesis to Device Applications, Springer, London, 2004.

75. H. Dai, J. Hafner, A. Rinzler *et al.*, *Nature* **384**, 147 (1998).

76. H. Dai, N. Franklin, and J. Han, *Appl. Phys. Lett.* **73**, 1508 (1998).

77. S. Wong, E. Jokselevich, C. M. Lieber *et al.*, *Nature* **394**, 52 (1998).

78. S. Wong, J. Harper, C. M. Lieber *et al.*, *J. Am. Chem. Soc.* **120**, 603 (1998).

79. J. Hafner, C. Cheung, C. M. Lieber *et al.*, *Nature* **398**, 761 (1999).

80. S. Tans, A. Verschueren, and C. Dekker, *Nature* **393**, 49 (1998).

81. R. Martel, T. Schmidt, H. R. Shea *et al.*, *Appl. Phys. Lett.* **73**, 2447 (1998).

82. H. Soh, C. Quate, H. Dai *et al.*, *Appl. Phys. Lett.* **75**, 627 (1999).

83. W. de Heer, A. Chaelain, and D. Vgart, *Science* **270**, 1179 (1995).

84. J. M. Bonard, J. P. Salvetat *et al.*, *Appl. Phys. Lett.* **73**, 918 (1998).

85. P. G. Collins and A. Zettl, *Appl. Phys. Lett.* **69**, 1969 (1996).

86. Y. Saito, K. Hamaguchi, K. Hata *et al.*, *Ultramicroscopy* **73**(N1-4), 1 (1998).

87. Q. Wang, A. Setlur, J. Lauerhaas *et al.*, *Appl. Phys. Lett.* **72**, 2192 (1998).

88. W. Z. Li, S. S. Xie, L. X. Qian *et al.*, *Science* **274**, 1707 (1996).

89. Z. F. Ren, Z. P. Huang, J. W. Xu *et al.*, *Science* **282**, 1105 (1998).

90. S. S. Fan, M. G. Chapline, N. R. Franklin *et al.*, *Science* **283**, 512 (1999).

91. Z. Pan, S. S. Xie, B. Chang *et al.*, *Nature* **394**, 631 (1998).

92. N. Wang, Z. K. Tang, G. D. Li, and J. S. Chen, *Nature* **408**, 50 (2000).

93. M. Endo and H. W. Kroto, *J. Phys. Chem.* **96**, 6941 (1992).

94. R. Saito, G. Dresselhaus, and M. S. Dresselhaus, *Chem. Phys. Lett.* **195**, 531 (1992).

95. J. C. Charlier, X. Blase, A. De vita *et al.*, *Appl. Phys.* **A68**, 267 (1999).

96. O. A. Louchev and Y. Sato, *Appl. Phys. Lett.* **72**(2), 194 (1994).

97. S. Iijima, P. M. Ajayan, and T. Ichihashi, *Phys. Rev. Lett.* **68**, 267 (1999).

98. C. H. Kiang and W. A. Goddard, *Phys. Rev. Lett.* **76**, 2515 (1996).

99. C. Laurent, E. Flahaut, A. Peighey *et al.*, *New J. Chem.* **22**, 1229 (1998).

100. L. Liu and S. S. Fan, *J. Am. Chem. Soc.* **123**, 11502 (2001).

101. Z. K. Tang, H. D. Sun, J. Chem, and G. Li, *Appl. Phys. Lett.* **73**(16), 2287 (1998).

102. S. Qiu and W. Pang, *Zeolites* **9**, 440 (1989).

103. Z. H. Yuan, H. Huang, S. S. Fan *et al.*, *Appl. Phys. Lett.* **78**, 3127 (2001).

104. Q. F. Zhan, Z. Y. Chen, D. S. Xue *et al.*, *Phys. Rev.* **B66**, 134436 (2002).

105. D. H. Qin, L. Cao, H. L. Li *et al.*, *Chem. Phys. Lett.* **358**, 484 (2002).

106. D. H. Qin, Y. Peng, H. L. Li *et al.*, *Chem. Phys. Lett.* **374**, 661 (2003).

107. V. M. Fedosyuk, O. I. Kasyutich, and W. J. Schwarzacher, *Magn. Mater.* **198**, 199, 246 (1999).

108. Y. W. Wang, G. W. Meng, L. D. Zhang *et al.*, *Chem. Phys. Lett.* **339**, 174 (2001).

109. Y. W. Wang, L. D. Zhang, G. W. Meng *et al.*, *J. Phys. Chem.* **B106**, 2502 (2002).

110. F. S. Li and L. Y. Ren, *Phys. Status Solidi*, **A193**, 196 (2002).

111. C. Z. Wang, G. W. Meng, L. D. Zhang *et al.*, *J. Phys. D: Appl. Phys.* **35**, 738 (2002).

112. Y. M. Lin, O. Rabin, S. B. Cronin *et al.*, *Appl. Phys. Lett.* **81**, 2403 (2002).

113. M. S. Sander, R. Gronsky, T. Sands *et al.*, *Chem. Mater.* **15**, 335 (2003).

114. Y. W. Wang, L. D. Zhang, G. W. Meng, X. S. Peng, Y. X. Jin, and J. Zhang, *J. Phys. Chem.* **B106**, 2502 (2002).

115. Y. Li, G. W. Meng, L. D. Zhang, and F. Phillip, *Appl. Phys. Lett.* **76**, 2011 (2000).

116. M. J. Zheng, L. D. Zhang, G. H. Li *et al.*, *Chem. Phys. Lett.* **363**, 123 (2002).

117. Y. C. Wang, I. C. Leu, and M. H. Hon, *J. Cryst. Growth* **237**, 564 (2002).

118. Y. C. Wang, I. C. Leu, and M. H. Hon, *J. Appl. Phys.* **95**, 1444 (2004).

119. B. B. Lakshmi, P. K. Dorbout, and C. R. Martin, *Chem. Mater.* **9**, 857 (1997).

120. L. E. Green, M. Law, P. D. Yang *et al.*, *Angew. Chem. Int. Edit.* **42**, 3031 (2003).

121. M. H. Huang, S. Mao, P. D. Yang *et al.*, *Science* **292**, 1897 (2001).

122. C. H. Liu, J. A. Zapien, S. T. Lee *et al.*, *Adv. Mater.* **15**, 838 (2003).

123. S. C. Lyu, Y. Zhang, C. J. Lee *et al.*, *Chem. Mater.* **15**, 3294 (2003).

124. M. J. Zheng, L. D. Zhang, G. H. Li, X. Y. Zhang, and X. F. Wang, *Appl. Phys. Lett.* **79**, 839 (2001).

125. H. F. Yang, Q. H. Shi, B. Z. Tian *et al.*, *J. Am. Chem. Soc.* **125**, 4724 (2003).

126. W. Han, S. Fan, Q. Li, and Y. Hu, *Science* **277**, 1287 (1997).

127. D. P. Yu, Z. G. Bai, Y. Ding, Q. L. Hang, H. Z. Zhang, J. J. Wang, Y. H. Zou, W. Q. Qian, G. C. Xiong, H. T. Zhou, and S. Q. Feng, *Appl. Phys. Lett.* **72**, 3458 (1998).

128. M. J. Zheng, G. H. Li, X. Y. Zhang, S. Y. Huang, Y. Lei, and L. D. Zhang, *Chem. Lett.* **13**, 3859 (2001).

129. J. Zhou, C. Y. Xu, X. M. Leu, C. S. Wang, C. Y. Wang, Y. Hu, and Y. T. Qian, *J. Appl. Phys.* **75**, 1835 (1994).

130. E. R. Leite, I. T. Weber, E. Long, and J. A. Varzla, *Adv. Mater.* **12**, 965 (2000).

131. A. Diehuez, A. R. Rdriguea, J. R. Morante, P. Nelli, L. Sangaletti, and G. Sberveglier, *J. Electrochem. Soc.* **146**, 3527 (1999).

132. Y. Lei, L. D. Zhang, G. W. Meng, G. H. Li, X. Y. Zhang, C. H. Liang, W. Chen, and S. X. Wang, *Appl. Phys. Lett.* **78**, 1125 (2001).

133. D. N. Furlong, D. Wells, and W. H. F. Sasse, *J. Phys. Chem.* **90**, 1107 (1986).

134. E. Joselevich and I. Willner, *J. Phys. Chem.* **98**, 7628 (1994).

135. L. Li, Y. W. Yang, G. H. Li, and L. D. Zhang, *Small* (2005), to be published.

136. L. Leontie, M. Caraman, A. Visinoiu, and G. I. Rusu, *Thin Solid Films* **473**, 230 (2005).

137. G. S. Wu, L. D. Zhang, B. C. Cheng, *et al.*, *J. Am. Chem. Soc.* **126**, 5976 (2004).

138. Y. Li, D. S. Xu, G. L. Guo *et al.*, *Chem. Mater.* **11**, 3433 (1999).

139. D. S. Xu, Y. J. Xu, G. L. Guo *et al.*, *Chem. Phys. Lett.* **325**, 340 (2000).

140. X. S. Peng, G. W. Meng, L. D. Zhang *et al.*, *Mater. Res. Bull.* **37**, 1369 (2002).

141. X. S. Peng, G. W. Meng, L. D. Zhang *et al.*, *J. Phys. D: Appl. Phys.* **34,** 3224 (2001).

142. D. Xu, X. Shi, G. L. Guo *et al.*, *J. Phys. Chem.* **B104**, 5061 (2000).

143. X. S. Peng, J. Zhang, L. D. Zhang *et al.*, *Chem. Phys. Lett.* **343**, 470 (2001).

144. W. B. Zhao, J. J. Zhu, H. Y. Chen *et al.*, *Scripta. Mater.* **50**, 1169 (2004).

145. X. S. Peng, J. Zhang, L. D. Zhang *et al.*, *J. Mater. Res.* **17**, 1283 (2002).

146. D. S. Xu, Y. G. Guo, G. L. Guo *et al.*, *J. Mater. Res.* **17**, 1711 (2002).

147. A. W. Zhao, G. W. Meng, L. D. Zhang *et al.*, *Appl. Phys.* **A76**, 537 (2003).

148. A. L. Prieto, M. S. Sander, M. S. Martin-Gonzalez *et al.*, *J. Am. Chem. Soc.* **123**, 7160 (2001).

149. M. S. Sander, A. L. Prieto, R. Gronsky *et al.*, *Adv. Mater.* **14**, 665 (2002).

150. Y. Huang, Q. Wei, C. M. Lieber, *et al.*, *Science* **291**, 630 (2001).

151. Y. H. Wang, C. H. Ye, L. D. Zhang *et al.*, *Appl. Phys. Lett.* **82**, 4253 (2003).

152. L. Motto, F. Billoudet, D. Thaudière, A. Naudon, and Mipileni, *Journal De Physique 111 France* **7**, 517 (1997).

153. D. Xu, D. Chen, Y. Xu, X. Shi, G. Guo, L. Gui, and Y. Tang, *Pure Appl. Chem.* **72**(1/2), 127 (2000).

154. A. E. Rakhshan, *J. Appl. Phys.* **81**, 7988 (1997).

155. C. Leppiler, P. Cowache, J. F. Guillemoes, N. Gibson, E. Ozsan, and D. Lincot, *Thin Solid Films* **361–361**, 118 (2000).

156. B. M. Basol, *Sol. Cells* **23**, 69 (1998).

157. C. Sella, P. Boncorps, and J. Vedel, *J. Electrochem. Soc.* **133**, 2043 (1986).

158. A. Sabady-Reitjes, L. M. Peter, M. E. Özsan, S. Dennison, and S. Webber, *J. Electrochem. Soc.* **140**, 2880 (1993).

159. G. S. Chen, L. D. Zhang, Y. Zhu, G. T. Fei, L. Li, C. M. Mo, and Y. Q. Mao, *Appl. Phys. Lett.* **75**, 2455 (1999).

160. G. Fasol, *Science* **272**, 1751 (1996).

161. F. A. Ponce and D. P. Boar, *Nature (London)* **386**, 351 (1997).

162. J. C. Zopler, R. J. Shul, A. G. Bala, R. G. Wilson, S. J. Pearton, and R. A. Stall, *Appl. Phys. Lett.* **68**, 2273 (1996).

163. Q. Chen, M. A. Khan, J. W. Wang, C. J. Sun, M. S. Shur, and H. Park, *Appl. Phys. Lett.* **69**, 794 (1996).

164. K. Kubota, Y. Kobayashi, and K. Fujimoto, *J. Appl. Phys.* **66**, 2984 (1989).

165. V. A. Joshkin, J. C. Roberts, F. G. Mcintosh, S. M. Redair, E. L. Piner, and M. K. Behbehani, *Appl. Phys. Lett.* **71**, 234 (1997).

166. Y. Xie, Y. Qian, W. Wang, S. Zhang, and Y. Zhang, *Science* **272**, 1926 (1996).

167. W. Han, S. Fan, Q. Li, and Y. Hu, *Science* **277**, 1287 (1997).

168. C. M. Balkas and R. F. Davis, *J. Am. Ceram. Soc.* **79**, 2309 (1996).

169. T. Azuhata, T. Sota, K. Suzuki, and S. Nakamura, *J. Phys.: Condens. Matter* **7**, L129 (1995).

170. D. P. Yu, Z. G. Bai, Y. Ding, Q. L. Hang, H. Z. Zhang, J. J. Wang, Y. H. Zou, W. Qian, G. C. Xiong, H. T. Zhou, and S. Q. Feng, *Appl. Phys. Lett.* **72**, 3458 (1998).

171. J. J. Trenier, K. M. Hickman, S. C. Goel, A. M. Viano, P. C. Gibbons, and W. E. Buhro, *Science* **270**, 1791 (1995).

172. C. J. Frosch and C. D. Thurmond, *J. Phys. Chem.* **62**, 611 (1958).

173. Y. H. Wang, C. G. Ye, G. Z. Wang, L. D. Zhang, Y. M. Liu, and Z. Y. Zhao, *Appl. Phys. Lett.* **82**, 4253 (2003).

174. D. A. Keen, S. Hull, W. Hayes, and N. J. G. Gardner, *Phys. Rev. Lett.* **77**, 4914 (1996).

175. M. Nagai and T. Nishino, *Solid State Ionics* **117**, 317 (1999).

176. Y. H. Wang, J. M. Mou, W. L. Cai, L. Z. Yao, and L. D. Zhang, *Mater. Lett.* **56**, 502 (2002).

177. Y. H. Wang, J. M. Mou, W. L. Cai, L. Z. Yao, and L. D. Zhang, *J. Mater. Res.* **16**, 990 (2001).

178. P. S. Kumar, P. B. Dayal, and C. S. Sunsndana, *Thin Solid Films* **357**, 111 (1999).

179. M. Nagai and T. Nishino, *Solid State Ionics* **53–56**, 63 (1992).

180. K. Tadanaga, K. Imai, M. Tatsumisago, and T. Minami, *J. Electrochem. Soc.* **147**, 4061 (2000).

181. M. Tatsumisago, K. Okuda, N. Itakura, and T. Minami, *Solid State Ionics* **121**, 193 (1999).

182. N. W. Ashcroft and N. D. Mermin, *Solid State Physics*, Sauders, Philadelphia, 1976, Chap. 25.

183. T. A. Mary, J. S. O. Ivans, T. Vogt, and A. W. Sleight, *Science* **272**, 90 (1996).

184. Y. Yamamura, N. Nakajima, and T. Tsuji, *Phys. Rev.* **B64**, 184109 (2001).

185. A. C. Baily and B. Yates, *J. Appl. Phys.* **41**, 5088 (1970).

186. W. G. Burgers, *Physica (Utrecht)* **1**, 561 (1934).

187. P. A. Madden, K. F. O'Sullivan, and G. Chiarott, *Phys. Rev.* **B45**, 10206 (1992).

188. O. Seok and D. W. Oxtoby, *Phys. Rev.* **B56**, 11485 (1997).

189. O. Seok and D. W. Oxtoby, *Phys. Rev.* **B58**, 5146 (1998).

190. J. Zhang, L. D. Zhang, X. F. Wang, *et al.*, *J. Chem. Phys.* **115**, 5714 (2001).

191. V. P. Menon and C. R. Martin, *Anal. Chem.* **67**, 1920 (1995).

192. A. Huczko, *Appl. Phys.* **A70**, 365 (2000).

193. M. Nishizawa, K. Mukai, C. R. Martin *et al.*, *J. Electrochem. Soc.* **144**, 1923 (1997).

194. Y. K. Zhou, J. Huang, and H. L. Li, *Appl. Phys.* **A76**, 53 (2003).

195. X. Y. Yuan, G. S. Wu, L. D. Zhang *et al.*, *Solid Stat. Commun.* **130**, 420 (2004).

196. X. Y. Yuan, T. Xie, G. S. Wu, Y. Lin, G .W. Meng, and L. D. Zhang, *Phys.* **E23**, 75 (2004).

197. W. Chen, S. L. Tang, M. Lu, and Y. W. Du, *J. Phys. Condens. Matter* **15**, 4623 (2003).

198. A. J. Yin, J. Li, W. Jian, A. J. Bennett, and J. M. Xu, *Appl. Phys. Lett.* **79**, 1039 (2001).

199. S. Park, S. Kim, S. Lee, Z. G. Khim, K. Char, and T. Hyeon, *J. Am. Chem. Soc.* **22**, 8581 (2000).

200. Y. W. Wang, G. W. Meng, C. H. Ling, G. Z. Wang, and L. D. Zhang, *Chem. Phys. Lett.* **343**, 174 (2001).

201. A. L. Prieto, M. S. Sander, M. S. Martin, R. Gronsky, T. Sands, and A. M. Stacy, *J. Am. Chem. Soc.* **123**, 7160 (2001).

202. C. P. Gibson and K. J. Putzer, *Science* **267**, 1338 (1995).

203. M. P. Pileni, B. W. Ninham, T. G. Krzywicki, J. Tanori, I. Lisiecki, and A. Filankembo, *Adv. Mater.* **11**, 1358 (1999).

204. M. Li, H. Schnablegger, and S. Mann, *Nature* **402**, 393 (1999).

205. Kadavanich and A. P. Alivisatos, *Nature* **404**, 59 (2000).

206. X. Y. Yuan, G. S. Wu, T. Xie, Y. Lin, and L. D. Zhang, *Nanotechnology* **15**, 59 (2004).

[illegible]

[illegible]

[illegible]

[illegible]

[illegible]

[illegible]

[illegible]

[illegible]

[illegible]

Chapter 6

Controlled Growth of Carbon Nanotubes

- Preparation, morphologies and structures of CNTs
- Long and continuous carbon nanotube yarns
- Controlled synthesis of single-walled carbon nanotubes
- Synthesis of double-walled carbon nanotubes (DWNTs)

Chapter 6

Controlled Growth of Carbon Nanotubes

6.1 Introduction

The carbon nanotube which is formed by rolling the graphite sheet has a seamless tube structure.[1] Figure 6.1 is the structure model, in which a_1 and a_2 are the unit vectors of the two-dimensional lattice of a nanolayer of graphite, and O is the origin. The OAB′B rectangle on the graphite sheet can be rolled up through the overlapping of O with A and B with B′ to form the carbon nanotube. Two ends of the carbon nanotube are often closed by half a fullerene sphere. The axis of this carbon nanotube is parallel to OB, and the circumference length is OA. This carbon nanotube can be represented by a vector C_h (the connective line of O and A).

$$C_h = na_1 + ma_2. \tag{6.1}$$

Here, both m and n are integers, and they can be used to characterize the structure of carbon nanotubes. The diameter d_1 of carbon nanotubes and the chiral angle Θ which is the angle formed from C_h and a_1 are also often used to characterize the structure of carbon nanotubes. The parameters, including C_h, m, n, Θ and d_l, can be related by the following expressions:

$$d_1 = \frac{C_h}{\pi} = \sqrt{3}a_{c-c}(m^2 + mn + n^2)^{\frac{1}{2}} \cdot \frac{1}{\pi}, \tag{6.2}$$

$$\Theta = \tan^{-1}\left[\sqrt{3}n/(2m+n)\right], \tag{6.3}$$

where, a_{c-c} is the neighboring carbon atom distance on the graphite sheet.

The carbon nanotubes include single-walled carbon nanotubes (SWNTs) and multi-walled carbon nanotubes (MWNTs). SWNTs are divided into three kinds. When $n = m$ and $\Theta = 30°$, the carbon nanotube (CNT) belongs to the armchair nanotube. When n or $m = 0$, the CNT is the zigzag nanotube. When $0° < \Theta < 30°$, the CNT is named as the chiral

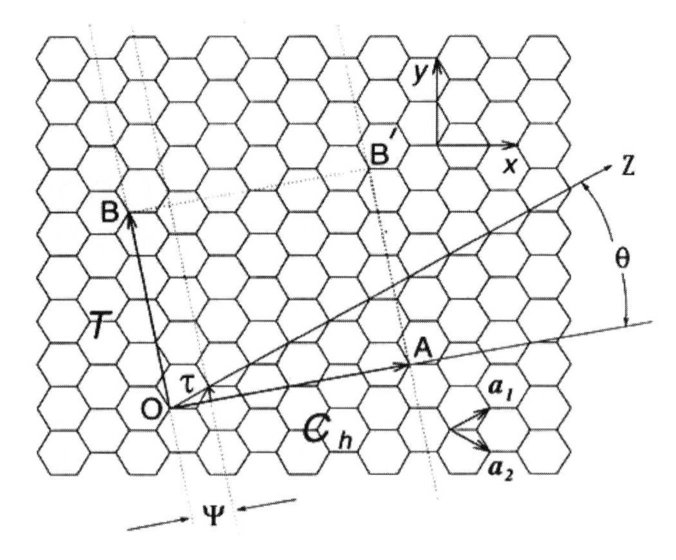

Fig. 6.1. Schematic representation of a unit cell OAB′B, which can form the CNT on a nanolayer of graphite.

nanotube. A MWNT is composed of several coaxial SWNTs, and these SWNTs have same structures or different structures.

Depending on the diameter and the helicity of the arrangement of graphite rings in the walls, the CNTs have been demonstrated to possess unusual electronic, phonic, magnetic, thermal and mechanical properties. For example, the less the diameter d_l is, the larger the difference between electronic state and sp^2, as a result the quantum effect becomes more obvious. Through theoretical calculating, Professor C. T. White *et al.*[2,3] indicate that when the difference between m and n is the integral times three, the CNTs are metallic. In other cases, the CNTs are semiconducting, and the width of the energy gap is directly proportional to $1/d_l$, regardless of their helicity. One-third of CNTs are metallic and two-thirds of CNTs are semiconducting. In 1998, through STM and STS (the electronic state density) experiments, Deker[4] and Lieber[5] *et al.* observed that the electronic transport properties of the individual CNT and its structure (the diameter d_l, and the chiral angle, Θ) were closely connected. Experimental results proved that the tiny change of the SWNT structure led to the change of carbon nanotube type, that is, the CNT sometimes is metallic and sometimes is semiconducting. Therefore, scientists hope to control the structure of CNTs during the preparation of CNTs. However, using present synthesis methods of CNTs, it is difficult to control the structure of CNTs. Now, the work on controlling the diameter of CNTs through the synthesis method of chemical vapor deposition on porous templates is an important progression. However, the controlling of the chiral angle, Θ, is still quite difficult. In addition, CNTs obtained by different synthesis methods are often a mixture of CNTs with different structures, including CNTs with different d_1 and Θ, SWNTs and MWNTs etc. Also, many products of carbon nanotubes are composed of random distribution nanotubes and their lengths are often shorter. The above-mentioned problems substantially affect the fundamental research and applications of carbon nanotubes. Therefore, finding out how to obtain pure SWNTs, pure MWNTs, long nanotubes and ordered arrays of CNTs with certain d_l and Θ is very important. Now, the work on the controlled synthesis of carbon nanotube structure presents an important progression. In this chapter, controlled synthesis, structures and morphologies of CNTs will be introduced.

6.2 Preparation, morphologies and structures of small diameter carbon nanotubes (CNTs)

Carbon nanotubes prepared by arc-discharge and laser evaporation possess perfect structure and very small defects. Therefore, most properties of CNTs were obtained through measuring these CNTs. The two ends of most of these CNTs are closed. For the small diameter CNTs, each end is generally closed by half a fullerene sphere. For the CNTs with the same structure, there are probably different kinds of semispheres to close their two ends. The (5, 5) CNT with an "armchair" structure and the (9, 0) CNT with a "zigzag" structure have respectively one kind of "fullerene semispheres" to close their ends. The (5, 5) and (9, 0) CNTs have less diameters, which are 0.688 and 0.715 nm, respectively. The number of "fullerene semisphere" structure increases rapidly with increase of the diameters of CNTs. For example, the (7, 7) CNTs with the diameter of 0.95 nm possess 87 kinds of "fullerene semispheres" to close their ends. Obviously, these fullerene semisphere structures are related to the diameter of CNTs. The larger the diameter is, the more kinds of fullerene semispheres are connected with the ends of CNTs.

These experimental results made by Xie *et al.* of the Institute of Physics, Chinese Academy of Sciences supposed that if the small "fullerene semispheres" were used as growth nuclei of CNTs. The small diameter CNTs should be obtained. Also, they supposed further to control the chiral of CNTs through controlling small nanotube diameters.

After CNTs were discovered, according to the close relation between CNTs and fullerene, it was inferred that two ends of the smallest CNTs are closed by C_{60}. During that time, no static fullerene with the diameter less than that of C_{60} was observed, and people thought that the smallest CNTs were the above-mentioned "armchair" type (5, 5) CNTs with the diameter of 0.688 nm and "zigzag" type (9, 0) CNTs with the diameter of 0.715 nm. In 1992, the Iijima group reported the CNTs with the diameter of 0.7 nm. People inferred that these CNTs are the (5, 5) or the (9, 0) nanotubes predicted theoretically. Until 1998, Professor Zettl synthesized successfully C_{36}.[6] Therefore, Xie *et al.* supposed the existence of the (0, 7) carbon nanotubes with the diameter less than that of C_{60}. They proposed that if bent graphite sheets are used as growth nuclei of the CNTs to prepare the CNTs, this will be energetically beneficial to the formation of CNTs with smaller diameters. Therefore, they used CNTs formed by rolling graphite sheets as precursors to synthesize CNTs with smaller diameters. As a result, they obtained successfully CNTs with the diameter of 0.5 nm. In January 2000, this result was published in Nature. In October 2000, Peng *et al.*[7] reported that they discovered CNTs with the diameter of 0.33 nm. In November of this year, Qin and Iijima *et al.*[8] and Wang *et al.*[9] reported the existence of CNTs with the diameter of 0.4 nm. In March 2004, a research group in Japan again pointed out that the diameter of the smallest CNTs is 0.3 nm. In 2002, Zhou *et al.*[10] obtained double-walled CNTs with the interval diameter of 0.4 nm. What is the smallest diameter of CNTs? Now, researchers still continue to study this subject.

6.2.1. Multi-walled carbon nanotubes (MWNTs)

In order to prepare the small diameter CNTs, Xie *et al.*[11] began to synthesize the CNTs as precursors of synthesizing the small diameter CNTs by arc-discharge. The anode and the cathode are graphite rods with diameters of 6 and 30 mm, respectively. The discharge chamber was first evacuated and then, the carrier gas (He or H_2) was pumped into the chamber to keep the gas pressure at 0.6 MPa. After the discharge started, the electric voltage between two electrodes was regulated to 20 V and the discharge current was controlled to 58 A (the current density was 160 A/cm^2). After the discharge was finished, the inner black core of a cathode deposit was taken out. These black products were composed of CNTs, amorphous carbon and carbon nanoparticles.

During the synthesis of the smallest CNTs ($d = 0.5$ nm), Xie *et al.*[11] used the same apparatus and conditions as described in Ref. 12, except for the anode.

The home-made arc-discharge apparatus, which consists of the discharge chamber, the current power supply, the vacuum system, the mechanical system, the monitored control system and the water-cooling system. Cooling water ran through the wall of the chamber, and the two electrodes were also water-cooled to enhance the cooling effect. A 30 mm diameter graphite disk of high density, high strength and spectral purity was used as the cathode, which was compactly mounted on a water-cooled copper holder. A graphite rod of spectral purity, 6 mm diameter, was used as the anode. The anode was prepared by boring a 3 mm diameter hole in the 6 mm diameter graphite rod. This hole was filled with a mixture of cobalt metal powder (5 at.% metal) and the above-mentioned black products. The anode can be fed at a steady rate between 0 and 3 cm/minute. Before the arc was initiated, the chamber was first vacuumed and then the helium was sent in until the gas pressure was 50–70 KPa. During the discharge, the arc current and voltage can be monitored and precisely controlled by related apparatus and the state of the arc can be visually observed through a window by using eye protection glasses.

Except for the arc initiation period, a stable discharge was sustained by steadily feeding the anode at a certain rate and setting the arc voltage and currents to certain parameters between 19 and 25 V and 60 and 80 A, respectively. The arc was mainly confined in the space between the opposite surfaces of the two electrodes and gave off glaring lights. By a visual examination of the arc, it was found that there was no erratic movement of the arc spots and the temperature was distributed homogeneously and constantly in the arc area. The whole cathode deposit was red hot and the anode was also red hot within a certain length at the discharging ends but the temperature on the two opposite surfaces was obviously higher. Under proper conditions, when the feeding of the anode was stopped, the interelectrode distance kept unchanged while the anode was being consumed and the cathode deposit was growing at the same time. In this "self-sustained" mode, an extremely stable and continuous discharge was achieved.

In the discharge, most of the graphite consumed from the anode was deposited on the cathode. The morphologies and microstructures of the deposit product in the center position of the cathode deposit rod were characterized by a high resolution transmission electron microscope (HRTEM, JEM 200-cx, operated at 200 kV).

Figure 6.2 shows an HRTEM image of the black products. It can be seen that the black products are MWNTs, and each CNT shows a uniform intershell spacing of 0.34 nm. The open arrow in Fig. 6.2 indicates the innermost

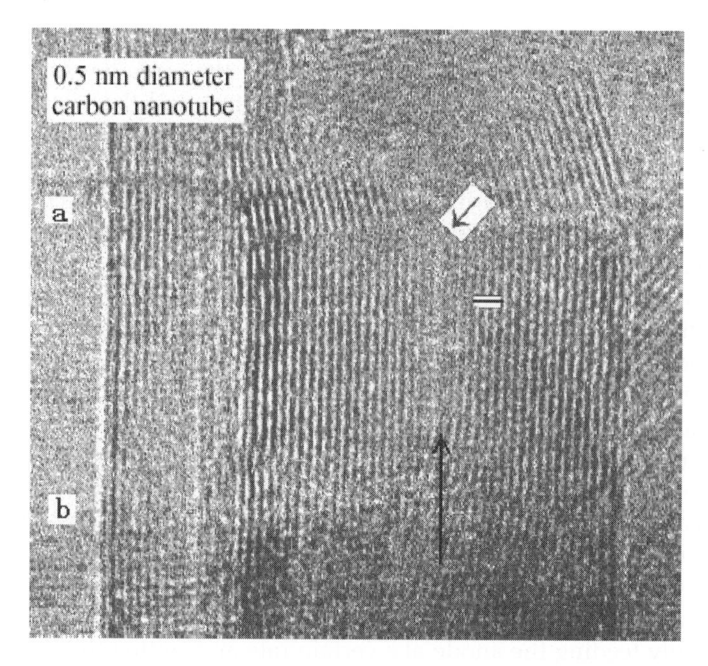

Fig. 6.2. HRTEM image of two nanotubes with outer diameters of 6 nm and 15 nm, respectively. The bottom arrow points to the innermost tube in the right nanotubes, which has a diameter of 0.5 nm. The other arrow which points to the tip of this tube, which is closed with half a C_{36} cage; scale bar: 1 nm.

nanotube, with a diameter of about 0.5 nm. This innermost nanotube is closed at the end by a half-circle, as indicated in Fig. 6.2. It is supposed that it is closed by half a C_{36} cage. Theoretical results predict that a zigzag nanotube with this diameter might have an (m, n) value of (6, 0) and a diameter of 0.47 nm or (7, 0) (0.55 nm).

The 0.5 nm diameter CNTs may grow from curved carbon fragments because the discharge anode was filled with carbon nanotubes. It can be supposed that the curved carbon fragments within the arc will require less energy to form carbon nanotubes. The use of the cobalt catalyst will result in curved fragments forming 0.5 nm diameter carbon nanotubes.

After the above work was published in November 2000, Iijima *et al.*[8] reported that they had discovered CNTs with the diameter of 0.4 nm. This small CNT is the innermost tube of the MWNT. They pointed out that these CNTs were seen in cathodic deposits produced by the arc-discharge of graphite rods in a hydrogen atmosphere without a metallic catalyst.[13]

Under these conditions, more CNTs (all multi-walled) are kept open as hydrogen etches away the capping atoms, a unique feature that helps maintain a favorable environment for smaller CNTs to form inside already-grown MWNTs. HRTEM images show that some of the smallest nanotubes were capped. The smallest CNTs with the diameters of 0.4 nm have an antichiral [3, 3] "armchair" structure. Iijima *et al.* proposed that these 0.4 nm nanotubes grow out of half of a C_{20} dodecahedron, in which the C–C bonding angle is 108°. Growth into a full nanotube is realized through a step-by-step mechanism on both the outer and inner surfaces.[14] The hydrogen atmosphere facilitates the formation of the halves of the C_{20} dodecahedron, stabilizing the structure by terminating certain dangling bonds with hydrogen, and keeps the inside of the nanotubes open so that after the confining outer shells have been formed, carbon species can enter the core to form the innermost shell.

At the same time when the above-described results were being reported, Wang *et al.*[9] reported that they obtained single-walled 0.4 nm carbon nanotube arrays.

6.2.2. Single-walled carbon nanotubes (SWNTs)

When Xie *et al.*[11] synthesized the MWNTs with the innermost tube of 0.5 nm diameter as described in Sec. 6.2.1, they observed the ash on the wall of the discharge chamber. As a result, in the ash they found the SWNTs. These nanotubes twined each other and formed the SWNT bundles, in which many catalyst particles were distributed, as shown in Figs. 6.3–6.5 which show HRTEM images of SWNT bundles and individual SWNTs, respectively. The SWNT bundle has a diameter of several tens of nm and is often curved. Figure 6.5(a) shows three small-diameter SWNTs, and "1", "2" and "3" in Fig. 6.5(a) represent SWNTs of the diameters of 0.7, 0.9 and 0.7 nm, respectively.

In Fig. 6.5(b), "2", "4" and "5" correspond to SWNTs of the diameter of 7 nm and "1" and "3" correspond to SWNTs of the diameters of 0.9 and 1.2 nm, respectively. Figure 6.5(c) displays the 0.5 nm diameter SWNTs.

6.2.3. Discussion and analysis[15]

The above results indicate that 0.5 nm diameter CNTs can be synthesized by the arc-discharge of the anode filled with CNTs. In the following, whether

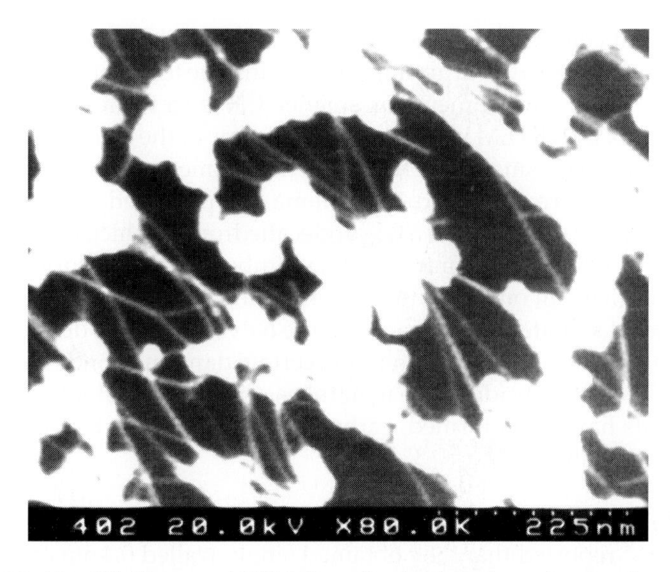

Fig. 6.3. The SEM image of SWNT bundles in the ash on the chamber wall.

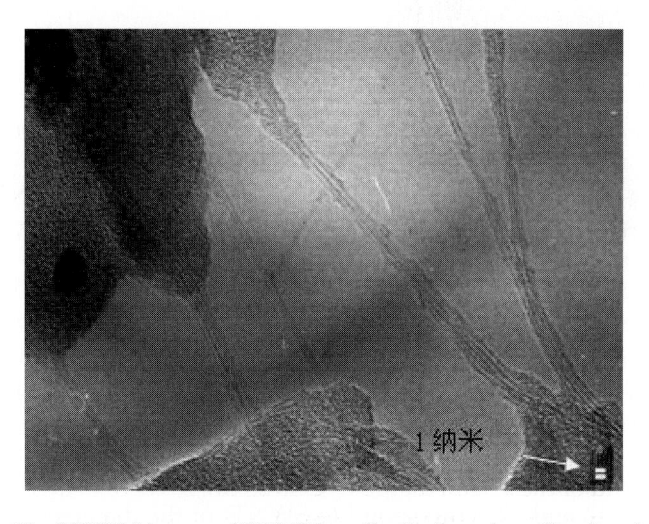

Fig. 6.4. The HRTEM image of SWNT bundles in the ash on the chamber wall.

the structure of the small diameter CNT can be controlled will be analyzed and explored.

Firstly, Xie *et al.* analyzed what the structure of the two end caps of a CNT is. The fact shows that there exist only two stable structure fullerence

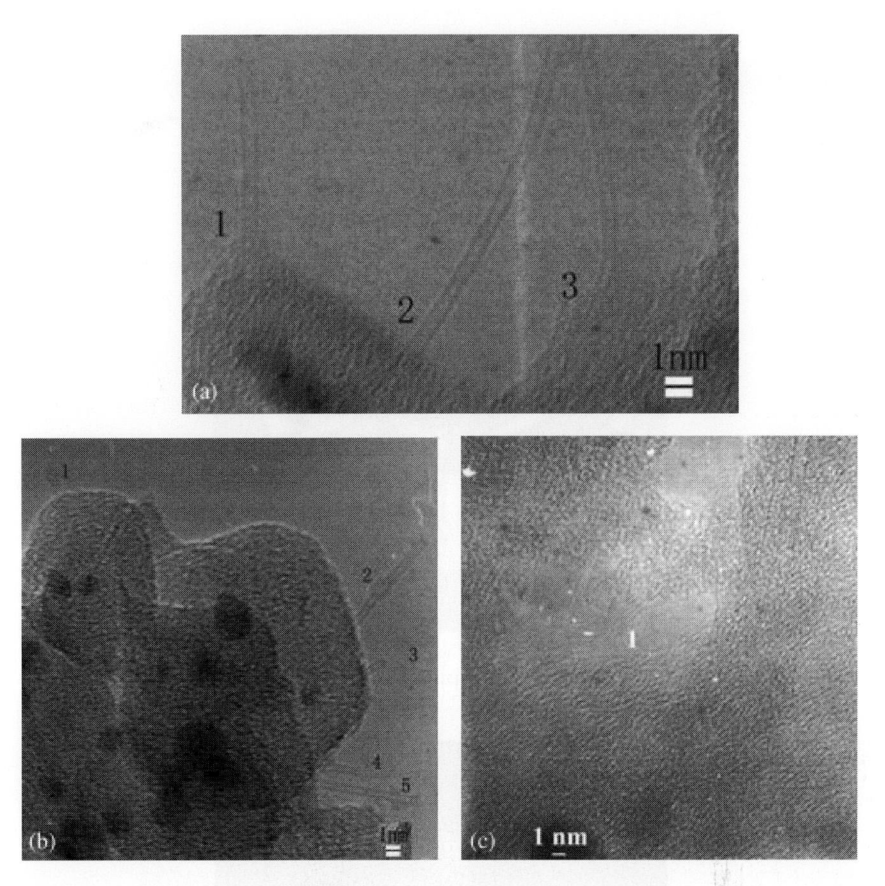

Fig. 6.5. (a), (b) and (c) are HRTEM images of individual SWNTs in the ash on the chamber wall.

with the diameter less than that of C_{60}. They are C_{36} and C_{20}. The diameters of C_{36} and C_{20} are 0.5 and 0.4 nm, respectively. Therefore, it can be supposed that each end of the 0.5 nm diameter CNT is closed by only half a C_{36} cage. Figure 6.6 gives the schematic picture of the C_{36} structure and Fig. 6.7 is the structure of the CNT, whose two ends are closed by C_{36}. It is clear that the C_{36} cage structure is not a sphere, but a spindle. A CNT is closed by the CNT connecting C_{36} with the direction perpendicular to the normals of six-member rings of the middle part of C_{36}. Therefore, the CNT has only one chiral. Namely, the CNT is the (6, 0) tube (Fig. 6.7). This means that the diameter and the chiral of CNTs can be controlled by the arc-discharge of the anode filled with CNTs.

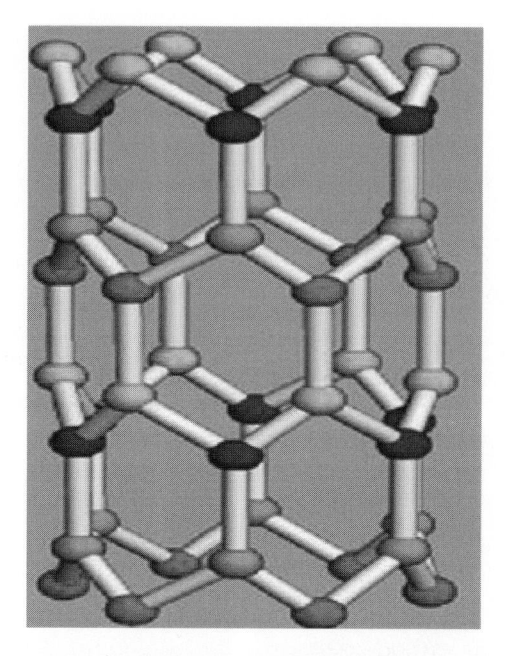

Fig. 6.6. The schematic picture of the C_{36} structure.

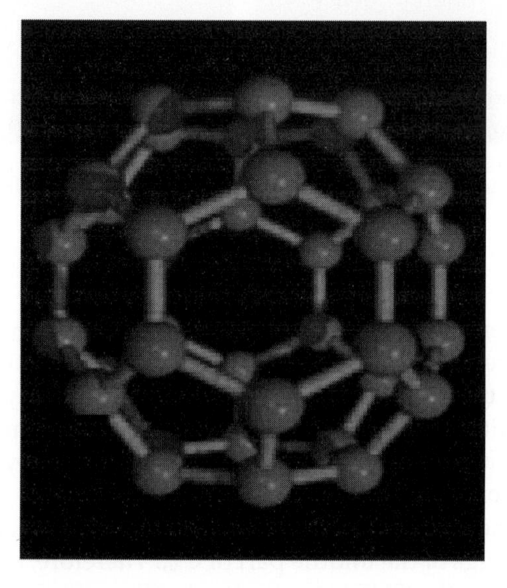

Fig. 6.7. The structure of the CNT, whose two ends are closed by C_{36}.

In the following, the growth mechanism of CNTs obtained by the arc-discharge of the anode filled with CNTs will be analyzed.

In order to explore the growth mechanism of CNTs, a graphite anode filled with graphite powder was used instead of the graphite anode filled with the CNTs to arc-discharge in the same experimental conditions. Results showed that no MWNTs with the inner diameter of 0.5 nm or the 0.5 nm diameter SWNTs were observed. Therefore, the growth mechanism of CNTs in the arc can be supposed as follows. When carbon in the anode entered the arc, two shape carbon groups existed: one is the curved carbon atom group and the other is the flat carbon atom group. Under reasonable conditions (the gas pressure, current etc.), the curved tube-like carbon sheets (the curved carbon atom group) possess lower energy. Therefore, they formed SWNTs in the arc.[16,17] Parts of SWNTs in the arc flowed out from the arc column and then reached the chamber wall. In this process, the carbon atom density in the carrier was lower and thus along the CNT's diameter direction, no growth appeared. This is why the CNTs on the chamber wall are still SWNTs. The Van der Waals forces between SWNTs led to the formation of the SWNT handles. Parts of SWNTs in the arc reached the cathode deposition rod and at this time, these SWNTs entered the carbon vapor of Maxwellian distribution.[18] In the carbon vapor, the SWNTs grew along their diameter direction. As a result, MWNTs formed.

The diameter of CNTs prepared using CNTs as the anode is less because the curved carbon atom sheet needs less energy forming the CNT in comparison with the flat carbon atom sheet. This is why 0.5 nm diameter CNTs can be produced by using the anode filled with CNTs during the arc-discharge.

In the following, according to the energy principle, the problem of the small diameter CNT formation will be further analyzed.

When a graphite sheet is rolled up to form a CNT, the energy change (ΔE) can be divided into two parts. One part is the increase of elastic energy, E_{bend}, caused by bending the graphite sheet. The other part is the decrease of energy, E_{bind}, caused by the graphite dangling bond descending, which is due to the formation of the closed structure of CNT. Whether the CNT can be formed is closely related to $E_{bend} + E_{bind}(= \Delta E)$.$E_{bend}$[16] and E_{bind} can be expressed as follows:

$$E_{bend} = C_1(1/D)^2 + C_2(1/D)^4. \tag{6.4}$$

Here, C_1 and C_2 equal 5.64 eV·Å^2 and 25.1 eV·Å, respectively,

$$E_{bind} = -b(1/D). \tag{6.5}$$

Here, b can be determined by using the following method. When $\Delta E < 0$, the CNTs can exist. Under the conditions of arc-discharge of the graphite rods, the 0.7 nm diameter CNTs may be found. It is supposed that in these conditions, $\Delta E = 0$. ΔE (0.7 nm) is expressed as follows:

$$\Delta E(0.7) = E_{bend} + E_{bind} = 5.64\left(\frac{1}{7}\right)^2 + 25.1\left(\frac{1}{7}\right)^4 - b\left(\frac{1}{7}\right) = 0, \tag{6.6}$$
$$b = 0.875\,\text{eV} \cdot \text{Å}.$$

Under the same conditions, for the 0.5 nm diameter CNT, ΔE (0.5 nm) $= E_{bend} + E_{bind} = 5.64\left(\frac{1}{5}\right)^2 + 25.1\left(\frac{1}{5}\right)^4 - 0.875\left(\frac{1}{5}\right) = 0.051 > 0$. Therefore, when CNTs are prepared by arc-discharge, using pure graphite as the carbon source, the 0.5 nm diameter CNTs cannot be obtained. However, when the CNTs are used as the carbon source, the CNTs can grow from the bended graphite sheets. These bended sheets can provide the above-mentioned energy (0.051). The bend degree is estimated as follows:

$$E_{bend} = \left[C_1\left(\frac{1}{D}\right)^2 + C_2\left(\frac{1}{D}\right)^4\right] \times \frac{5}{D} = 0.051, \tag{6.7}$$
$$D = 10.5\,\text{Å}.$$

The bend degree of the graphite sheet, which forms the 0.5 nm diameter CNT, is $\frac{5}{10.6} \times 2\pi$ $(=0.94\pi)$. Therefore, the shape of this graphite sheet approximates to a semicircle.

6.3 Very long carbon nanotubes and continuous carbon nanotube yarns (fibers)

The carbon nanotubes (CNTs) have unique characteristics and many potential applications. For instance, they can be used to make single electron transistors, field effect transistors, atomic force microscope probes, field emission electron guns, gas sensors and nanotemplates etc. Now, most applications belong to the application under the microscopic scale. Discovering how to use the unusual mechanical, thermoconducting and electric properties etc. of carbon nanotubes in the macroscopic scale is a challenge to researchers. There are two routes to solve this problem. One way is through the synthesis of CNTs with macroscopic lengths. Many researchers[19–21] obtained very long carbon nanotubes. But it is difficult to synthesize carbon nanotubes of any length. The other way is to form yarns by connecting CNTs. In 2000, Vigolo *et al.*[22] reported that by using a polymer solution, they aligned and connected dispersed SWNTs to form continuous ribbons or yarns. In 2002, Fan *et al.*[23] reported that continuous carbon nanotube yarns were drawn out from superaligned arrays of carbon nanotubes. In 2004, Li *et al.*[24] published a paper, in which they reported continuously spun fibers consisting of pure carbon nanotube fibers directly spun from an aerogel formed during synthesis by chemical vapor deposition. These results are beneficial to applications of CNTs in the macroscopic scale.

6.3.1. Very long carbon nanotubes[19]

Xie *et al.* reported that pyrolysis of acetylene over iron/silicon substrates was an effective method with which to produce very long, multi-walled carbon nanotubes that reached about 2 mm in length.

To obtain long carbon nanotubes (CNTs), the substrates were prepared by a sol-gel process using the following technique. Tetraethoxysilane (10 ml) with 1.5 M iron nitrate aqueous solution (15 ml) and ethyl alcohol (10 ml) were mixed by magnetic stirring for 20 minutes. A few drops of concentrated hydrogen fluoride (0.4 ml) were added into this mixture, which was stirred for another 20 minutes. The mixture was then dropped onto a quartz plate to form a film 30–50 nm thick. After gelatin, the film was dried overnight at 80°C, during which time the gel cracked into small pieces of substrate of area 5–20 mm^2. The substrates were calcined at 450°C for 10 h under vacuum and then reduced at 500°C for 5 h in a flow of 9% H_2/N_2 under 180 torr.

At this stage, large quantities of iron/silica nanoparticles, which acted as the catalyst for nanotube growth, formed evenly on all surfaces of the substrates. The substrates were put in a furnace. Subsequently, CNTs were produced on the substrates in a flow of 9% acetylene in nitrogen at 600°C for 1–48 h under 180 torr.

Figure 6.8 displays the low-magnification electron microscope image of aligned CNT (carbon nanotube) arrays (Fig. 6.8(a)). The lengths of the nanotube arrays increased with growth time, and reached about 2 mm (Fig. 6.8(b)) after 48 h of growth.

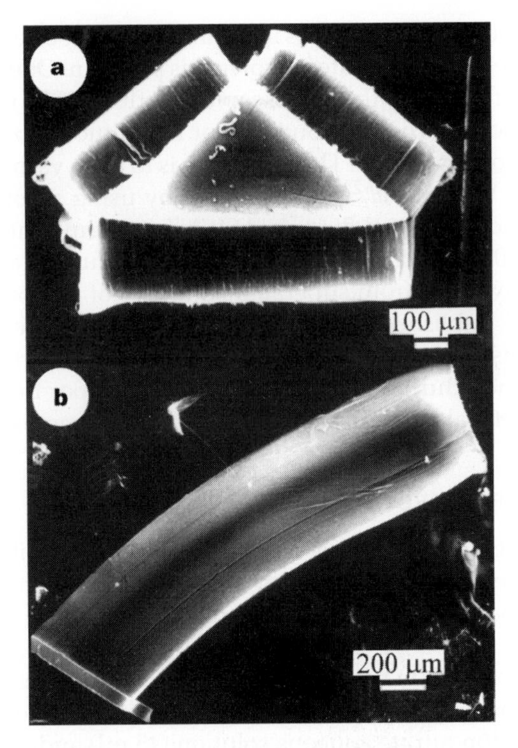

Fig. 6.8. Low-magnification scanning electron microscope images of aligned carbon nanotube arrays. (a) Top view of a sample after 5 h of growth. The triangular substrate is covered with nanotube arrays growing outwards perpendicularly from every surface of the substrate. (b) A sample after 48 h of growth. Only one nanotube array remains after stripping off other arrays from the substrate.

High-magnification scanning electron microscope images (Fig. 6.9(a)) indicate that the CNTs grow outwards separately and perpendicularly from the substrate to form an array. The external diameter of CNTs is about 20–40 nm, with a spacing of about 100 nm between the tubes.

High-resolution transmission electron microscope observations (Fig. 6.9(b)) show that the multi-layered graphite tubes (10–30 layers) are coated with a small amount of amorphous carbon on their periphery owing to the low growth temperature (600°C) and long growth time.

The results of resistivity measurements demonstrate that the resistivity of some thinner nanotube bundles cleaned from the arrays reaches 10^{-2} to

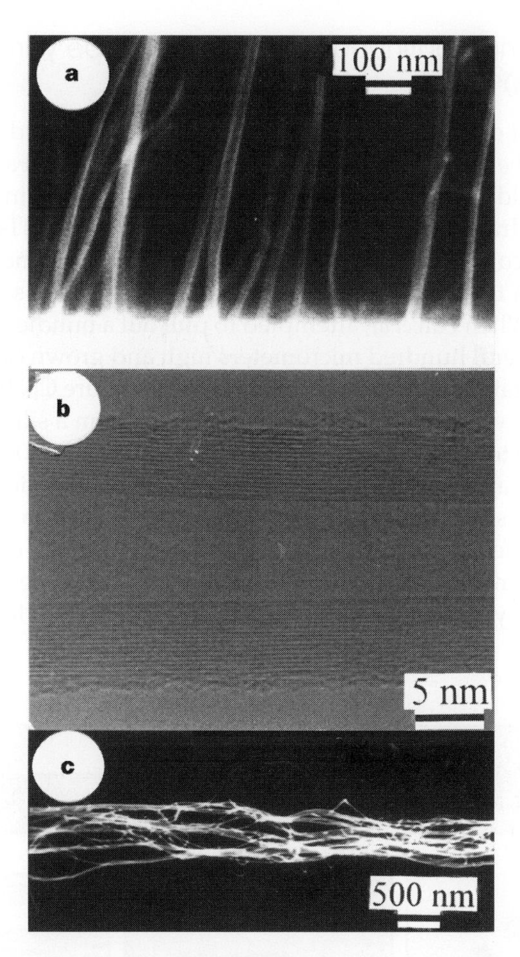

Fig. 6.9. Scanning and transmission electron microscope images of nanotubes after 48 h of growth. (a) High-magnification scanning electron microscope image of an aligned carbon nanotube array. (b) High-resolution transmission electron microscope image of a carbon nanotube. (c) Scanning electron microscope image of a thin nanotube bundle cleaved from a nanotube array along the growth direction of the nanotubes. The bundle contains about 20 nanotubes.

$10^{-3}\Omega$ cm at room temperature. This resistivity is an order of magnitude larger than the value reported previously[25,26] because of the presence of larger numbers of defects in these very long nanotubes. These results indicate that these very long nanotubes do indeed grow continuously without any interruption within the nanotube arrays.

6.3.2. Spinning continuous carbon nanotube yarns (fibers)[23,24]

The creation of continuous yarns made out of CNTs would enable macroscopic nanotube devices and structures to be constructed. Fan *et al.*[23] found that CNTs could be self-assembled into yarns of up to 30 cm in length simply by being drawn out from superaligned arrays of CNTs, and that the strength and conductivity of these yarns could be enhanced by heating them at a high temperature. These continuous CNT yarns were obtained occasionally. When Fan *et al.* attempted to pull out a bundle of CNTs from a CNT array several hundred micrometers high and grown on a silicon substrate, they obtained instead a continuous yarn of pure CNTs (Fig. 6.10(a)). This process is very similar to drawing a thread from a silk cocoon, corresponding here to the CNT array. Figure 6.10(b) shows a 100 μm high, free standing CNT array held by adhesive tape. The indentation at the top of the array marks the region that is being turned into a yarn 30 cm length and 200 μm wide. Continuous yarns can only be drawn out from superaligned arrays in which the CNTs are aligned parallel to one another and are held together by Van der Waals interactions to form bundles (Fig. 6.10(c)).

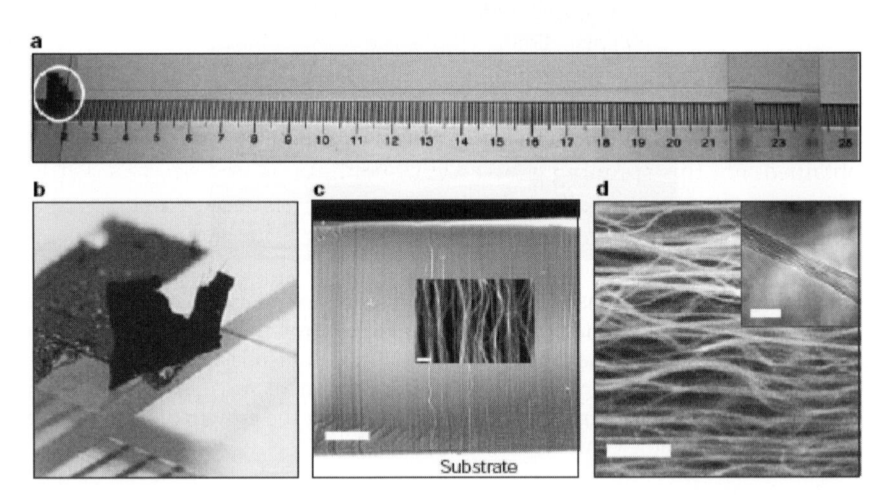

Fig. 6.10. Carbon nanotube yarns. (a), (b) A carbon nanotube yarn being continuously pulled out from a free standing carbon nanotube array (a) which is shown enlarged in (b) (roughly $\times$ 8 magnification). (c) Scanning electron microscope (SEM) images of a carbon nanotube array grown on a silicon substrate, showing the superalignment of carbon nanotubes (scale bars: 100 μm; inset, 200 nm). (d) SEM image of the yarn in (a); inset: transmission electron microscope (TEM) image of a single thread of the yarn (scale bars: 500 nm; inset: 100 nm).

The yarns usually appear as thin ribbons, which are composed of parallel threads (Fig. 6.10(d)). The size of the yarn can be controlled by the tip size of the tool that is used to pick up the yarn. The smaller the tip, the thinner the yarn.

In the following, direct spinning of CNT fibers from chemical vapor deposition synthesis will be introduced.[24] The key requirements for continuous spinning are the rapid production of the highly pure nanotube to form an aerogel in the furnace hot zone and the forcible removal of the product from reaction by continuous wind-up. Therefore, Li *et al.* used ethanol as the carbon source, and 0.23 to 2.3 wt.% ferrocene and 1.0 to 4.0 wt.% thiophene were dissolved in ethanol. Then, the solution was injected at 0.08 to 0.25 ml/minute from the top of the furnace into a hydrogen carrier gas that flowed at 400 to 800 ml/minute, with the furnace hot zone in the range of 1050 to 1200°C. The aerogel occurred very soon after the introduction of the solution (the precursors). It was then stretched by the gas flow into the form of a sock, elongating downwards along the furnace axis. The aerogel was continuously drawn from the hot zone by winding it onto a rotating rod. The authors evaluated several possible wind-up protocols with different geometries and spinning temperatures. In one of these protocols, the spindle was aligned at about 25° to the furnace axis and is rotated about that axis at 90 rpm. The device penetrated the hot zone to capture the aerogel before it reached the cool zone. When the area covered by this spindle was 12% of the cross-section of the furnace tube, the majority of the nanotubes produced could be captured and continuous spinning was obtained. For this spinning geometry the continuous fibers have a degree of twist and they were collected either at the top of the spindle or along its length. Fibers were unwound from the spindle and wound up onto another rod. For the second wind-up geometry, the spindle was rotated normal to the furnace axis outside the hot zone, at a position where the temperature was about 100°C. The advantage of this geometry is that well-defined, thin, unentangled films could be obtained when the wind-up speed was close to the velocity of the gas. With an increasing wind-up speed, the thin films separated into discrete threads that were wound simultaneously onto the spindle.

For the spun fibers, MWNTs or SWNTs could be controlled. When the concentration of thiophene in ethanol was 1.5 to 4.0 wt.%, the hydrogen flow rate was 400 to 800 ml/minute and the synthesis temperature was 1100 to 1180°C, MWNTs formed. When the concentration of thiophene in ethanol, the hydrogen flow rate and the synthesis temperature were about 0.5 wt.%, about 1200 ml/minute and 1200°C, respectively, SWNTs formed.

In order to achieve a clean product, it is important that the final stage in the reaction chain to produce carbon is finely balanced, so as to discourage the formation of non-fibrous particles with the nanotubes. Experiments proved that a high hydrogen flow rate was found to suppress carbon formation, whereas the removal of hydrogen led to the precipitation of particulate carbon rather than nanotubes.

The alignment, purity and structure of the fibers obtained from ethanol-based reactions were characterized by electron microscopy, Raman spectroscopy and thermogravimetric analysis. In MWNT fibers, the nanotube diameters were 30 nm, with an aspect ratio of about 1000. They contained 5 to 10 wt.% iron. The quality of alignment of the nanotubes was measured from transforms of scanning electron microscope images. The full-width of the inter-nanotube interference peak measured around the azimuthal circle was about 11°. The SWNT fibers contained more impurities than the MWNT fibers. The diameters of SWNTs were between 1.6 and 3.5 nm. The SWNTs were organized in bundles with lateral dimensions of 30 nm. Raman spectra show the typical radial breathing modes, with peaks at 180, 243 and 262 cm^{-1} with a 514.5 nm excitation laser.

6.4 Controlled synthesis of single-walled carbon nanotubes

Besides laser ablation and arc-discharge, chemical vapor deposition (CVD) of the pyrolysis of C-containing compounds such as CO, CH_4, and C_2H_4 etc. on nanometallic particles is also used to prepare CNTs. In past years, people developed the floating catalyst CVD to prepare SWNTs. This synthesis method is a continuous growth CNT method. Namely, CNTs are continuously produced in the flowing reaction gas.[20,27–30] Cheng *et al.* prepared large numbers of SWNTs by using the floating catalyst CVD method through the pyrolysis of the mixture of benzene-dicyclopentadienyl iron-thiophene at 1200°C. After that, Rao *et al.*[27] Nikolaev *et al.*[28] and Bladh *et al.*[29] used a similar method to prepare SWNTs through the pyrolysis of other C-containing gases such as C_2H_4, CO and CH_4 etc. at 800 to 1200°C. The research results show that CO is one better carbon source because the SWNTs made of CO are cleaner and there are less amorphous carbon covers on the surfaces of CNTs.[28] Inversely, when other hydrocarbons are used to prepare SWNTs, there are usually more amorphous carbon covers on the surface of SWNTs because these hydrocarbons such as C_2H_4, C_6H_6 easily show a rapid pyrolysis. These surface covers are difficult to be cleaned. However, under a certain reaction temperature and pressure, the decomposition rate of CO is much lower than that of hydrocarbons. Therefore, during the preparation of SWNTs, a large flow of CO (1000 ~ 2000 cm^3/minute) should be used, that is, the carbon transform efficiency of CO is lower. Xie *et al.*[31] used C_2H_2 as a precursor to successfully prepare pure SWNTs.

6.4.1. Preparation of pure single-walled carbon nanotubes

The preparation apparatus of pure SWNTs[37] is shown in Fig. 6.11. The experimental reactor is a quartz tube with an inner diameter of 30 mm. The dicyclopentadienyl iron is used as the catalyst. Firstly, dicyclopentadienyl sublimed in the first furnace. The sublimation temperature of dicyclopentadienyl is $60 \sim 90$°C. The sublimated dicyclopentadienyl was then carried by the mixture gas of argon (1200 cm^3/minute) and C_2H_2 ($3 \sim 10$ cm^3/minute) into the second furnace, in which the SWNTs formed. The second furnace temperature was controlled at 750 to 1200°C. The pressure in the quartz reactor was kept at about 0.1 MPa. A water-cooling collector was assembled at the end of the quartz tube. The formed SWNTs were carried by the gas flow to the water-cooling collector, resulting in the formation of the film. The SEM observation of the film shows that the film production contains mainly small diameter fibers, which are twined with each other, and some small particles. The HRTEM images indicate that these small diameter fibers are SWNT bundles or individual SWNTs, and these small particles are iron particles which are capped by several carbon layers (Fig. 6.12). These SWNTs and particles are not capped by amorphous carbon. This is because the low carbon source bias pressure was exactly controlled, resulting in no remainder self-pyrolysis carbon depositing onto the surfaces of the products.

The C_2H_2 bias pressure substantially affects the SWNT growth. Xie *et al.*[31] found that when the carbon source bias pressure was higher than

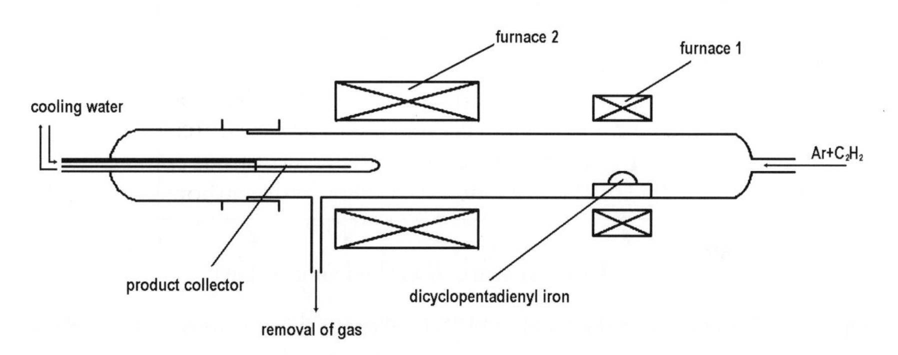

Fig. 6.11. The floating catalyst CVD experiment apparatus.

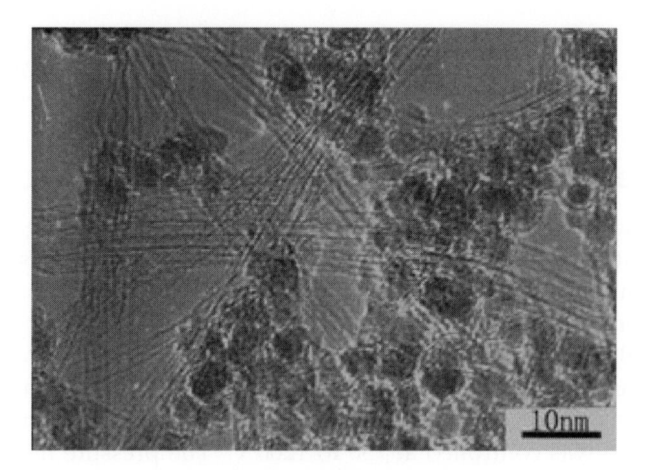

Fig. 6.12. The HRTEM image of SWNTs.

12 torr, almost no SWNT deposition appeared on the surface of the collector. Figure 6.13 demonstrates the change of the yield of SWNTs with the carbon source bias pressure. From this figure, it can be seen that when the bias pressure of the carbon source is lower than 5.0 torr, the yield of SWNTs increases with the increase of the carbon bias pressure. However, when the carbon bias pressure exceeds 5.0 torr, the yield of SWNTs begins to decrease. This is because too much self-pyrolysis carbon was produced and thus the

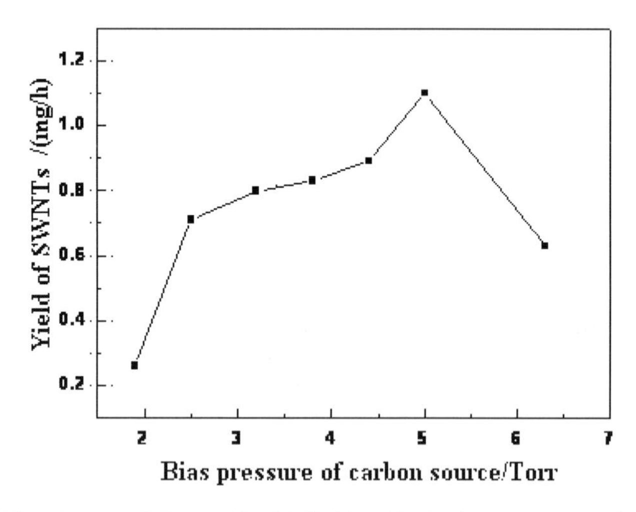

Fig. 6.13. The change of the yield of SWNTs with the bias pressure of the carbon source.

catalyst particles were capped by the remaining pyrolysis carbon, resulting in the loss of the catalyst activity. When Hafier *et al.*[31] prepared SWNTs using the catalysis pyrolysis of C_2H_4 or CO, their experimental results also proved that the growth rate of SWNTs is controlled by the carbon source quantity provided by the catalyst particles.

The growth temperature of SWNTs is also an important parameter. According to literature reports, when hydrocarbons are used as the precursors and the floating catalysis method is used to prepare SWNTs, the growth temperature of SWNTs is generally 1100 to 1200°C. Xie *et al.* found that by using C_2H_2 as the precursor, the SWNTs could be synthesized in a wider temperature range (750 to 1200°C). Moreover, different reaction temperatures affected the yield of SWNTs. Table 6.1 lists the relation between the reaction temperature and the yield of SWNTs. At a temperature lower than 800°C, only fewer products adhere to the collector. But, when the reaction temperature is higher than 1100°C, the main product is grey amorphous carbon. The optimization temperature range for SWNT growth is 900 to 1000°C. This is different from other researcher's reports, that suggest that the higher the reaction temperature (up to 1200°C) is, the higher the yield of SWNTs.[27,28] Why do Xie's experimental results at a high temperature show the lower yield? This can be analyzed as follows. The self-pyrolysis rate of C_2H_2 is accelerated at a high temperature and thus the rate that the carbon source provides to the catalyst particles is altered. As a result, more amorphous carbon is produced and fewer SWNTs are formed.

Through the change of the sublimation temperature of dicyclopentadienyl iron, the numbers of iron particles capped by carbon in the products, which are mainly composed of pure SWNTs and iron particles capped by carbon, can be controlled. Under the lower sublimation temperature of dicyclopentadienyl iron (the temperature of the first furnace), the quantity

Table 6.1. The relation between the SWNT yield and the growth temperature.

Temperature/°C	Dicyclopentadienyl iron/($\times 10^{-5}$ mol/h)	C_2H_2 bias pressure/Torr	Yield/(mg/h)
800	4.68	5.0	0.16
850	4.68	5.0	0.36
900	3.74	5.0	1.2
1000	4.68	5.0	1.10
1100	4.33	5.0	0.72

Table 6.2. The relation between the yield of SWNT films and the quantity used of dicyclopentadienyl iron.

Quantity used of dicyclopentadienyl iron/ ($\times 10^{-5}$ mol/h)	Growth temperature/°C	C_2H_2 bias pressure/Torr	Yield/(mg/h)
11.29	900	6.3	4.40
8.55	900	6.3	3.00
5.91	900	6.3	1.52
3.74	900	6.3	1.20

used of dicyclopentadienyl iron decreases and hence iron particles can be substantially decreased so that purer SWNTs can be obtained, as shown in Table 6.2.

From the HRTEM images, the diameter distribution of SWNTs was measured. The results are shown in Fig. 6.14. It is clear that the average diameter of SWNTs is 1.1 nm. The maximum diameter is less than 2 nm. The 0.7 nm diameter SWNTs are occasionally observed.

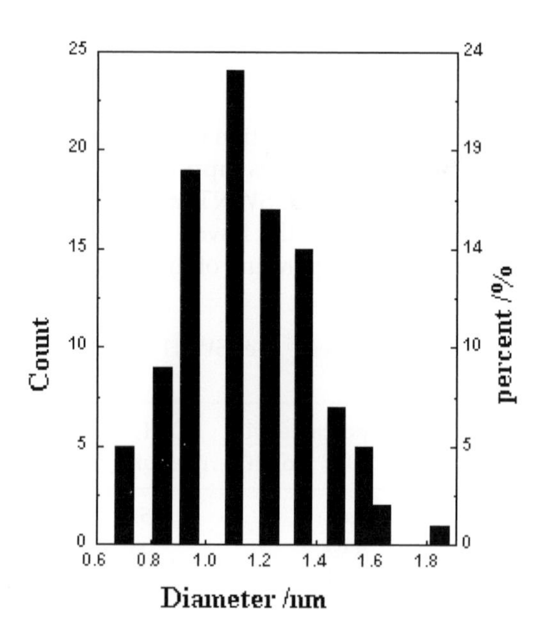

Fig. 6.14. The tube diameter distribution measured from HRTEM images.

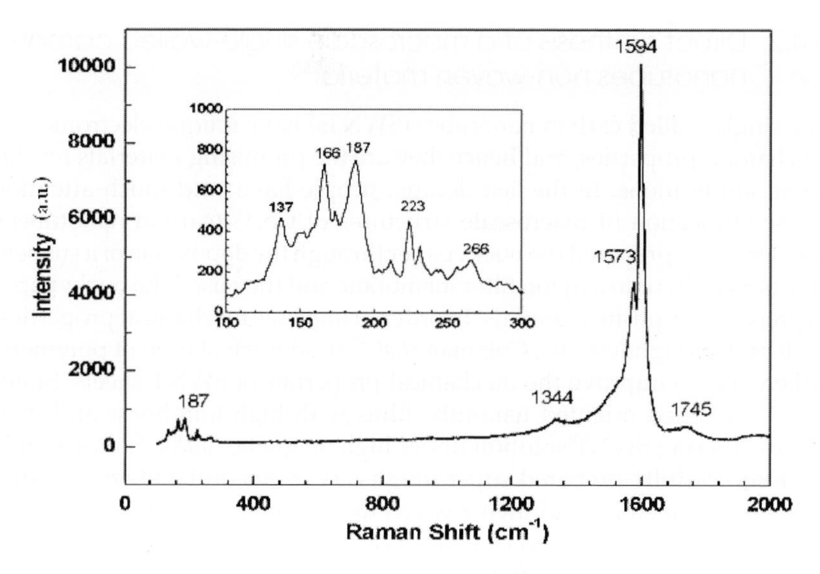

Fig. 6.15. The Raman spectra of typical SWNTs.

Figure 6.15 displays the Raman spectra of the products. The laser wavelength is 514.5 nm. There are two group feature peaks of SWNTs in Fig. 6.15. This shows that the as-received products are SWNTs. The inset gives the first group peaks located at the low frequently region, including the peaks located at 137, 166, 187, 223 and 266 cm^{-1}. They are the radial breathing modes of SWNT, and their frequency shift is related to the diameter of SWNTs. This relation is

$$\omega = \frac{223.75}{d}. \tag{6.8}$$

Here, ω is the frequency shift of breathing modes, and d is the diameter of SWNTs. According to expression (6.8), the diameters of SWNTs are located in the range of 0.84 nm (266 cm^{-1}) to 1.63 nm (137 cm^{-1}) for as-received SWNTs. This result is consistent with the diameters obtained by HRTEM observations. The second group character peaks are the tangential stretch-shrinkage modes located at 1573 cm^{-1} and 1549 cm^{-1}, corresponding to G mode of graphite. The small size effect of SWNTs leads the G mode to split into two peaks. The intensity of the D peak located at 1344 cm^{-1} is weaker. This means that the quantity of amorphous carbon in the products is very small. This result is also consistent with that of the HRTEM observations.

6.4.2. Direct synthesis of a macroscale single-walled carbon nanotubes non-woven material[36]

The single-walled carbon nanotubes (SWNTs) have unique electronic and mechanical properties, and hence they are the promising materials for different applications. In the last decade, people have paid much attention to the fabrication of macroscale structures of CNTs (carbon nanotubes). Rinzler *et al.*[32] prepared the buckypaper through the deposition of a suspension of SWNTs onto a nylon filter membrane and they used the buckypaper to prepare composite materials. In order to improve mechanical properties, such as the toughness etc., Coleman *et al.*[33] used intercalation of polymeric adherences to improve the mechanical properties of SWNT sheets. Smith *et al.*[34] obtained oriented nanotube films with high toughness and density by filtering SWNT solution under high magnetic fields. Dalton *et al.*[35] have successfully prepared super-tough carbon-nanotube fibers by using the spinning method and then weaved these fibers into textiles. Li *et al.*[24] spun continuous fibers and ribbons of carbon nanotubes directly from the chemical vapor deposition (CVD) synthesis zone of a furnace using a liquid source of carbon and an iron nanocatalyst. Fan *et al.*[23] pulled out a bundle of carbon nanotubes from a carbon nanotube array, obtaining continuous carbon nanotube yarns. Last year, Xie *et al.*[36] reported the direct synthesis of a large-scale SWNT non-woven material with compact structure by an optimum floating CVD technique. In the next paragraph, we will introduce the preparation and characterization of SWNT non-woven materials.

Figure 6.16 show the preparation setup. It is a two-stage furnace system fitted with a special quartz tube structure. An inserted tube with a small diameter of 10 mm is embedded in an outer quartz tube with the diameter of 35 mm, as shown in Fig. 6.16. The dicyclopentadienyl iron (ferrocene)

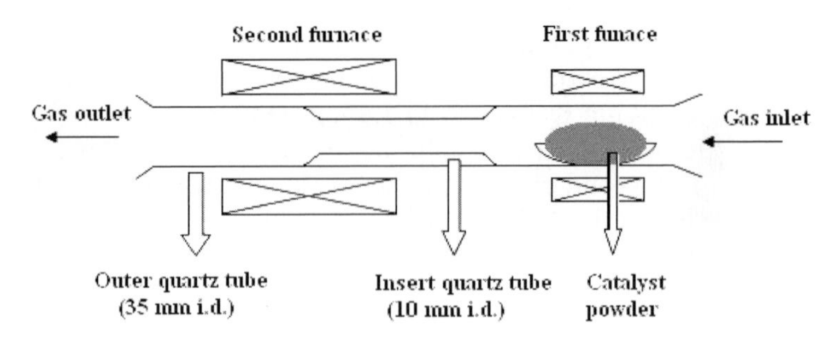

Fig. 6.16. Schematic of the floating catalyst CVD setup for preparing SWNT non-woven materials.

and sulfur powder with a molar ratio of 20 : 1 were mixed uniformly and sublimed in the first furnace at 80–95°C. This sublimed catalyst powder then was carried by the mixture of the flow argon and methane, whose speeds were 1500 sccm and $1 \sim 3$ sccm, respectively, into the second furnace through a narrow inserted quartz tube (Fig. 6.16). The reaction temperature of the second furnace was 1100°C. The pressure in the quartz tube was kept constant at 1 atm. After reaction for 6–9 h, large pieces of nanotube non-woven material were peeled off from the wall of the quartz zone. Then, the nanotube non-woven material was heated at 400–500°C for 48 h in air, so that the Fe catalyst particles and other carbon impurities were oxidized, followed by immersing the nanotube non-woven material into concentrated HCl (38%) to dissolve the iron oxides. After that, the nanotube non-woven material was repeadly washed with deionized water until the pH value of the product reached about 7.0.

Scanning electron microscopy (SEM), transmission electron microscopy (TEM) and micro-Raman spectroscopy were used to characterize the as-received nanotube non-woven material.

Figure 6.17 shows an optical image of the as-received non-woven material with the area of several tens of square centimeters.

Scanning electron microscopy (SEM) images of the non-woven material are shown in Figs. 6.18(a)–6.18(c). From Fig. 6.18(a), it can be seen that the

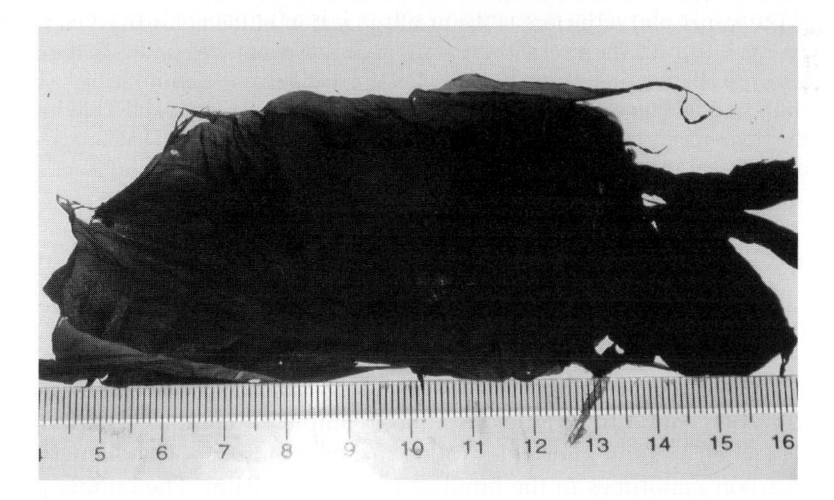

Fig. 6.17. Optical micrograph of a large non-woven material with area of about 40 cm^2.

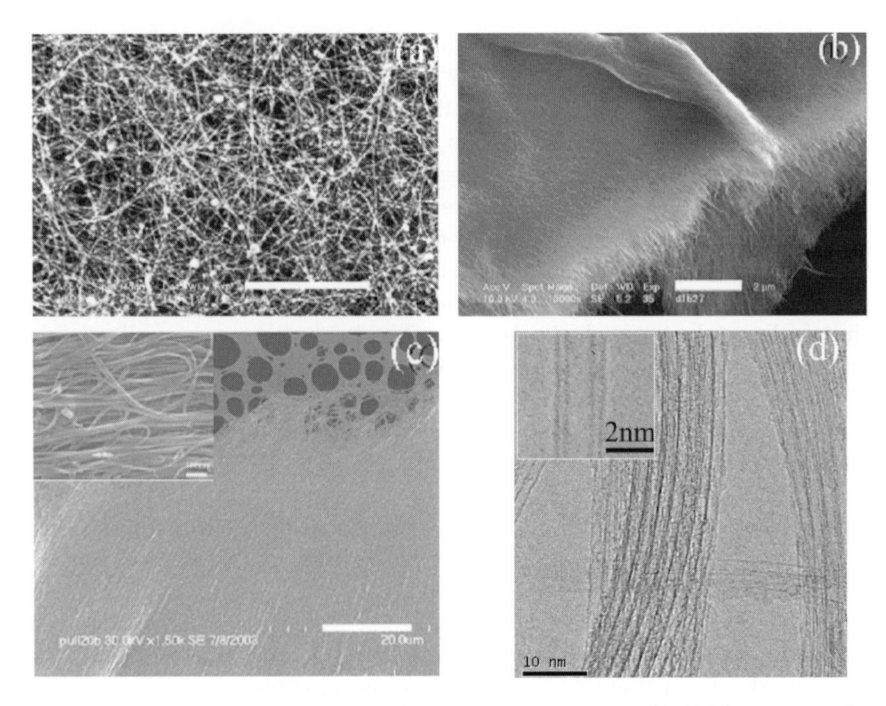

Fig. 6.18. SEM and TEM pictures of non-woven materials. (a) SEM image of the as-grown SWNT non-woven material (scale bar = 1 μm). (b) SEM image of the purified SWNT non-woven material, showing an area near the torn sample edge (scale bar = 2 μm). (c) SEM image of the purified non-woven material after failure. The upper part of picture is the breakage region of the non-woven material (scale bar = 10 μm). The inset shows details of carbon nanotube bundles in the fracture region aligned preferentially under loading (scale bar = 100 nm), which show the bundle diameter (~ 30 nm) and interstitial space (~ 20 nm). (d) HRTEM image of the non-woven material (scale bar = 10 nm). The inset is a HRTEM image of a double-walled carbon nanotube (scale bar = 2 nm).

as-received non-woven material is composed of highly-entangled carbon nanotube bundles, decoated with very small diameter nanoparticles. The non-woven material is of compact structure.

The high resolution TEM image (Fig. 6.18(d)) shows that these entangled carbon nanotube bundles are almost completely composed of SWNTs with a small amount of double-walled carbon nanotubes. The diameter of the carbon nanotubes in the bundles is about 1 ~ 2 nm. The bundle possesses a diameter of about 30 nm, which are firmly entangled to form the non-woven material. The nanoparticles in the non-woven material are iron catalysts encapsulated by several graphite layers. Figure 6.18(b) shows no iron particles in the non-woven material and energy-dispersive X-ray spectrometry (EDXS) cannot detect the iron particle existence in the purified SWNT non-woven material. This implies that after purification, most of the iron particles in the SWNT non-woven material were effectively removed.

The Raman spectra of SWNT non-woven materials are shown in Fig. 6.19. It can be observed that the Raman spectra consist of two main groups of peaks (Figs. 6.19(a) and 6.19(b)). The first group is composed of the radial breathing modes. According to expression (6.8) described in 6.4.1 $\left(\omega = \frac{223.75}{d}\right)$, the Raman peak at 207 cm^{-1} in Fig. 6.19(b) suggests that nanotubes with a diameter of 1.1 nm are dominant in the as-received non-woven material. After purification, the appearance of a radial breathing mode peak at 143 cm^{-1} implies that the number of nanotubes with larger diameters is increased. This is because small diameter tubes are

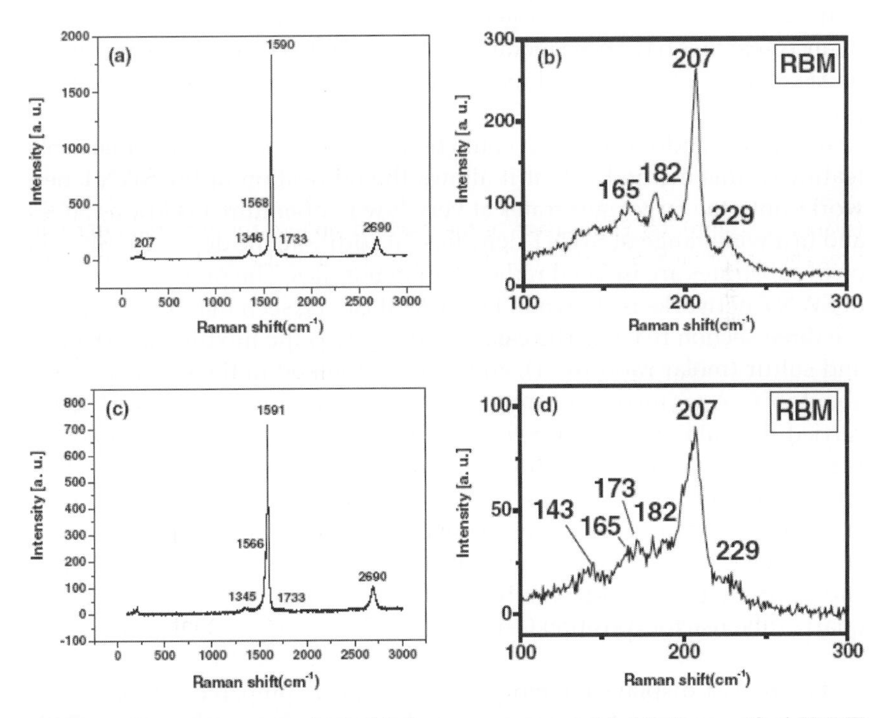

Fig. 6.19. Raman spectra of (a), (b) the as-grown and (c), (d) the purified SWNT non-woven material, obtained with an excitation wavelength of 514.5 nm. Graphs (b), (d) show details of the peak distribution in the low frequency range (RBM).

less stable than larger ones during oxidization. In the second group with high frequency (Figs. 6.19(a) and 6.19(c)), the peaks at 1566 and 1591 cm^{-1} are related to the E_g graphite mode. The low intensity ratio of the *D*-band (1346 cm^{-1}) to the tangential mode *G*-band (1590 cm^{-1}) suggests the good graphitization of the as-received SWNT non-woven material. Furthermore, compared to the as-received non-woven material, the relatively lower intensity of the *D*-band (Fig. 6.19(c)) indicates that the purified non-woven material is almost free of amorphous carbon.

6.4.3. Synthesis of random networks of single-walled carbon nanotubes

The single-walled carbon nanotubes (SWNTs) are a promising class of materials for future functional devices and computing systems.[37–40] Therefore, fabricating the building block of SWNTs and then integrating them into various devices are inevitably involved. It is necessary to spread SWNTs without bundles onto the target substrate. In order to reach this goal, many researches have been made.[39–44] We will introduce the research work of Xie's group.[45] They developed a simple CVD method with which the preparation of isolated SWNTs and the assembly of them into the two-dimensional random networks could be achieved in one step. A remarkable feature of this method is that it allows the fabrication of the SWNT networks onto the target substrates at very low temperature (as low as 50°C) and in a wide range of SWNT densities. In addition, the deposited SWNTs on Si substrates are isolated rather than in bundles. The preparation setup of SWNT networks is shown in Fig. 6.20. It can be seen that this setup has the three-section reactor. The catalyst source was the mixture of ferrocene and sulfur (molar ratio 16 : 1), and it was sublimed in the first furnace at about 55°C. A gas mixture of argon (300 $\sim$ 2000 sccm) and acetylene (1 sccm) carried the sublimed catalyst through a narrow connection tube into the reaction zone in the second furnace (position B in Fig. 6.20). The reaction temperature of position B was 1100°C and the system pressure was held at 1 atm. In this case, the temperature of region C–D could range between 50 and 300°C. As a result, the SWNT random networks were formed on Si substrates that were horizontally preplaced in the outer-end section of the quartz tube reactor (position C–D).

Figure 6.21 displays the morphology of the as-prepared samples. This image was obtained by a tapping-mode atomic force microscope (TM-AFM). Obviously, some line-like structures extend out from the white spots, intersecting each other and forming a continuous network. These lines

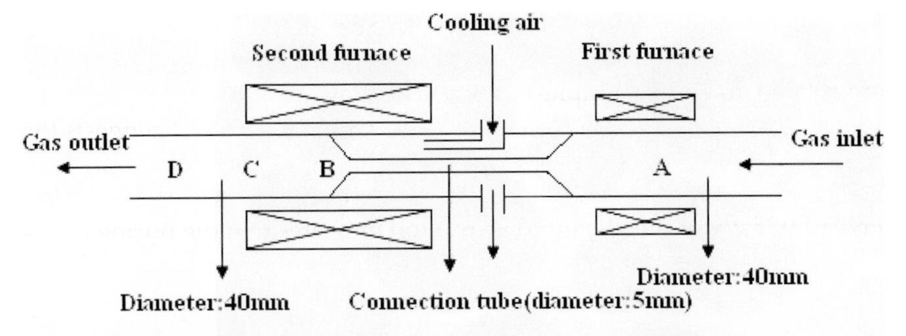

Fig. 6.20. Schematic diagram of the floating catalyst CVD apparatus for preparing random SWNT networks.

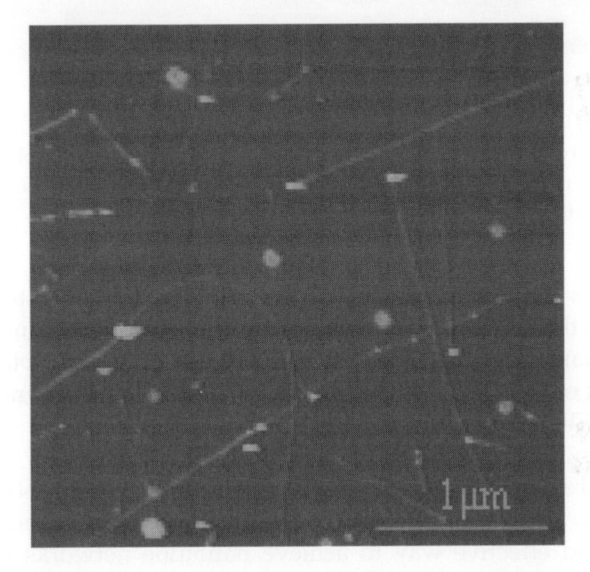

Fig. 6.21. Topography image of the nanotube network recorded by a tapping-mode AFM. About 95% of the nanotubes, determined by their height measurement, are below 3.0 nm in diameter.

change from several to dozens of micrometers in length except for a few extending to more than 50 μm. The height measurements on these samples through the cross-section method indicated that most of the nanotubes ($> 95\%$) have diameters ranging from 0.7 to 3.0 nm. To further characterize the networks, these line-like structures have been collected on carbon-coated copper grids during the deposition process. Figure 6.22(a) gives a typical SEM image of the "grid" sample. It can be observed that the lines span the grid hold and interconnect or intercross each other. The HRTEM

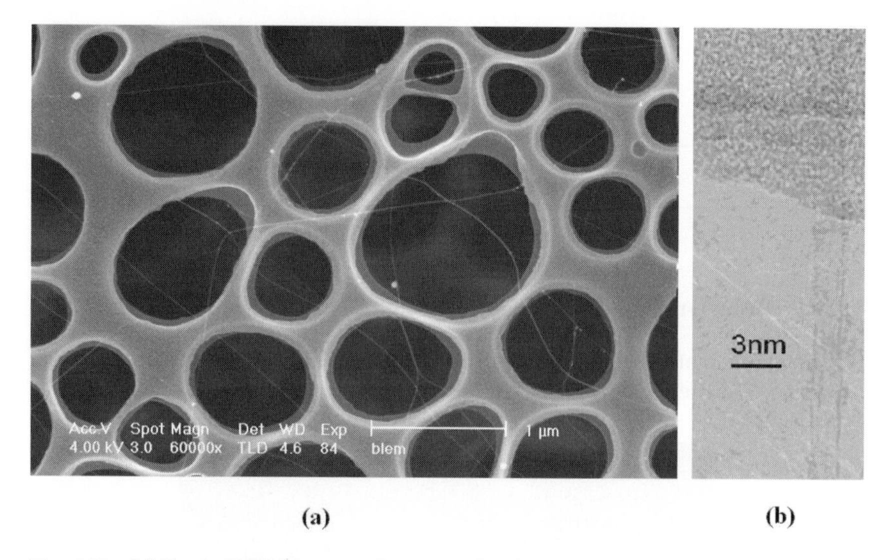

Fig. 6.22. (a) Typical SEM image of a nanotube-deposited grid sample prepared by collecting the nanotubes onto the copper grid during the preparation process. (b) TEM of the grid sample, with an individual SWNT shown.

observations (in Fig. 6.22(b)) confirm that these line-like structures are SWNTs and the white spots are catalyst particles. The SWNT in the figure is 1.7 nm in diameter. According to the above results, we can obtain the conclusion that most of these lines occurring on the Si substrate are individual or isolated SWNTs.

There are several important features about the SWNT networks obtained by Xie's group. Firstly, the technique described here provides a simple and effective way to achieve nanotube networks (or isolated SWNTs). The whole production process contains only one step so that it avoids the troublesome process of pre-preparing the definite catalyst on the target substrate before the nanotube synthesis. Also, the SWNTs synthesized with this method are nearly free of defects. Secondly, the deposition position of the nanotube network is far away from their formation zone. This prevents the nanotube networks and the target substrates from being exposed at a high temperature. Therefore, it is beneficial to using different substrates for future various applications. Thirdly, the density of the SWNT networks on the Si substrate can be readily adjusted in a wide range by changing the deposition position (changing the deposition temperature) and process duration. Finally, it is extremely convenient to characterize the properties of the as-received nanotubes with HRTEM. Namely, during the preparation process of the nanotube network, the copper grid is placed in the deposition position to obtain the sample for HRTEM observations. However, the SWNT networks obtained by this preparation technique also have shortcomings, which are mainly the difficulty in controlling the deposited positions of SWNTs and obtaining pure SWNTs (some double-walled carbon nanotubes existed in SWNTs).

6.5 Synthesis of double-walled carbon nanotubes (DWNTs)

For many applications, it is highly desirable to prepare some special type of carbon nanotubes (CNTs) with high quality, for example double-walled CNTs (DWNTs). Recently, DWNTs have been attracting increasing attention.[46–50] The main reason is their ability to promote insight into the interaction between grapheme layers in MWNTs (multi-walled carbon nanotubes). As we know, preparing pure DWNTs is not easy.[49,51,52] In the next paragraph, we will introduce the production of cleaner DWNTs reported by Xie's group.[53]

The experimental setup for the synthesis of DWNTs is the same as that described in 6.4.3 (Fig. 6.20). Namely, a three-section quartz tube is mounted in a dual furnace system: the head section, the middle thin transition tube and the end section. Researchers found that a different diameter ratio of the transitive section to the end section had an apparent effect on the purity of DWNTs. Therefore, they have fixed the diameters of the head and end sections of the quartz tube reactor while changing the diameter of the transitive section to 20, 15 and 5 mm (the samples prepared in them are marked as S20, S10 and S5). The preparation process of DWNTs in this floating catalyst system is as follows. A quartz boat containing the catalyst mixture of ferrocene and sulfur was placed in the center of the head section of the reactor. Then, the catalyst sublimed at 60 ~ 90°C. This was carried by the flow of argon (300 ~ 2400 sccm) and acetylene (1 ~ 20 sccm) into the second furnace through the transitive tube. During the process, the catalyst decomposed and formed the sulfur-contaminated iron nanoparticles. The DWNTs nucleated and grew on these catalyst nanoparticles at 1100°C and at 1 atm. As a result, the film-like products were deposited on the inner wall of the exit part of the reactor. Then, the film-like deposits were peeled off for characterizing.

Figure 6.23 shows SEM images for the samples prepared at various transition section diameters. It is clear that the samples consist of filaments and some impurities. These filament HRSEM images indicated that they were

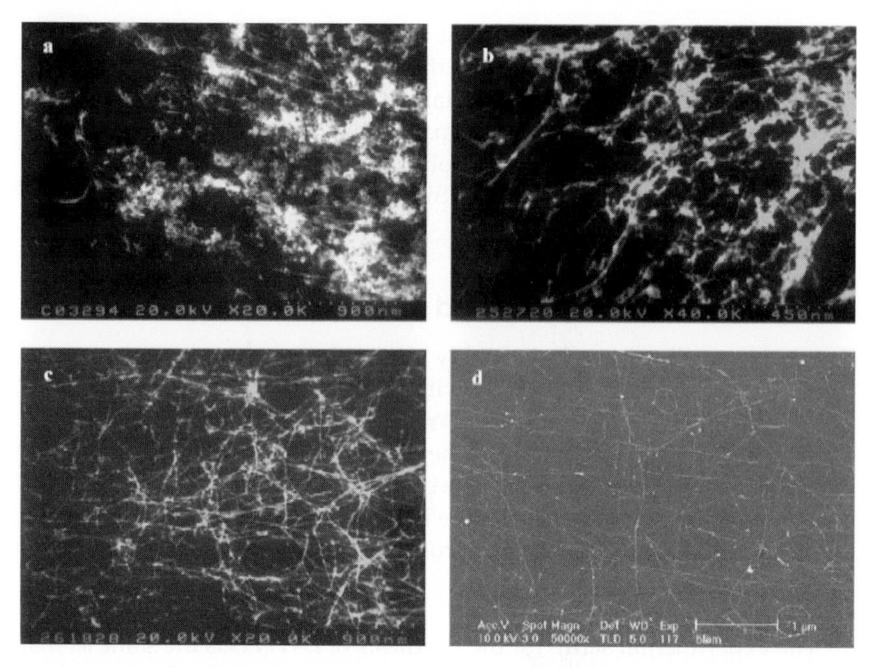

Fig. 6.23. SEM images for samples (a) S20, (b) S10 and (c) S5, showing the morphology changes of the DWNT samples with the transition tube diameters reduced. (d) DWNT samples on the Si substrate obtained with the transition tube of less than 5 mm diameter, described in detail in the text.

made up of DWNT bundles or individual DWNTs whose HRTEM images are given in Figs. 6.24 and 6.25. The rest were SWNTs with impurities; no three- or more-walled tubes were observed. The filaments can reach several tens of micrometers in length and several to several dozens of nanotubes in width. HRTEM images revealed that DWNTs have outer diameters ranging from about $1 \sim 3$ nm and inner diameters from 0.4 to 2.1 nm. Most of the impurity particles in the samples were iron catalysts overcoated by a graphitic layer, and amorphous carbon, which was confirmed by XRD analysis and HRTEM characterization.

From Fig. 6.23, it can be seen that the diameter of the transition tube has an obvious influence on the morphology of DWNT samples, and with the tube diameter reduced from 20 to 5 mm, the impurities greatly decrease. Table 6.3 list the average values of EDX spectroscopy measurements on samples S20, S10 and S5 as complementary evidence of product purity. With the diameter of the transition tube reduced from 20 to 5 mm, the corresponding catalytic iron content in the samples decreased from about

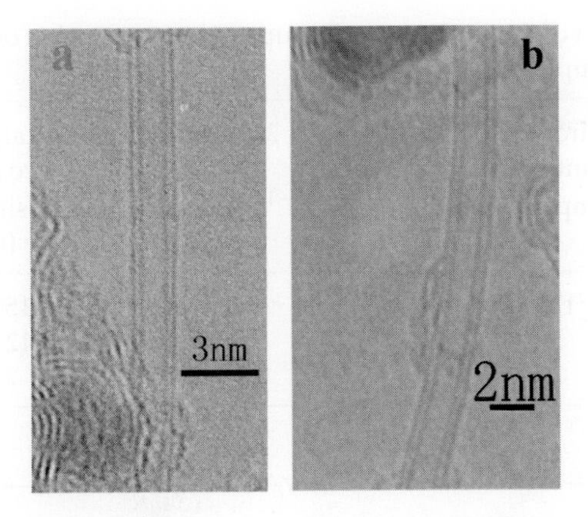

Fig. 6.24. HRTEM images of individual DWNTs.

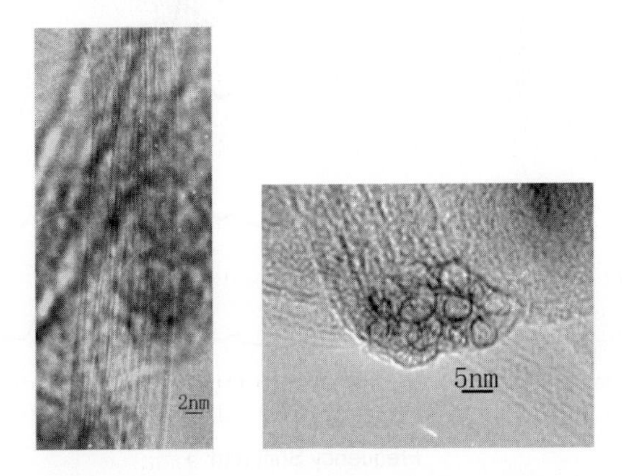

Fig. 6.25. HRTEM images of DWNT bundles.

12 to 2 mol%. Figure 6.23(d) shows that when the diameter of the transition section is less than 5 mm, there is less amorphous carbon and iron particles in the samples, which were deposited on Si substrates, than in S5, and DWNTs are very clean.

The resonant Raman spectra of samples S20, S10 and S5 are shown in Fig. 6.26. The Raman spectrum of S5 shows the largest intensity because of its higher concentration of DWNTs relative to S20 and S10. Also, all the samples have nearly same peak positions including RBM and *G* bands. This indicates that the DWNTs prepared with various transition tube diameters possess the same diameter distribution.

Table 6.3. Effect of the experimental setup on the preparation of DWNTs at the sublimation temperature of 70°C.

Samples	Iron content in samples (%)	Sublimed ferrocene $\times 10^{-4}$(mol h^{-1})	Production rate (mg h^{-1})	Catalyst size in transition tube (nm)	Catalyst size in outer-end tube (nm)
S20	12	3.9	14.7	15	80
S10	5	0.9	4.3	12	50
S5	2	0.3	0.2	7	20

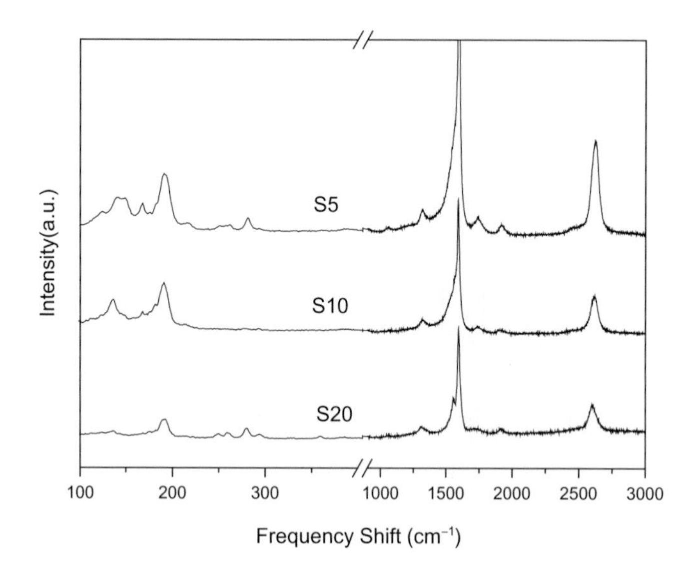

Fig. 6.26. Raman spectra for samples S20, S10 and S5 at room temperature.

In order to investigate what parameters affect the DWNT purity in the floating process, Xie's group conducted some experiments. They found that by changing the catalyst sublimation temperature, they could adjust the amount of sublimed catalyst, which played an important role in determining the purity of DWNT samples. When the sublimation temperature increased from 60 to 90°C, more catalyst could be sublimed, resulting in the samples containing more impurities. However, when the sublimation temperature was below 60°C, little product was deposited on the inner wall of the reactor. Some other experimental parameters such as the relative molar value of ferrocene to sulfur, the reaction gas flow and the cobaltocene as catalyst precursors were investigated. The results show that changing the above parameters had very small effects on the purity of DWNTs. It should be noted that the above processes did not involve all the important experiment parameters.

Bibliography

1. S. Iijima, *Nature* **354**, 56 (1991).
2. J. W. Mintmire, B.I. Dunlap, and C.T. White, *Phys. Rev. Lett.* **68**, 631 (1992).
3. N. Hamada, S. Sawada, and A. Oshiyama, *Phys. Rev. Lett.* **68**, 1579 (1992).
4. H. W. G. Wildoer, I. C. Venema, A. G. Rinzler *et al.*, *Nature* **391**, 59 (1998).
5. T. W. Odom, J. L. Huang, C. M. Lieber *et al.*, *Nature* **391**, 62 (1998).
6. C. Piskoti, J. Yarger, and A. Zttl, *Nature* **393**, 771 (1998).
7. L. M. Peng, Z. L. Zhang, Z. Q. Xue, Q. D. Wu, Z. N. Gu, and D. G. Pettifor, *Phys. Rev. Lett.* **85**, 3249 (2000).
8. L. C. Qin, X. L. Zhao, K. Hirahara, Y. Miyamoto, Y. Ando, and S. Iijima, *Nature* **408**, 50 (2000).
9. N. Wang, Z. K. Tang, G. D. Li, and J. S. Chen, *Nature* **408**, 50 (2000).
10. Z. P. Zhou, L. J. Ci, X. H. Chen, D. S. Tang, X. Q. Yan, D. F. Liu, Y. X. Liang, H. J. Yuan, W. Y. Zhou, G. Wang, and S. S. Xie, *Carbon* **41**, 337 (2003).
11. L. F. Sun, S. S. Xie, W. Liu, W. Y. Zhou, Z. Q. Liu, D. S. Tang, G. Wang, and L. X. Qian, *Nature* **403**, 384 (2000).
12. B. H. Chang, S. S. Xie, W. Z. Li, C. Y. Wang, W. Y. Zhou, L. X. Qian, and L. Gan, *Sci. China* **41A**, 431 (1998).
13. X. Zhao, M. Ohkohchi, M. Wang, S. Iijima, T. Ichihashi, and Y. Ando, *Carbon* **35**, 775 (1997).
14. S. Iijima, P. M. Ajayan, and T. Ichihashi, *Phys. Rev. Lett.* **69**, 3100 (1992).
15. S. S. Xie, S. S. Fan, Y. T. Qian *et al.*, Preparation, characterization and properties (Chap. 2) in Nanomaterials and Nanostructure eds. L. D. Zhang and S. S. Xie, Chemical Publisher, Beijing, China (in Chinese).
16. S. Sawada and N. Hamada, *Solid State Commun.* **83** (11), 917 (1992).

17. N. Bourgeois and L. A. Bursill, *Int. J. Mod. Phys.* **B10** (5), 563 (1996).

18. T. W. Ebbesen ed., *Carbon nanotubes: Preparation and properties*, CRC press, *New York*, 1997 p. 169.

19. Z. W. Pan, S. S. Xie, B. H. Chang, C. Y. Wang, L. Lu, W. Liu, M. Y. Zhou, and W. Z. Li, *Nature* **394**, 631 (1998).

20. H. M. Cheng, F. Li, X. Sun, S. D. M. Brown, M. A. Pimenta, A. Marucci, G. Dresselhaus, and M. S. Dresselhaus, *Chem. Phys. Lett.* **289**, 602 (1998).

21. H. W. Zhu, C. L. Xu, D. H. Wu, B. Q. Wei, R. Vajtai, and P. M. Ajayan, *Science* **296**, 884 (2002).

22. B. Vigolo, A. Penicaud, C. Coulon, C. Sauder, R. Pailler, C. Journet, P. Bernier, and P. Poulin, *Science* **290**, 1331 (2000).

23. K. L. Jiang, Q. Q. Li, and S. S. Fan, *Nature* **419**, 801 (2002).

24. Y. L. Li, I. A. Kinloch, and A. H. Windle, *Science* **304**, 276 (2004).

25. M. Proctor, P. Yeo, and A. Lack, *The Natural History of Pollination*, Harper Collins, London, 1996.

26. N. M. Waser, L. Chittka, M. V. Price, N. M. Williams, and J. Ollerton, *J. Ecology* **77**, 1043 (1996).

27. B. C. Satishkumar, A. Govindaraj, R. Sen, and C. N. R. Rao, *Chem. Phys. Lett.* **293**, 47 (1998).

28. P. Nikolaev, M. J. Bronikowski, R. K. Bradley, F. Rohmund, D. T. Colbert, K. A. Smith and R. E. Smalley, *Chem. Phys. Lett.* **313**, 91 (1999).

29. K. Bladh, L. K. Falk, and F. Rohmund, *Appl. Phys.* **A70**, 317 (2000).

30. M. Endo and T. Koyama, *Japanese Patent* **60**, 32818 (1985).

31. L. Ci, S. S. Xie, D. S. Tang, X. Q. Yan, Y. B. Li, Z. Q. Liu, X. P. Zou, W. Y. Zhou, and G. Wang, *Chem. Phys. Lett.* **349**, 191 (2001); J. H. Hafner, M. J. Bronikowski, B. R. Azamian, *et al.*, *Chem. Phys. Lett.* **296**, 195 (1998).

32. (a) A. G. Rinzler, J. Liu, P. Nikolaev, C. B. Huffman, F. J. Rodriguez-Macias, P. J. Boul, A. H. Lu, D. Heymann, D. T. Colbert, R. S. Lee, J. E. Fischer, A. M. Rao, P. C. Eklund, and R. E. Smalley, *Appl. Phys. A-Mater. Sci. Process.* **67**, 29 (1998). (b) R. H. Baughman, C. X. Cui, A. A. Zakhidov, Z. Iqbal, J. N. Barisci, G. M. Spinks, G. G. Wallace, A. Mazzoldi, D. D. Rossi, A. G. Rinzler, Q. Jaschinski, S. Roth, and M. Kertesz, *Science* **284**, 1340 (1999).

33. J. N. Coleman, W. J. Blau, A. B. Dalton, E. Munoz, S. Collins, B. G. Kim, J. Razal, M. Selvidge, G. Vieiro, and R. H. Baughman, *Appl. Phys. Lett.* **82**, 1682 (2003).

34. B. W. Smith, Z. Benes, D. E. Luzzi, J. E. Fischer, D. A. Walters, M. J. Casavant, J. Schmid, and R. E. Smalley, *Appl. Phys. Lett.* **77**, 666 (2000).

35. A. B. Dalton, S. Collins, E. Mounoz, J. M. Razal, V. H. Ebron, J. P. Ferraris, J. N. Coleman, B. G. Kim, and R. H. Baughman, *Nature* **423**, 703 (2003).

36. L. Song, L. Ci, L. Lv, Z. P. Zhou, X. Q. Yan, D. F. Liu, H. J. Yuan, Y. Gao, J. X. Wang, L. L. F. Liu, X. W. Zhao, X. Dou, W. Zhou, G. Wang, C. Wang, and S. Xie, *Adv. Mater.* **16**(17), 1529 (2004).

37. S. J. Tans, A. R. M. Verschueren, and C. Dekker, *Nature* **393**, 49 (1998).

38. T. Rueckes, K. Kim, E. Joselevich, G. Y. Tseng, C. L. Cheung, and C. M. Lieber, *Science* **289**, 94 (2000).

39. P. C. Colline, M. S. Amold, and P. Avouris, *Science* **292**, 706 (2000).

40. A. Bachtold, P. Hadley, T. Nakanishi, and C. Dekker, *Science* **294**, 1317 (2001).

41. J. Kong, H. T. Soh, A. M. Cassell, C. F. Quate, and H. J. Dai, *Nature* **395**, 878 (1998).

42. J. H. Hafnet, C. L. Cheung, T. H. Oosterkamp, and C. M. Lieber, *J. Phys. Chem. B* **105**, 743 (2001).

43. E. S. Snow, J. P. Novak, P. M. Campbell, and D. Park, *Appl. Phys. Lett.* **82**, 2145 (2003).

44. A. M. Cassell, G. C. McCool, H. T. Ng, J. E. Koehne, B. Chen, J. Li, J. Han, and M. Meyyappan, *Appl. Phys. Lett.* **82**, 817 (2003).

45. Z. P. Zhou, L. J. Ci, L. Song, X. Q. Yan, D. F. Liu, H. J. Yuan, Y. Gao, J. X. Wang, L. F. Liu, W. Y. Zhou, G. Wang, and S. S. Xie, *J. Phys. Chem. B* **108**, 10751 (2004).

46. S. Bandow, G. Chen, G. U. Sumanasekera, R. Gupta, M. Yudasaka, S. Iijima, and P. C. Eklund, *Phys. Rev. B* **66**, 075416 (2002).

47. Y. K. Kwon, and D. Tománek, *Phys. Rev. B* **58**(24), R16001 (1998).

48. K. Tanaka, H. Aoki, H. Ago, T. Yamabe, and K. Okahara, *Carbon* **35**, 121 (1997).

49. J. L. Hutchison, N. A. Kiselev, E. P. Krinichnaya, A. V. Krestinin, R. O. Loutfy, A. P. Morawsky, V. E. Muradyan, E. D. Obraztsova, J. Sloan, S. V. Terekhov, and D. N. Zakharov, *Carbon* **39**, 761 (2001).

50. S. Bandow, M. Takizawa, K. Hirahara, M. Yudasaka, and S. Iijima, *Chem. Phys. Lett.* **337**, 48 (2001).

51. W. C. Ren, F. Li, J. A. Chen, S. Bai, and H. M. Cheng, *Chem. Phys. Lett.* **359**, 196 (2003).

52. W. Z. Li, J. G. Wen, M. Sennett, and Z. F. Ren, *Chem. Phys. Lett.* **368**, 299 (2003).

53. Z. P. Zhou, L. J. Ci, L. Song, X. Q. Yan, D. F. Liu, H. J. Yuan, Y. Gao, J. X. Wang, L. F. Liu, W. Y. Zhou, G. Wang, and S. S. Xie, *Carbon* **41**, 2607 (2003).

Chapter 7

Synthesis of Inorganic Non-carbon Nanotubes

Chapter 7

Synthesis of Inorganic Non-carbon Nanotubes

7.1 Introduction

The nanotube is one general kind of quasi-one-dimensional nanomaterial with a typical diameter smaller than 200 nm and length ranging from a hundred nanometers to micrometers, and with a hollow interior. The cross-section of the nanotubes varies from circular, rectangular, square, hexagonal, triangular, to irregular, depending on the specific crystalline structure of the materials. The first example of an inorganic nanotube is the carbon nanotube, discovered in 1991 by Iijima.[1] From then on, efforts have been made to synthesize inorganic nanotubes. In the early years, materials with layered structures similar to graphite were examined, and first demonstrated by Tenne *et al.* in 1992 for WS_2[2] and MoS_2,[3] and by Zettl *et al.* in 1995 for BN.[4] Later on, the inorganic nanotube family expanded to include more and more members, including metal chalcogenides, halides, oxides, pure metals, and so on. Recently, inorganic nanotubes have also been synthesized for materials of non-layered structures, such as GaN, ZnO, AlN, and so on. The progress of the discovery of inorganic nanotubes and the enrichment of the inorganic nanotube family has been reviewed by several researchers.[5–13]

In fact, the early report concerning inorganic nanotubes may be traced back to 1955, when Hofer *et al.* reported nanotubules of carbon.[14] In 1960, Bacon reported nanotubules of carbon, and proposed the scroll model of the formation of the tubules.[15] In 1967, Yada reported the nanotube structure for chrystotite asbestos.[16] In 1969, Hess *et al.* reported nanotubes of quasi-graphitic carbon.[17] Later on, Kaito *et al.* reported β-SnS_2 nanotubes.[18]

The zeal for investigating inorganic nanotubes has been stimulated by the novel physical properties and new physical concepts for this new structure.[19] Taking the carbon nanotube as an example, its fabulous physical and chemical properties enable numerous applications ranging from field emission, sensor, to electronic circuit, and so on.[20–22] Other inorganic nanotubes have also been predicted or demonstrated to display novel properties. For example, WS_2 nanotubes show ultrahigh shock wave resistance and excellent performance under high load.[23] NbS_2 and $NbSe_2$ nanotubes have been predicted to possess superconducting properties.[24,25] Inorganic nanotubes are also believed to have stronger quantum confinement than nanowires of similar diameter due to the hollow interior.[26,27]

This chapter reviews the synthesis of inorganic nanotubes according to material structure, namely, layered and non-layered structure. The growth mechanisms for these two kinds of materials are very different. Since the synthesis of carbon nanotubes has appeared in many handbooks and has been detailed in Chap. 6, it will not be treated in depth in this chapter.

7.2 Synthesis of inorganic nanotubes

Inorganic materials can be generally classified into three categories: (i) two-dimensional materials, which are those materials linked by ionic or covalent bonds and are extended in two dimensions to form a lamellar structure, with the layers stacked parallel and linked by Van der Waals forces to form a three-dimensional structure, such as graphite, WS_2, and BN; (ii) quasi-two-dimensional materials, which are those materials linked by ionic or covalent bonds and are extended in one dimension to form a platelet structure, and the platelets are stacked side by side by Van der Waal's forces to form a lamellar structure, and the lamellar structures are stacked parallel and linked by Van der Waal's forces to form a three-dimensional structure, such as Bi_2S_3, Sb_2S_3, Se, and Te; (iii) three-dimensional materials, which are those materials linked by ionic or covalent bonds and extended in three dimensions, such as ZnO and GaN.

Materials in the first category all possess the graphite-like layered structure, and are explored most extensively. They mainly form nanotubes with layered walls, namely, multi-walled nanotubes. Materials in the second category have been less explored, due much to the difficulty in controlling the formation kinetics of the nanotubes. These materials either form layered walls such as Bi_2S_3, or non-layered walls such as Se and Te. Materials in the last category have attracted attention only recently, especially for the important wide band gap semiconductors such as ZnO and GaN. These materials all form nanotubes of non-layered walls.

Besides the structural distinction among the three categories of nanotubes, the formation mechanisms also differ greatly, which will be elaborated in the following sections.

7.2.1. Inorganic nanotubes based on two-dimensional structures

The materials in this category mainly include transition metal chalcogenides and halides, boron nitride and its derivatives, rare earth metal oxides and their derivatives. This section will be divided into subsections according to the above subcategories.

7.2.1.1. *Inorganic nanotubes based on graphite (carbon nanotubes)*

Carbon nanotube was first discovered by Iijima in the arc discharge deposit of graphite anode in 1991.[1] The nanotubes are a multi-walled structure. The distance of the neighboring layer is 0.34 nm, the same as in graphite. Iijima proposed a curving model to account for the formation of carbon nanotubes, which is the basis for the explanation of nanotube formation from two-dimensional structures.

It is noteworthy that carbon nanotube formation is efficient only when a suitable catalyst, often a transition metal, is employed.

7.2.1.2. *Inorganic nanotubes based on transition metal chalcogenides and halides*

Tenne *et al.* analyzed the structure of graphite and found out its essential two-dimensional nature. He also found out the structural similarity between graphite and WS_2, MoS_2, and other layered materials. If the curving model for the carbon nanotube is correct, then it may function also for the latter materials. Tenne *et al.* did obtain WS_2 and MoS_2 nanotubes in subsequent investigations.[2,3] Following their pioneering work, researchers in the world carried out more extensive investigations to find other layered materials for the formation of nanotubes.

The first example in this subcategory is WS_2. The WS_2 nanotubes were produced by heating thin tungsten films in an atmosphere of hydrogen sulphide.[2] Later, WS_2 nanotubes were also synthesized by the reduction of WO_3 with H_2, and then treated with H_2S,[28] pyrolysis of precursors,[29] and so on.[30,31]

The molecular model of WS_2 is shown in Fig. 7.1(a), where the layered structure is demonstrated. An HRTEM image of a WS_2 nanotube is shown in Fig. 7.1(b). It is clear that the nanotube is multi-walled. The spacing between two neighboring walls is close to 0.62 nm, corresponding to the separation between two layers of bulk WS_2.

Lately, gram scale production of high-purity WS_2 nanotubes has been achieved by a fluidized-bed reactor.[32]

The second example is MoS_2. Its structure is similar to that of WS_2. The MoS_2 nanotubes were synthesized by a similar method to that of WS_2,

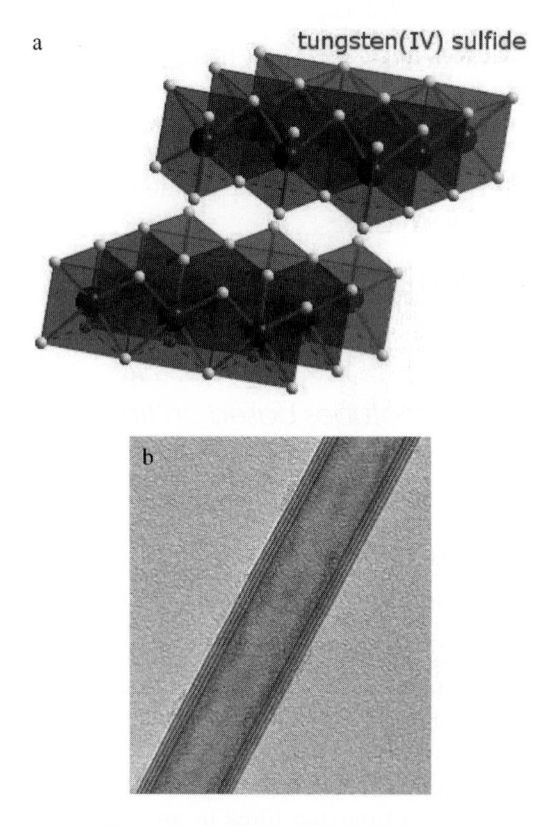

Fig. 7.1. (a) Schematic illustration of layered structure of WS_2. (b) HRTEM image of a WS_2 nanotube, where the multi-walled structure is clear.[11]

namely, reduction of MoO_3 with H_2, and then treated with H_2S.[3] MoS_2 nanotubes were also synthesized by other groups.[33,34]

The molecular model of MoS_2 is shown in Fig. 7.2(a), where the layered structure is demonstrated. An HRTEM image of a MoS_2 nanotube is shown in Fig. 7.2(b). It is clear that the nanotube is also multi-walled. The spacing between two neighboring walls is close to 0.62 nm, the same to that of WS_2 nanotubes, corresponding to the separation between two layers of bulk MoS_2.

Recently, W and Mo selenide nanotubes have also been produced.[35] $MoS_{2-x}I_y$ subnanometer nanotubes and doped nanotubes have also been synthesized.[36–38]

Nanotubes of TiS_2 were produced by heating TiS_2 nanoparticles.[39] $TiSe_2$ nanotubes were synthesized by direct reaction of Ti with Se powder.[40]

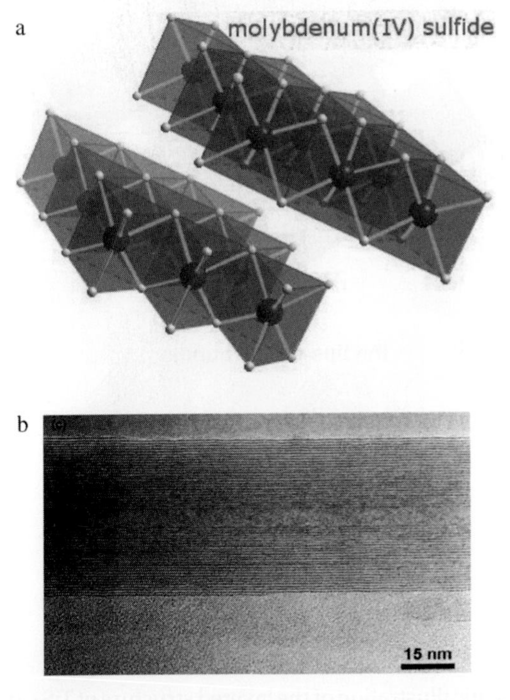

Fig. 7.2. (a) Schematic illustration of layered structure of MoS_2. (b) HRTEM image of a MoS_2 nanotube.[33]

The structure of TiS_2 is shown in Fig. 7.3(a), where the layered structure is demonstrated. An HRTEM image of a TiS_2 nanotube is shown in Fig. 7.3(b). It is clear that the nanotube is multi-walled. The spacing between two neighboring walls is close to 0.63 nm, corresponding to the 10% enlargement of the separation between two layers of bulk TiS_2. An HRTEM image of a $TiSe_2$ nanotube is shown in Fig. 7.3(c). The spacing between two neighboring walls is close to 0.60 nm, corresponding to the separation between two layers of bulk $TiSe_2$.

ReS_2 nanotubes were produced by H_2S treating of the Re precursor.[41] An HRTEM image of a ReS_2 nanotube is shown in Fig. 7.4. It is clear that the nanotube is multi-walled. The spacing between two neighboring walls is close to 0.62 nm, corresponding to the separation between two layers of bulk ReS_2.

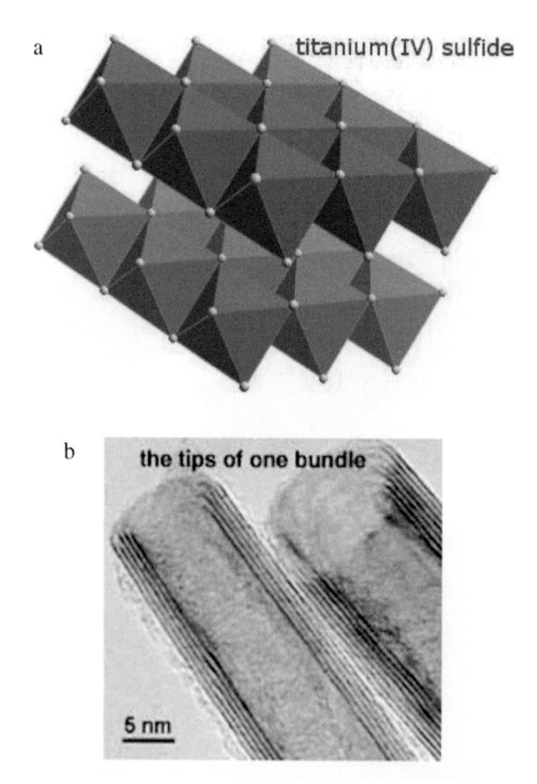

Fig. 7.3. (a) Schematic illustration of the layered structure of TiS_2. (b) HRTEM image of a TiS_2 nanotube.[39] (c) HRTEM image of a $TiSe_2$ nanotube.[40]

c

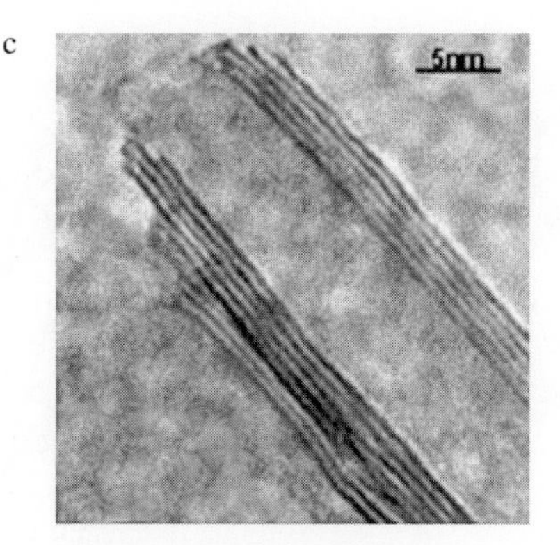

Fig. 7.3. (*Continued*)

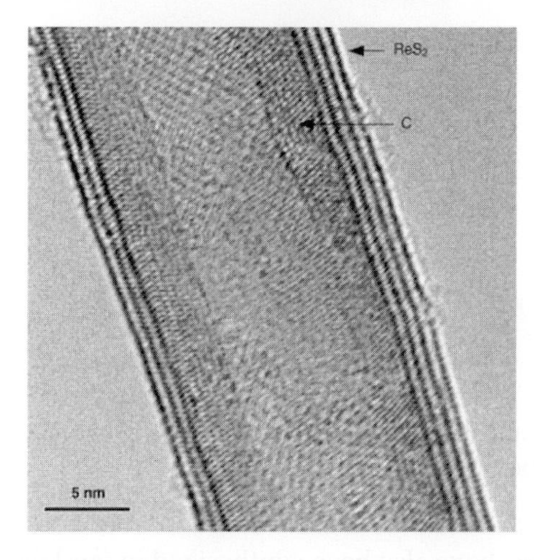

Fig. 7.4. HRTEM image of a ReS_2 nanotube.[41]

ZrS_2, HfS_2, NbS_2, TaS_2, and InS nanotubes have also been synthesized.[42–45]

SnS_2 nanotubes were also synthesized.[18,46] Hong *et al.* synthesized SnS_2 nanotubes by laser ablation of SnS_2 pellets.[46] The structure of SnS_2 is shown in Fig. 7.5(a), where the layered structure is demonstrated. An HRTEM image of an SnS_2 nanotube is shown in Fig. 7.5(b). It is clear that the nanotube is multi-walled. The spacing between two neighboring walls is close to 0.59 nm, corresponding to the separation between two layers of bulk SnS_2.

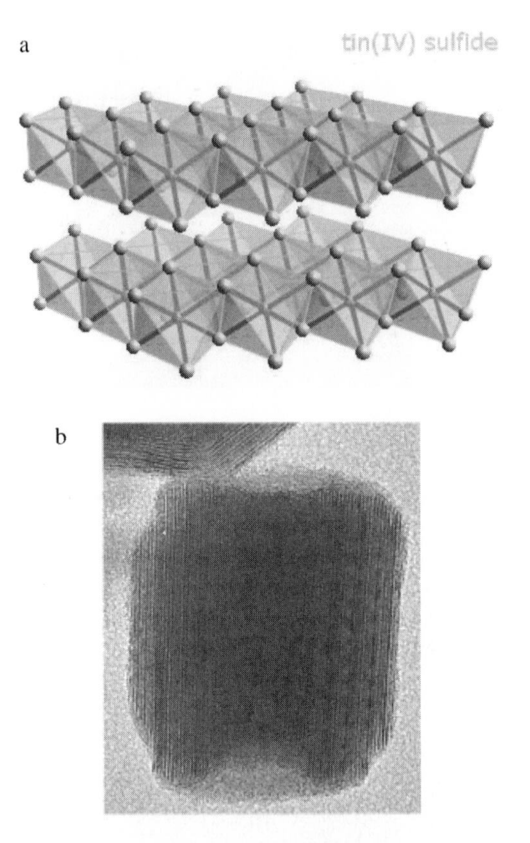

Fig. 7.5. (a) Schematic illustration of layered structure of SnS_2. (b) HRTEM image of an SnS_2 nanotube.[46]

SnS_2 nanotubes have also been synthesized in our laboratory by physical evaporation of SnS_2 nanopowders in the atmosphere of sulfur. Figure 7.6 exhibits the general morphologies of the as-synthesized SnS_2 nanotubes.

Bi_2Te_3 nanotubes were synthesized in our laboratory by a two-step approach. The first step is the synthesis of Bi_2Te_3 nanoplatelets by a sonochemical method. $Bi(NO_3)_3 \cdot 5H_2O$ and elemental Te were used as starting materials. The mixture of the materials were sonicated in aqueous solution for 1 h. The morphology of the as-prepared Bi_2Te_3 nanoplatelets is shown in Fig. 7.7(a), where the platelet structure is demonstrated. The as-synthesized Bi_2Te_3 nanoplatelets were annealed at 350 °C under the protection of Ar. The nanoplatelets were converted to nanotubes after the annealing. The nanotubes were shown in Fig. 7.7(b). An HRTEM image of a Bi_2Te_3 nanotube

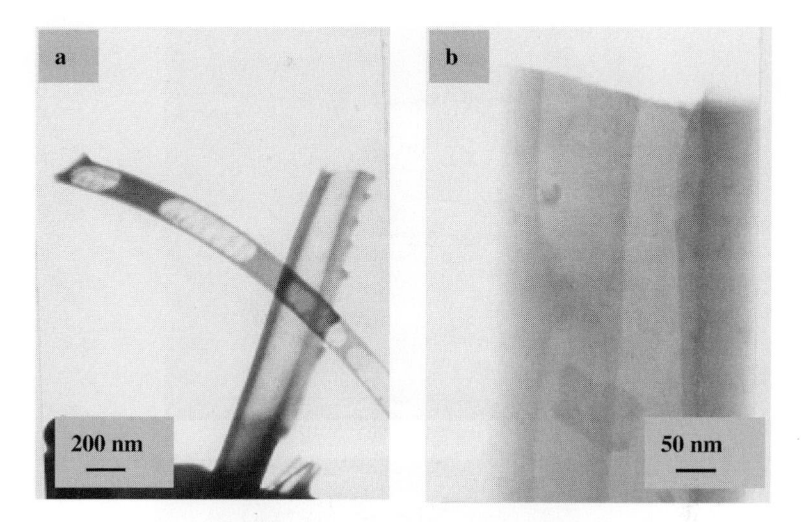

Fig. 7.6. TEM image of SnS_2 nanotubes.

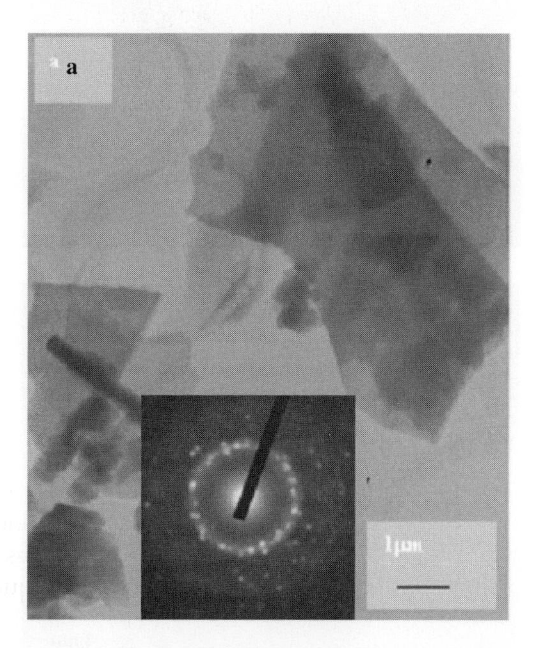

Fig. 7.7. (a) TEM image of Bi_2Te_3 nanoplatelets, (b) and (c) SEM images of Bi_2Te_3 nanotubes, (d) HRTEM image of a Bi_2Te_3 nanotube.

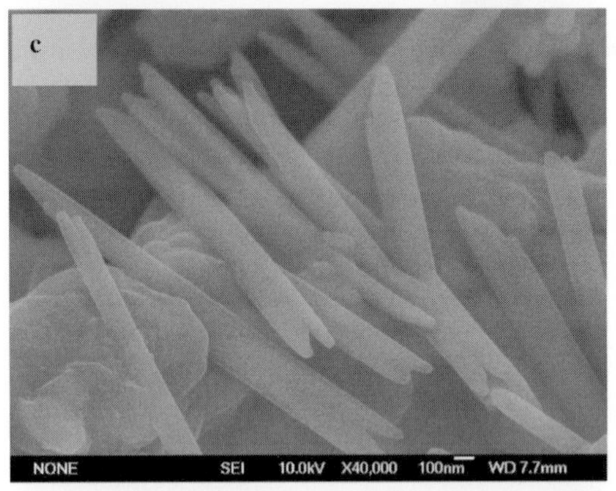

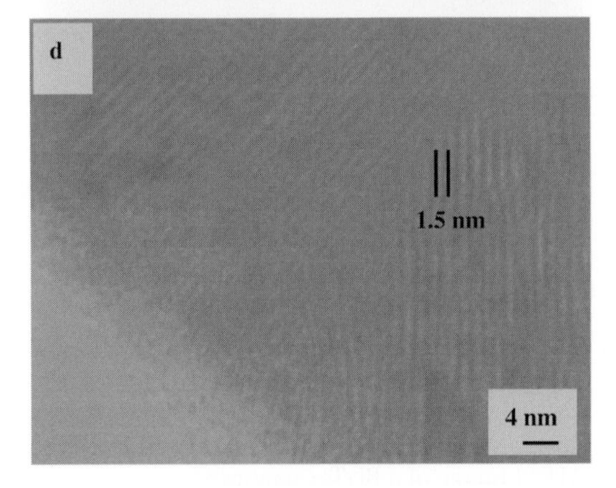

Fig. 7.7. (*Continued*)

is shown in Fig. 7.7(c). It is clear that the nanotube is polycrystalline. The spacing between two neighboring lattice fringes is close to 1.5 nm, corresponding to the separation between two layers of bulk Bi_2Te_3.

Even though Bi_2Te_3 is also a layered material, in the present study, it seems that the nanotube formation does not follow the complete roll-up of single lamellar layer mechanism as described earlier. Instead, Bi_2Te_3 nanotubes may be formed by the formation of nanoloops and through the oriented growth mechanism, which will be described later.

$NiCl_2$ nanotubes were synthesized by laser ablation of hydrated $NiCl_2$.[47,48] An HRTEM image of a $NiCl_2$ nanotube is shown in Fig. 7.8. It is clear that the nanotube is multi-walled. The spacing between two neighboring walls is close to 0.58 nm, corresponding to the separation between two layers of bulk $NiCl_2$.

7.2.1.3. *Inorganic nanotubes based on boron nitride and the derivatives*

Zettl *et al.* also noticed the progress of the synthesis of inorganic nanotubes, and realized the synthesis of BN nanotubes.[4] BN nanotubes have also been synthesized by other groups lately.[49–51]

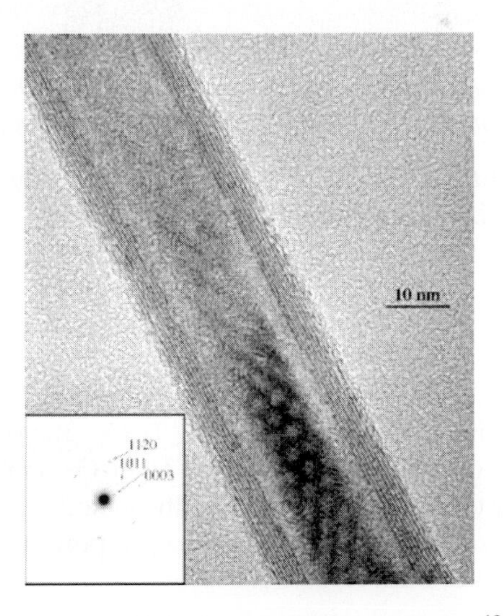

Fig. 7.8. TEM image of a $NiCl_2$ nanotube.[48]

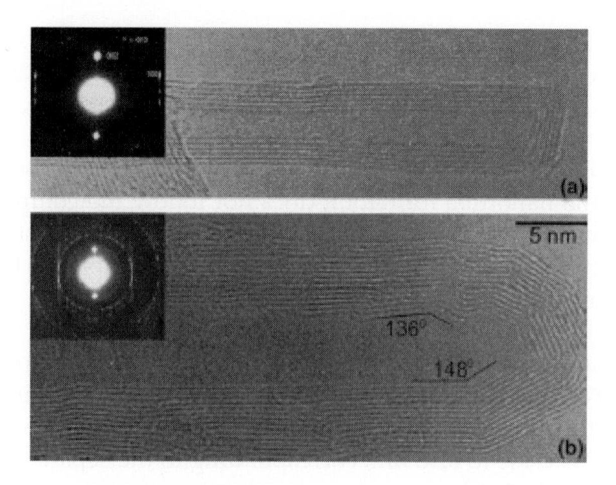

Fig. 7.9. HRTEM image of BN nanotubes.[50]

BN nanotubes were synthesized in a plasma arc discharge apparatus. An HRTEM image of a BN nanotube is shown in Fig. 7.9. It is clear that the nanotube is multi-walled. The spacing between two neighboring walls is close to 0.33 nm, corresponding to the separation between two layers of bulk BN.

The derivatives of BN such as BC_xN nanotubes have also been synthesized.[52,53]

7.2.1.4. *Inorganic nanotubes based on rare earth and transition metal oxides and their derivatives*

Nanotubes including VO_x, Y_2O_3, Y_2O_2S, $Eu(OH)_3$, $Dy(OH)_3$, and so on, have been synthesized by a hydrothermal approach.[54–59]

A TEM image of an $Er(OH)_3$ nanotube is shown in Fig. 7.10(a). An HRTEM image of an $Eu(OH)_3$ nanotube is shown in Fig. 7.10(b).

$H_2Ti_3O_7$ nanoscrols were formed by treating TiO_2 with concentrated NaOH.[60] $H_2Ti_3O_7$ and $Na_xH_{2-x}Ti_3O_7$ nanoscrols were also synthesized.[61–63] An HRTEM image of an $H_2Ti_3O_7$ nanotube is shown in Fig. 7.11. It is clear that the nanotube is multi-walled. The spacing between two neighboring walls is close to 0.78 nm, corresponding to the separation between two layers of bulk $H_2Ti_3O_7$.

Nanotubes of Bi and Sb have also been produced.[64,65]

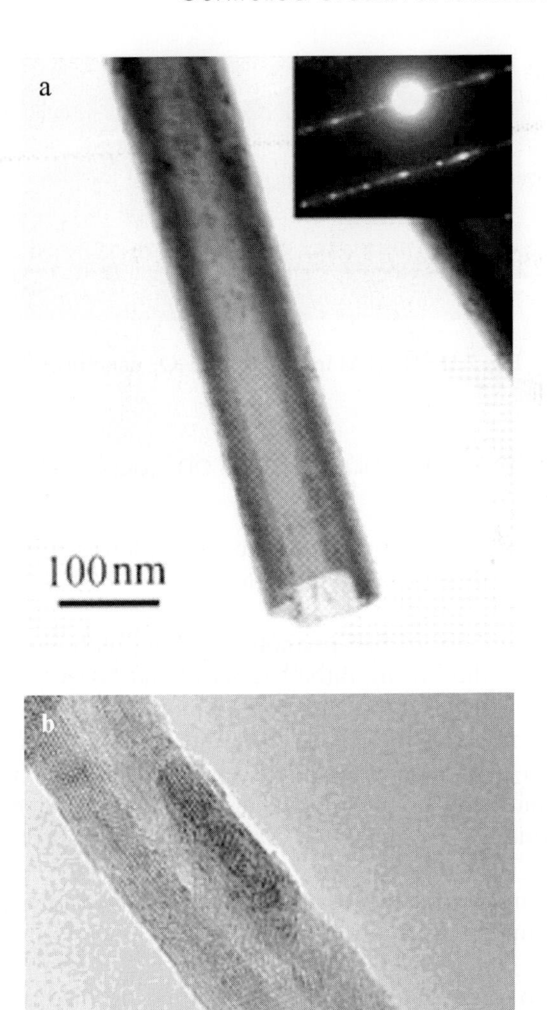

Fig. 7.10. (a) TEM image of an $Er(OH)_3$ nanotube; (b) HRTEM image of an $Eu(OH)_3$ nanotube.[56]

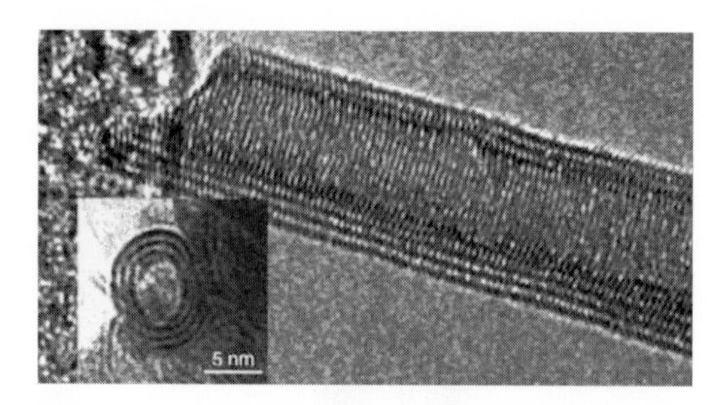

Fig. 7.11. HRTEM image of $H_2Ti_3O_7$ nanotubes.[60]

7.2.2. Inorganic nanotubes based on quasi-two-dimensional structures

Nanotubes from this category have been less explored. Bi_2S_3, Sb_2O_3, Sb_2S_3, Se, and Te nanotubes belong to this category.

Pan *et al.* synthesized Bi_2S_3 nanotubes along with nanoparticles by a solvothermal method using $Bi(NO_3)_3{\cdot}5H_2O$ and NH_2CSNH_2 as starting materials.[66] Ota *et al.* synthesized Bi_2S_3 nanotubes by a micelle-template assisted method using $Bi(NO_3)_3{\cdot}5H_2O$ and triethanol amine as starting materials.[67] Bi_2S_3 nanotubes had also been synthesized in our laboratory in 2002 by chemical vapor deposition using Bi_2S_3 nanopowders and sulfur as starting materials.[68] The quasi-two-dimensional structure of Bi_2S_3 is shown in Fig. 7.12(a). TEM images of Bi_2S_3 nanotubes are displayed in Fig. 7.12(b). An HRTEM image of a Bi_2S_3 nanotube is displayed in Fig. 7.12(c). The layered structure is clear.

In order to investigate how Bi_2S_3 nanotubes formed, we synthesized Bi_2S_3 nanoplatelets by a sonochemical method similar to the Bi_2Te_3 case. Then the as-synthesized Bi_2S_3 nanoplatelets were annealed at 400°C under the protection of Ar. We did obtain Bi_2S_3 nanotubes by this method. The Bi_2S_3 nanoplatelets are shown in Fig. 7.13(a). In Fig. 13(b), Bi_2S_3 nanotubes are displayed. From this image, Bi_2S_3 nanotubes at different formation stages can be observed. The arrows in the figure point to some nanotubes with incomplete closure. A conclusion can be drawn from this observation that Bi_2S_3 nanotubes in this study were formed by a mechanism involving the simultaneous roll-up and oriented growth processes.

Sb_2O_3 nanotubes were synthesized in several groups. Sb_2O_3 nanotubes have been synthesized in our laboratory by chemical vapor deposition

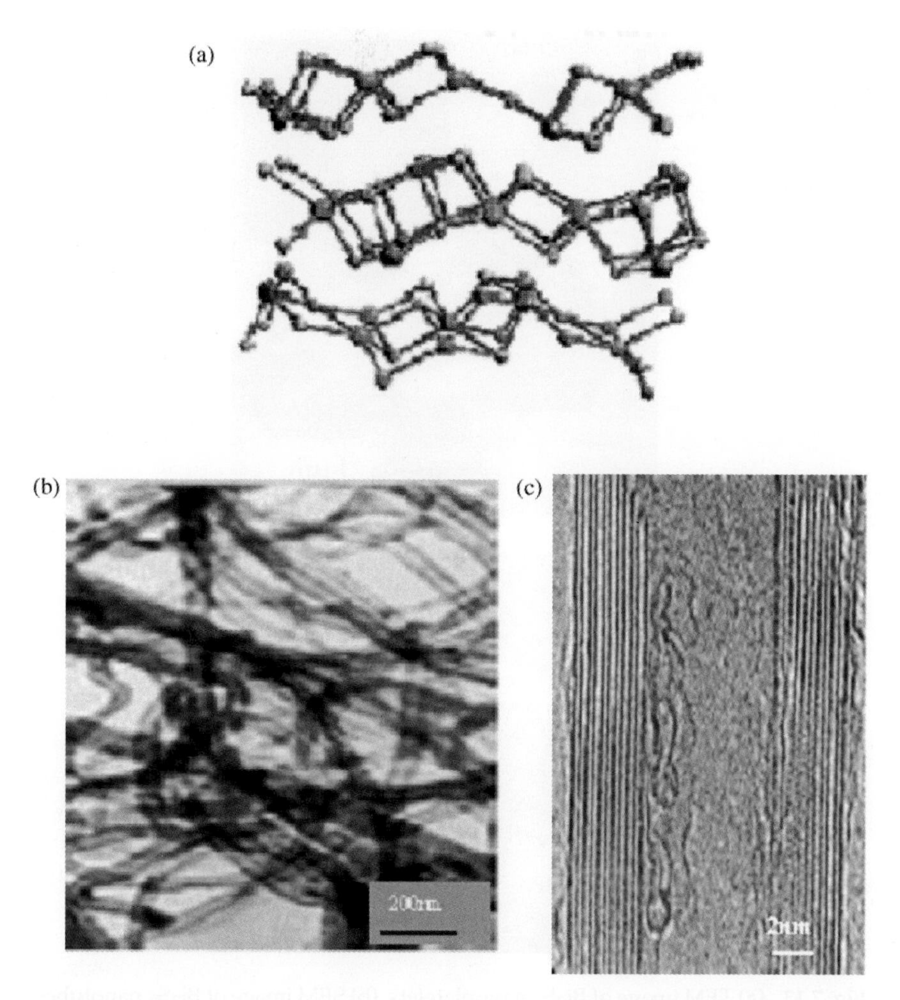

Fig. 7.12. (a) Schematic illustration of the layered structure of Bi_2S_3. (b) TEM image of Bi_2S_3 nanotubes. (c) HRTEM image of a Bi_2S_3 nanotube; the multi-walled structure is clear.[68]

using Sb_2S_3 nanopowders as the starting material and the process of simultaneous oxidation.[69] Zhang *et al.* in our laboratory synthesized Sb_2O_3 nanotubes by surfactant-assisted solvothermal method using $SbCl_3$, CTAB (cetyl trimethylammonium bromide), and $NaBH_4$ as starting materials.[70] The layered structure of Sb_2O_3 is shown in Fig. 7.14(a). The SEM image of Sb_2O_3 nanotubes are displayed in Fig. 7.14(b), where the arrows mark the openings of the nanotubes. Figure 7.14(c) is a TEM image of an Sb_2O_3

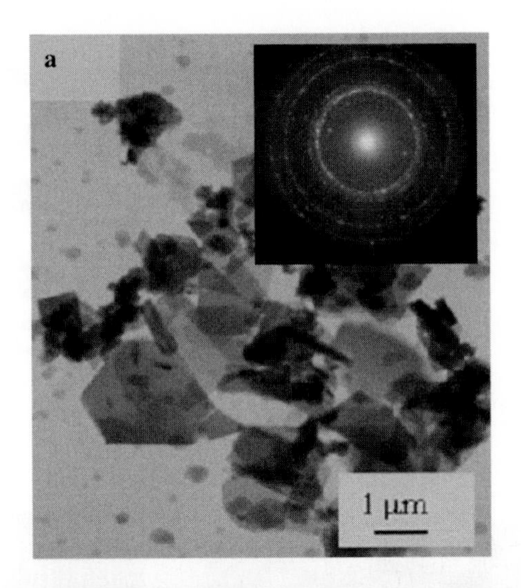

Fig. 7.13. (a) TEM image of Bi_2S_3 nanoplatelets. (b) SEM image of Bi_2S_3 nanotubes; the arrows point to the incomplete closure of the nanotubes.

nanotube, where an arrow points to the breakage of the nanotube into platelets. TEM images of solution-route synthesized Sb_2O_3 nanotubes were shown in Fig. 7.15. Sb_2O_3 nanotubes may be formed by the mechanism similar to Bi_2S_3 nanotubes, namely, the simultaneous roll-up and oriented growth processes, which is evidenced from Fig. 7.15(c), where the incomplete closure of the nanotube is revealed.

Sb_2S_3 nanotubes were synthesized similar to the method we used to synthesized Bi_2S_3 nanotubes.[71] Yang *et al.* synthesized Sb_2S_3 nanotubes

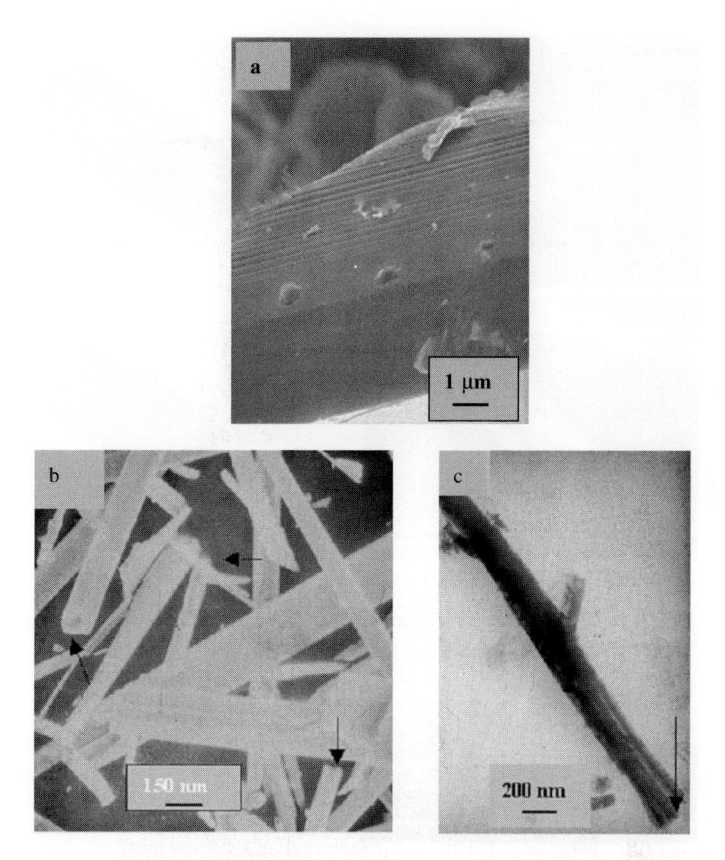

Fig. 7.14. (a) SEM image of the layered structure of Sb_2O_3. (b) SEM image of Sb_2O_3 nanotubes; the arrows mark the openings of the nanotubes. (c) TEM image of an Sb_2O_3 nanotube; the arrow points to the breakage of the nanotube into platelets.[69]

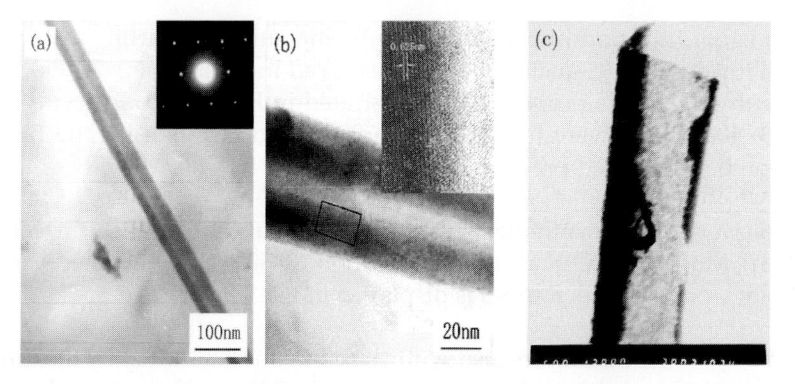

Fig. 7.15. (a) and (b) TEM images of Sb_2O_3 nanotubes. (c) TEM image of a Sb_2O_3 nanotube; the incomplete closure of the nanotube is clear.[70]

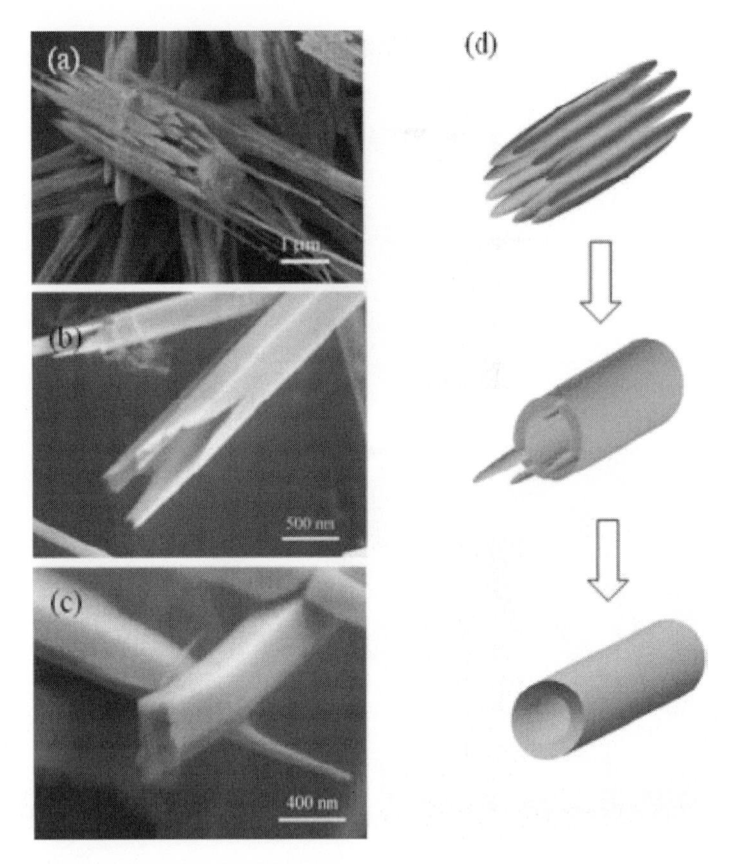

Fig. 7.16. (a)–(c) SEM images of Sb_2S_3 nanotubes. (d) Proposed growth model of Sb_2S_3 nanotubes.[71]

by a vapor transport method using Sb_2S_3 and sulfur as starting materials. SEM images of Sb_2S_3 nanotubes are displayed in Figs. 7.16(a)–7.16(c). The growth model was proposed and illustrated in Fig. 7.16(d). It seems that Sb_2S_3 nanotubes were formed by the similar mechanism for the Bi_2Te_3 nanotubes described previously.

Bezverkheyy synthesized sulfur microtubes by heat treating $(NH_4)_2Mo_2S_{12}{\cdot}2H_2O$ and acetone solution at 50°C.[72] The general morphology of sulfur microtubes is displayed in Fig. 7.17.

Sulfur microtubes were also synthesized in our laboratory by the direct evaporation and deposition of sulfur powder. The sulfur microtubes are

Fig. 7.17. TEM image of S microtubes.[72]

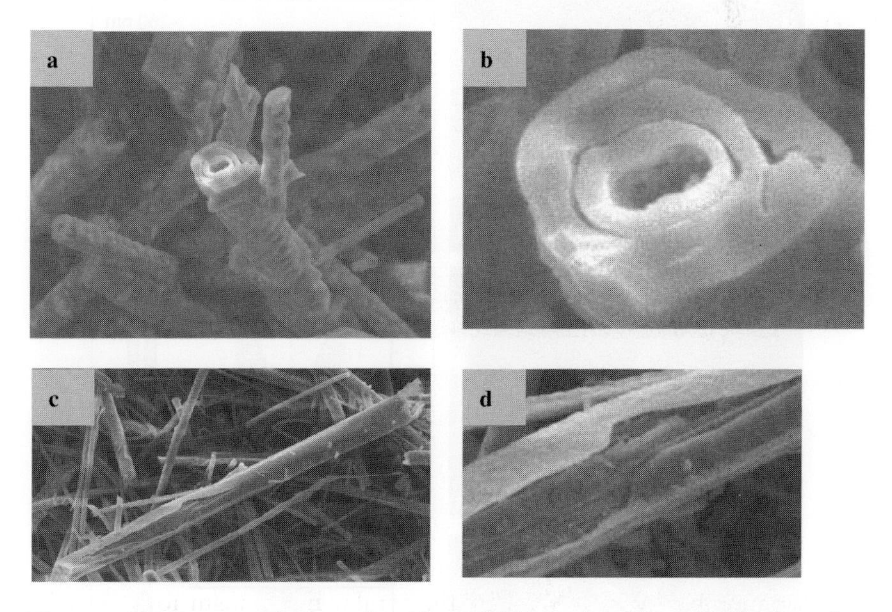

Fig. 7.18. SEM image of S microtubes. The multi-layered and branched S microtubes are displayed.

shown in Figs. 7.18(a)–7.18(d). Some sulfur microtubes have branches on the stem (Fig. 7.18(b)), and some sulfur microtubes show a multi-layered structure (Fig. 7.18(b)). In Fig. 7.18(d), the sulfur microtube breaks on the surface, and layered structures are evident.

Te[73–75] and Se[76,77] nanotubes were synthesized in several groups. Mayers *et al.* synthesized Te nanotubes by refluxing ethylene glycol and orthotelluric acid at 179°C.[73] The TEM images of the Te nanotubes are shown in Figs. 7.19(a)–7.19(c). They proposed a concentration depletion mechanism to account for the formation of the nanotubes as illustrated in the figure below. Although it seems distinct from other mechanisms for nanotube formation, this mechanism is still similar to what has been proposed for the formation of Bi_2Te_3 nanotubes earlier.

Mo *et al.* synthesized Te nanotubes along with nanoribbons by a hydrothermal method using Na_2TeO_3 and aqueous ammonia as starting materials.[74] They proposed a rolling mechanism of nanoribbons to form nanotubes.

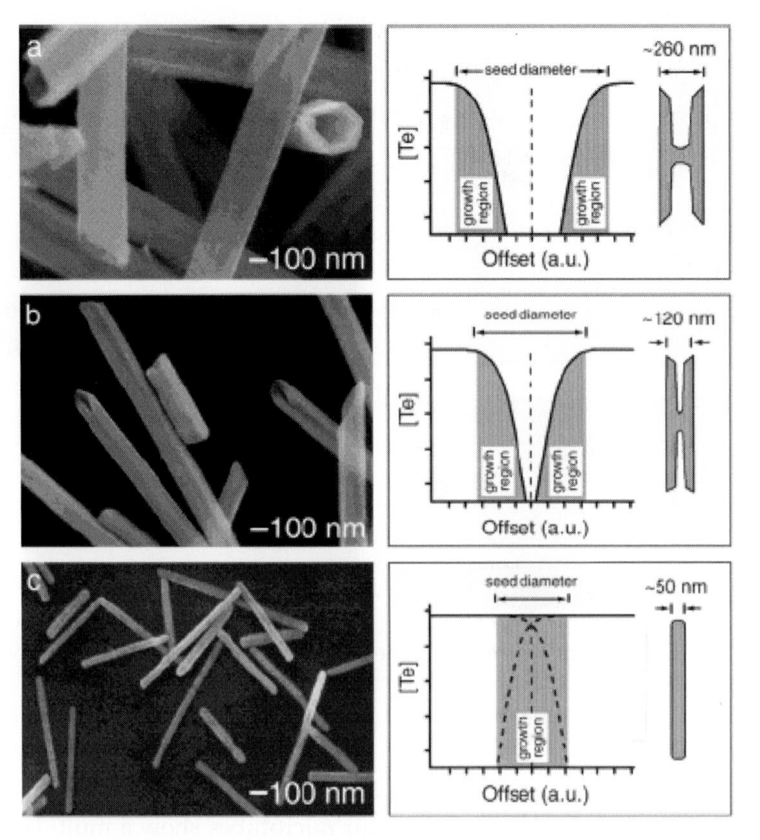

Fig. 7.19. (a)–(c) TEM images of Te nanotubes. The images in the right illustrate the proposed growth mechanism of the nanotubes.[73]

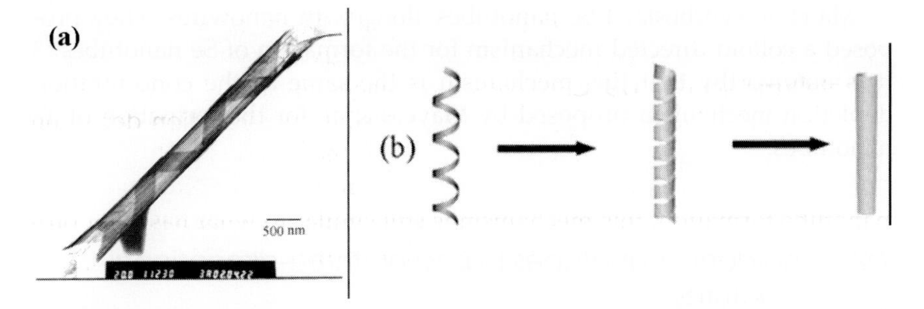

Fig. 7.20. (a) TEM image of an as-formed Te nanotube. (b) Schematic illustration of the roll-up and seaming mechanism.[74]

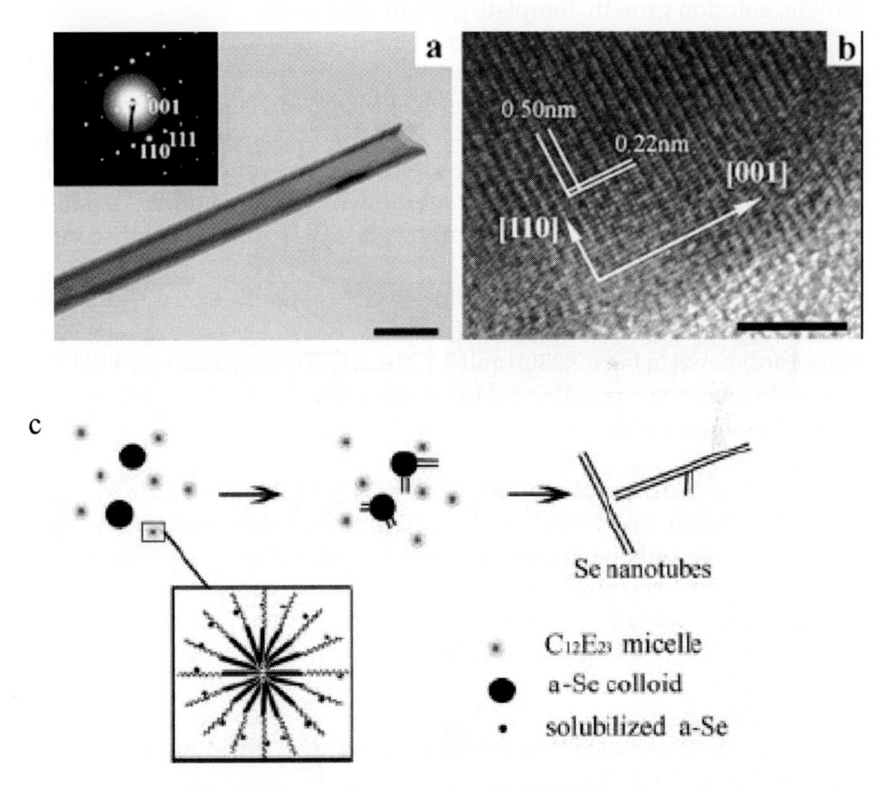

Fig. 7.21. (a) TEM image of a Se nanotube; (b) HRTEM image of a Se nanotube; (c) schematic illustration of the formation mechanism of the Se nanotubes.[76]

Ma *et al.* synthesized Se nanotubes along with nanowires. They proposed a colloid directed mechanism for the formation of Se nanotubes.[76] It is noteworthy that this mechanism is the same as the concentration depletion mechanism proposed by Mayers *et al.* for the formation of Te nanotubes.

7.2.3. Inorganic nanotubes based on three-dimensional structures

Nanotubes in this category are distinct such that they all function as a single part. They can be synthesized by a wide range of methods, including high-temperature vapor phase growth, vapor-liquid-solid catalyzed growth, solution growth, template growth, and so on.

GaN nanotubes were synthesized by Goldberger *et al.* by using ZnO nanorods as the removable template.[78] TEM and HRTEM images of GaN nanotubes are shown in Figs. 7.22(a)–7.22(c). The GaN nanotubes are essentially single crystals judged from the HRTEM images and the corresponding SAED pattern. Previously, they synthesized silica nanotubes using Si nanorods as the template by oxidation and selective removal of the Si core.[79]

Hu *et al.* converted Ga_2O_3 nanowires to GaN nanotubes by nitridization of the nanowires using NH_3.[80] TEM and HRTEM images of the GaN nanotubes are shown in Figs. 7.23(a) and 7.23(b). It is apparent from the HRTEM image and the corresponding SAED pattern that the GaN nanotubes are single crystalline entity.

Hu *et al.* grew cubic phase rectangular GaN nanotubes by the reaction of Ga_2O_3, carbon, and NH_3.[81] GaN nanotubes are shown in Figs. 7.24(a) and 7.24(b). The GaN nanotubes in the SEM image (Fig. 7.24(a)) shows the rectangular cross-section. The GaN nanotubes are cubic phase, different from the other reports described earlier. The authors proposed that the stress due to the rectangular structure contributes to the stabilization of this metastable phase.

AlN nanotubes have been synthesized by nitriding aluminum powder.[82]

InP nanotubes were synthesized by Barkkers *et al.* by a vapor-liquid-solid method.[83] Figures 7.25(a) and 7.25(b) show the TEM and HRTEM images of InP nanotubes. The formation of the InP nanotubes was assisted

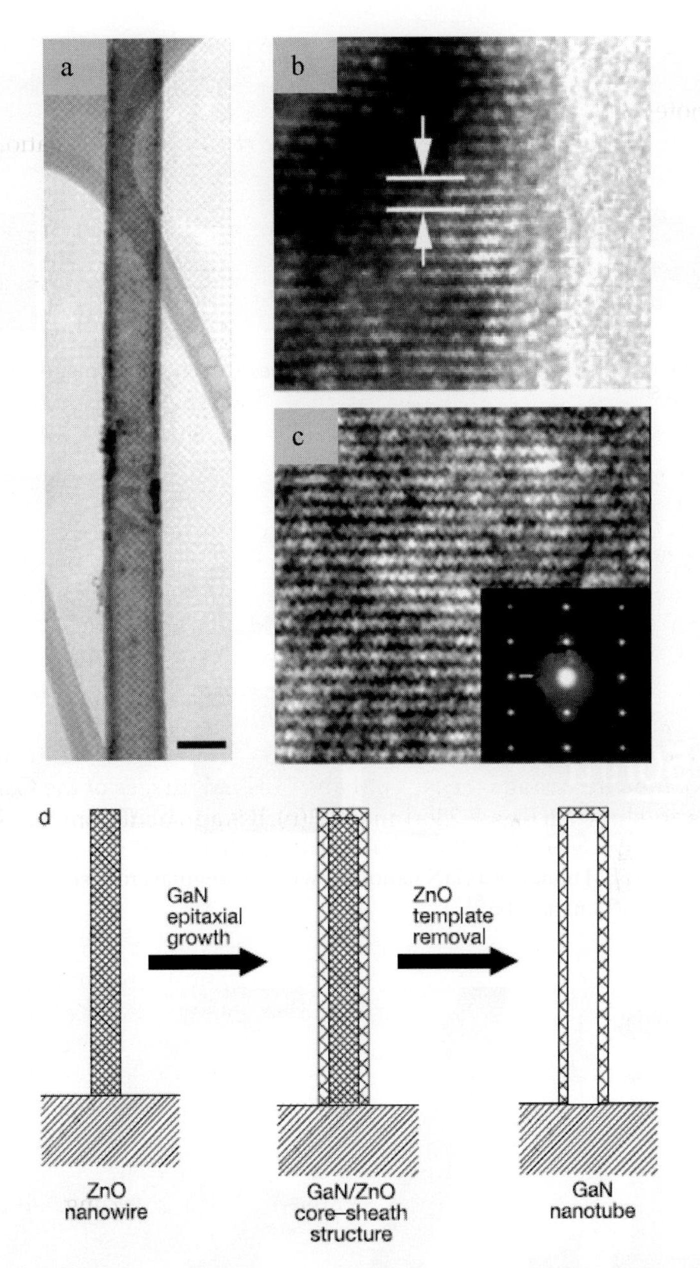

Fig. 7.22. (a) TEM image of a GaN nanotube, (b) and (c) HRTEM images of GaN nanotubes. (d) Schematic illustration of the formation of GaN nanotubes using ZnO nanorods as the removable template.[78]

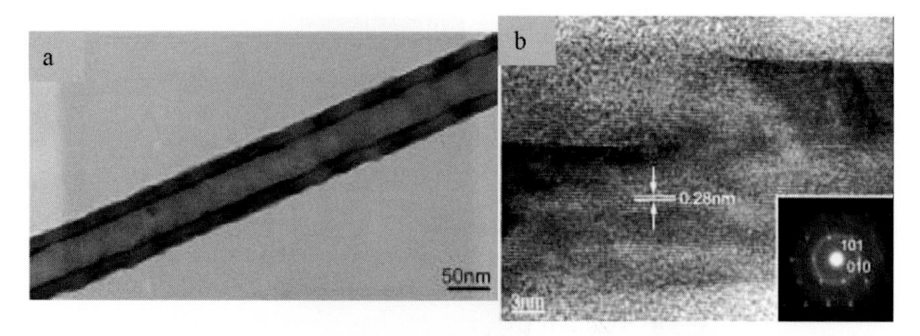

Fig. 7.23. (a) TEM image of a GaN nanotube. (b) HRTEM image of a GaN nanotube.[80]

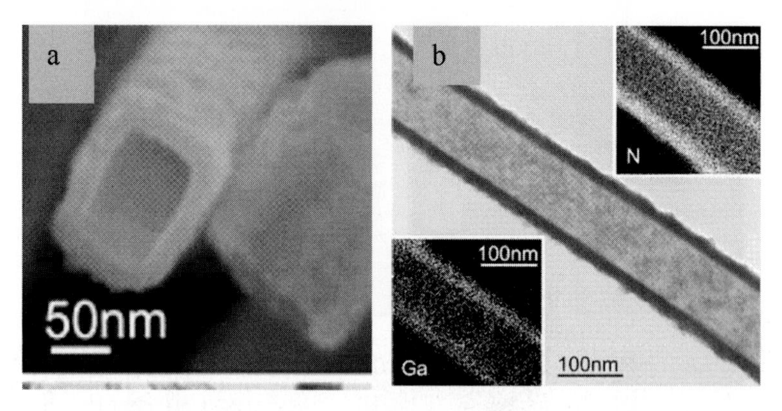

Fig. 7.24. (a) SEM image of a GaN nanotube with rectangular cross-section. (b) TEM image of a GaN_2 nanotube.[81]

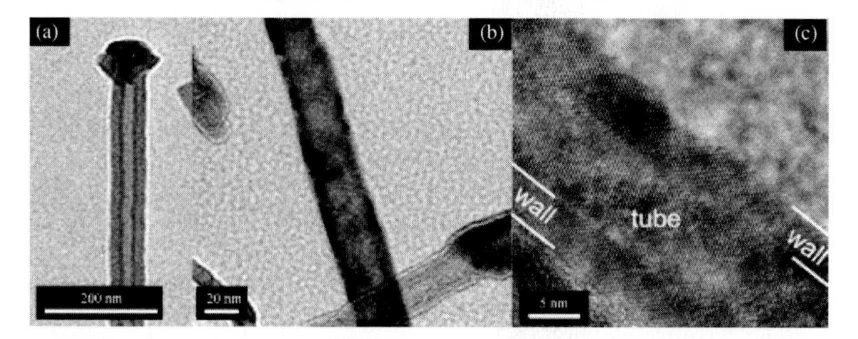

Fig. 7.25. (a) and (b) TEM images of InP nanotubes. (c) HRTEM image of an InP nanotube.[83]

by the metal particles on the tip of the nanotubes with controlling the diffusion and growth rates.

ZnO nanotubes could by synthesized by high-temperature growth[84] and solution growth methods.[85,86]

Yu *et al*. synthesized ZnO nanotube arrays by oxidation of zinc foils with formamide solution. ZnO nanotube arrays are shown in Fig. 7.26.

Sun *et al*.[87] synthesized multi-walled metal nanotubes by using Ag nanowires as the sacrificial template. They synthesized Au/Ag and Pd/Ag multi-walled nanotubes by galvanic reactions of Ag and $HAuCl_4$, $Pd(NO_3)_2$ solutions, respectively. The Au/Ag and Pd/Ag multi-walled nanotubes are shown in Figs. 7.27(a) and 7.27(b), respectively. Figure 7.27(c) is a schematic illustration of the formation process of the nanotubes.

By templating porous alumina membrane, nanotubes of various inorganic materials were synthesized in our laboratory. Wu *et al*. synthesized Eu_2O_3 nanotubes,[88] Wang synthesized Cu nanotubes,[89] Xiao synthesized beaded WO_3 nanotubes,[90] just to name a few.

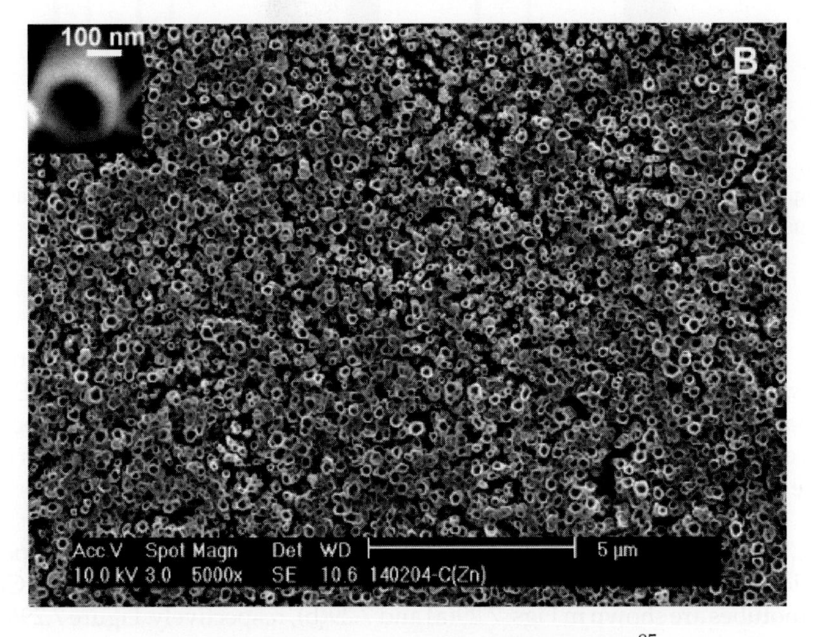

Fig. 7.26. SEM image of a ZnO nanotube array.[85]

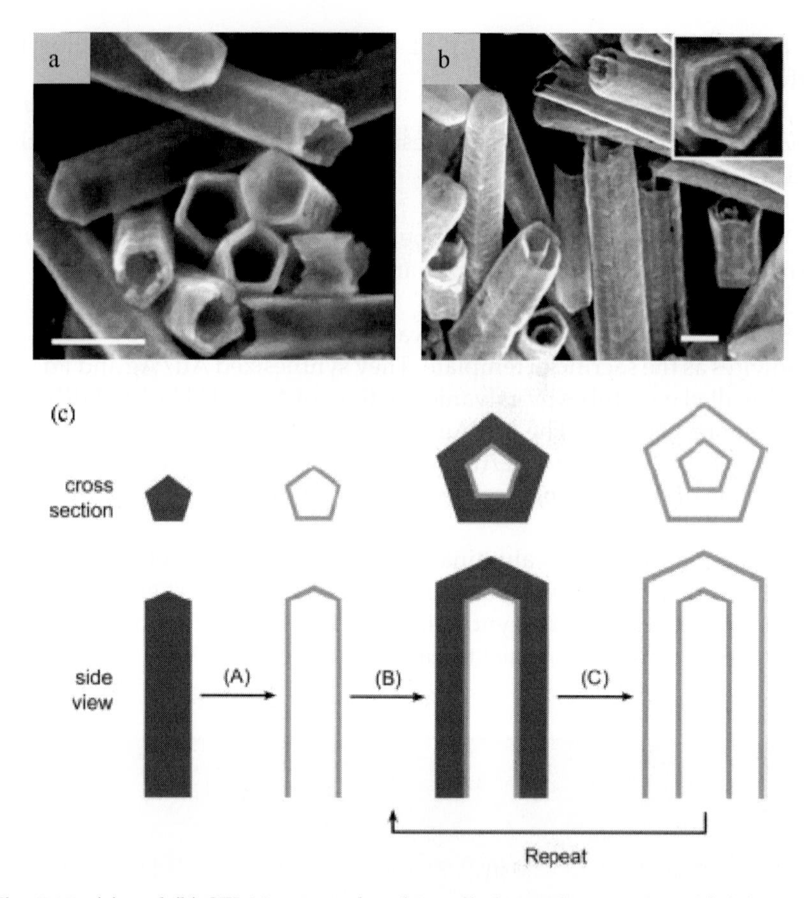

Fig. 7.27. (a) and (b) SEM images of multi-walled metal nanotubes. (c) Schematic illustration of the formation mechanism of the nanotubes.[87]

Eu_2O_3 nanotubes were synthesized in porous alumina template by the *in situ* hydrolysis of $Eu(NO_3)_3$ with urea at 80°C.[88] The SEM and TEM images of Eu_2O_3 nanotubes are shown in Figs. 7.28(a) and 7.28(b), respectively. Figure 7.28(c) is a schematic illustration of the formation process of the nanotubes.

Wang *et al.* synthesized Cu nanotube arrays by an electrochemical deposition method using $CuSO_4{\cdot}5H_2O$ and H_3PO_4 as electrolytes.[89] The Cu nanotubes are shown in Figs. 7.29(a) and 7.29(b), respectively. Figure 7.29(c) is a schematic illustration of the formation process of the nanotubes.

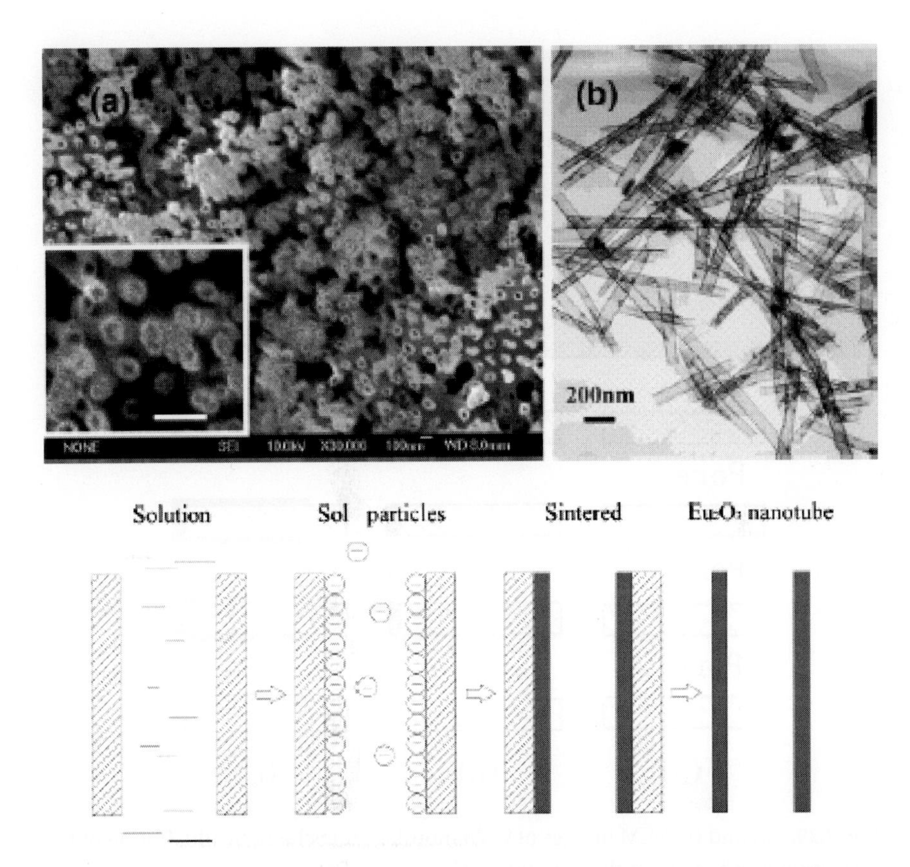

Fig. 7.28. (a) SEM, and (b) TEM images of Eu_2O_3 nanotubes. (c) Schematic illustration of the formation mechanism of the nanotubes.[88]

Xiao *et al.* synthesized porous $W_{18}O_{49}$ beaded nanotubes by a sol-gel method.[90] The $W_{18}O_{49}$ beaded nanotubes are shown in Fig. 7.30(a). Figure 7.30(b) is a schematic illustration of the formation process of the nanotubes.

7.2.4. Formation mechanisms of inorganic nanotubes

The mechanisms for the formation of inorganic nanotubes are divergent. The most understood is for the third category, where the nanotubes form by templating removable nanowires or nanoholes. Nanotubes in this category could also be formed by the *in situ* conversion of nanowires or nanorods by nitridation or sulfidation[80,81] or galvanic replacement.[87]

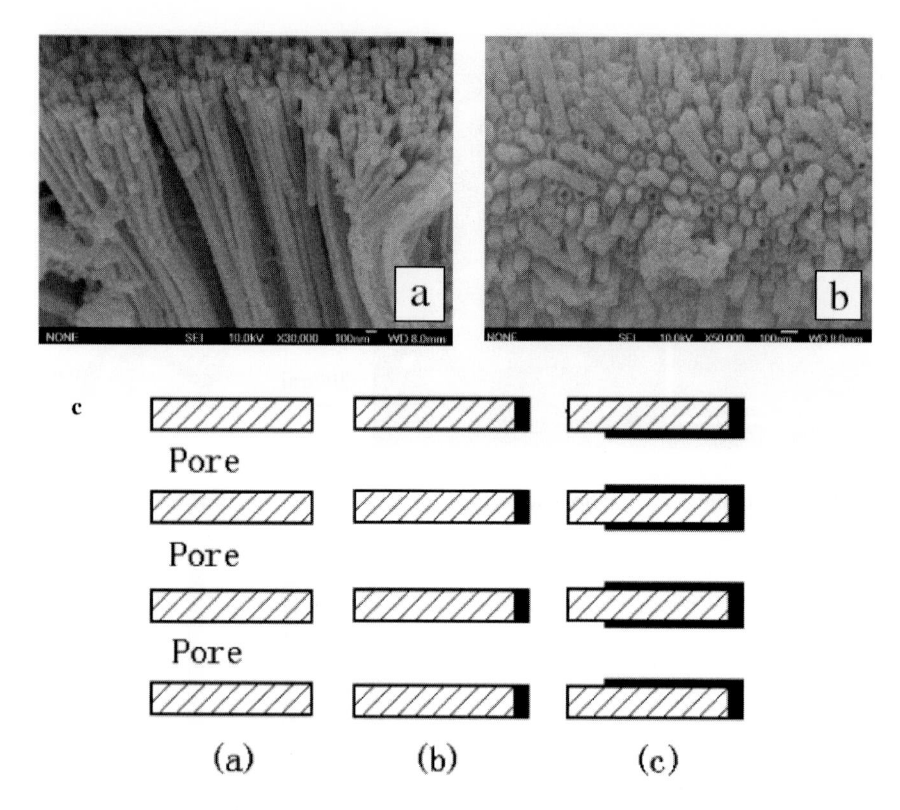

Fig. 7.29. (a) and (b) SEM images of Cu nanotubes. (c) Schematic illustration of the formation mechanism of the nanotubes.[89]

Nanotubes in the first category could also be synthesized by the *in situ* conversion method similarly.[28,32] The more general mechanism for the formation of inorganic nanotubes in the first category is believed to be the roll-up and seaming model. Li *et al.* pyrolysized the lamellar precursor of WS_4^{2-}, and they did find the formation of WS_2 nanoscrols from the rolling of the precursor.[91] Therefore, they proved this rolling mechanism was also operative to the formation of WS_2 nanoscrols. This roll-up mechanism is also operative for specific materials in the second category, and Bi_2S_3 nanotubes are believed to form according to this mechanism. The most observed formation mechanism for the second category is, however, the formation of nanoloop and the oriented growth mechanism which is the case for Se, Te, and Sb_2S_3 nanotubes. The nanotubes in the first category, such as Bi_2Te_3 and $H_2Ti_3O_7$, could also be formed by this mechanism.

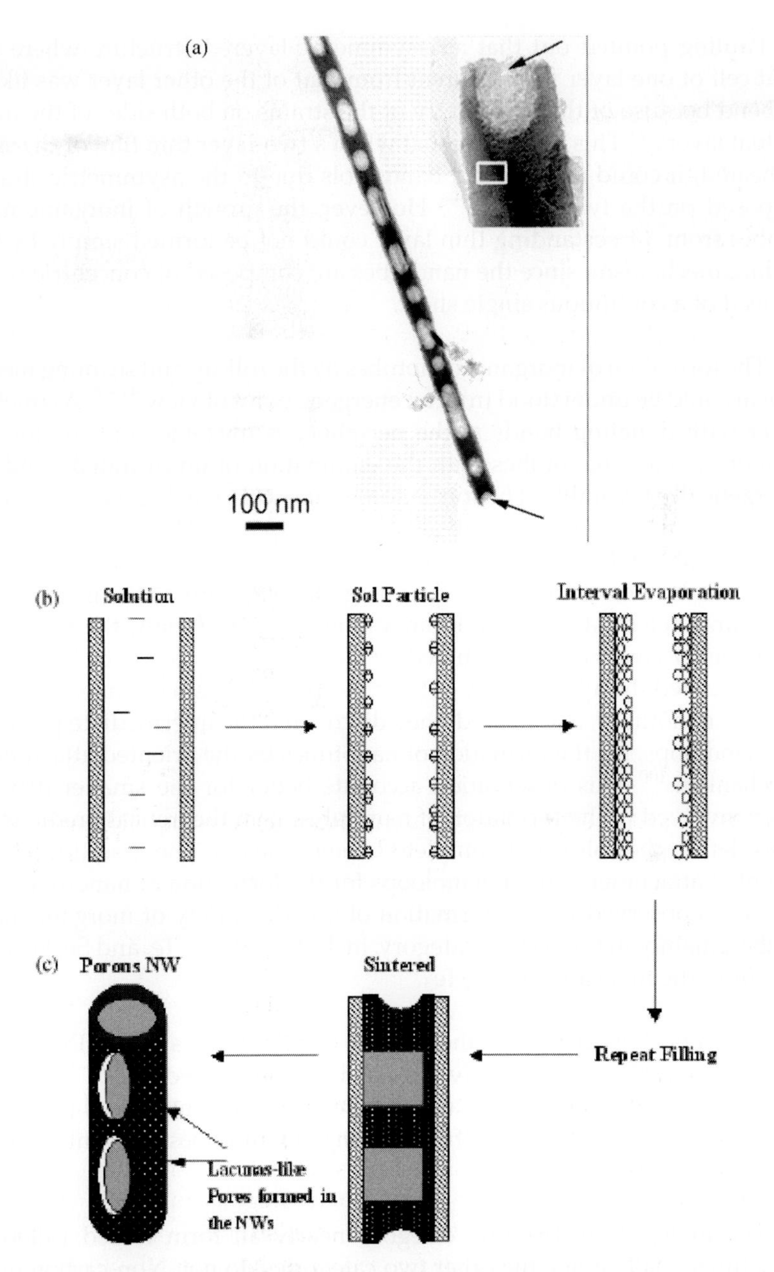

Fig. 7.30. (a) and (b) SEM images of Cu nanotubes. (c) Schematic illustration of the formation mechanism of the nanotubes.[90]

Pauling pointed out that an asymmetric-layered structure where the unit cell of one layer was different from that of the other layer was likely to bend because of the asymmetry of the strains on both sides of the individual layers.[92] This is the case that when a two-layer thin film of Si/SiGe is heated, it could roll to form nanoscrols due to the asymmetric strains imposed on the two sides.[93–96] However, the growth of inorganic nanotubes from a freestanding thin layer could not be formed simply by the rolling mechanism, since the nanotubes are composed of concentric walls instead of a continuous single sheet.

The formation of inorganic nanotubes by the roll-up and seaming mechanism could be understood from an energetic point of view.[97–99] A lamellar layer with dangling bonds at the periphery is unstable, and the formation of seamless nanotubes with the elimination of unsaturated bonds is energetically favorable. The formation of nanotubes will generate defects in the structure, and strain will also be built up in it. The closed structure will be stable when the energy decrease due to the elimination of dangling bonds over-compensates the energy increase due to the inclusion of structural defects and strain. From a kinetic point of view, the roll-up of the lamellar layers may be initiated by thermal stress. However, Kukovecz *et al.* doubted the applicability of the roll-up mechanism of lamellar sheets to form nanotubes, and indeed, they discovered the intermediate product, the nanoloops, for the formation of nanotubes by the oriented attachment mechanism.[100] This observation accounts better for the smaller driving force involved in the formation of nanotubes than theoretical predictions, considering the roll-up of complete lamellar layers. The assertion of the oriented attachment on the nanoloops for the formation of nanotubes has also been observed for the formation of a wide variety of inorganic nanotubes, mainly in the second category, including Sb_2S_3, Te, and Se, besides one from the first category, Bi_2Te_3.

Theoretical predictions of the stability of nanotubes, such as BN, P, B, Si, and so on have been extensively carried out in the last decade.[101–104] Unfortunately, nanotubes of pure P and B have not been synthesized in a laboratory so far. The existence of freestanding Si nanotubes is a controversial issue.

The materials in the first category nearly all form closed fullerene structures,[105–111] while the other two categories do not. Non-carbon inorganic nanotube and fullerene structures are generally faceted. The faceting of the structure is to relieve the inherent strains in roll-up and bending of the layered structure to the closed and partially-closed nanotubes and fullerenes.[112–114] An example of the inorganic nanotubes with a rectangular cross-section is shown in the SEM images in Fig. 7.31. Carbon nanotubes and fullerenes lack the faceting morphology. It is interesting that carbon nanotubes with triangular cross-sections could be formed by templating a porous alumina membrane with triangular holes (Fig. 7.32).[115] It should also be noted that BN and C could form horns, cones, and baboons, in additional to nanotubes and fullerene structures.[116–122]

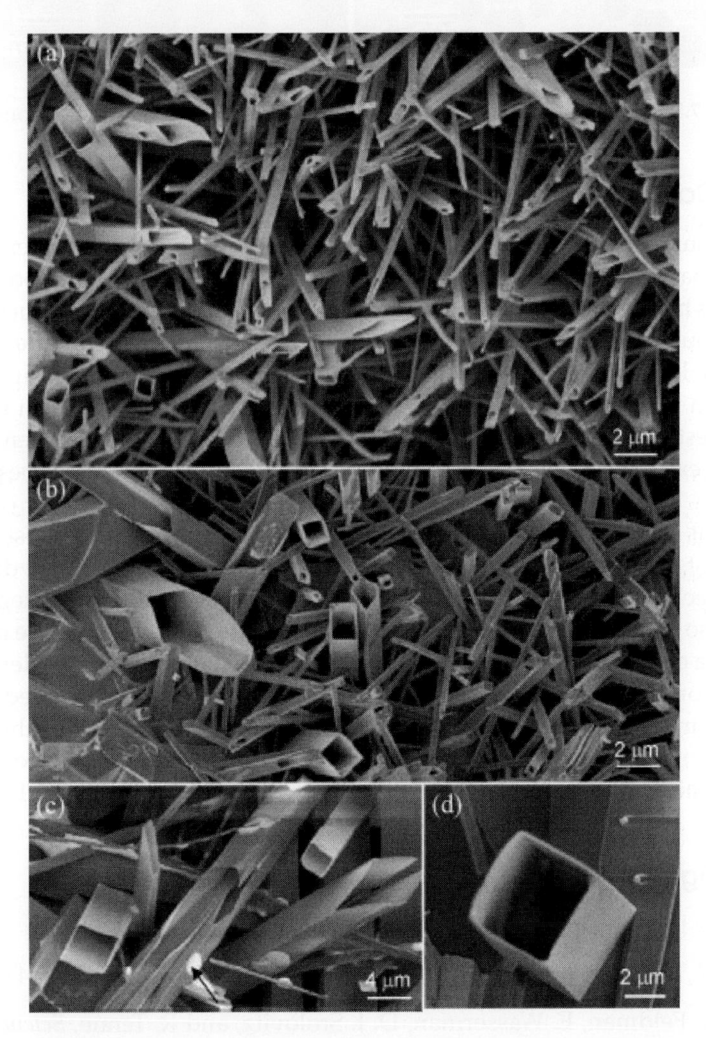

Fig. 7.31. SEM images of mullite microtubes with rectangular cross-sections.[113]

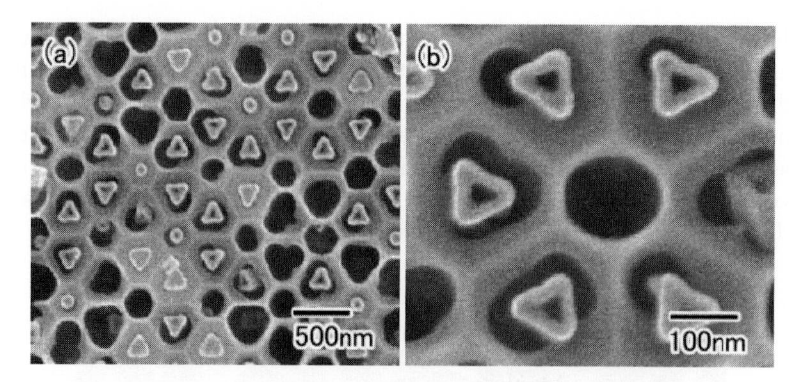

Fig. 7.32. SEM images of carbon nanotubes with triangular cross-sections.[115]

7.3 Concluding remarks

Inorganic nanotubes are an interesting family of quasi-one-dimensional nanomaterials. Synthesis of inorganic nanotubes other than carbon nanotubes has been carried out extensively, however, large-scale production of high-purity nanotubes has only been achieved for BN and WS_2. The reason for the low productivity and low purity is the poor understanding of the growth mechanism and formation process of the nanotubes. Even for the simplest roll-up mechanism, scientists have not arrived at a consensus as to the extent of its functions in the formation of inorganic nanotubes.

This chapter aimed to review the inorganic nanotubes synthesized in the last decade according to the structure of the materials, and discuss the mechanisms for the formation of the nanotubes in different categories. We also intended to include as many realistic examples as possible and to draw a clear picture which could be easily followed up by the readers. It is noteworthy that this field is developing very rapidly. What is deemed novel today may become outdated one year later. It is our hope that the brief review in this chapter would benefit those who are interested in exploring inorganic nanotubes, both for the synthesis and for the growth mechanism.

Bibliography

1. S. Iijima, *Nature* **354**, 56 (1991).
2. R. Tenne, L. Margulis, M. Genut, and G. Hodes, *Nature* **360**, 444 (1992).
3. Y. Feldman, E. Wasserman, D. J. Srolovitz, and R. Tenne, *Science* **267**, 222 (1995).

4. N. G. Chopra, R. J. Luyken, K. Cherrey, V. H. Crespi, M. L. Cohen, S. G. Louie, and A. Zettl, *Science* **269**, 966 (1995).
5. W. Tremel, *Angew. Chem. Int. Ed.* **38**, 2175 (1999).
6. R. Tenne and A. K. Zettl, *Topics Appl. Phys.* **115**, 83 (2001).
7. G. R. Patzke, F. Krumeich, and R. Nesper, *Angew. Chem. Int. Ed.* **41**, 2446 (2002).
8. R. Tenne, *Eur. J. Chem.* **8**, 5296 (2002).
9. A. L. Ivanovskii, *Russian Chem. Rev.* **71**, 175 (2002).
10. C. R. N. Rao and M. Nath, *Dalton Trans.* 1 (2003).
11. R. Tenne, *Angew. Chem. Int. Ed.* **42**, 5124 (2003).
12. M. Remskar, *Adv. Mater.* **16**, 1497 (2004).
13. G. S. Zakharova, V. L. Volkov, V. V. Ivanovskaya, and A. L. Ivanovskii, *Russian Chem. Rev.* **74**, 651 (2005).
14. L. J. E. Hofer, E. Sterling, and J. T. McCartney, *J. Phys. Chem.* **59**, 1153 (1955).
15. R. Bacon, *J. Appl. Phys.* **31**, 283 (1960).
16. K. Yada, *Acta Cryst.* **23**, 704 (1967).
17. W. M. Hess, L. L. Bann, and G. C. McDonald, *Rubber Chem. Technol.* **42**, 1209 (1969).
18. C. Kaito, Y. Saito, and K. Fujita, *J. Cryst. Growth* **94**, 967 (1989).
19. D. Keller, *Nature* **384**, 111 (1996).
20. P. Ball, *Nature* **414**, 142 (2001).
21. M. S. Dresselhaus, *Nature* **432**, 959 (2004).
22. H. Xu, *Nat. Mater.* **4**, 649 (2005).
23. Y. Zhu, T. Sekine, K. S. Brigatti, S. Firth, R. Tenne, R. Rosentsveig, H. W. Kroto, and D. R. M. Walton, *J. Am. Chem. Soc.* **125**, 1329 (2003).
24. G. Seifert, H. Terrones, M. Terrones, and T. Frauenheim, *Solid State Commun.* **115**, 635 (2000).
25. M. Nath, S. Kar, A. K. Raychaudhuri, and C. R. N. Rao, *Chem. Phys. Lett.* **368**, 690 (2003).

26. V. V. Ivanovskaya, A. N. Enyashin, A. A. Sofronov, Yu. N. Makurin, N. I. Medvedeva, and A. L. Ivanovskii, *Solid State Commun.* **126**, 489 (2003).

27. Y. Oshima, A. Onga, and K. Takayanagi, *Phys. Rev. Lett.* **91**, 205503 (2003).

28. A. Rothschild, G. L. Frey, M. Homyonfer, R. Tenne, and M. Rappaport, *Mater. Res. Innovat.* **3**, 145 (1993).

29. J. Chen, S. Li, F. Gao, and Z. Tao, *Chem. Mater.* **15**, 1012 (2003).

30. E. B. Mackie, D. H. Galvan, E. Adem, S. Talapatra, G. L. Yang, and A. D. Migone, *Adv. Mater.* **12**, 495 (2000).

31. Y. Zhu, W. K. Hsu, N. Grobert, B. He, M. Terrones, H. Terrones, H. W. Kroto, and D. R. M. Walton, *Chem. Mater.* **12**, 1190 (2000).

32. A. Rothschild, J. Sloan, and R. Tenne, *J. Am. Chem. Soc.* **122**, 5169 (2000).

33. W. K. Hsu, B. Chang, Y. Zhu, W. Han, H. Terrones, M. Terrones, N. Grobert, A. K. Cheetham, H. W. Kroto, and D. R. M. Walton, *J. Am. Chem. Soc.* **122**, 10155 (2000).

34. M. Nath, A. Govindaraj, and C. N. R. Rao, *Adv. Mater.* **13**, 283 (2001).

35. M. Nath and C. N. R. Rao, *Chem. Commun.* 2236 (2001).

36. M. Remskar, A. Mrzel, Z. Skraba, A. Jesih, M. Ceh, J. Demsar, P. Stadelmann, F. Levy, and D. Mihailovic, *Science* **292**, 479 (2001).

37. M. Nath, K. Mukhopadhyay, and C. N. R. Rao, *Chem. Phys. Lett.* **352**, 163 (2002).

38. W. K. Hsu, Y. Zhu, N. Yao, S. Firth, R. J. H. Clark, H. W. Kroto, and D. R. M. Walton, *Adv. Funct. Mater.* **11**, 69 (2001).

39. J. Chen, Z. Tao, and S. Li, *Angew. Chem. Int. Ed.* **42**, 214 (2003) 7.

40. J. Chen, Z. Tao, S. Li, X. Fan, and S. Chou, *Adv. Mater.* **15**, 1379 (2003).

41. M. Brorson, T. W. Hansen, and C. J. H. Jacobsen, *J. Am. Chem. Soc.* **124**, 11582 (2002).

42. M. Nath and C. N. R. Rao, *J. Am. Chem. Soc.* **123**, 4841 (2001).

43. M. Nath and C. N. R. Rao, *Angew. Chem. Int. Ed.* **41**, 3451 (2002).

44. Y. Q. Zhu, W. K. Hsu, H. W. Kroto, and D. R. M. Walton, *J. Phys. Chem.* **B106**, 7623 (2002).

45. J. A. Hollingsworth, D. M. Poojary, A. Clearfield, and W. E. Buhro, *J. Am. Chem. Soc.* **122**, 3562 (2000).

46. S. Y. Hong, R. P. Biro, Y. Prior, and R. Tenne, *J. Am. Chem. Soc.* **125**, 10470 (2003).

47. Y. R. Hacohen, E. Grunbaum, R. Tenne, J. Sloan, and J. L. Hutchison, *Nature* **395**, 336 (1998).

48. Y. R. Hacohen, R. P. Biro, E. Grunbaum, Y. Prior, and R. Tenne, *Adv. Mater.* **14**, 1075 (2002).

49. J. Cumings and A. Zettl, *Chem. Phys. Lett.* **316**, 211 (2000).

50. C. Tang, Y. Bando, and T. Sato, *Chem. Phys. Lett.* **362**, 185 (2002).

51. P. Cai, L. Chen, L. Shi, Z. Yang, A. Zhao, Y. Gu, T. Huang, and Y. Qian, *Solid State Commun.* **133**, 621 (2005).

52. M. Terrones, A. M. Benito, C. M. Digo, W. K. Hsu, O. I. Osman, J. P. Hare, D. G. Reid, H. Terrones, A. K. Cheetham, K. Prassides, H. W. Kroto, and D. R. M. Walton, *Chem. Phys. Lett.* **257**, 576 (1996).

53. M. Terrones, D. Golberg, N. Grobert, T. Seeger, M. R. Reyes, M. Mayne, R. Kamalakaran, P. Dorozhkin, Z. Dong, H. Terrones, M. Ruhle, and Y. Bando, *Adv. Mater.* **15**, 1899 (2003).

54. M. Niederberger, H. J. Muhr, F. Krumeich, F. Bieri, D. Gunther, and R. Nesper, *Chem. Mater.* **12**, 1995 (2000).

55. M. Yada, M. Mihara, S. Mouri, M. Kuroki, and T. Kijima, *Adv. Mater.* **14**, 309 (2002).

56. X. Wang and Y. Li, *Chem. Euro. J.* **9**, 5627 (2003).

57. X. Wang, X. Sun, D. Yu, B. Zou, and Y. Li, *Adv. Mater.* **15**, 1442 (2003).

58. A. Xu, Y. Fang, L. You, and H. Liu, *J. Am. Chem. Soc.* **125**, 1494 (2003).

59. Y. Fang, A. Xu, L. You, R. Song, J. C. Yu, H. Zhang, Q. Li, and H. Liu, *Adv. Funct. Mater.* **13**, 955 (2003).

60. G. Du, Q. Chen, R. Che, Z. Yuan, and L. Peng, *Appl. Phys. Lett.* **79**, 3702 (2001).

61. Q. Chen, W. Zhou, G. Du, and L. Peng, *Adv. Mater.* **14**, 1208 (2002).

62. X. Sun and Y. Li, *Chem. Euro. J.* **9**, 2229 (2003).

63. Y. Lan, X. Gao, H. Zhu, Z. Zheng, T. Yan, F. Wu, S. P. Ringer, and D. Song, *Adv. Funct. Mater.* **15**, 1310 (2005).

64. Y. Li, J. Wang, Z. Deng, Y. Wu, X. Sun, D. Yu, and P. Yang, *J. Am. Chem. Soc.* **123**, 9904 (2001).

65. H. Hu, M. Mo, B. Yang, M. Shao, S. Zhang, Q. Li, and Y. Qian, *New J. Chem.* **27**, 1161 (2003).

66. D. Pan, S. Zhang, G. Li, Y. Chen, and J. Hou, *Inter. J. Nanosci.* **1**, 187 (2002).

67. J. R. Ota and S. K. Srivastava, *Nanotechnology* **16**, 2415 (2005).

68. C. Ye, G. Meng, Z. Jiang, Y. Wang, G. Wang, and L. Zhang, *J. Am. Chem. Soc.* **124**, 15180 (2002).

69. C. Ye, G. Meng, L. Zhang, G. Wang, and Y. Wang, *Chem. Phys. Lett.* **363**, 34 (2002).

70. Y. Zhang, G. Li, and L. Zhang, *Chem. Lett.* **33**, 334 (2004).

71. J. Yang, Y. C. Liu, H. M. Lin, and C. C. Chen, *Adv. Mater.* **16**, 713 (2004).

72. I. Bezverkheyy, P. Afanasiev, C. Marhic, and M. Danot, *Chem. Mater.* **15**, 2119 (2003).

73. B. Mayers and Y. Xia, *Adv. Mater.* **14**, 279 (2002).

74. M. Mo, J. Zeng, X. Liu, W. Yu, S. Zhang, and Y. Qian, *Adv. Mater.* **14**, 1658 (2002).

75. Z. He, S. Yu, and J. Zhu, *Chem. Mater.* **17**, 2785 (2005).

76. Y. Ma, L. Qi, J. Ma, and H. Cheng, *Adv. Mater.* **16**, 1023 (2004).

77. H. Zhang, D. Yang, Y. Ji, X. Ma, J. Xu, and D. Xue, *J. Phys. Chem. B* **108**, 1179 (2004).

78. J. Goldberger, R. He, Y. Zhang, S. Lee, H. Yan, H. J. Choi, and P. Yang, *Nature* **422**, 599 (2003).

79. R. Fan, Y. Wu, D. Li, M. Yue, A. Majumdar, and P. Yang, *J. Am. Chem. Soc.* **125**, 5254 (2003).

80. J. Hu, Y. Bando, D. Golberg, and Q. Liu, *Angew. Chem. Int. Ed.* **42**, 3493 (2003).

81. J. Hu, Y. Bando, J. Zhan, F. Xu, T. Sekiguchi, and D. Golberg, *Adv. Mater.* **16**, 1465 (2004).

82. Q. Hu, Z. Hu, X. Wang, Y. Lu, X. Chen, H. Xu, and Y. Chen, *J. Am. Chem. Soc.* **125**, 10176 (2003).

83. E. P. A. M. Barkkers and M. A. Verheijin, *J. Am. Chem. Soc.* **125**, 440 (2003).
84. X. Kong, Y. Ding, and Z. L. Wang, *J. Phys. Chem. B* **108**, 570 (2004).
85. H. Yu, Z. Zhang, M. Han, X. Hao, and F. Zhu, *J. Am. Chem. Soc.* **127**, 2378 (2005).
86. Q. Li, V. Kumar, Y. Li, H. Zhang, T. J. Marks, and R. P. H. Chang, *Chem. Mater.* **17**, 1001 (2005).
87. Y. Sun and Y. Xia, *Adv. Mater.* **16**, 264 (2004).
88. G. Wu, L. Zhang, B. Cheng, T. Xie, and X. Yuan, *J. Am. Chem. Soc.* **126**, 5976 (2004).
89. Y. Wang, C. Ye, X. Fang, and L. Zhang, *Chem. Lett.* **33**, 166 (2004).
90. Z. Xiao, L. Zhang, X. Tian, and X. Fang, *Nanotechnology* **16**, 2647 (2005).
91. Y. Li, X. Li, R. He, J. Zhu, and Z. Deng, *J. Am. Chem. Soc.* **124**, 1411 (2002).
92. L. Pauling, *Proc. Nat. Acad. Sci.* **16**, 578 (1930).
93. V. Ya. Prinz, V. A. Seleznev, A. K. Gutakovsky, A. V. Chehofskiy, V. V. Preobrazhenskii, M. A. Putyato, and T. A. Gavrilova, *Physica E* **6**, 828 (2000).
94. O. G. Schmidt and K. Eberl, *Nature* **410**, 168 (2001).
95. V. Ya. Prinz, D. Gruetzmacher, A. Beyer, C. David, B. Ketterer, and E. Deckardt, *Nanotechnology* **12**, 399 (2001).
96. Yu. V. Nastaushev, V. Ya. Prinz, and S. N. Svitasheva, *Nanotechnology* **16**, 908 (2005).
97. G. Seifert, T. Kohler, and R. Tenne, *J. Phys. Chem. B* **106**, 2497 (2002).
98. S. Zhang, L. Peng, Q. Chen, G. Du, G. Dawson, and W. Zhou, *Phys. Rev. Lett.* **91**, 256103 (2003).
99. D. J. Srolovitz, S. A. Safran, M. Homyonfer, and R. Tenne, *Phys. Rev. Lett.* **74**, 1779 (1995).
100. A. Kukovecz, M. Hodos, E. Horvath, G. Radnoczi, Z. Konya, and I. Kiricsi, *J. Phys. Chem. B* **109**, 17781 (2005).
101. A. Rubio, J. L. Corkill, and M. L. Cohen, *Phys. Rev. B* **49**, 5081 (1994).
102. G. Seifert and E. Hernandez, *Chem. Phys. Lett.* **318**, 355 (2000).
103. I. Boustani, *Phys. Rev. B* **55**, 16426 (1997).

104. R. Zhang, S. T. Lee, C. K. Law, W. K. Lee, and B. K. Teo, *Chem. Phys. Lett.* **364**, 251 (2002).

105. M. Nath, C. N. R. Rao, R. P. Biro, A. A. Yaron, and R. Tenne, *Chem. Mater.* **16**, 2238 (2004).

106. Y. Mastai, M. Homyonfer, A. Gedanken, and G. Hodes, *Adv. Mater.* **12**, 1010 (1999).

107. Y. Feldman, A. Zak, R. P. Biro, and R. Tenne, *Solid State Sci.* **2**, 663 (2000).

108. Y. R. Hacohen, R. P. Biro, Y. Prior, S. Gemming, G. Seifert, and R. Tenne, *Phys. Chem. Chem. Phys.* **5**, 1644 (2003).

109. R. P. Biro, N. Sallacan, and R. Tenne, *J. Mater. Chem.* **13**, 1631 (2003).

110. J. A. Ascencio, M. P. Alvarez, L. M. Molina, P. Santiago, and M. J. Yacaman, *Surf. Sci.* **526**, 243 (2003).

111. X. Wang and Y. Li, *Angew. Chem. Int. Ed.* **42**, 3497 (2003).

112. L. Yin, Y. Bando, J. Zhan, M. Li, and D. Golberg, *Adv. Mater.* **17**, 1972 (2005).

113. X. Kong, Z. L. Wang, and J. S. Wu, *Adv. Mater.* **15**, 1445 (2003).

114. X. Chen, J. Ma, Z. Hu, Q. Wu, and Y. Chen, *J. Am. Chem. Soc.* **127**, 7982 (2005).

115. T. Yanagishita, M. Sasaki, K. Nishio, and H. Masuda, *Adv. Mater.* **16**, 429 (2004).

116. T. Yanagishita, M. Sasaki, K. Nishio, and H. Masuda, *Adv. Mater.* **16**, 429 (2004).

117. C. Tang, M. Lamy de la Chapelle, P. Li, Y. Liu, H. Dang, and S. Fan, *Chem. Phys. Lett.* **342**, 492 (2001).

118. L. Bourgeois, Y. Bando, W. Han, and T. Sato, *Phys. Rev. B* **61**, 7686 (2000).

119. C. Zhi, Y. Bando, C. Tang, D. Golberg, R. Xie, and T. Sekiguchi, *Appl. Phys. Lett.* **87**, 063107 (2005).

120. K. Murata, K. Kaneko, F. Kokai, K. Takahashi, M. Yudasaka, and S. Iijima, *Chem. Phys. Lett.* **331**, 14 (2000).

121. G. Zhang, X. Jiang, and E. Wang, *Science* **300**, 18 (2003).

122. Z. Sun, Z. Liu, J. Du, Y. Wang, B. Han, and T. Mu, *J. Phys. Chem. B* **108**, 9811 (2004).

Chapter 8
Novel Properties of Nanomaterials

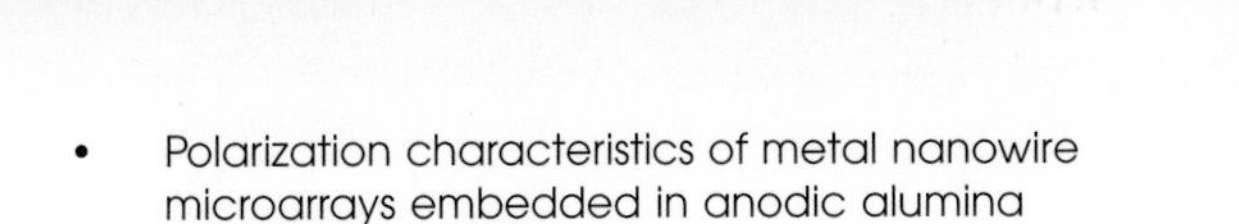

- Polarization characteristics of metal nanowire microarrays embedded in anodic alumina membrane templates
- Electronic and magnetic properties of Bi-based nanowire arrays
- Thermal expansion properties of nanowire arrays

Chapter 8

Novel Properties of Nanomaterials

8.1 Introduction

Nanomaterials and nanostructures have attracted steadily growing interest due to their fascinating properties. Recently, some novel properties that are associated with nanometer-sized dimensions have been found.[1] The best-established examples include size-dependent excitation,[2] quantized or ballistic conductance,[3] Coulomb blockade or single electron tunneling,[4] metal-insulator transition,[5,6] photoluminescence and lasing properties,[7] photochemical sensing,[8] piezoelectric properties,[9] dielectric properties[10] etc.

In this chapter, we mainly present the polarization characteristics of metal nanowire microarrays embedded in anodic alumina membrane templates, electronic and magnetic properties of Bi-based nanowire arrays, and thermal expansion properties of nanowire arrays.

8.2 Polarization characteristics of metal nanowire microarrays embedded in anodic alumina membrane templates

8.2.1. Introduction

Polarization plays a key role in optical devices such as isolators, modulators, and switches. With the development of optical techniques, the miniaturization and highly efficient performance of the devices could be explored in much depth. In these cases, microarray polarizing elements have received growing attention owing to their potential applications in optical communications, optical integrated circuits, and orientation detections.

In contrast to the conventional wire grids fabricated by evaporation[11,12] and lithography[13–16] methods, metallic nanowire arrays can be synthesized through a facile template-based synthesis approach[17–20] rather than complicated and expensive techniques. As early as in 1987, Miyagi *et al.* predicted the possibility of fabricating a kind of polarizer using the technique of anodic oxidation of aluminium and sealing process.[21] Subsequently, Satio and co-workers fabricated a novel wire-grid polarizer made of anodic alumina film filled with metallic columns, which indicated anisotropic optical properties.[22] However, owing to the immature anodization technique of aluminium and inherent difficult electro-deposition process of the metal, an obvious non-uniform distribution of the optical loss at a certain wavelength was discovered.

In an attempt to improve the polarization properties, ordered arrays of metal nanowires embedded in a anodic alumina membrane (AAM) template have been synthesized successfully.[23–25] The template can work as an efficient wire-grid polarizer at the near infrared wavelengths. On the other hand, we analyze the mechanism of the polarization in detail and explore the size-dependent polarization characteristics of the arrays.

8.2.2. Optical measurement

Before optical measurement, a portion of the metal/AAM sample was sliced off and the cross-sectional surfaces were polished until the length (L) became as thin as about 10 μm (see Fig. 8.1). The thickness (d) of the sample was about 60 μm (see Fig. 8.1). The prepared sample was stuck onto a SiO_2 glass slide for mechanical support during the optical measurement. The optical loss of the sample was evaluated by a spectrophotometer (Cary 5E) and the experimental setup was shown in Fig. 8.2. The sample was

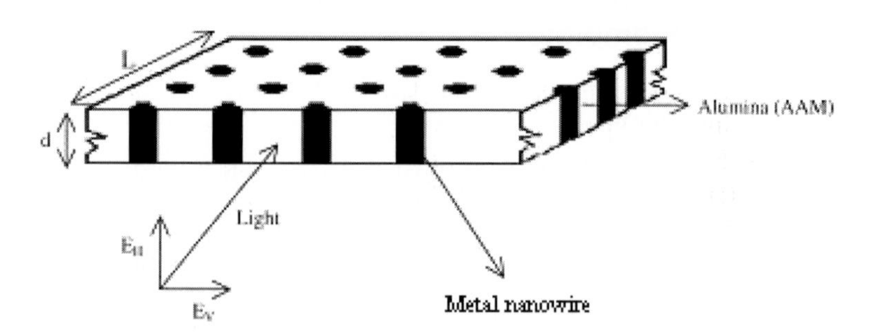

Fig. 8.1. Schematic illustration of the Cu/AAM sample.[23]

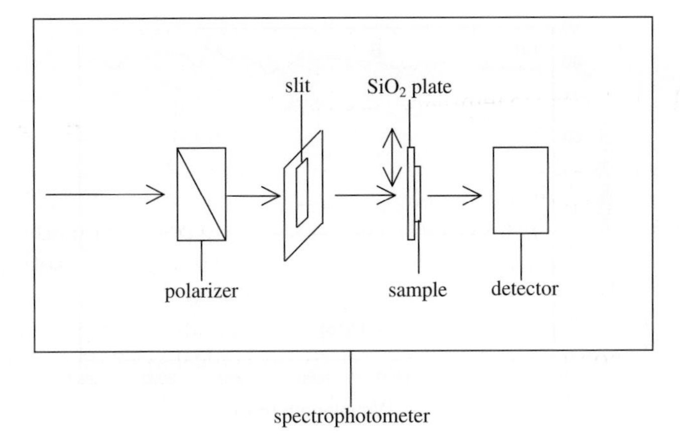

Fig. 8.2. The experimental setup used for the evaluation of the optical loss. The sample and the SiO_2 glass slide can be moved up and down (see the *double arrow*) by a micromechanical stage.[23]

positioned on a rotating stage so that the direction of polarization might be horizontal (H polarization) or vertical (V polarization) to the nanowires. The infrared light (from 1.0 to 2.2 μm) transmitted through a polarizer prism and a narrow slit and then reached the sample at normal incidence.

8.2.3. Polarization characteristics

8.2.3.1. *Cu/AAM*

The preparation and detailed structural characterization of the anodic alumina membrane (AAM) and metal nanowire microarrays embedded in anodic alumina membrane templates (Cu/AAM) can be seen in Chap. 5 and Ref. 23.

The optical transmission of the Cu/AAM sample (Fig. 8.3(a)) was obtained using the spectrophotometer (Fig. 8.2). The light transmitted through the SiO_2 glass slide without a sample on it was also measured as a reference (curve B in Fig. 8.3(a)). Curves H and V denote the transmittance of horizontal and vertical incident light, respectively. It is seen that the curve H is amplified eight times for clear display. The transmission difference between H and V demonstrates prominent optical anisotropy of the sample, i.e., optical transmission increases significantly when the direction of polarization is changed from horizontal to vertical in the nanowires.

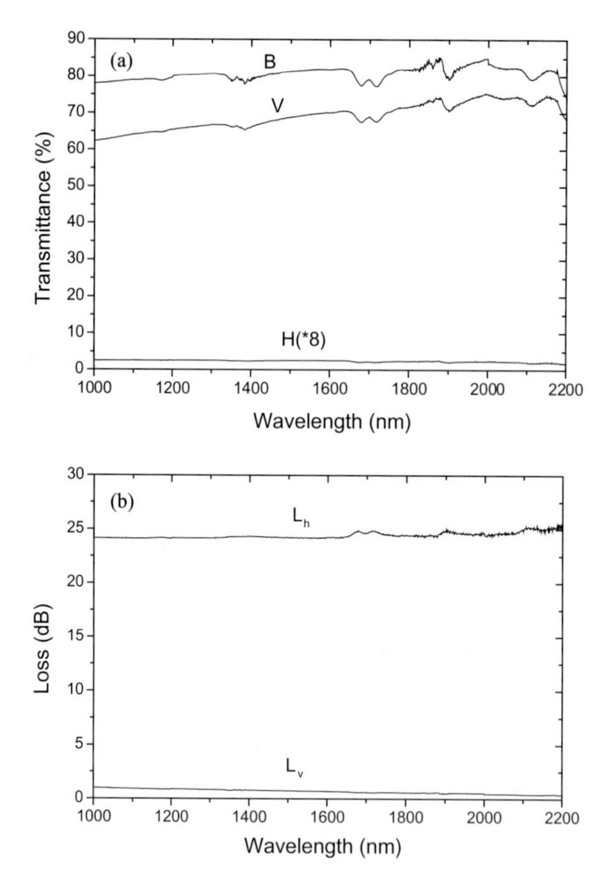

Fig. 8.3. (a) Transmission and (b) optical loss of copper nanowire arrays embedded in the AAM measured at the wavelength region 1 μm ~ 2.2 μm. Curve B expresses the transmission of incident light that only passes through the silica substrate; curves H and V express the transmission of incident light that passes through both silica and sample, horizontal and vertical to the channels, respectively. The curve H is amplified eight times for clear display. In the (b) curves, L_h and L_v express the extinction ratio and insertion loss calculated from (a), respectively.[23]

Figure 8.3(b) shows the optical loss of the sample calculated from the transmittance. Curve L_h (or L_v) was calculated from the light intensity ratio of B to H (or V). Therefore, curve L_h presents the extinction ratio (i.e. the loss of horizontal polarization), and curve L_v presents the insertion loss (i.e. the loss of vertical polarization). In the whole applied wavelength region, there is much difference in optical loss between L_h and L_v. For example, at the wavelength of 2.1 μm, the insertion loss is only 0.38 dB and the extinction ratio is as large as 25 dB. With the increase of the wavelength, the extinction ratio increases, whereas the insertion loss decreases. This is in agreement with the theoretical prediction.[26]

Curves L_h and L_v in Fig. 8.3(b) reveal the polarizing function of the sample, i.e. 24 to 32 dB for the extinction ratio and 0.5 dB for average insertion loss in the wavelength region 1 μm $\sim$ 2.2 μm, respectively. The extinction ratio is smaller than the expected theoretical values.[26] The possible reasons for the discrepancies are as follows: variation of the nanowire density, and oxidation of the metal nanowires. The large insertion loss of the sample seems to originate mainly from the scattering on the rough sample surfaces.

To investigate the homogeneity of the transmission of the sample, we measured the optical transmission of the sample from different positions by moving the sample up and down slightly with a micromechanical stage (see double arrows in Fig. 8.2). The results (Fig. 8.4) reveal that the transmissions of V and H are slightly different from their counterparts shown in Fig. 8.3(a). The inhomogeneity of the transmission of the sample is due to the non-uniformity of the sample thickness.

To evaluate the polarizer of Cu nanowire arrays embedded in the AAM for practical application, we compare it with two popular commercial polarizers, i.e., the conventional wire grid and the Glan–Thompson

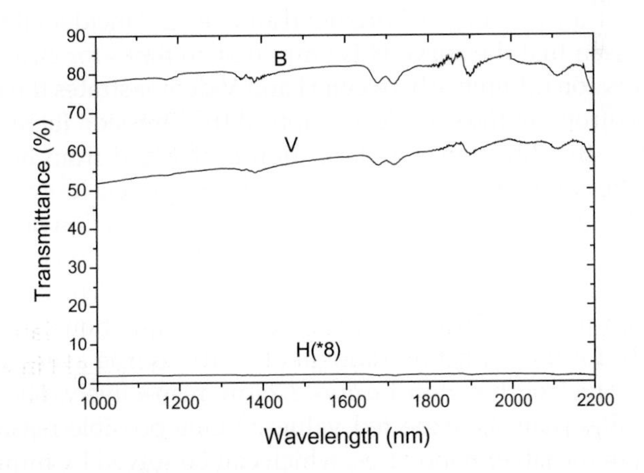

Fig. 8.4. Optical transmission of the Cu/AAM sample measured at different positions compared with Fig. 8.3.[23]

prism. The extinction ratio and the insertion loss in Fig. 8.3(b) show a more efficient polarizer than those of the conventional wire grids,[27–29] because the performance of a wire grid polarizer is determined by the wire size. To attain a high extinction ratio and low insertion loss, the thickness and spacing of wires must be sufficiently smaller than the wavelength of the incident light. Since the conventional wire grids are fabricated by photolithography, the wire size cannot be made small enough, and consequently this results in the poor performance in comparison with copper nanowires embedded in the AAM. The Glan–Thompson prism has an excellent polarization property from visible to near infrared. However, it is made of calcite crystals and is rather expensive. Although the polarization of the sample is inferior to the Glan–Thompson prism at present, the Cu/AAM polarizer has an advantage for practical applications from the viewpoint of fabrication cost.

8.2.3.2. *Ag/AAM*

The fabrication and detailed structural characterization of silver nanowire arrays embedded in the AAM are similar to those of the copper nanowire arrays.[24]

The optical transmission of the Ag/AAM sample is displayed in Fig. 8.5(a). The light intensity transmitted through the SiO_2 glass plate without a sample on it was also measured as a reference (curve B). Curves H and V denote the transmittance of horizontal and vertical incident light, respectively. It is seen that the curve H is amplified 40 times for clearer display. The transmission difference between H and V demonstrates the prominent optical anisotropy of the sample, i.e. optical transmission increases significantly when the direction of polarization is changed from horizontal to vertical in the nanowires. Figure 8.5(b) shows the optical loss of the sample calculated from the transmittance. We see that the extinction ratio increases when the wavelength becomes longer, whereas the insertion loss decreases. This is in agreement with the theoretical prediction.[26] The curves L_h and L_v presented in Fig. 8.5(b) reveal the polarizing function of the sample, i.e. 25–26 dB for the extinction ratio and 0.77 dB for the average insertion loss in the wavelength region $1\,\mu m \sim 2.2\,\mu m$, respectively. The extinction ratio is smaller than the expected values.[26] One possible reason may be the defects in the silver nanowires, which can be solved by improving the electrochemical deposition process. Other possible reasons for the discrepancies may be as follows: variation of the nanowire density, and oxidation

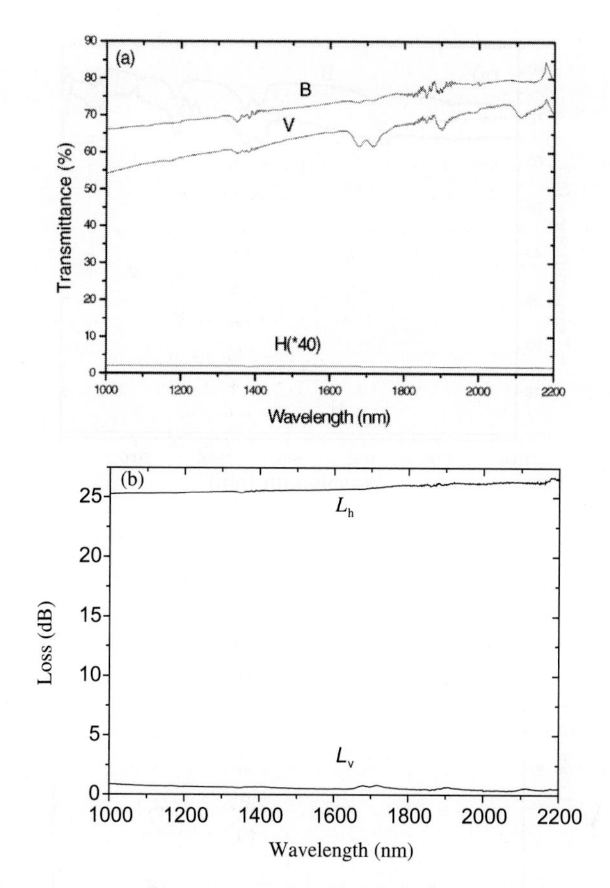

Fig. 8.5. (a) Transmission and (b) optical loss of silver nanowire arrays embedded in the AAM.[24]

of the metal Ag nanowires. The large insertion loss of the sample seems to originate mainly from the scattering on the rough sample surfaces.

8.2.3.3. *Pb/AAM*

Similarly, Pb nanowire arrays embedded in the AAM prepared in a sulfuric acid solution have been synthesized by DC electro-deposition.[25] Figure 8.6 shows the transmittances and optical losses of laminated Pb nanowire arrays embedded in the AAM. The transmission difference between H and V demonstrates prominent optical anisotropy of the laminated sample. From Fig. 8.6, we can observe that the extinction ratio is 17–18 dB/cm^2 and the mean insertion loss is 0.4 dB/cm^2 at the wavelengths. However, the

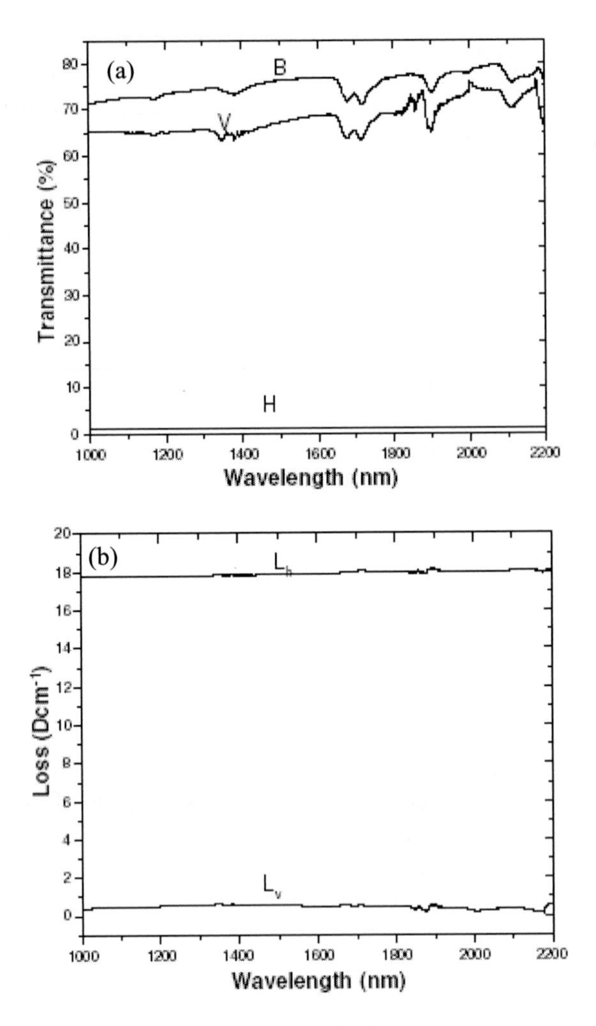

Fig. 8.6. (a) Transmission and (b) optical loss of the laminated lead nanowire arrays embedded in the AAM.[25]

extinction ratio is smaller than expected, probably owing to the leak light that is transmitted through the adhesive layers.

8.2.4. Theoretical calculation

Polarization properties of the metal nanowire microarrays embedded in the AAM are theoretically investigated to explore the mechanism of the polarization. We not only analyze the origin of the polarization but also demonstrate the size dependence of the polarization property. High extinction ratios and low insertion losses can be expected by properly modulating the diameter and spacing according to the optical loss spectra. Furthermore, a design of nanowire grid polarizer at an optimum dimension in the near and mid infrared wavelength region is presented.

8.2.4.1. *Theory model*

A schematic illustration of the ordered nanostructure is depicted in Fig. 8.7.[30] The nanostructure contains hexagonally-packed microarrays of copper nanowires parallel to each other, and the nanowires are $2a$ in diameter and b in spacing.

Compared with the experiment,[23] the typical values of $2a$ and b are 90 nm and 130 nm, respectively, and λ represents the wavelength of the incident waves in the near infrared region (e.g. $1.0 \sim 2.2\,\mu$m), thus, we have $a, b << \lambda$. The wavenumbers k_1 and k_2 can be written as $k_1 = 2\pi n_1/\lambda$ and $k_2 = 2\pi n_2/\lambda$, where n_1 and n_2 correspond to the refractive indexes of alumina and metal, respectively. The refractive index of metal can be expressed as $n_2 = n_m - ik$,[31] where n_m and k are the real and imaginary parts of the complex refractive index, respectively. Moreover, n_m determines a phase lag of the light wave traveling through the metal, and k (the so-called extinction coefficient) describes an attenuation of the intensity of the light wave. E_s and E_p are the electric fields of the incident waves, which correspond to light polarized parallel (s polarization) and perpendicular

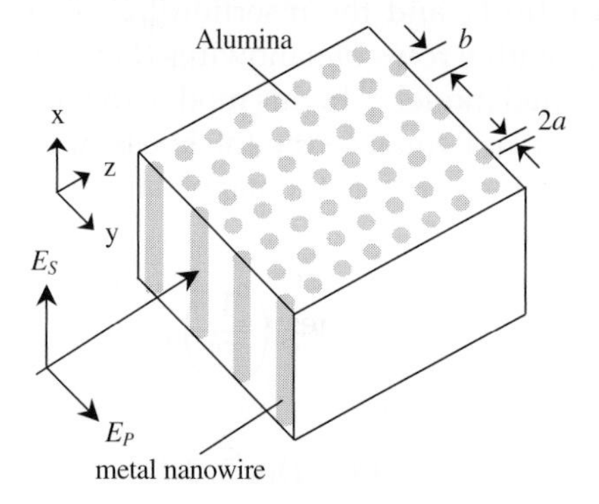

Fig. 8.7. Schematic illustration of a microarray of metal nanowires embedded in the AAM.[30]

(p polarization) to the nanowire axis. As for the incident light wave propagating normal to the nanowires, the optical losses of a row of the nanowires can be determined by considering a coupling effect of adjacent nanowires,[32] If we consider a row of the nanochannels of the AAM as fully filled with metal nanowire, then the transmission coefficient can be given by

$$t = 1 - \left[1 - \frac{ib\eta_1}{\pi k_1 a^2}\right]^{-1} - \left[1 + \frac{ik_1 b(\eta_0 - \log 2)}{\pi}\right]^{-1}, \tag{8.1}$$

where η_0 and η_1 are parameters dependent on the polarization direction. One can obtain η_0 and η_1 using the boundary conditions. For s polarization, the parameters take the form

$$\eta_{0s} = \frac{J_0(k_2a) - (k_2a)J_0'(k_2a)\log(\pi a/b)}{(k_1a)^2 J_0(k_2a)/2 + (k_2a)J_0'(k_2a)}, \tag{8.2}$$

$$\eta_{1s} = \frac{J_1(k_2a) + (k_2a)J_1'(k_2a)}{J_1(k_2a) - (k_2a)J_1'(k_2a)}. \tag{8.3}$$

For p polarization, one has

$$\eta_{0p} = \frac{(k_2a)J_0(k_2a)/(k_1a)^2 - J_0'(k_2a)\log(\pi a/b)}{(k_2a)J_0(k_2a)/2 + J_0'(k_2a)}, \tag{8.4}$$

$$\eta_{1p} = \frac{(k_2a)J_1(k_2a) + (k_1a)^2 J_1'(k_2a)}{(k_2a)J_1(k_2a) - (k_1a)^2 J_1'(k_2a)}, \tag{8.5}$$

where J_0, J_1 are Bessel functions of 0 and 1 orders, and J_0', J_1' are the corresponding derivatives, respectively.

The extinction ratio L_s and the insertion loss L_p of the hexagonally-ordered microarray with n rows of nanowires (Fig. 8.8) can be determined by utilizing a simplified model.[30] In this model, the nanowires are uniform in length (e.g. 60 μm) and equal to the thickness of the template, and the nanowire diameter is the same as the nanochannel diameter. The optical loss α can be expressed as

$$\alpha = \frac{1}{d}\log\left(\frac{1}{T^n}\right), \tag{8.6}$$

where

$$d = (n-1)(\sqrt{3}/2)b. \tag{8.7}$$

L_s and L_p can be defined by $(\alpha_s - \alpha_p)d$ and $\alpha_p d$, respectively. Here α_s and α_p correspond to the optical losses for s polarization and p polarization. We

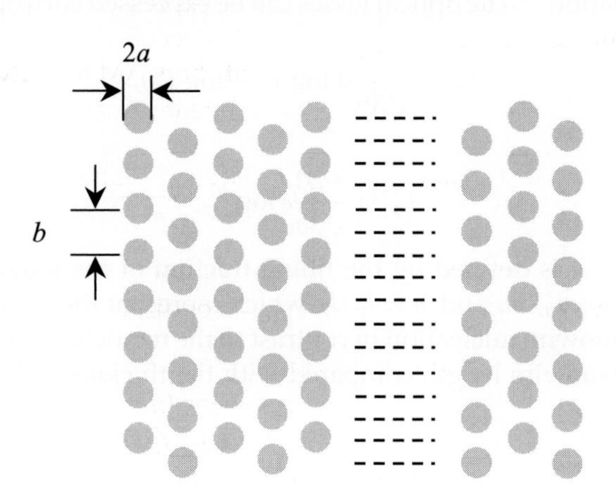

Fig. 8.8. Schematic drawing of the microarray with n rows of nanowires.[30]

chose $d = 10\,\mu$m to match with the experimental data.[23] In this case the number of rows $n >> 1$. L_s and L_p now reduce to

$$L_s = \frac{-20}{\sqrt{3}b}(\log T_s - \log T_p) \tag{8.8}$$

and

$$L_p = \frac{-20}{\sqrt{3}b} \log T_p. \tag{8.9}$$

T_s and T_p follow that

$$T_s = |t_s|^2; \qquad T_p = |t_p|^2, \tag{8.10}$$

where t_s and t_p can be obtained from Eqs. (8.1)–(8.5), which describe the transmission with a row of the nanowires for s polarization and p polarization, respectively.

In fact, full filling (100%) of the nanochannels is still a significantly challenge via the self-assembly approach,[33] which requires that the nanowire nucleation must occur simultaneously in all of the nanochannels, as well as the growth rate of the nanowires must be uniform to ensure that the nanowires are uniform in length and equal to the thickness of the template. Considering the filling inhomogeneity in fabricating nanowire arrays, a filling fraction of the nanochannels should be incorporated into this theoretical calculation.[34] The optical losses can be expressed correspondingly as

$$L_s = \frac{-20}{\sqrt{3}b}\delta(\log T_s - \log T_p), \tag{8.11}$$

and

$$L_p = \frac{-20}{\sqrt{3}b}\delta \log T_p, \tag{8.12}$$

where $\delta = \delta_m \delta_l$ is defined by the filling fraction of the nanochannels. In addition, $\delta_m = m_p/m_f$ and $\delta_l = l_p/l_f$, which represent the average percentage of the nanowire nucleation in contrast to the nanochannel number, and that of the nanowire length compared with the thickness of the template, respectively.

On the basis of Eqs. (8.8)–(8.9) or (8.11)–(8.12), the optical loss spectra of the nanostructures can be numerically simulated.[30,34] In order to investigate the polarization properties of the nanostructures in a wide range of wavelengths, which covers and extends beyond that in the experiment,[23,24] the refractive indexes n_1 of alumina are obtained from Refs. 35 and 36, and the refractive indexes n_2 of metals are taken from the compilations of Lynch and Hunter.[37]

8.2.4.2. *Numerical simulation*

Figure 8.9 shows the spectra of the microarray with the nanowire diameter $2a$ of 90 nm and spacing b of 130 nm according to Eqs. (8.9) and (8.10). It can be seen that L_s increases while L_p decreases with increasing wavelength. Furthermore, L_s is much larger than L_p especially at the long wavelengths. Our theoretical spectra are in agreement with the previous experiment,[23] whereas L_s is larger for our theoretical value than the experimental result. This discrepancy may imply that the density and crystallization of the copper film used in the theoretical simulations are superior to copper nanowires on the basis of the template-based synthesis approach. The results also suggest that a larger fraction of the copper nanowires reflects and absorbs the incident waves by comparison to the experiment, which gives rise to larger L_s for the copper nanowire microarrays as discussed in the following sections.

The origin of L_s and L_p may be analyzed in terms of the following discussion. For an s-polarized wave propagating along the z direction, driven by the field E_s along the nanowire axis, the conduction electrons in the

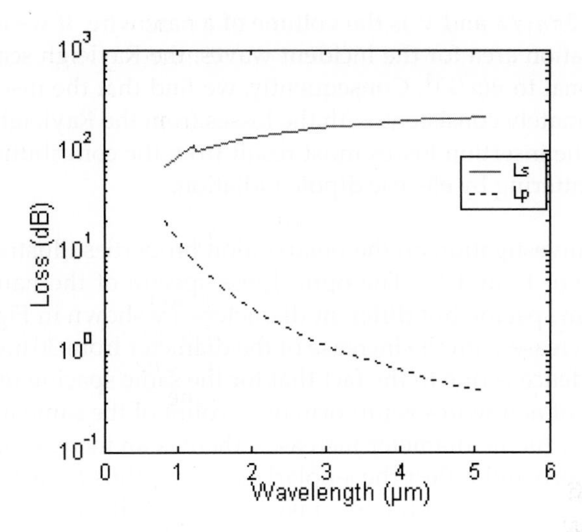

Fig. 8.9. The optical loss spectra of the ordered microarray of copper nanowires of 90 nm in diameter and 130 nm in spacing.[30]

metal lose energy both by electron-phonon and electron-lattice defect interactions, and in this case the light wave is absorbed. In addition, Fresnel reflection occurs at discrete copper-alumina interfaces. Since the extinction coefficient k increases with the increase of the wavelength, the absorption increases correspondingly. Moreover, the Fresnel reflection increases due to the increase of the reflectivity R_F[38] for the s-polarized wave with the wavelength at normal incidence. Therefore, the strong absorption and Fresnel reflection lead to the large L_S.

In contrast, for the thin nanowires of tens of nanometers in diameter, the electrons driving by the field E_p are not free to move very far in the direction perpendicular to the nanowire axis, thus the corresponding field component of a p-polarized wave is essentially unaltered. In addition, one notes that the loss decreases with the increase of the wavelength for p polarization. Since the reflection of the p-polarized wave is the same as that of the s-polarized wave at normal incidence,[38] L_p does not originate from the strong Fresnel reflection at the interfaces. Because the diameter of the nanowires is very small compared with the wavelengths, in terms of Rayleigh approximation,[39] the average scattering cross-section per cylindrical nanowire can be given by

$$C_{sca} = 2k_1^4 V^2 / 3\pi, \tag{8.13}$$

where $k_1 = 2\pi n_1/\lambda$ and V is the volume of a nanowire. If we consider the same irradiation area for the incident waves, the Rayleigh scattering loss is proportional to $n(a/\lambda)^4$. Consequently, we find that the insertion losses are approximately consistent with the losses from the Rayleigh scattering. Therefore, the insertion losses must result from the contribution from the Rayleigh scattering by electric dipole radiation.

Further investigation on the polarization properties illustrates the size dependence of L_s and L_p. The optical loss spectra of the nanostructures with uniform spacing but different diameters are shown in Fig. 8.10. Both L_s and L_p increase with the increase of the diameter from 30 nm to 100 nm. This dependence is due to the fact that for the same spacing (e.g. 130 nm), the number of nanowires is uniform on account of the same incident area of the waves. As the diameter increases, there is an increasing amount of metal to absorb and reflect the s-polarized wave, thus L_s increases. When the metal with increasing fraction takes part in the Rayleigh scattering for p polarization, L_p enhances.

The spectra of the nanostructures with uniform diameter but different spacing are illustrated in Fig. 8.11. Both L_s and L_p decrease with the spacing increasing from 70 nm to 130 nm. In an analogous way, for the same

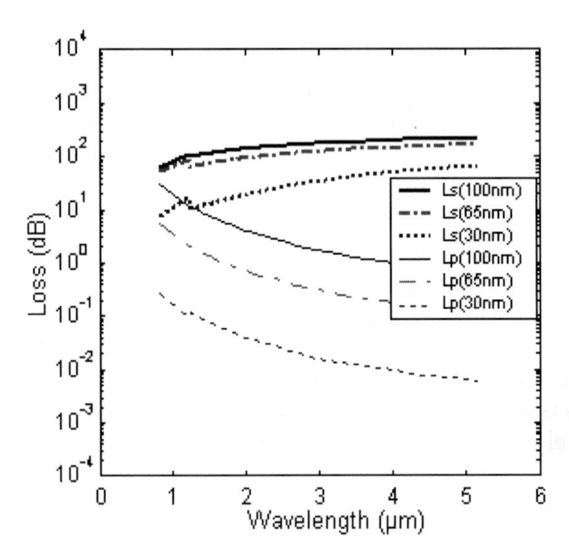

Fig. 8.10. The optical loss spectra of the ordered microarrays of copper nanowires with uniform nanowire spacing of 130 nm but different nanowire diameters of 30 nm, 65 nm, and 100 nm.[30]

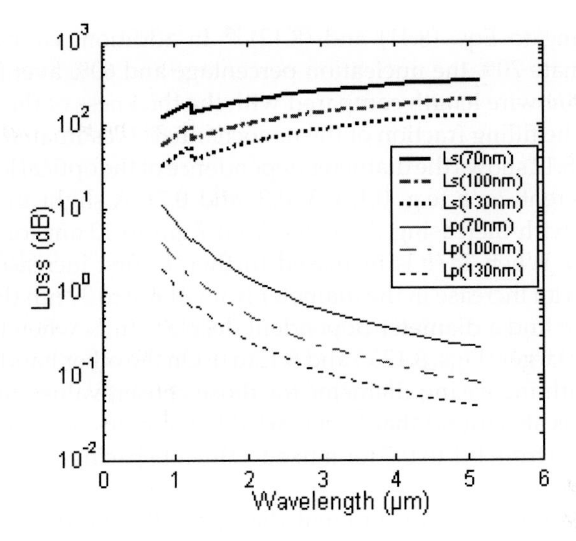

Fig. 8.11. The optical loss spectra of the ordered microarrays of copper nanowires with uniform nanowire diameter of 50 nm but different spacing of 70 nm, 100 nm, and 130 nm.[30]

diameter (e.g. 50 nm), the number of nanowires decreases as the spacing increases, which gives rise to the result that a decreasing amount of the metal takes part in the absorption and reflection, and in this respect L_s decreases. Similarly, L_p also decreases due to the decreasing Rayleigh scattering from the metal.

The strong size dependence of L_s and L_p is observed in Figs. 8.10 and 8.11. Comparing the curves, we see that it is difficult to synchronously obtain large L_s and small L_p. For instance, if the ratio of $2a$ and b is very small, L_p is exceedingly low, however, L_s decreases a lot. On the other hand, if the ratio of $2a$ and b is close to 1, L_s is very high. Nevertheless, L_p is also quite large. In the present nanostructures, the ratio of $2a$ to b at the optimized dimensions is around 0.5 at the wavelengths investigated, and in this case comparatively large L_s and relatively small L_p can be obtained.

It is worthy to note that both L_s and L_p clearly exhibit the same size dependence of polarization characteristics by varying independently one quantity (e.g. diameter or spacing) described above. Nevertheless, the current interest is in how to increase L_s and synchronously diminish L_p in the spectra range. We simulate L_s and L_p with different ratios of $2a$

to b according to Eqs. (8.11) and (8.12).[34] In addition, we can approximately estimate 70% the nucleation percentage and 60% average percentage of the nanowire length compared with the thickness of the template.[34] As a result, the filling fraction of the nanochannels is estimated to be about 42%. Figure 8.12 shows the diameter dependence of the optical losses choosing the different $2a/b$ (e.g. 0.1, 0.3, 0.5, and 0.7). A slight increase in L_s is observed with increasing diameter from 7 nm to 13 nm for $2a/b$ of 0.1 (Fig. 8.12(a)). When $2a/b$ is increased further, L_s first increases and then reduces with an increase in the diameter from 21 nm to 39 nm (Fig. 8.12(b)). However, we find a diameter-dependent decrease in L_s when the value of $2a/b$ is 0.5 or larger (Figs. 8.12(c) and 8.12(d)). On the other hand, L_p slightly enhances with increasing diameter for those chosen values of $2a/b$. Furthermore, it is also found that L_p remarkably increases with an increase in ratio of $2a$ to b from 0.1 to 0.7 for a fixed value of spacing.

The optical losses with different diameters for those selected values of $2a/b$ can be analyzed as follows. Since both $2a$ and b display a strong influence on the optical loss for s-polarization, different contributions of the loss from those quantities will give rise to the different L_s. As $2a/b$ is very small (e.g. 0.1), the contribution from the diameter possibly becomes dominant as compared to that from the number (e.g. in inverse proportion to the spacing), which results in a slight increase in L_s with increasing diameter. However, the contribution from the number becomes significant when $2a/b$ is larger (e.g. 0.5 and 0.7). In this case, the diameter-dependent decrease is attributed to the reduction in the number of nanowires. For a mid $2a/b$ (e.g. 0.3), the contributions from the diameter and the number do not show distinct difference. The contribution from the diameter perhaps plays a dominant role at the shorter wavelengths. Nevertheless, that from the number may become increasingly prominent at the longer wavelengths. In fact, this case is very complicated for the qualitative analysis because it is not enough to interpret the reason by only considering the size rather than the wavelength according to the quantitative description demonstrated in the theory section.

On the other hand, if we take into account the same irradiation area for the incident waves, the Rayleigh scattering loss L_R is given by

$$L_R \propto m_p \delta_l^2 (a/\lambda)^4, \tag{8.14}$$

and the number of the nanowires in the hexagonal arrays (Fig. 8.8) can be given by

$$m_p = 2\delta_m A/(\sqrt{3}b^2). \tag{8.15}$$

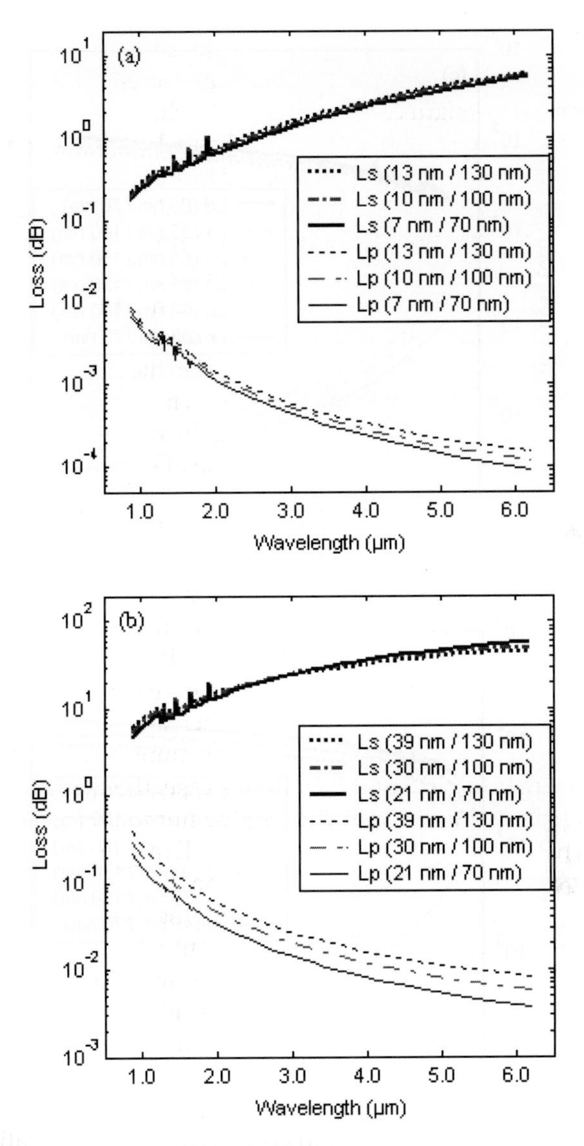

Fig. 8.12. Diameter evolution of optical loss spectra of the microarrays of silver nanowires when selecting the different ratios of $2a$ to b: (a) $2a/b = 0.1$, (b) $2a/b = 0.3$, (c) $2a/b = 0.5$, and (d) $2a/b = 0.7$.[34]

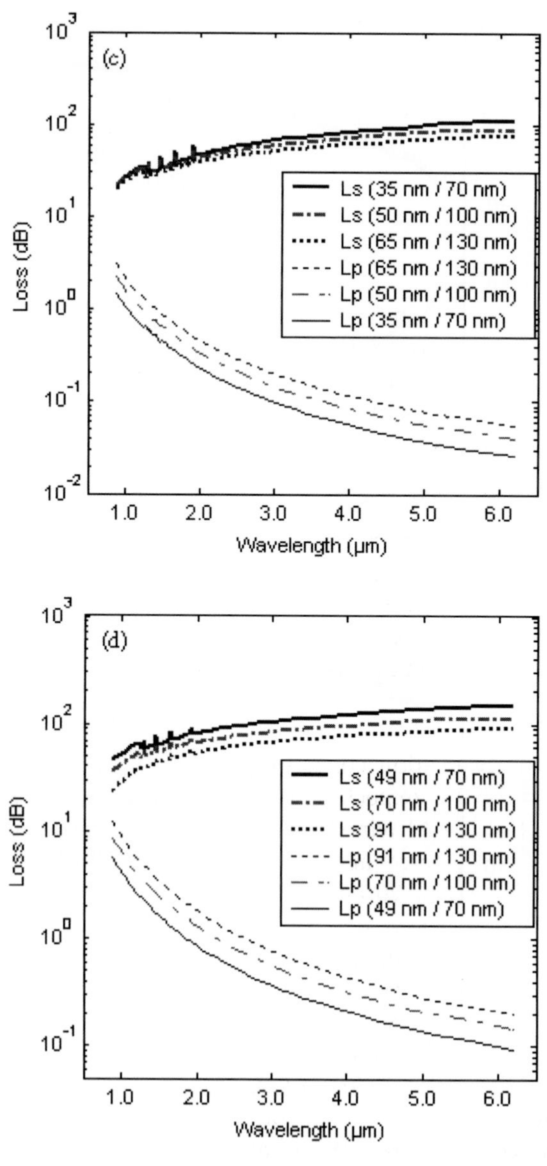

Fig. 8.12. (*Continued*)

Equation (8.14) now has the form

$$L_R \propto \frac{A}{2\sqrt{3}}\delta_m\delta_l^2(2a/b)^2(a^2/\lambda^4). \tag{8.16}$$

When $2a/b$ is fixed, δ_m and δ_l are invariable, we have

$$L_R \propto (a^2/\lambda^4). \tag{8.17}$$

Hence, L_p increases when the diameter increases for a fixed value of $2a/b$ (Figs. 8.12(a)–8.12(d)). Since the number of the nanowires is uniform for the same spacing, the scattering loss can be evaluated to be proportional to a^4/λ^4. Consequently, L_p obviously enhances with increasing ratio of $2a$ to b for the same spacing.

On the basis of the analysis of the diameter-dependent polarization characteristics for the different values of $2a/b$, we see that a larger L_s and a smaller L_p can be obtained by a proper choice of $2a/b$ (e.g. 0.5) as shown in Fig. 8.12(c). In particular, L_s increases and L_p reduces with the decreasing diameter from 65 nm to 35 nm. Therefore, the large L_s and small L_p can be simultaneously obtained by reducing the diameter of the nanowires for the selected $2a/b$.

An important question concerning the design goal of the nanowire grid polarizer is the expected tolerances of the polarization properties. We note that an Al wire-grid polarizer fabricated by electron-beam lithography has exhibited an extinction ratio of 30 dB and an insertion loss of 0.97 dB (loss of 20%) at an 800 nm wavelength band. Recently a nanowire-grid polarizer designed by a nanoimprint lithography and electron-beam evaporation process can achieve a high extinction ratio of 40 dB at the near infrared spectra range (e.g. C band).[40] Zhou *et al.* have successfully designed an in-fiber polarizer by use of 45° tilted fiber Bragg grating (TFBG) structures and they have obtained an extinction ratio of 33 dB near 1550 nm.[41] Compared with tolerances of those polarizers designed by a lithography approach, an expected manufacturing tolerance is an extinction ratio of 40 dB and an insertion loss 0.5 dB for silver nanowire microarrays embedded in the AAM template by the self-assembly strategy (e.g. at the telecommunication wavelength of 1550 nm), in this respect >40 dB extinction ratio and <0.5 dB insertion loss are the expected tolerances at a longer wavelength. Since a slight increase of L_p and a remarkable rising of L_s with increasing filling fraction of the nanochannels are observed in Fig. 8.13, we can achieve the

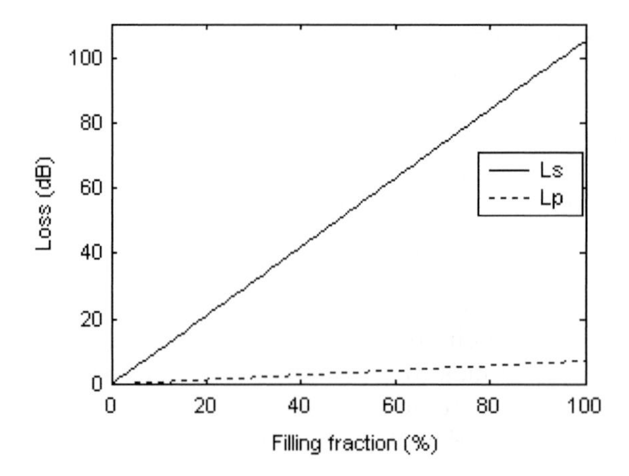

Fig. 8.13. Optical losses with filling fraction of the nanochannels of silver nanowire microarray at a telecom wavelength of 1550 nm.[34]

expected design goal for the fabricating tolerance by enhancing the filling fraction when selecting 0.5 for the ratio of $2a$ to b and 35 nm for the diameter.

8.2.5. Conclusion

Metal nanowire arrays transmit p-polarized waves while strongly attenuate s-polarized waves in the near and mid-infrared wavelength region, which means they possess efficient polarization properties. The optical losses are found to be strongly dependent on the diameter, spacing, and the ratio of diameter to spacing for s and p polarizations. We can simultaneously achieve large extinction ratios and small insertion losses by selecting an appropriate $2a/b$ as well as by reducing the diameter. It is predicted that the sizes (e.g. $2a/b$ of 0.5 and $2a$ of 35 nm) may be expected to be the optimal parameters for the design of nanowire grid polarizer at the near and mid-infrared wavelengths.

8.3 Electronic and magnetic properties of Bi-based nanowire arrays

Efficient solid-state energy conversion is based on the Peltier effect for cooling and the Seebeck effect for power generation calls for materials with high electrical conductivity, high Seebeck coefficient, and low thermal conductivity. However, the improvement of thermoelectric materials requires a detailed knowledge of all the factors that determine their properties. After 30 years of slow progress, thermoelectric material research experienced a resurgence, inspired by the developments of new concepts and theories to engineer electron and phonon transport in nanostructures. In nanostructures, quantum and classical size effects provide opportunities to tailor the electron and phonon transport through structural engineering. Quantum wells, quantum wires, and quantum dots have been employed to change the band structure, energy levels, and density of states of electrons, and have led to improved energy conversion capability of charge carriers compared to those of their bulk counterparts. Interface reflection and scattering of phonons in these nanostructures have been utilized to reduce the heat conduction. Recently, increases in the thermoelectric figure of merit based on size effects for either electrons or phonons have been demonstrated.

Bismuth, with a rhombohedral crystal lattice structure, is a semimetal with a small effective electron mass, long carrier mean free path, highly anisotropic Fermi surface and small energy overlap (about 38 meV at 77 K) between the L-point conduction band and the T-point valence band, which can lead to a semimetal-semiconductor transition in bismuth nanowires with diameter decreasing to a certain value (about 60 nm at 77 K).[42,43] Theoretical calculations have indicated that Bi and its related alloys are promising candidates for low-dimensional thermoelectric materials.[44–46] A single nanowire cannot carry a high enough current for thermoelectric applications, thus the synthetic methods that yield high-quality nanowire arrays are of particular interest. Many types of TE material nanowire arrays, such as Bi, Sb, $Bi_{1-x}Sb_x$, $Bi_{2-x}Sb_xTe_3$, $Bi_2Te_{3-y}Se_y$, have been fabricated by different techniques in the anodic alumina membranes (AAM) with uniformly sized, high-density and high aspect ratio nanochannels.[47–51]

8.3.1. Bi nanowire arrays[52]

The resistance was measured using a low-frequency AC bridge with excitation voltages varying from 20 μV to 2 mV. The excitation voltage level did not affect the results. In zero magnetic fields, the temperature dependence of the resistance of Bi nanowires with various diameters of 20, 50, and 70 nm is shown in Fig. 8.14. The data are presented normalized to 100 K, and are denoted by $R(T)/R(T = 100\,\text{K})$. It is easy to see from Fig. 8.14, that the temperature dependence is quite different between the three samples. When T increases from 4.2 to 100 K, $R(T)/R(T = 100\,\text{K})$ for the 50 nm sample decreases, while $R(T)/R(T = 100\,\text{K})$ increases for the 70 nm sample. For the 20 nm sample, the temperature dependence of resistance first

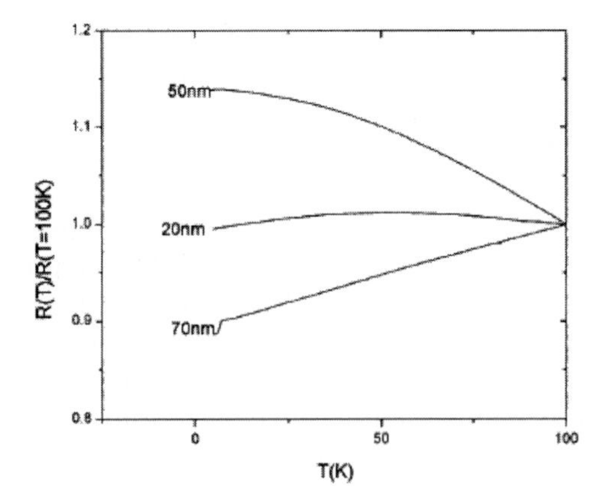

Fig. 8.14. Temperature dependence of the zero-field resistance of Bi nanowires of different diameters, normalized to the resistance at 100 K.[52]

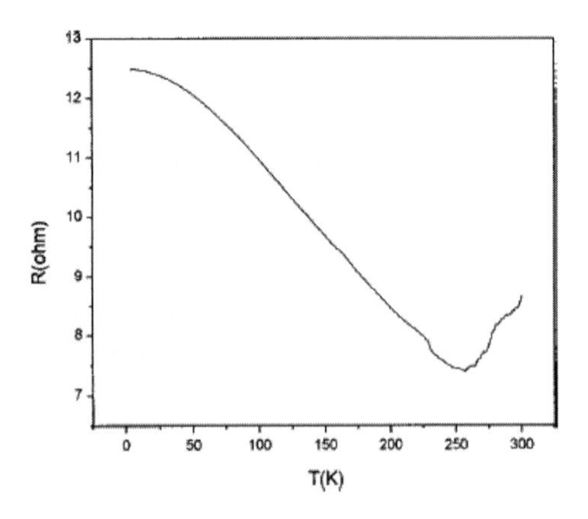

Fig. 8.15. Temperature dependence of the zero-field resistance of Bi nanowires with 50 nm diameter in 4.2–300 K.[52]

increases, then decreases. Figure 8.15 shows the temperature dependence of resistance of 50 nm Bi nanowires, where a minimum at 258 K is observed. A maximum at 50 K for the 20 nm sample is shown in Fig. 8.16.

For the 70 nm sample, the contribution of localization effects to the resistance is not apparent, and the carrier mobility dominates, thus a metallic-like positive temperature coefficient of resistance (TCR) is observed. With the diameter decreasing, the carrier mobility is suppressed by scattering at the wire boundary and acoustic phonon scattering. For the 50 nm samples, in higher temperatures ($T > 258$ K), the change of mobility exerts a

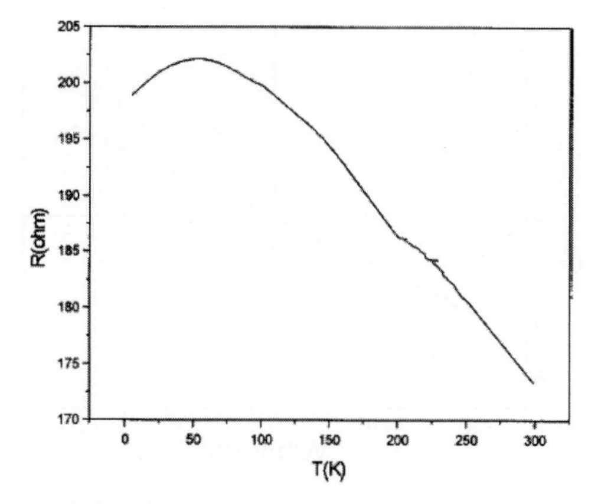

Fig. 8.16. Temperature dependence of the zero-field resistance of Bi nanowires with 20 nm diameter in 4.2–300 K.[52]

decisive influence on the temperature variation of resistance. With the temperature decreasing to 258 K, the contribution of the increasing of mobility decreases due to the scattering at the wire boundary, resulting in a minimum at 258 K. In temperatures below 258 K, the change of carrier density dominated, resulting in a negative TCR. When the diameter decreases to about 20 nm and the temperature drops below a certain value T_{max}, the localization effects are very apparent. The decisive factor in the temperature variation of the resistance of Bi nanowires is the strong growth of the mobility. The maximum of the resistance for 20 nm Bi wires can be explained on the basis of this. Thus, the minimum and maximum on the temperature dependence of the resistance of the Bi wires appear as a result of competition between the influence of the carrier density and their mobility, which have different characters and opposite signs.

8.3.2. Bi-Bi homogeneous nanowire junction[53]

Figure 8.17 shows the typical *I–V* plots for a Bi M-S junction nanowire array with two segments of 100 nm and 35 nm in diameter and a Bi nanowire array with the diameter of 100 nm. The *I–V* curve of the junction array shows intrinsic nonlinear and asymmetric behavior at room temperature, but does not pass through the zero point (due to a charging effect in the apparatus; see curve (b) in Fig. 8.17). To assess the effect of contacts, we performed similar measurements on the arrays of straight Bi nanowires, and found as expected that the *I–V* curve of the straight Bi nanowire arrays is linear and almost symmetric (curve (a) in Fig. 8.17), which provides further

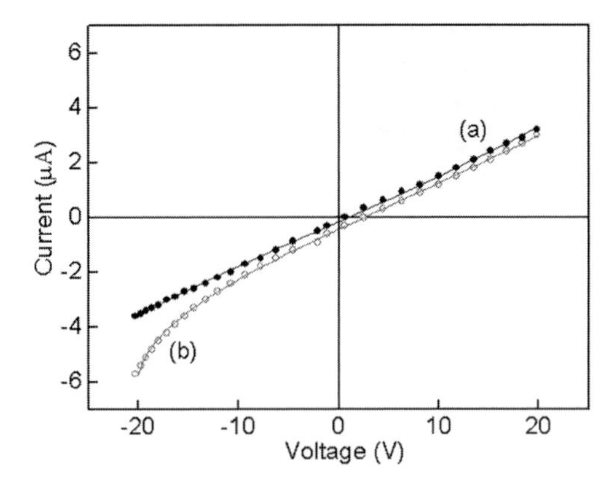

Fig. 8.17. *I–V* curves of (a) Bi metal nanowire array and (b) Bi M-S junction nanowire array with a diameter of 100 nm in the AAM.[53]

evidence that the observed nonlinear behavior is an inherent property of the Bi M-S junction nanowire arrays. The higher value of the conductance for the junction nanowires array is due to the larger area in measuring the *I–V* characteristics as compared with that of a metallic nanowire array.

8.3.3. Y-segment Bi nanowire array[54]

The current–voltage *I–V* characteristic of the parallel *Y*-branched Bi nanowires (NWs) embedded in the AAO template was measured at room temperature, as illustrated in Fig. 8.18(a) (inset). The result reveals that the *I–V* curve (Fig. 8.18(a)) is nonlinear and asymmetrical about the zero bias. This characteristic is analogous to the Schottky junction between a metal and a degenerate *n*-type semiconductor. The possible energy diagram for the device is shown in Fig. 8.18(b) (inset). The observed characteristics can be explained by assuming that the semiconductor is degenerate at room temperature with conduction electrons available from the bottom of the conduction band E_c to the Fermi energy E_F. The *I–V* curve shows a kink close to the zero bias on the positive side. This is highlighted on the conductance curve, which was derived from the *I–V* curve by numerical differentiation (Fig. 8.18(b)). The conductance curve goes through a minimum at around 110 meV. This energy is then the difference between E_c and E_F. The forward bias region up to 1.4 eV is characteristic of carrier injection processes, as the metal crosses the forbidden energy gap in the semiconductor. The actual band gap of the semiconductor channel is complicated to extract, given the fact that the tunneling barrier shape changes due to

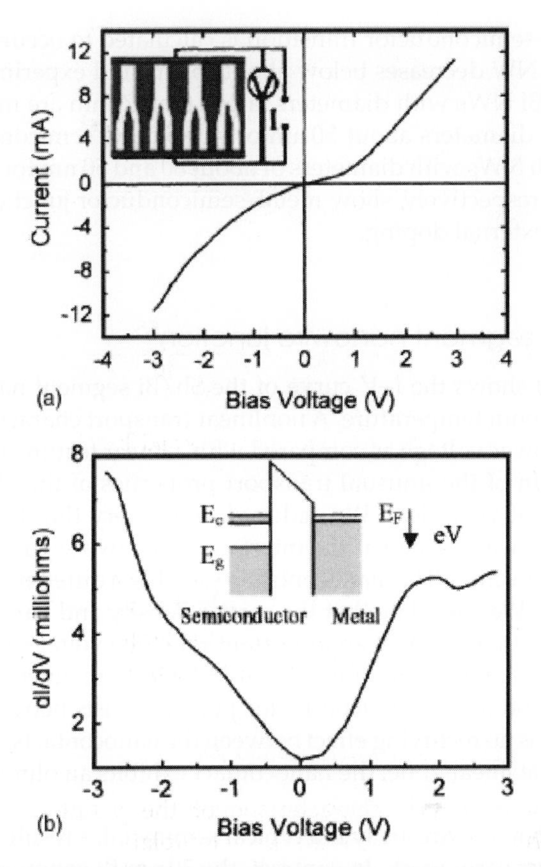

Fig. 8.18. (a) *I–V* characteristics of parallel *Y*-branched Bi NWs embedded in the AAO template. The inset is a schematic illustration of the *I–V* measurement for the *Y*-branched Bi NWs. (b) Conductance curve, showing a minimum at about 110 meV, which is the Fermi-level energy difference, measured from the conduction-band edge of the semiconductor. Inset: Energy band diagram of the semimetal and degenerate semiconducting Bi NW.[54]

image charges. Other factors to be considered are surface charging at the interface. The curve shows a peak at 1.9 eV, suggesting the unavailability of additional conduction states with increase in bias. The reverse bias is dominated by electron tunneling from the semimetal Bi into the conduction band of the degenerate semiconductor Bi NW channel. The kink at −1.9 eV in the reverse bias region of the conductance is notable, as it indicates variation in conduction states in the metallic part of the NW, which may be responsible for the small peak at 1.9 eV. Previous studies revealed that semimetal-to-semiconductor transition is calculated to occur as the diameter of the Bi NW decreases below about 65 nm, and experimental results showed that Bi NWs with diameters larger than 70 nm are metallic, while Bi NWs with diameters about 50 nm or smaller are semiconductive. Our *Y*-branched Bi NWs with diameters of about 80 and 50 nm for the stem and the branches respectively, show metal–semiconductor junction properties without any external doping.

8.3.4. Bi-Sb segment nanowire junction[55]

Figure 8.19(a) shows the *I–V* curve of the Sb/Bi segment nanowire junction array at room temperature. A nonlinear transport characteristic can be observed at lower voltage as compared with a linear feature at higher voltage. The origin of the unusual transport properties of the Sb-Bi segment nanowire is not very clear. We deduced that a very thin transitional (or alloying) layer was formed at the interface of the two segment nanowires during the deposition. This nanocontact region has a different electric transport property. We speculate that the carrier density and band structure of the nanocontact region play an important role. The voltage remains small until the breakdown of the transitional layer at high current. This result indicates that no metal/semiconductor junction exists between Sb and Bi because there is no rectifying effect between the nanocontacts, and at higher voltage, i.e. the linear zone, the nanocontact exhibits an ohmic contact.

A 40 nm Sb nanowire array is a typical semimetal, a result that has been reported in our early work. In contrast, the 30 nm Bi nanowire array is an

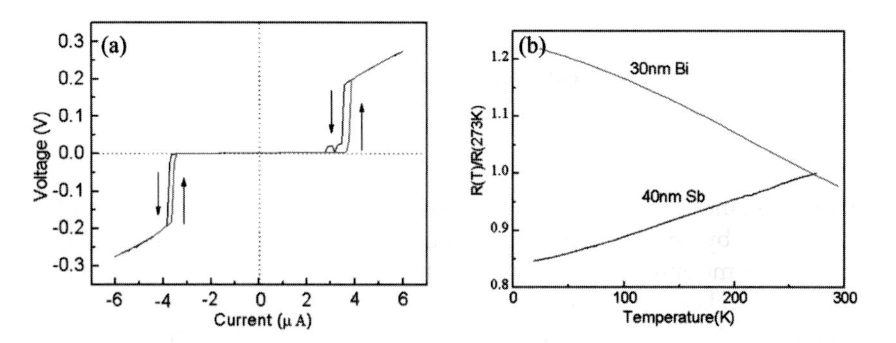

Fig. 8.19. (a) *I–V* characterization curve of the Sb/Bi segment nanowire nanojunction array at room temperature, and (b) temperature dependence of the resistance ratio $R(T)/R(273\,\mathrm{K})$ of Sb and Bi nanowire arrays at zero magnetic field.[55]

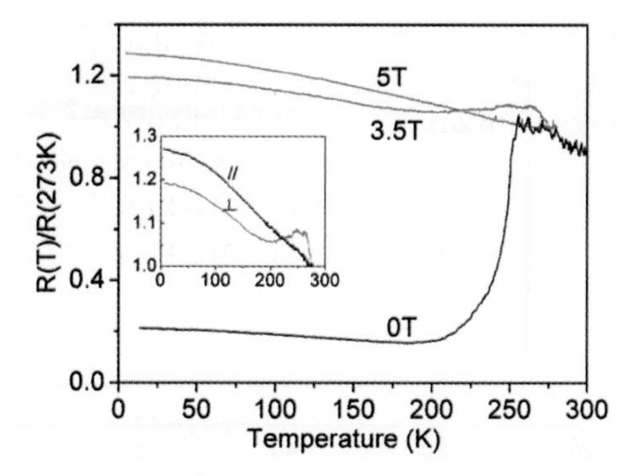

Fig. 8.20. Temperature dependence of the resistance ratio $R(T)/R(273\,\text{K})$ of Sb/Bi segment nanowire nanojunction array at different magnetic fields (B $\perp$ the axis of the nanowires).[55]

anomalous metallic semimetal, since its resistance increases with decreasing temperature (Fig. 8.19(b)). The decreasing carrier numbers for both the electrons and holes play a dominated role in determining the temperature behavior of the Bi nanowires. Figure 8.20 shows an unusual temperature dependence of the resistance with increasing temperature with and without magnetic fields (the measurements were made in the linear region of the *I–V* curve). Without a magnetic field, the resistance of the Sb/Bi junction array slightly decreases with increasing temperatures from 4 K to 210 K. When the temperature is greater than 210 K, the resistance rises abruptly, and above 250 K, the resistance again slightly decreases. The exact mechanism of this switching feature is presently not understood. A magnetic field of 3.5 T greatly weakens the effect, and a 5 T field totally eliminates the switching behavior (Fig. 8.20).

8.4 Thermal expansion properties of nanowire arrays

8.4.1. AgI nanowire arrays[56]

The AgI nanowire arrays were prepared in the ordered porous alumina membrane by an electrochemical method. *In situ* high-temperature X-ray diffraction measurements show that the nanowire arrays possess hexagonal close-packed structure (β-AgI) at 293 K, orienting along the (002) plane, whereas at 473 K, the nanowire arrays possess a body-centered cubic structure (α-AgI), orienting along the (110) plane, as shown in Fig. 8.21, which indicates that the AgI nanowires have a transition from the β to α phase when the temperature increases from 293 to 433 K. As presented in Ref. 57,

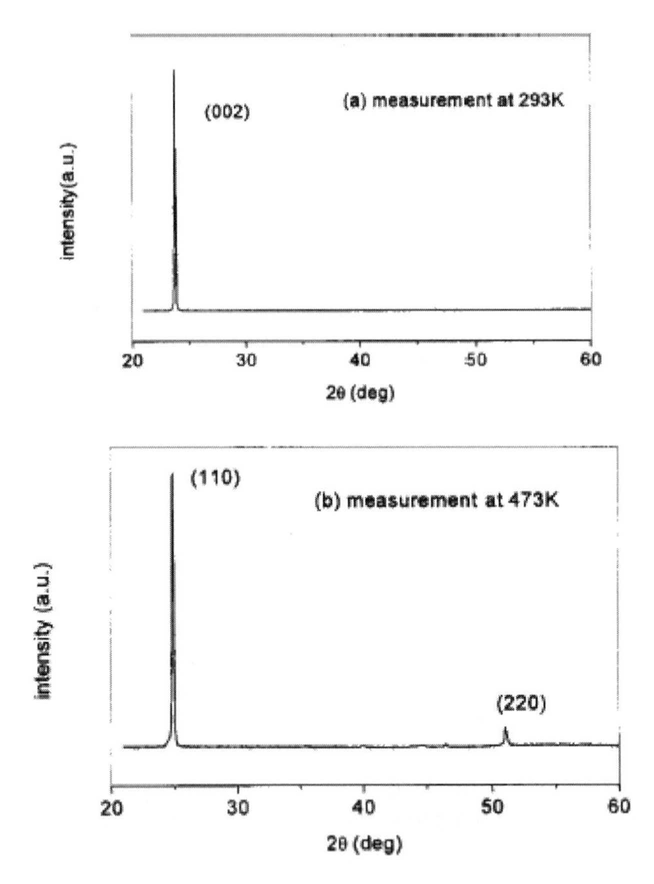

Fig. 8.21. XRD patterns of the AgI nanowire arrays, oriented along the (002) plane.[56]

AgI undergoes a phase transition from β to α phase at 420 K. However, in our experiment the phase transition occurs at above 433 K at least, which clearly indicates that the AgI nanowire arrays in the alumina membrane have a higher phase transition temperature.

Furthermore, with increase of temperature, the peak positions of the (002) plane of β-phase shift toward higher angle directions (see Fig. 8.22(a)), suggesting a negative expansion property. From the X-ray diffraction pattern and Bragg equation, $2d \sin \theta = \lambda$ (d, θ and λ are the interplanar spacing, diffraction angle and X-ray incidence wavelength, respectively), the lattice parameter d can be calculated. The lattice parameters of β-AgI as a function of the temperature were shown in Fig. 8.22(b), which clearly shows the variation of the lattice constants with temperature. The

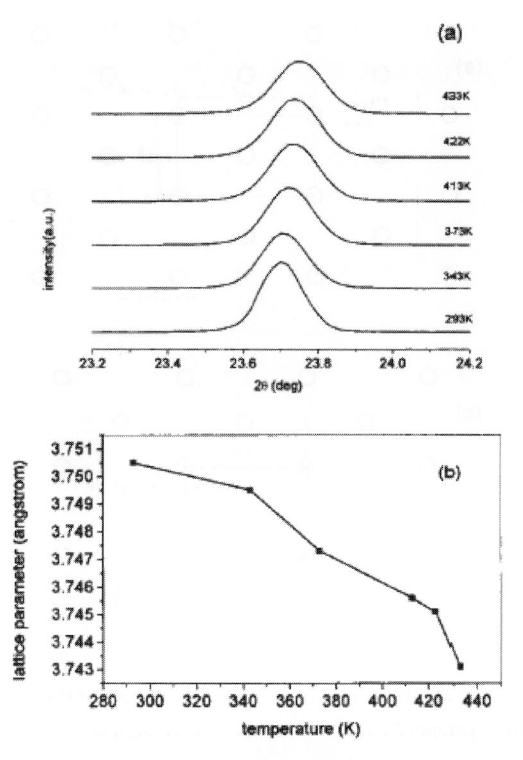

Fig. 8.22. (a) XRD patterns of the (002) plane of the β-AgI measuring from 293 to 433 K. (b) Temperature dependence of the lattice parameters of AgI nanowires from 293 to 433 K.[56]

close-packed structure transforms into body-centered cubic structure in the AgI nanowires, which can be interpreted according to the Burgers mechanism. In Fig. 8.23, for the transition from the (002) plane of hexagonal close-packed structure to the (110) plane of body-centered cubic structure, the iodine ions will have a glide, implying that it will expand along the y direction, resulting in the distance between the iodine ions changing from 4.59 to 5.04 Å, and contracting along the x direction (see Fig. 8.23(a)), resulting in the distance between the iodine ions changing from 7.95 to 7.12 Å. Meanwhile, the plane spacing contracted from 3.75 Å of the (002) plane of the β-phase to 3.53 Å of the (110) plane of the α-phase. Due to the fact that AgI nanowires are located in the channels of the alumina membrane, the expansion in the radial direction will be hindered, resulting in a rising phase transition temperature.

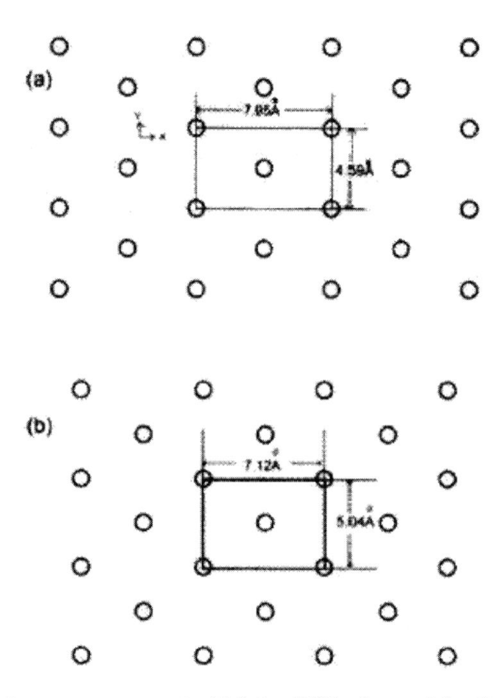

Fig. 8.23. Iodine ions arrangement of (a) the (002) plane of the β phase and (b) the (110) plane of the α phase.[56]

8.4.2. Bi nanowire arrays[58]

To investigate the thermal expansion behavior of the Bi nanowire arrays with different diameters, *in situ* high-temperature X-ray diffraction (XRD) was performed in the temperature range from 300 to 500 K under a high vacuum atmosphere. Temperatures were kept constant at each point for 20 minutes before measurement, and scans were carried out for $20° < 2\theta < 80°$. XRD patterns (Fig. 8.24) reveal that the Bi nanowire arrays have a high growth orientation along the (110) plane.

The lattice parameters of the (110) plane of Bi nanowires with different diameters (10, 20, 40, 60 and 80 nm) were studied, and the results are shown in Fig. 8.25. The lattice parameter firstly increases with increasing temperature and then decreases at a certain temperature for Bi nanowires with the diameter in the range 20–60 nm, and there is a critical transition temperature, T_c, at which the temperature coefficient of the lattice parameter changes from positive to negative (Figs. 8.25(a)–8.25(c)). The T_c shifts to high temperatures with the increase of the diameter from 20 nm to 60 nm.

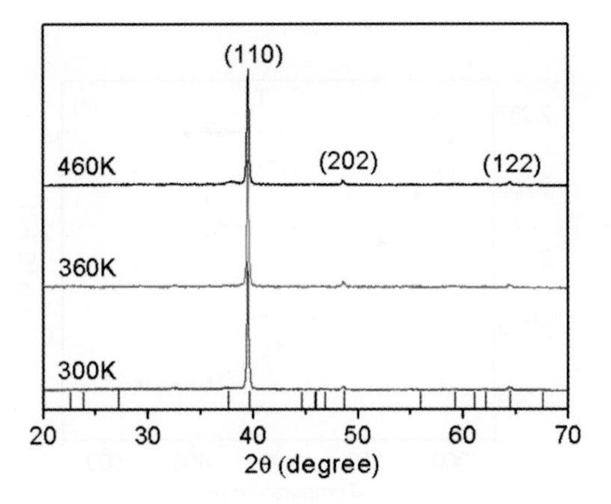

Fig. 8.24. XRD patterns of Bi nanowire arrays with the diameter of 10 nm at different temperatures.[58]

The lattice parameter of Bi nanowires with a diameter of 10 nm always decreases with increasing temperature and there is not a transition temperature, while that with a diameter of 80 nm always increases. These results indicate that Bi nanowires with a very small diameter might possess a negative thermal expansion behavior even in a room temperature region. It is also worthy to note that: (1) the lattice parameter of the (110) plane of the Bi nanowire is larger than that of the Bi bulk material (2.2730 Å). Compared with the bulk, the decreased or increased lattice constants were observed in different nanomaterials. The intrinsic reasons are different depending on the processing histories and surface structure of the nanomaterials. (2) The lattice parameter of the Bi nanowires increases with the increasing diameter of the nanowires.

The positive thermal expansion is commonly observed in most bulk material, which can be understood by accounting for the effects of the anharmonic lattice potential on the equilibrium lattice separations and is characterized by the Gruneisen parameter. In most cases, negative thermal expansions, mainly originating from a structural related phase transition, have been observed among anisotropic systems in only a narrow temperature range. A negative thermal expansion coefficient has been observed in AgI nanowires, when the structure changes from hexagonal close-packed to body-centered cubic due to the restriction of the AAM wall on the nanowires. Theoretical calculation based on Gruneisen's law with certain simple assumptions indicated that highly anisotropic bulk material Bi has either positive or negative thermal expansion behavior depending on

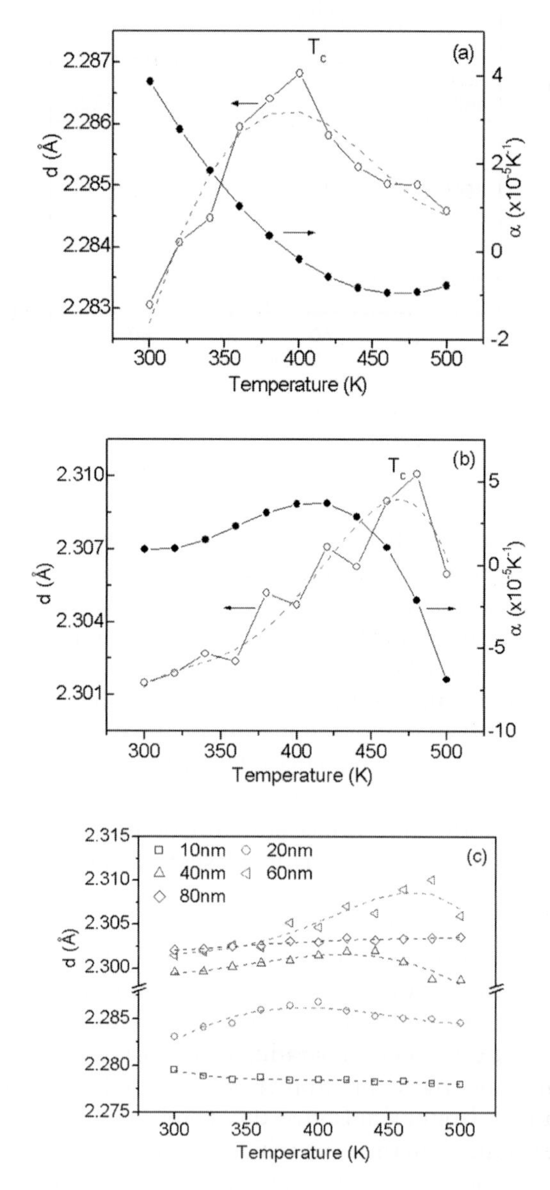

Fig. 8.25. Temperature dependences of lattice parameter, d (open symbols) of Bi nanowires with different diameters (the dashed line is the fourth-order polynomial fitting). (a) 20 nm, (b) 40 nm, and (c) 10–80 nm.[58]

crystallographic directions. In the present study, a transition of the thermal expansion coefficient from positive at low temperatures to negative at high temperatures has been observed in freestanding rhombohedral (110) Bi nanowires (20–60 nm) within the AAM and without structural change over the whole temperature range studied. Our results proved from experiment that the lattice thermal contraction is an intrinsic property of Bi nanowires. A similar phenomenon was also observed in nanoparticles, in which the effect of the valence electron energy on the equilibrium lattice positions was introduced. In terms of the minimum energy principle, a transition of thermal expansion coefficient from positive to negative was anticipated at a certain temperature T_C, which varied with the diameters of nanoparticles. As for nanowires, where discrete levels are separated by only a few meV, the electronic effects become significant even at ordinary temperatures, and the influence of the valence electron potential on the equilibrium lattice separation in nanowires thus should be taken into consideration. Lattice shrinking results in an increase in the level separation at elevated temperatures, which, on the one hand, reduces the number of electrons occupying the excited states as dictated by the Fermi–Dirac factor and, on the other hand, raises the thermal energy of individual electrons in the excited states. These two factors compete delicately to achieve a lower electronic potential energy that results in the transition from thermal expansion to thermal contraction. Other factors, such as defects, surface stress and finite-size modified lattice potential, might be significant in determining the thermal behaviors of nanowires.

8.4.3. Cu nanowire arrays[59]

The Cu nanowires were prepared in alumina membranes with ordered pore arrays by an electrochemical deposition. X-ray diffraction observations show that the nanowires are oriented along the (110) direction. The thermal expansion coefficients of the nanowires were measured by *in situ* high-resolution X-ray diffraction in the temperature range from 298 to 693 K. From Fig. 8.26, we can find the following features: (1) the lattice parameters of the Cu nanowires are slightly increased with respect to the equilibrium lattice constants for bulk Cu; (2) the thermal expansion coefficient of the Cu nanowires is much smaller than that of the bulk Cu.

Decreased or increased lattice parameters were observed in different nanomaterials.[60,61] The intrinsic reason for the decreased or increased lattice parameters could be different depending on the processing histories of the nanomaterials and surface structure. In our experiment, the Cu

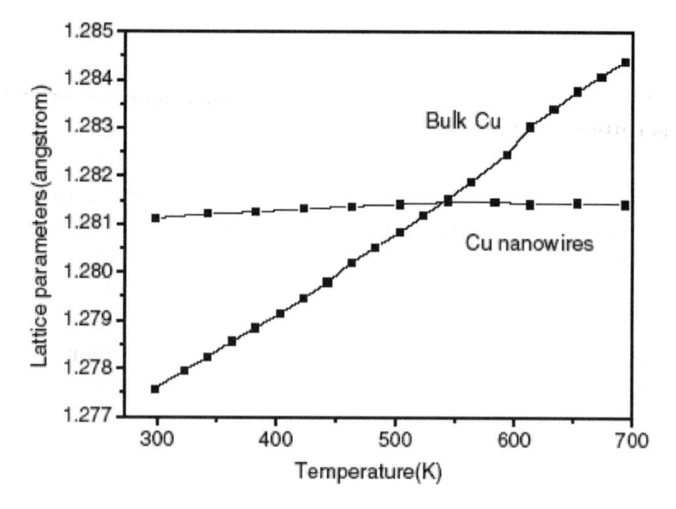

Fig. 8.26. Temperature dependences of lattice parameters of the Cu nanowires and bulk Cu.[59]

nanowires were prepared by an electrochemical deposition method. The defects that induced a stress field that resulted in the lattice expansion,[62] such as single vacancies and vacancy clusters, could exist on the grain boundaries or surfaces of the nanowires.

Why is the thermal expansion coefficient, α, of the Cu nanowires much smaller than that of bulk Cu? As we know, the thermal expansion depends mainly on the nonlinear crystal lattice vibration. The more obvious the vibration, the larger the α is. The contribution of the nonlinear crystal lattice vibration in the grain boundary component is apparently larger than that in the grain component because of random atomic arrangement in grain boundaries, which arises from the existence of a large number of vacancies and vacancy clusters in them. For one-dimensional Cu nanowires, the grain diameter is about 40 nm, which is obtained from X-ray diffraction peaks by using Scherrer's formula, and is almost the same as the nanowire's. This implies that for every Cu nanowire, all the grain boundaries almost cross with the nanowire surface. With the temperature increasing, the vacancies and vacancy clusters in nanowires move easily along the grain boundaries onto the surfaces of the nanowires and disappear quickly. As a result, atoms in grain boundaries are rearranged rapidly, and cause the increase of the order degree of the grain boundary structure. This makes the nonlinear vibration contribution in grain boundaries to α decrease quickly. Moreover, the nonlinear vibration of the crystal lattice in grain boundaries becomes the main contribution to α. The quick decrease of the grain boundary contribution to α causes the Cu nanowires to have a very small α. For Cu nanocrystalline bulk samples with a large volume fraction of grain boundaries, it is difficult for vacancies and vacancy clusters to move fast along the grain boundary towards the sample surface. Therefore, with increasing temperature, the nonlinear vibration of the crystal lattice becomes more and more obvious in both grain boundaries and grains. This gives rise to highly thermal expansions. This is why the nanocrystalline bulk has a high α. As we know, the conventional Cu bulks contain very few grain boundaries. The nonlinear vibration of the crystal lattice in grains is the main contribution to α with elevating temperature. The nonlinear vibration of the crystal lattice in grains with very small vacancies and vacancy clusters becomes more and more obvious because the atom rearrangement process in grains is not caused by the annihilation of defects. Conversely, as mentioned above, for the Cu nanowires with a large volume fraction of grain boundaries, the grain boundary, atom rearrangement process becomes very rapid with the increase of temperature due to fast annihilation of vacancies and vacancy clusters on the nanowire surface. The nonlinear vibration of the crystal lattice in grain boundaries becomes so weak that their contribution to thermal expansion decreases substantially. At the same time, the nanowire grain boundary is the main contribution to α. Therefore, the α of Cu nanowires becomes lower than that of conventional Cu bulks.

Bibliography

1. P. D. Yang, Y. Y. Wu, and R. Fan, *Inter. J. Nanosci.* **1**, 1 (2002).
2. C. B. Murray, C. R. Kagan, and M. G. Bawendi, *Annu. Rev. Mater. Sci.* **30**, 545 (2000).
3. K. K. Likharev and T. Claeson, *Sci. Am.* June, 80 (1992).
4. K. K. Likharev, *IBM J. Res. Develop.* **32**, 144 (1988).
5. G. Markovich, C. P. Collier, S. E. Henrichs, F. Remacle, R. D. Levine, and J. R. Heath, *Acc. Chem. Res.* **32**, 415 (1999).
6. Y. N. Xia, P. D.Yang, Y. G. Sun, Y. Y. Wu, B. Mayers, B. Gates, Y. D. Yin, F. Kim, and H. Q. Yan, *Adv. Mater.* **15**, 353 (2003).
7. M. H. Huang, S. Mao, H. Feick, H. Q. Yan, Y. Y. Wu, H. King, E. Weber, R. Russo, and Y. P. Yang, *Science* **292**, 1897 (2001).
8. H. Kind, H. Yan, M. law, B. Messer, and Y. P. Yang, *Adv. Mater.* **14**, 158 (2002).

9. X. Y. Kong and Z. L. Wang, *Nano Lett.* **3**, 1625 (2003).
10. X. S. Fang, C. H. Ye, L. D. Zhang, and T. Xie, *Adv. Mater.* **17**, 1661 (2005).
11. G. R. Bird and M. Parrish, *J. Opt. Soc. Am.* **50**, 886 (1960).
12. J. B. Young, H. A. Graham, and E.W. Peterson, *Appl. Opt.* **4**, 1023 (1965).
13. H. Tamada, T. Doumuki, T. Yamaguchi, and S. Matsumoto, *Opt. Lett.* **22**, 419 (1997).
14. J. Guo and D. Brady, *Opt. Eng.* **36**, 2268 (1997).
15. G. P. Nordin, J. T. Meier, P. C. Deguzman, and M. W. Jones, *J. Opt. Soc. Am. A* **16**, 1168 (1999).
16. J. Guo and D. Brady, *Appl. Opt.* **39**, 1486 (2000).
17. C. R. Martin, *Science* **266**, 1961 (1994).
18. K. Nielsch, F. Müller, A. P. Li, and U. Gösele, *Adv. Mater.* **12**, 582 (2000).
19. G. Yi and W. Schwarzacher, *Appl. Phys. Lett.* **74**, 1746 (1999).
20. A. Huczko, *Appl. Phys.* **A70**, 365 (2000).
21. M. Miyagi, Y. Hiratani, T. Taniguchi, and S. Nishida, *Appl. Opt.* **26**, 970 (1987).
22. M. Saito, M. Kirihara, T. Taniguchi, and M. Miyagi, *Appl. Phys. Lett.* **55**, 607 (1989).
23. Y. T. Pang, G. W. Meng, Y. Zhang, Q. Fang, and L. D. Zhang, *Appl. Phys. A* **76**, 533 (2003).
24. Y. T. Pang, G. W. Meng, Q. Fang, and L. D. Zhang, *Nanotech.* **14**, 20 (2003).
25. Y. T. Pang, G. W. Meng, L. D. Zhang, Y. Qin, X. Y. Gao, A. W. Zhao, and Q. Fang, *Adv. Funct. Mater.* **12**, 719 (2002).
26. M. Saito and M. Miyagi, *Appl. Opt.* **28**, 3529 (1989).
27. J. P. Auton, *Appl. Opt.* **6**, 1023 (1967).
28. P. K. Cheo and C. D. Bass, *Appl. Phys. Lett.* **18**, 565 (1971).
29. J. P. Auton and M. C. Hutley, *Infrared Phys.* **12**, 95 (1972).
30. J. X. Zhang, L. D. Zhang, C. H. Ye, M. Chang, Y. G. Yan, and Q. F. Lu, *Chem. Phys. Lett.* **400**, 158 (2004).

31. H. van de Hulst, Light Scattering by Small Particles, John Wiley & Sons Inc., New York, 1957.
32. M. Saito and M. Miyagi, *J. Opt. Soc. Am. A* **6**, 1895 (1989).
33. A. L. Prieto, M. S. Sander, M. Martin-Gonzalez, R. Gronsky, T. Sands, and A. M. Stacy, *J. Am. Chem. Soc.* **123**, 7160 (2001).
34. J. X. Zhang, Y. G. Yan, X. L. Cao, and L. D. Zhang, *Appl. Opt.* **45**, 297 (2006).
35. M. L. Lang and W. L. Wolfe, *Appl. Opt.* **22**, 1267 (1983).
36. F. Gervais, Aluminum oxide (Al_2O_3), in Handbook of Optical Constants of Solids II, E. D. Palik, eds., Academic Press, New York, 1991, pp. 761–775.
37. D. W. Lynch and W. R. Hunter, in Handbook of Optical Constants of Solids, E. D. Palik, ed., Academic Press, New York, 1985, p. 283.
38. M. Born and E. Wolf, Principles of Optics, Pergamon, Oxford, 1980.
39. A. A. Kokhanovsky, Optics of Light Scattering Media: Problems, and Solutions, Springer, Chichester, 2001.
40. J. J. Wang, W. Zhang, X. Deng, J. Deng, F. Liu, P. Sciortino, and L. Chen, *Opt. Lett.* **30**, 195 (2005).
41. K. Zhou, G. Simpson, X. Chen, L. Zhang, and I. Bennion, *Opt. Lett.* **30**, 1285 (2005).
42. Z. B. Zhang, X. Z. Sun, M. S. Dresselhaus, and J. Y. Ying, *Appl. Phys. Lett.* **73**, 1589 (1998).
43. Z. B. Zhang, J. Y. Ying, and M. S. Dresselhaus, *J. Mater. Res.* **13**, 1745 (1998).
44. L. D. Hicks and M. S. Dresselhaus, *Phys. Rev.* **B47**, 16631 (1993).
45. Y. M. Lin, O. Rabin, S. B. Cronin, and J. Y. Ying, *Appl. Phys. Lett.* **81**, 2403 (2002).
46. Y. M. Lin, X. Z. Sun, and M. S. Dresselhaus, *Phys. Rev.* **B62**, 4610 (2000).
47. J. Heremans, C. M. Thrush, Y. M. Lin, Z. Zhang, M. S. Dresselhauls, and J. F. Mansfield, *Phys. Rev.* **B61**, 2921 (2000).
48. Y. Zhang, G. H. Li, Y. C. Wu, B. Zhang, W. H. Song, and L. D. Zhang, *Adv. Mater.* **14**, 1227 (2002).

49. Y. Peng, D. H. Qin, R. J. Zhou, and H. L. Li, *Mater. Sci. Eng.* **B77**, 246 (2000).

50. M. Martín-González, A. L. Prieto, R. Gronsky, T. Sands, and A. M. Stacy, *Adv. Mater.* **15**, 1003 (2003).

51. M. Martín-González, G. J. Snyder, A. L. Prieto, R. Gronsky, T. Sands, and A. M. Stacy, *Nano Lett.* **3**, 973 (2003).

52. X. F. Wang, J. Zhang, H. Z. Shi, Y. W. Wang, G. W. Meng, X. S. Peng, and L. D. Zhang, *J. Appl. Phys.* **89**, 3847 (2001).

53. L. Li, Y. Zhang, G. H. Li, W. H. Song, and L. D. Zhang, *Appl. Phys. A* **80**, 1053 (2005).

54. Y. T. Tian, G. W. Meng, S. K. Biswas, P. M. Ajayan, S. Sun, and L. D. Zhang, *Appl. Phys. Lett.* **85**, 967 (2004).

55. Y. Zhang, L. Li, and G. H. Li, *Nanotech.* **16**, 2096 (2005).

56. Y. H. Wang, C. H. Ye, G. Z. Wang, and L. D. Zhang. *Appl. Phys. Lett.* **82**, 4253 (2003).

57. K. Tadanaga, K. Imai, M. Tatsumisago, and T. Minami, *J. Electrochem. Soc.* **147**, 4061 (2000).

58. L. Li, Y. Zhang, Y. W. Yang, X. H. Huang, G. H. Li, and L. D. Zhang, *Appl. Phys. Lett.* **87**, 031912 (2005).

59. Y. H. Wang, J. J. Yang, C. H. Ye, X. S. Fang, and L. D. Zhang, *Nanotech.* **15**, 1437 (2004).

60. Y. H. Zhao, K. Zhang, and K. Lu, *Phys. Rev.* **B56**, 14322 (1997).

61. R. Banerjee, E. A. Sperling, G. B. Thompson, and H. L. Fraser, *Appl. Phys. Lett.* **82**, 4250 (2003).

62. W. Qin, Z. H. Chen, P. Y. Huang, and Y. H. Zhuang, *J. Alloys Compounds* **44**, 1915 (2001).

Chapter 9
Applications

- Sensors
- Field emission of carbon nanotubes and its application
- Light polarization
- Light-bulb filaments made of carbon nanotube yarns
- Electronic and optoelectronic nanoscale devices

Chapter 9

Applications

9.1 Introduction

Nanomaterials have extensive applications in many fields due to their unique properties. In the aspects of improving products of conditional industry, protecting the environment, saving energy and applying advanced science and technology, some achievements have been made. In this chapter, we introduce mainly applications of one-dimensional nanomaterials in some advanced technique fields, such as nanosensors (biosensors, gas sensors and chemical sensors), field emission of carbon nanotubes and field emission devices, light polarization devices of one-dimensional materials, light-bulb filaments made of carbon nanotube yarns, electronic and optoelectronic nanoscale devices.

9.2 Sensors

Nanosensors play a major role in semiconductor processing, medical diagnosis, environmental sensing, and national security. Due to their higher sensitivity and high resolving power, nanosensors have potential applications in microfluid detection, such as harmful gas molecules, individual bacteria and viruses etc. The detection principle is that interaction of target analytes with nanodetection devices produces physical and chemical property changes such as electric, optical, magnetic, mass, pH value etc. changes. In the next paragraph, we will introduce mainly the research progress of nanosensors, including gas sensors, biosensors and chemical sensors.

9.2.1. SnO_2 gas sensors

As we know, the electric properties of SnO_2, especially nano-SnO_2, are very sensitive to many gases such as CO, O_2, NO_2, and C_2H_5OH. Recently, many researchers study the gas sensors of SnO_2 nanowires or/and nanoribbons.

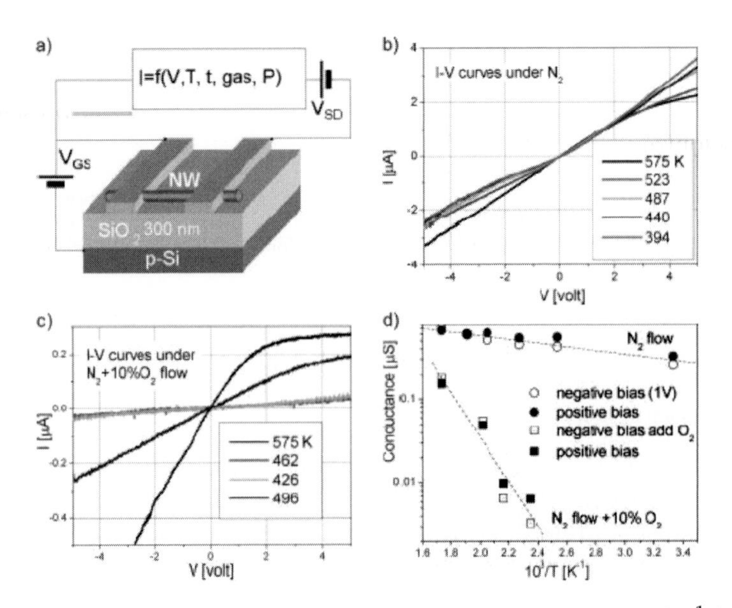

Fig. 9.1. The SnO_2 nanowire gas sensor and the gas sensing test results.[1] (a) The SnO_2 nanowire gas sensor; *I*–*V* characteristics for the SnO_2 nanowire in (b) an inert environment; and (c) an oxidizing environment. Nanowires were preconditioned for 2 h at 525 K under flowing dry N_2 before each set of measurements. The results shown were collected by first raising the nanowires to the highest temperature indicated, then decreasing the temperature incrementally and allowing thermal equilibrium to be established before the conductance was measured.

We firstly introduce CO and O_2 sensors of SnO_2 nanowires reported by Kolmakov *et al.*[1] The SnO_2 nanowire sensor was prepared as follows. SnO_2 nanowires were deposited on a SnO_2/Si substrate, outfitted with vapor-deposited Au/Ti electrodes that were contacted by microscope-guide, micro-positioned contacting probes, as shown in Fig. 9.1(a).

Current–voltage (*I*–*V*) and conductance measurements were made on isolated individual nanowires. The conductance and *I*–*V* characteristics of nanowires depend mainly on the temperature and the ambient gas partial pressure. Figure 9.1(b) shows the *I*–*V* curve in the absence of oxygen. It can be seen that the nanowire is a fairly good conductor. When oxygen is introduced into the gas cell, in which the nanowires are located, the nanowire becomes an insulator (Fig. 9.1(c)). This dramatic change from a conductor to insulator is closely related to the nature of the species adsorbed on the nanowire's surface because these adsorbed gases can cause a change in the nanowire's bulk electronic properties.

In the absence of an oxidant, for example, in an inert atmosphere, the low resistance and its weak temperature dependence can be seen (Fig. 9.1(b)). Kolmakov *et al.* indicate that this behavior is possibly due to the presence of easily ionized oxygen vacancies, V_O, in the nano SnO_2 that remain ionized over a wide temperature range. When annealed at high temperatures in an inert atmosphere, the surface of SnO_2 loses many of the lattices $O^{2-}_{(s)}$ and/or ionosorbed $O^{-}_{(s)}$ species and oxygen vacancies are formed. These processes can be expressed by Eqs. (9.3) and (9.4).

$$O^{2-}_{(s)} \rightarrow \frac{1}{2}O_{2(g)} + V_O \tag{9.1}$$

$$V_O \rightarrow V_O^{+} + e^{-} \tag{9.2}$$

$$V_O^{+} \rightarrow V_O^{++} + e^{-} \tag{9.3}$$

$$O^{-}_{(s)} \rightarrow \frac{1}{2}O_{2(g)} + e^{-} \tag{9.4}$$

Here the s and g refer to surface and gas, respectively. V_O is a vacancy with two trapped electrons.

As a result, the SnO_2 nanowire changes from the surface acceptor state to the surface donor state.

When the SnO_2 nanowire is exposed in oxygen, the nanowire recreates the surface acceptor (runs reactions (9.3) and (9.4) backwards) and hence the nanowire conductance decreases and restores the temperature dependence of the conductance to the exponential form typical of intrinsic semiconductors. At this time, the activation energy presents an obvious change (Fig. 9.1(d)). This is the basic reason for SnO_2 nanowires with oxygen sensing.

When combustible gases (CO) react with pre-adsorbed oxygen species like $O^{-}_{(s)}$ or $O^{2-}_{(s)}$ to form CO_2, the following reaction takes place:

$$CO_{(g)} + O^{-}_{(s)} \rightarrow CO_2 + e^{-} \tag{9.5}$$

The steady-state surface oxygen concentration decreases and a few electrons are donated back into the bulk in an increased conductivity. Kolmakov *et al.* indicate that the electron exchange between the surface states and the bulk take place within a surface layer whose thickness is the order of the Debye length, λ_D. The SnO_2 nanowire radius is the order of λ_D, or less than λ_D. This means that in air, the SnO_2 nanowire is depleted of its conduction

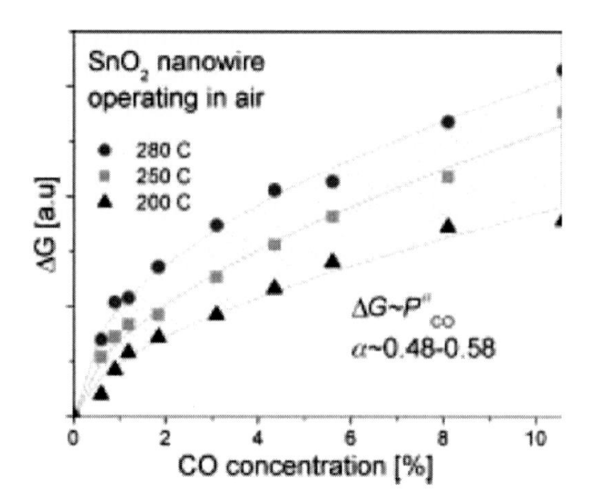

Fig. 9.2. The change in conductance of an individual SnO_2 nanowire as a function of CO concentration at different temperatures.[1]

electrons almost all of which are captured by surface acceptors and hence is a non-conductive state under an oxidizing ambient. The conductance, ΔG, of the SnO_2 nanowire sharply increases when a reducing gas (CO) is introduced into the gas cell. Figure 9.2(a) displays effect of CO concentration on conductance at different temperatures. The data was fitted to $\Delta G = P^a$. P is the CO partial pressure and a is about 0.5 over a wide range of CO concentrations.

Comini *et al.*[2] investigated CO sensing, NO_2 sensing and ethanol sensing of SnO_2 nanobelt sensors. The SnO_2 nanobelts have a width to thickness ratio of about 5 : 1 and rutile-like tetragonal structures. For electrical measurements, SnO_2 nanobelts were deposited onto alumina 3 mm × 3 mm blank substrates. Different deposition densities were obtained through the control of the growing parameters, including temperature, gradients, and distance from source material. The substrate with SnO_2 nanobelts was then equipped with a platinum meander on the backside, acting as a heater and a temperature sensor. Platinum interdigitated contacts were sputtered onto the nanobelt layer.

For a gas-sensing test, a constant flux of synthetic air at a flow of 0.3l per minute was the gas carrier. The desired concentration of pollutants such as CO, or NO_2 or ethanol was mixed with the synthetic air. The measurement was carried out in a sealed chamber at 20°C under controlled humidity. The sensor was biased by 1 V.

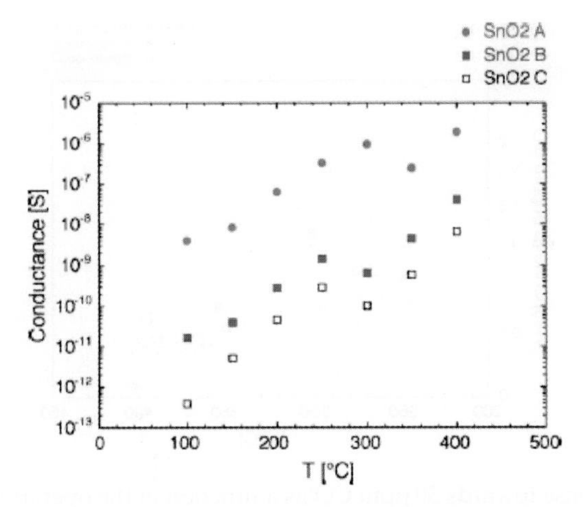

Fig. 9.3. Conductance of tin oxide nanobelts with different densities exposed to a humid synthetic air flux of 300 l per minute, pH 50% at 20°C.[2]

The conductance of nanobelts with different densities as a function of the operating temperature is shown in Fig. 9.3. Samples SnO_2 A, SnO_2 B, SnO_2 C have a decreasing density of nanobelts and give a consequent decrease in the conductance. The conductance of sample B and sample C show a minimum at 300°C and sample A possesses a minimum at higher temperatures. This phenomenon is attributed to the maximum oxygen coverage.

In the following, the response of a sensor is defined as the normalized variation of resistance for an n-type semiconductor and oxidizing gases ($S = \Delta R/R_0$) while for reducing gases it is defined as the normalized variation of conductance for an n-type semiconductor ($S = \Delta G/G_0$).

Figure 9.4 indicates the response towards 30 ppm CO for three different densities of SnO_2 nanobelts as a function of operating temperature with 50% pH at 20°C. From this figure, it can be observed that for each sample, there exists a maximum of the response at a temperature corresponding to the maximum oxygen coverage for the layers (300 for SnO_2 B and C and 350 for the other layer). This behavior is consistent with the reaction mechanism proposed by previous authors, for example, Kolmakov *et al.* That is dependent on the oxygen coverage. Also, the density of the samples affects the response.

For this SnO_2 nanobelt sensor, Comini *et al.*[2] also investigated NO_2 sensing. In the case of NO_2, there are proofs of the reactions directly

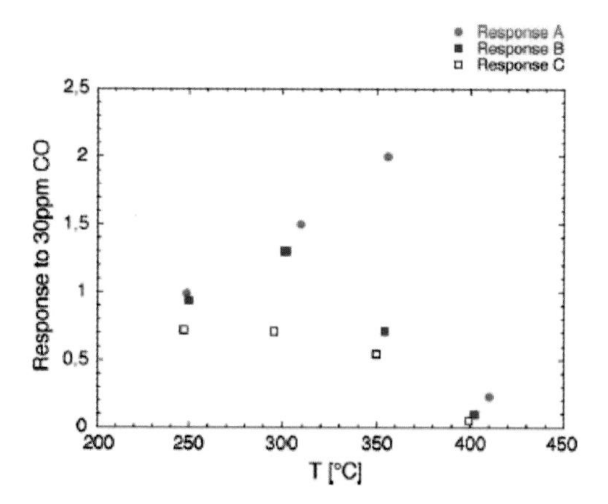

Fig. 9.4. Response towards 30 ppm CO as a function of the operating temperature with 50% pH at 20°C.[2]

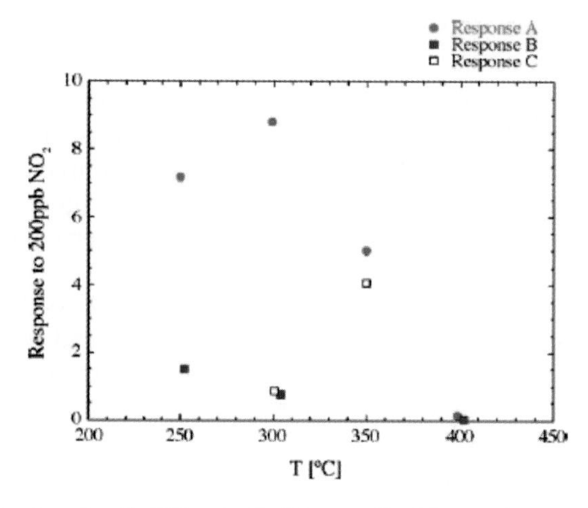

Fig. 9.5. Response towards 200 ppb of nitrogen dioxide as a function of the operating temperature with 50% pH at 20°C.[2]

with the semiconductor surface other than with the oxygen chemisorbed at the surface. Tamaki *et al.*[3–5] found that the adsorbates originating from NO_2 are essentially the same as those for NO, since the NO_2 molecule dissociates easily over the SnO_2 surface. Figure 9.5 displays the variation of the response toward 200 ppb of NO_2 as a function of the operating temperature. Clearly, the conductance decreases rapidly. The maximum of the response appears at relatively low temperatures, due to the particular reaction mechanism of NO_2 with SnO_2, not related to the oxygen chemisorbed species. Also, the density of nano-SnO_2 belts affects the response remarkably.

The ethanol sensing of SnO_2 nanobelt sensors has been investigated by Comini *et al.*[2] A scheme for the interpretation of the transformation of ethanol to aldehyde through catalytic interaction with a metal oxide semiconductor surface has been proposed by Idriss.[6] The reaction process is as follows:

$$\begin{aligned} CH_3CH_2OH + O(s) + M(s) &\rightarrow CH_3CH_2O\text{-}M(ads) + O\text{-}H(ads) \\ &\rightarrow CH_3CHO + H\text{-}M(ads) + O\text{-}H(ads) \end{aligned} \tag{9.6}$$

where (s) and (ads) are a surface site or an adsorbed species, respectively. Comini *et al.* indicate that the residual hydrogen from ethanol adsorption generally desorbs as H_2O or as H_2. H_2O can originate from the recombination of M-H and OH adsorbed from the environment. This product then desorbs, leaving an oxygen vacancy and a partially-reduced metal, which in the presence of gas phase O_2 is re-oxidized. Because ethanol exhibits a reducing behavior, for the *n*-type semiconductor the conductance increases with the gas introduction. Comini *et al.* experimentally investigated sensitivity at a concentration of ethanol below 200 ppm. Figure 9.6 shows the response of the SnO_2 nanobelts towards 10 ppm of ethanol as a function of the operating temperature with 50% RH at 20°C. Obviously, the increase of

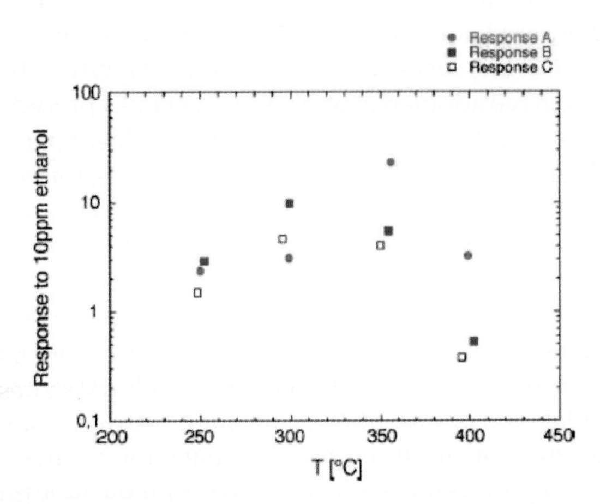

Fig. 9.6. Response towards 10 ppm of ethanol as a function of the operating temperature with 50% RH at 20°C.[2]

electric conductance is observed. The maximum of the response is between 300 and 350°C for all the layers and there exists a difference in the response between the different layers.

Recently, Cai *et al.*[7] reported that they prepared the SnO_2 nanostructured mono- and multi-layered ordered porous film as the gas sensors, and measured the gas (ethanol gas) sensing properties. The main contents are as follows.

Cai *et al.*[7] constructed nanostructured SnO_2 gas sensors based on two-dimensional colloidal crystal templates. The fabrication procedure is as follows.

The colloidal monolayer floating in the $SnCl_4$ solution is directly picked up by a commercially available ceramic tube (2 mm in outer diameter and 5 mm in length) with a relatively rough surface. After subsequent heat treatments, the SnO_2-ordered porous film grows directly on the surface of the gas-sensor substrate (ceramic tube). Figure 9.7 shows a schematic depiction, a photograph, and a microstructure of an as-fabricated gas sensor. There are two gold electrodes at the ends of the ceramic tube, and a Pt resistance heater is placed inside the tube to control the temperature. The ordered porous film can cover the entire surface of the tube and connect the two electrodes. However, as shown in Fig. 9.7(d), there are many gaps (or breaks) in the pore walls (or skeletons) because of mismatched thermal expansion coefficients of SnO_2 and the ceramic tube. These breaks could adversely influence the conductivity of the gas sensor. The resistance of a mono-layered ordered porous film on the tube is higher than 30 MΩ, which leads to no signal while measuring the sensing properties. To increase the conductivity, we have adopted a layer-by-layer (LbL) procedure. By using a tube which has already been covered with a monolayer porous film as the substrate, and repeating the above-mentioned process, Cai *et al.* obtained a second ordered porous layer on the tube. The first layer can act as a buffer layer. Using this procedure, there are fewer gaps in the pore walls of the second layer. Importantly, the breaks in the first layer can be remedied during the formation of the second layer due to the presence of the precursor solution. Using this process, the microstructures, and hence the conductivity of the gas sensors can be improved considerably. Also, researchers can fabricate multi-layered porous films on the tube with any number of layers, as required for practical applications. They have found that the resistance of nanostructured porous films with four layers on the tube is below 3 MΩ, which is an acceptable value for conventional gas-sensing measurements. Figure 9.8(a) shows the surface morphology of a gas sensor consisting of

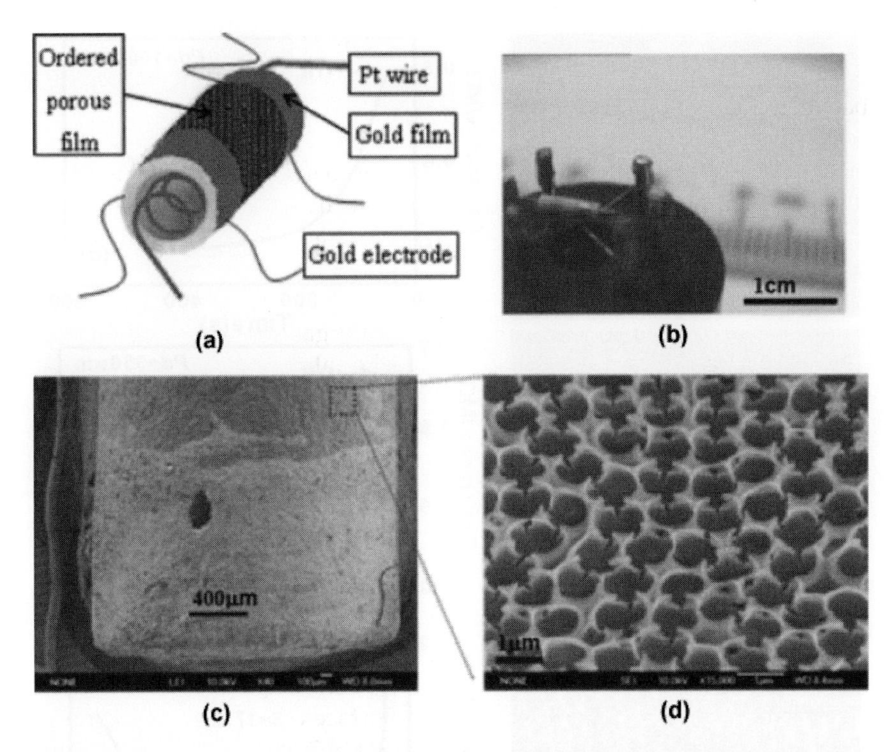

Fig. 9.7. A gas sensor fabricated with a SnO_2-ordered porous monolayer film on a ceramic tube. (a) A sketch of the gas-sensor structure; (b) a photograph of a final gas sensor; (c) a low-magnification field-emission scanning electron microscopy (FESEM) image of the gas-sensor surface; and (d) magnified image of the area marked in (c).[7]

four ordered porous layers of SnO_2 on the ceramic tube, prepared using a 1000 nm diameter PS sphere colloidal monolayer. To increase the specific surface area of the film, PS spheres with smaller diameters can be used. Figures 9.8(b) and 9.8(c) show corresponding sensors with four layers of SnO_2 on the tube, fabricated using 350 and 200 nm PS spheres, respectively. The surface morphologies look quite homogeneous, compared with the monolayer porous film (Fig. 9.7(d)). Obviously, such gas sensors with multi-layered porous films cannot be fabricated with three-dimensional (3D) colloidal crystals.[8,9]

The gas sensing properties of the four-layered porous films (sensors) have been measured in a custom-built experimental setup.[10] The sensor response to a 100 ppm ethanol gas at 300°C for these three types of

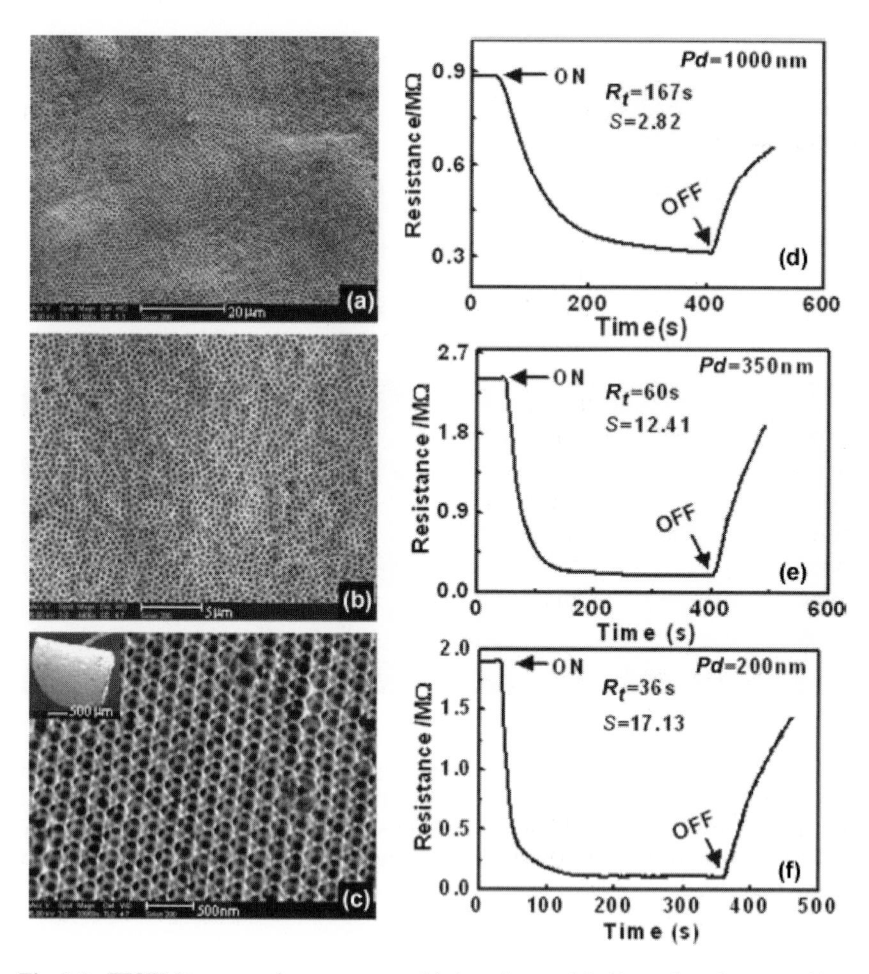

Fig. 9.8. FESEM images of gas sensors with four-layered SnO_2-ordered porous films and their corresponding sensing properties in ethanol gas. The sensors have been fabricated with (a) 1000 nm, (b) 350 nm, and (c) 200 nm diameter PS sphere colloidal monolayers. The inset in Fig. 9.4(c) is a low-magnification image of a part of the gas sensor. Graphs (d)–(f) show the corresponding measured resistance changes versus time for 100 ppm ethanol gas at 300°C. "ON" means injection of gas into the test chamber and "OFF" means venting of the gas from the chamber [*Pd* is the pore (or PS sphere) diameter, *Rt* is the response time, and *S* is the sensitivity].[7]

nanostructured multi-layered porous films are shown in Figs. 9.8(d)–9.8(f). Here, the response time is defined as the time required for the resistance of the sensor to change to 90% of the saturation value after exposure to the test gas. The gas sensitivity is defined as $S = R_{air}/R_{gas}$, where R_{air} and R_{gas} are the resistances of the sensor in air and ethanol gas, respectively. It can be seen that the response time decreases and the sensitivity increases with a reduction in the size of the PS spheres in the colloidal monolayer (and thus in the pore size). The response times are about 167, 60, and 36 s for sensors from 1000, 350, and 200 nm PS spheres, respectively. The corresponding sensitivities are 2.8, 12.4, and 17.1, respectively. This implies that we can very easily control the response time and sensitivity of such nanostructured sensors by changing the size of the PS spheres. In addition, it should be mentioned that such an LbL method leads to good reproducibility and uniformity in the microstructure and thickness of the films.

As for gas-sensing properties, the increased sensitivity and decreased response time with reduction of the PS sphere size can be attributed to the increased surface areas. For SnO_2 gas sensors, the changes in resistance are mainly induced by the adsorption and desorption of gas molecules from the surfaces of the sensing structures. In an air environment, oxygen molecules are generally chemisorbed onto the surface of SnO_2 as O_2^-, O^-, and O^{2-}, which exist in an equilibrium state:[10]

$$O_2 \Leftrightarrow O_2^- \Leftrightarrow O^- \Leftrightarrow O^{2-} \quad (9.7)$$

Because of the adsorbents, some electrons in SnO_2 are captured,[11] and hence the porous films show a higher resistance. When the film is exposed to reductive ethanol gas at 300°C, the surface oxygen species react with the gas, decreasing the surface concentration of oxygen ions, and hence increasing the conductivity of the SnO_2-ordered porous films.[12,13] Obviously, films with a higher surface area (or smaller pore size) can adsorb more molecules, and thus possess a higher concentration of oxygen ions. When such films are exposed to the gas, more adsorbed oxygen is desorbed, which leads to a higher variation in resistance, or higher sensitivity. Further experiments have shown that the conductance or sensitivity is not changed significantly when the number of the layers is increased above four.

In summary, SnO_2 nanostructures, including nanowires, nanobelts and nanostructured ordered porous mono- and multi-layered films can be used as CO, O_2, NO_2 and ethanol gas sensors with very good electrical responses to gases, comparable to the 3D counterpart and even superior to the 3D counterpart.

The gas sensing properties of the SnO_2 sensors depend on the surface reaction [The gas reacts directly with the semiconductor (for example,

SnO_2) surface or with the oxygen chemisorbed at the surface.] that leads to a change in the electric properties (resistance or conductance).

9.2.2. Biosensors

Biosensors are one very important kind of sensor. They possess potential for broad applications in practical biotechnological and medical applications in various bio-detection systems. Recently, many researchers began to design biosensors using nanomaterials such as nanoparticles, nanowires and nanoribbons etc. due to these sensors exhibiting high sensitivity to pathogenic bacteria, viruses etc. In the next paragraph, several typical biosensors will be introduced.

9.2.2.1. *Nanodevices for electrical detection of single viruses*[14]

Viruses are among the most important causes of human disease and an increasing concern as agents for biological warfare and terrorism. Rapid, selective, and sensitive detection of viruses is central to implementing an effective response to viral infection. Now, established methods for viral analysis have not obtained rapid detection at a single virus level and often require a relatively high level of sample manipulation that is inconvenient for infectious materials. The ability to detect rapidly, directly, and selectively individual virus particles, is very important to human health care because it could enable diagnosis at the earliest stages of virus replication within a body system. Lieber *et al.*[14] converted an array of nanowire-based field effect transistors (FETs) into a virus detector by coating the surfaces of the transistors with antibody receptors. The following describes simply this virus detector preparation process. Silicon nanowires were synthesized by chemical vapor deposition with gold particles of 20 nm diameter as catalysts, silane as the reactant, and diborane as the *p*-type dopant with a B/Si ratio of 1 : 4000. The arrays of Si nanowire devices were defined by using photolithography with Ni metal contacts[15] on silicon substrates with a 600 nm thick oxide layer. The metal contacts to the nanowires were isolated by subsequent deposition of about 50 nm thick Si_3N_4 coating. The space between source-drain electrodes (active sensor area) was 2 μm. Lieber *et al.* used a two-step procedure to covalently link antibody receptors to the surfaces of the Si nanowire devices. Firstly, the devices were reacted with a 1% ethanol solution of 3-(trimethoxysilyl) propyl aldehyde for 30 minutes, washed with ethanol, and then heated at 120 for 15 minutes. Secondly, mAb receptors, anti-hemagglutinin for influenza A and anti-adenovirus group III, were coupled to the aldehyde-terminated nanowire surfaces by reaction with 10–100 μg/ml antibody in a pH 8, 10 mM phosphate buffer solution containing 4 mM sodium cyanoborohydride. The surface density of the antibody was controlled by varying the reaction time from 10 minutes to 3 h. Unreacted aldehyde surface groups were subsequently passivated by reaction with ethanolamine under similar conditions. Device arrays for multiplexed experiments were made in the same way except that different antibody solutions were spotted on the different regions of the array.

During viral sensing experiments, virus samples were delivered to the nanowire device arrays by using fluidic channels.

Figure 9.9 gives the schematic of two Si nanowire devices and detection of single viruses.[14] When a virus particle binds to the antibody receptor on a nanowire device, the conductance of that device should change from the baseline value, and when the virus unbinds, the conductance should return to the baseline value. For a *p*-type nanowire, the conductance should decrease (increase) when the surface charge of the virus is positive (negative).[16] The conductance of the second nanowire does not change because no virus binds to the antibody receptor on this nanowire device during this same time period. Modification of different nanowires within the array with receptors specific for different viruses provides a means for simultaneous detection of multiple viruses. Therefore, Lieber *et al.*[14] have used arrays of individual addressable Si nanowire field-effect transistors to detect viruses. Nanowire elements within the arrays were functionalized with the same or different virus-specific antibodies as receptors for selective binding. In the following, we will introduce the virus detection results given by Lieber *et al.*

Figure 9.10(a) shows the change of conductance with time, which was recorded simultaneously from two nanowires in the same device array modified with antibodies specific for the influenza A virus.[14] It can be seen that when a solution containing about 100 virus particles per μl is delivered to the sensor elements, the conductance presents discrete changes. These experiments have several features. First, the magnitude and duration of the conductance drops are nearly the same for a given nanowire. Second, an excess of free antibodies added to the viral solution eliminates the well-defined conductance changes. Third, the discrete conductance changes are uncorrected for the two nanowire devices in the micro-fluidic channel. Fourth, the frequency of the discrete conductance drops is directly proportional to the number of virus particles in solution.

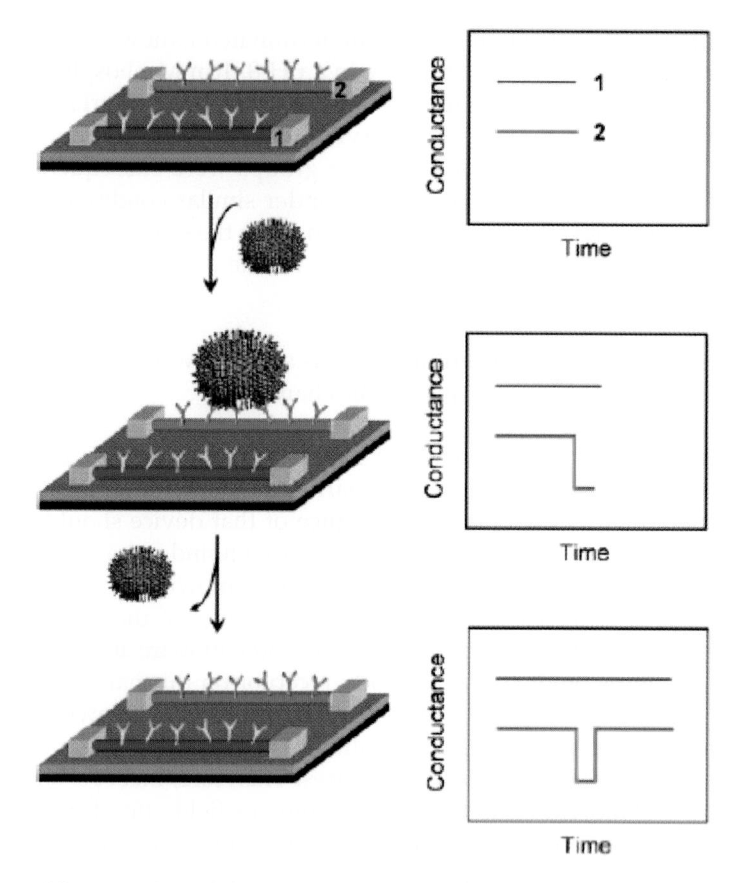

Fig. 9.9. Nanowire-based detection of single viruses. (Left) Schematic shows two nanowire devices, 1 and 2, where the nanowires are modified with different antibody receptors. Specific binding of a single virus to the receptors on nanowire 2 produces a conductance change. (Right) Characteristic of the surface charge of the virus only in nanowire 2. When the virus unbinds from the surface, the conductance returns to the baseline value.[14]

Lieber *et al.* measured the change of the discrete conductance with pH at constant ionic strength to probe viral surface charge, as shown in Fig. 9.10(b). The results show that the discrete conductance changes decrease and then increase in magnitude but with the opposite sign as the pH increases from 5.5 to 8 with a point of zero conductance change appearing between pH 6.5 and 7.0. The *p*-type nanowire devices show a reduced (increased) conductivity upon binding of single influenza A viruses at pH $< 7 (\geq 7)$.

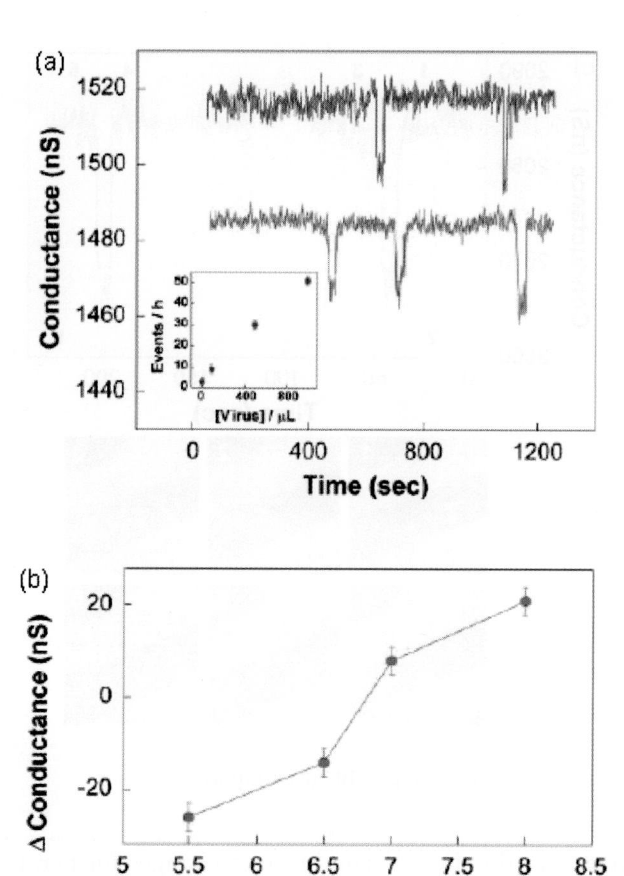

Fig. 9.10. Selective detection of single viruses. (a) Conductance versus time data recorded simultaneously from two silicon nanowires elements, red and blue plots, within a single device array after introduction of an influenza A solution. (Inset) Frequency of single virus events as a function of virus solution concentration. (b) Conductance changes associated with single influenza A virus binding/unbinding as a function of solution pH. (c) Conductance (upper) and optical (lower) data recorded simultaneously versus time for a single silicon nanowire device after introduction of the influenza A solution. Combined bright-field and fluorescence images correspond to time points 1–6 indicated in the conductance data; the virus appears as a red dot in the images. The solid white arrow in image 1 highlights the position of the nanowire device, and the dashed arrow indicates the position of a single virus. Images are $8 \times 8\ \mu$m. All measurements were performed with solutions containing 100 viral particles per μl.[14]

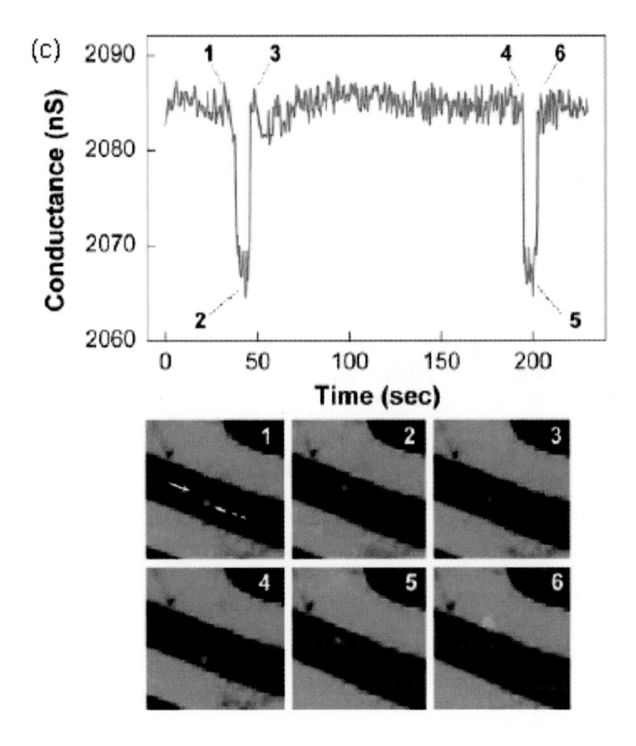

Fig. 9.10. (*Continued*)

To characterize the discrete conductance changes further, Lieber *et al.* have carried out simultaneous electrical and optical measurements. They collect conductance, fluorescence, and bright-field data from a single nanowire device with fluorescently labeled viruses. Clearly, each discrete conductance change corresponds to a single virus binding to and unbinding from the nanowire. Figure 9.10(c) shows clearly that the conductance at points 1, 3, 4 and 6 corresponds to a virus (a red dot) unbinding from the nanowire, and the conductance at points 2 and 5 corresponds to a virus binding to the nanowire.

Lieber *et al.* also found that the density of specific antibodies affects the binding/unbinding properties. When the density of specific antibodies is very high, sequential binding of virus particles without unbinding is observed, as shown in Fig. 9.11(a). On the other hand, a low density of specific antibodies show discrete conductance changes caused by the interaction of single viruses with an average duration, 1.1 ± 0.3 s. This value is about 20 times shorter than that of Fig. 9.10, where an intermediate antibody density was used. In Fig. 9.11(b), the event 2 corresponds to the

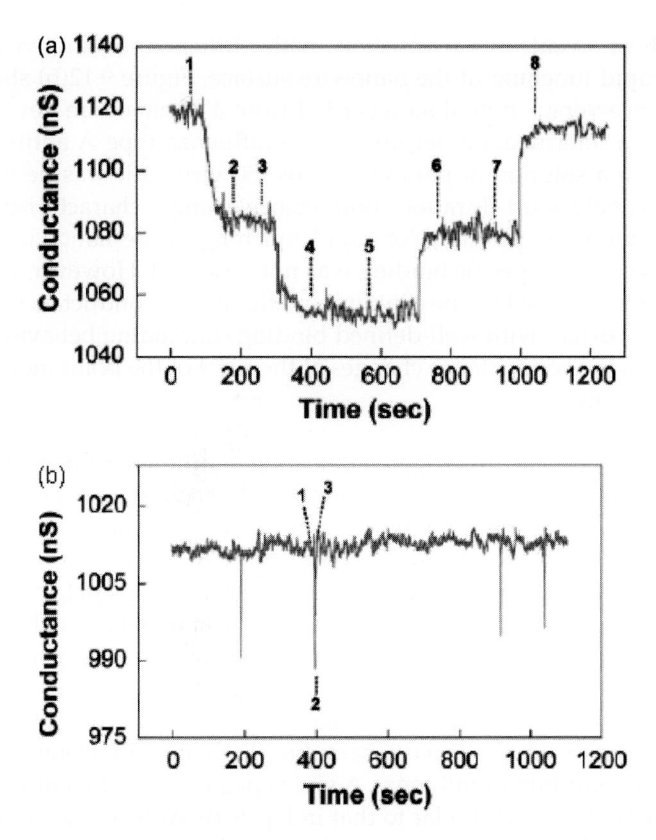

Fig. 9.11. Single virus binding selectivity. (a) The conductance versus time curve recorded from a single nanowire device with a high density of anti-influenza type A antibody. Influenza A solution was added before point 1, and the solution was switched to a pure buffer between points 4 and 5 on the plot. (b) The conductance versus time curve recorded from a single nanowire device with a low density of anti-influenza type A antibody. Measurements were made by using a solution containing 100 viral particles per μl.[14]

above case. In addition, Fig. 9.11(b) shows that other events have a time scale of 0.4 $\pm$ 0.1 s that is characteristic of the diffusion of the virus past and/or rapid touching of the nanowire surface. These short events can be readily distinguished on the basis of temporal behavior from selective binding exhibited by devices with a comparable density of specific antibody.

Delivery of a solution containing influenza (100 virus/μl) to a nanowire device functionalized with a high density of anti-adenovirus group III antibodies, which have no specificity against influenza A, only exhibited discrete conductance changes of short duration, 0.4 $\pm$ 0.1 s (Fig. 9.12(a)).

These short events are consistent with diffusion of the virus past and/or rapid touching of the nanowire surface. Figure 9.12(b) shows the conductance versus time data recorded from a Si nanowire device modified with a intermediate density of anti-influenza type A antibody. Initially, when a solution of paramyxovirus (50 virus/μl) was delivered to the device, only short-duration conductance changes characteristic of diffusion of the virus past and/or rapid touching of the nanowire surface was observed and specific binding was not observed. However, when the solution was changed to one containing influenza A, conductance changes became consistent with well-defined binding/unbinding behavior (see in Fig. 9.12(b), the conductance changes at the right of the point indicated by the black arrow).

Lastly, Lieber *et al.* investigated detection of different viruses at the single particle level by modifying the nanowire device surface in an array with antibody receptors specific either for influenza A (nanowire 1) or adenovirus (nanowire 2). Simultaneous conductance measurements were made. Introduction of adenovirus, which is negative at the given pH, to the device array yields positive conductance changes for nanowire 2 with a time of 16 ± 6 s which is similar to the selective binding/unbinding in Fig. 9.10. Shorter, $\sim$0.4 s duration positive conductance changes are also observed for nanowire 1 (Fig. 9.12(c)). These changes are characteristic of a charged virus diffusing past and rapidly sampling the nanowire element. On the other hand, addition of influenza A yields negative conductance changes for nanowire 1. This is similar to that in Fig. 9.10. At this time, nanowire 2 exhibits short duration negative conductance changes, which correspond to the diffusion of influenza A viral particles past the nanowire device. If a mixture of influenza A and adenovirus was introduced to the array, selective binding/unbinding responses for influenza A and adenovirus can be detected simultaneously by nanowire 1 and nanowire 2, respectively.

The experimental results obtained by Lieber *et al.* indicate that the single viruses can be detected directly with high selectivity, including parallel detection of different viruses, in electrical measurements using antibody functionalized nanowire field-effect transistors.

9.2.2.2. *Nanoelectromechanical devices for detection of viruses*[17]

Llic *et al.*[17] have used arrays of chemically functionalized, surface micromachined polycrystalline silicon cantilevers to measure binding events of the

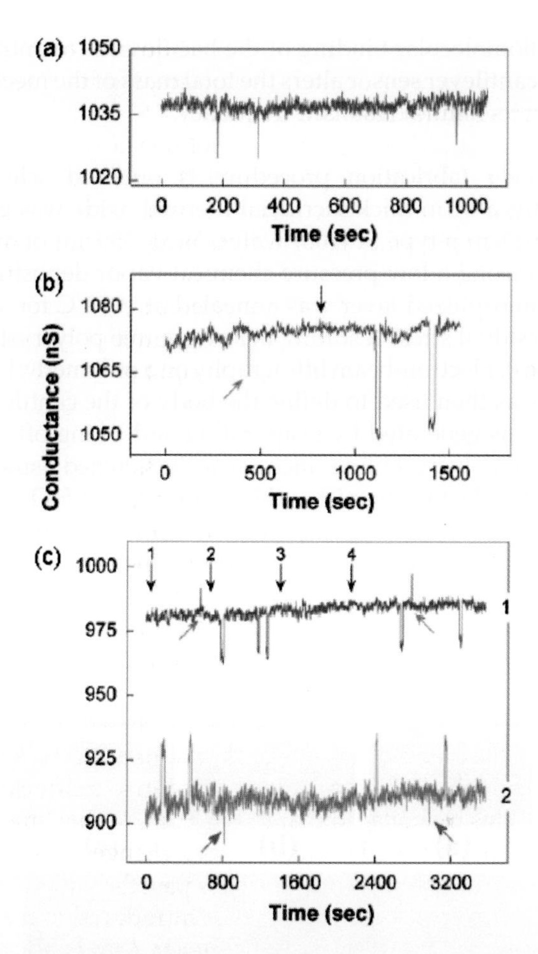

Fig. 9.12. Selective and multiplexed single virus detection. (a) Conductance versus time curve recorded from a silicon nanowire device after introduction of influenza A virus solution; the device had a high surface coverage of anti-adenovirus group III antibody. (b) Conductance versus time data recorded from a silicon nanowire device modified with an intermediate density for anti-influenza type A antibody. Initially, a solution of paramyxovirus (50 virons per μl) was delivered to the device, and at the point indicated by black arrow the solution was changed to one containing influenza A (50 virons per μl). (c) Conductance versus time data recorded simultaneously from two silicon nanowires elements; one nanowire (nanowire 1) was modified with anti-influenza type A antibody (blue data), and the other (nanowire 2) was modified with anti-adenovirus group III antibody (red data). Black arrows 1–4 correspond to the introduction of adenovirus, influenza A, pure buffer, and a 1 : 1 mixture of adenovirus and influenza A, where the virus concentrations were 50 viral particles per μl in phosphate buffer (10 μm, pH 6.0). Small red and blue arrows in B and C highlight conductance changes corresponding to diffusion of viral particles past the nanowire and not specific binding.[14]

baculovirus. Biomolecular binding of the baculovirus to antibody-treated regions of the cantilever sensor alters the total mass of the mechanical oscillator, changing its natural resonant frequency.

The cantilever fabrication procedure is outlined schematically in Fig. 9.13. Firstly, a 1 μm thick sacrificial thermal oxide was grown on the surface of a 10 Ωcm *p*-type Si (100) wafer. Next, 150 nm of amorphous Si was deposited using a low pressure chemical vapor deposition (LPCVD) process. The amorphized layer was annealed at 1050°C for 15 minutes to alleviate the residual stress resulting in a stress-free polycrystalline silicon film (Fig. 9.13(a)). Electron-beam lithography on a polymethyl methacrylate bilayer resist was then used to define the body of the cantilever. A metallic etch mask was generated by evaporating and lifting off a 30 nm thick chromium layer. Polycrystalline silicon was then etched using reactive ion etching in a CF_4 plasma chemistry (Fig. 9.13(b)). The SiO_2 was removed with hydrofluoric acid, leaving suspended cantilever beams above a Si substrate (Fig. 9.13(c)). Figure 9.13(d) shows an oblique angle scanning electron

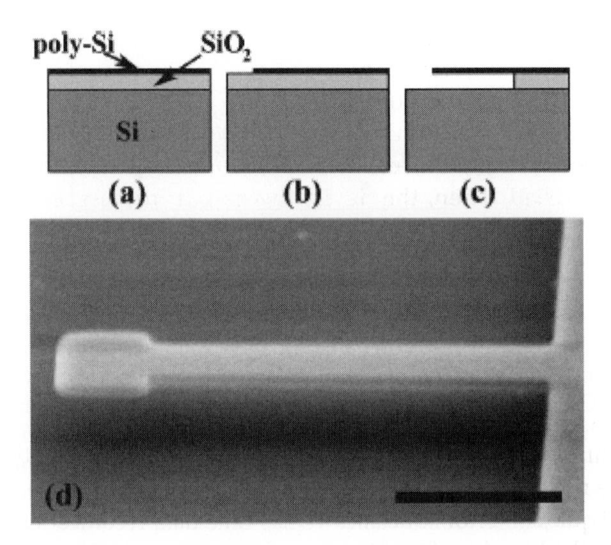

Fig. 9.13. Fabrication sequence of the nanomechanical oscillators. (a) 1 mm thermal oxidation and LPCVD deposition of the polycrystalline silicon device layer; (b) lithographic definition of the oscillator; (c) sacrificial silicon dioxide is removed in hydrofluoric acid; and (d) scanning electron micrographs of released cantilever oscillators with $l = 6\,\mu$m, $w = 0.5\,\mu$m, $t = 150$ nm with a 1μm $\times$ $1\,\mu$m paddle. Scale bar corresponds to $2\,\mu$m.[17]

micrograph of a released cantilever oscillator of length $l = 6\,\mu\text{m}$, width, $w = 0.5\,\mu\text{m}$, thickness, $t = 150\,\text{nm}$, with a $1\,\mu\text{m} \times 1\,\mu\text{m}$ paddle.

This device can be modeled as a harmonic oscillator with mass m_{osc} and resonant frequency f_o. The estimated minimum detectable surface mass Δm loading is given by

$$\Delta m = K\left[\frac{1}{(f_o - \Delta f)^2} - \frac{1}{f_o^2}\right], \tag{9.8}$$

where K is the spring constant and Δf is the frequency shift. It is assumed that the flexural rigidity is unchanged due to subsequent additional mass loading, and $\Delta f \ll f_o$-first-order expansion of Eq. (9.8) yields

$$\frac{\Delta m}{\Delta f} = \frac{2m_{\text{osc}}}{f_o}. \tag{9.9}$$

The frequency spectrum of the oscillator with $l = 6\,\mu\text{m}$ was measured, as shown in curve 1 of Fig. 9.14(a). Thus substrates containing oscillator arrays were immersed into a solution of AcV1 antibodies for 1 h and then washed in de-ionized water and nitrogen dried. The immobilization of the AcV1 antibodies increased the total mass of the oscillators. As a result, the resonant frequency decreased (the curve 1 shifts to curve 2 of Fig. 9.14(a)). Devices were then immersed in a buffer solution with baculovirus concentrations ranging between 10^5 and 10^7 pfu/ml for 1 h. Following the immobilization of baculovirus, devices were rinsed in de-ionized water and nitrogen-dried. Then, the devices were put into a vacuum chamber and evacuated to a pressure of $\times 10^{-6}$ torr. The experimental conditions were similar to those during measure curve 1. The frequency spectrum of the devices was measured (curve 3 of Fig. 9.14(a)). From Eq. (9.9), the minimum detectable masses for the $l = 6.8$, and $10\,\mu\text{m}$ oscillator were 0.41, 0.52, and 0.96 attograms/H_z, respectively. Figure 9.14(b) displays the frequency shift with the baculovirus concentration for oscillators with $l = 6$, 8, and $10\,\mu\text{m}$. At the lower concentration of the spectrum (10^5 pfu/ml), considering a weight of a single baculovirus as $\sim 1.5 \times 10^{-5}$ g, it is estimated that the average number of bound virus particles is 6. This implies that with a mechanical quality factor of about 10^4, the fabricated oscillators are capable of detecting the binding of a single baculovirus.

Figure 9.14(c) shows that a buffer solution without baculovirus was used to evaluate the effect of non-specific absorption of the buffer solution to the antibody layer on the frequency shift. Results indicate a small frequency shift change and even no shift in the frequency.

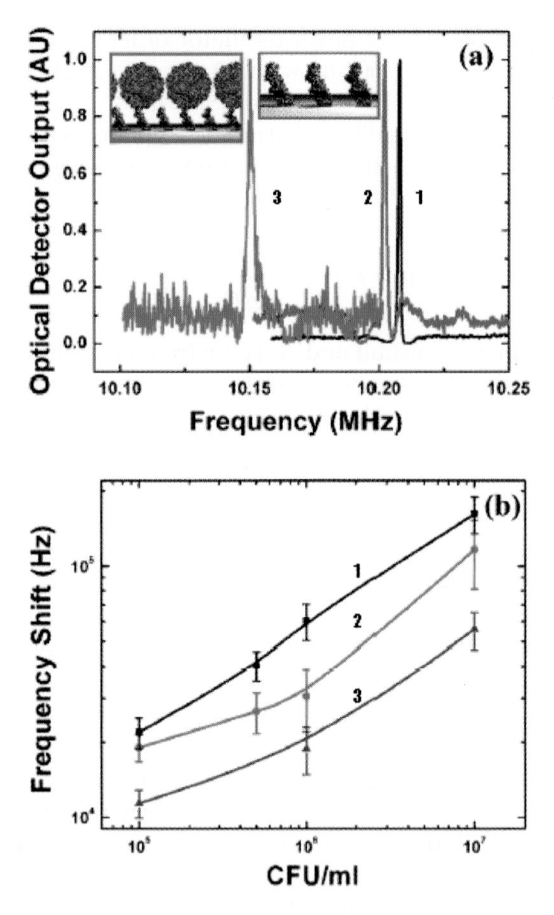

Fig. 9.14. Frequency spectra and sensitivity of the nanoelectromechanical cantilevers. (a) Measured frequency spectra of the as-fabricated $l = 6\,\mu$m oscillator (1), with additional antibody (2) and baculovirus (3) mass loading. Insets schematically depict binding of AcV1 to the substrate (left) and immobilization of baculovirus (right). (b) Frequency shift as a function of the baculovirus concentration for 6 μm (1), 8 μm (2), and 10 μm (3), long cantilevers. (c) Frequency spectra of control measurements with $l = 8\,\mu$m cantilevers (1), with antibodies (2), a buffer solution without baculovirus (3) and a buffer solution containing baculovirus of 10^8 pfu/ml concentration (4).[17]

In summary, arrays of surface AcV1 antibody-coated polycrystalline silicon nanomechanical cantilever beams can be used to detect binding from various concentrations of baculoviruses in a buffer solution. The $0.5\,\mu$m $\times$ $6\,\mu$m cantilevers have mass sensitivities in the order of 10^{-19} g/H_2, enabling the detection of an immobilized AcV1 antibody

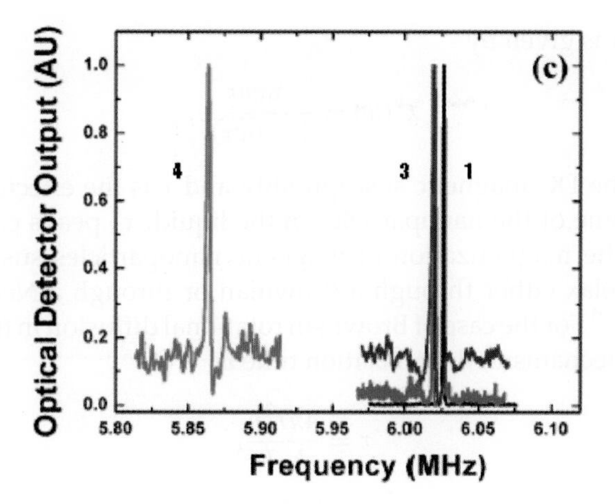

Fig. 9.14. (*Continued*)

monolayer corresponding to a mass of about 3×10^{-15} g. These devices can detect the mass of single virus particles bound to the cantilever.

9.2.2.3. *Biological magnetic sensors*[18]

The biological and medical applications of nanomagnetic particles have attracted recent interest due to the successful development of different fabrication techniques that combine fine magnetic nanoparticles with biologically relevant coatings.[19] This coating can have a biological recognition function for specific target molecules through ligand receptor and antigen-antibody bindings. Magnetic nanoparticles have many applications. Recently, many efforts focus on developing magnet-based biodetection platforms. Many of these sensor schemes rely on detecting the stray magnetic field of the biomagnetic particles upon binding them to a magnetic-field sensor using the target as a link. One weakness of all of the sensing schemes proposed by researchers is that they do not permit discrimination between several different targets that may have similar biological binding affinity. Chung *et al.*[18] experimentally demonstrate a substrate-free magnetic sensor scheme based on the changes of dynamic magnetic properties of magnetic nanoparticles suspended in liquids. This sensor allows distinguishing between several possible targets with similar binding affinity. Chung *et al.* indicate that the magnetic response of nanoparticles suspended in a liquid to a small alternating magnetic field with frequency w can be expressed by a complex magnetic susceptibility x. The imaginary part, x'', of x is given by

$$x''(w) = \frac{x_0 w\tau}{1 + (w\tau)^2}. \tag{9.10}$$

Here x_0 is the DC magnetic susceptibility and τ is the effective magnetic relaxation time of the nanoparticles in the liquid. x'' peaks correspond to $w = \tau^{-1}$. The magnetization of magnetic nanoparticles suspended in a liquid can relax either through a Brownian or through a Neel relaxation mechanism.[20] For the case of Brownian rotational diffusion in the dominant relaxation mechanism, the relaxation time is

$$\tau = \frac{4\pi r^3 \eta}{k_B T}, \tag{9.11}$$

where η is the dynamic viscosity of the liquid, r is the hydrodynamic radius of the biomagnetic nanoparticles, and T is the temperature. According to expressions (9.10), (9.11) and $w = \tau^{-1}$ at the peak of $x''(w)$, it can be known that if the hydrodynamic radius increases due to the binding of target molecules to the biomagnetic nanoparticles, the Brownian relaxlation time increases and the frequency for the peak of AC x decreases.

Connolly *et al.*[21] demonstrate that the dominant relaxation of the two relaxation mechanisms depends on the particle size. Neel relaxation is the dominant mechanism for particles less than 20 nm while the Brownian mechanism is dominant for larger particle diameters with a single domain structure. Chung *et al.*[18] give an approximate guide for the needed stable single domain magnetic nanoparticle diameters (Fig. 9.15).

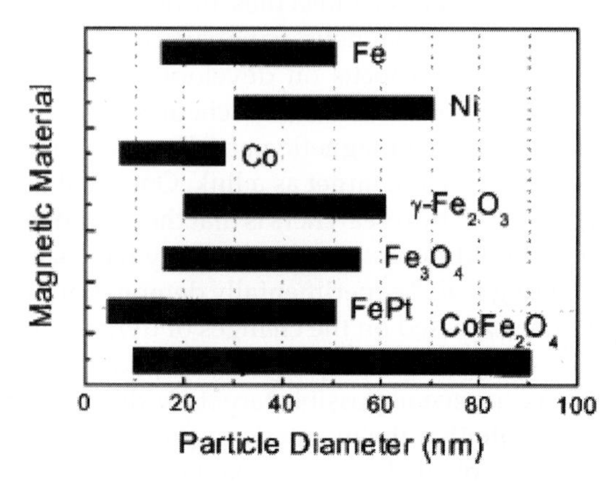

Fig. 9.15. Particle diameters for stable single domain magnetic nanoparticles.[18]

Chung *et al.*[18] investigated the magnetic characterization of avidin-coated magnetic (Fe_3O_4) nanoparticles in an aqueous solution. The particle concentration in the fluid is ~6 mg/ml (~2×10^{15} particles/ml). The dried particle has about 10 nm diameter magnetic core which is covered by an avidin-coated shell that is 20–30 nm thick. During measuring magnetizations, the nanoparticle sample was diluted with phosphate buffer saline (PBS, pH = 7.0). The experimental results show that the room temperature hysteresis curve of avidin-coated magnetic nanoparticles presents paramagnetic behavior, which is caused by single domain nanoparticles rotating freely in the liquid to be aligned with the external magnetic field.

Figure 9.16 shows imaginary part of the AC magnetic susceptibility, x'', as a function of frequency for avidin-coated magnetic particles suspended in a PBS buffer solution. It can be observed that at 300 K, a peak in the AC susceptibility appears at 210 Hz. When the PBS buffer solution was cooled to 250 K, which is below its freezing point, the frequency peak disappeared. This result implies that the low frequency peak at 300 K arises from rotational diffusive Brownian relaxation of the magnetization, and does not

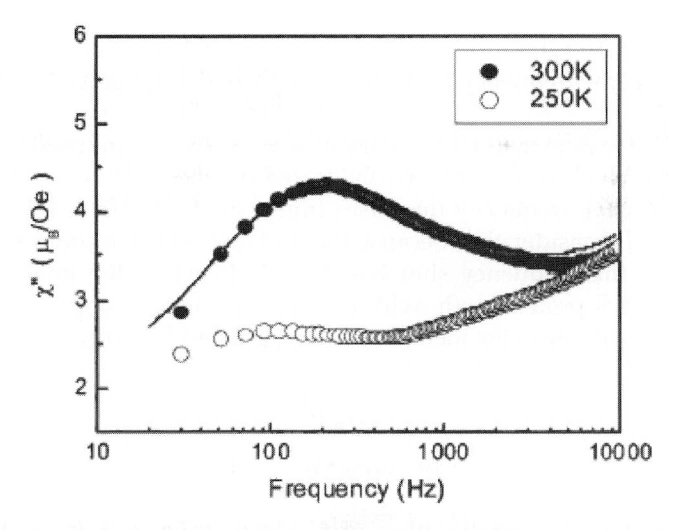

Fig. 9.16. Imaginary part of the AC magnetic susceptibility as a function of frequency for an avidin-coated magnetic particle suspended in a PBS buffer solution. Solid symbols are measured at 300 K, and open symbols at 250 K, which is below the freezing point of the carrier liquid. The solid line is from a convolution of Eq. (9.1) with a Gaussian-sized distribution of the nanoparticles and using the 250 K data as background.[18]

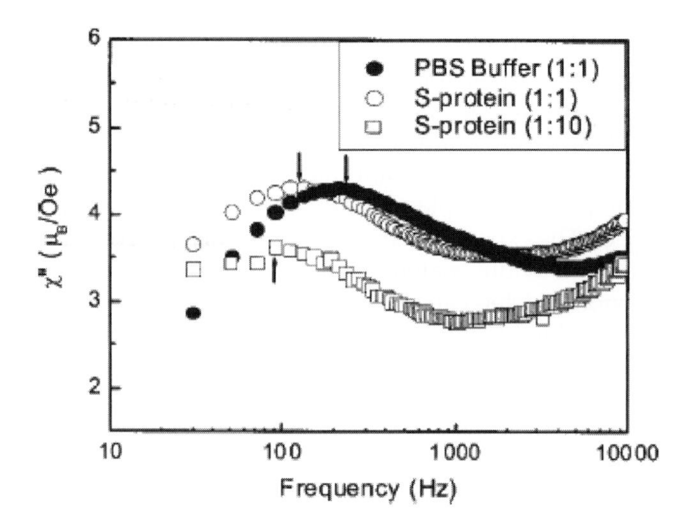

Fig. 9.17. Imaginary part of the AC magnetic susceptibility of an avidin-coated magnetic particle before (solid) and after (open) binding to S protein. Different concentrations of S protein (open circles and squares) result in similar frequency shifts.[18]

arise from Neel relaxation. If the magnetic relaxation is due to the Neel mechanism, the frequency peak should not have disappeared at 250 K.

In order to demonstrate the sensing scheme, the biotinylated S protein (6.3 μM) was added into the avidin-coated magnetic nanoparticles. As a result, the peak frequency decreases from 210 to 120 Hz (see Fig. 9.17). Chung *et al.* consider that because biotinylated S protein does not have magnetics, this frequency shift has to be induced by the interaction of biotinylated S protein with avidin-coated magnetic nanoparticles. This interaction will cause the increase of nanoparticles' hydrodynamic radius. The TEM observations show that the diameter of the avidin-coated magnetic nanoparticles is about 50 nm. After S protein binding, the particle diameter is about 60 nm. When ten times more S proteins were added to the sample, the frequency was further reduced.

Figure 9.18 displays the plots of x'' versus frequency. It can be seen after binding to biotinylated T7 bacteriophage, the frequency peak of x'' disappears. Chung *et al.* indicated that the magnetic nanoparticles are immobilized upon binding to the T7 bacteriophage. This is because of the avidin-coated magnetic nanoparticle aggregation after adding the biotinylated T7 bacteriophage particles. Therefore, this aggregation actually

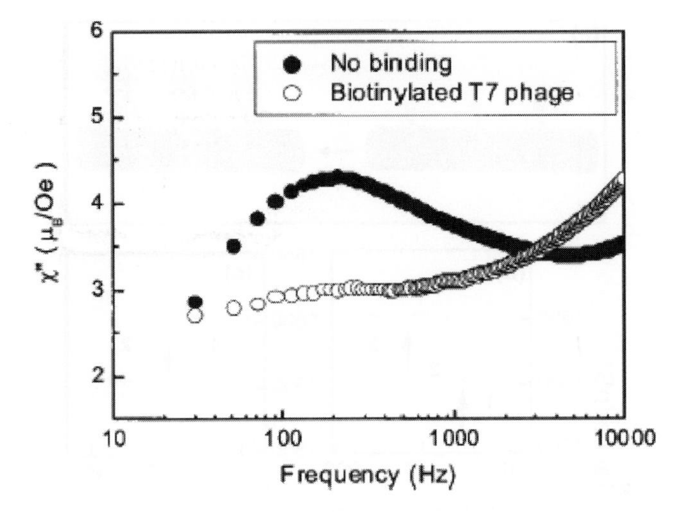

Fig. 9.18. Imaginary part of the AC magnetic susceptibility of an avidin-coated magnetic particle before (solid) and after (open) binding to biotinylated T7 bacteriophage.[18]

immobilizes the particles, and thus the rotational motion of the particles in the aggregate is blocked. This is why the frequency peak caused by Brownian relaxation disappears.

These results demonstrate that by using the frequency peak of the AC susceptibility, people can monitor the attachment of the targeted molecules to magnetic nanoparticles, and thus a biosensor scheme based on the Brownian relaxation of magnetic nanoparticles in a liquid is provided.

9.2.2.4. *Biotin-modified Si nanowire nanosensors for detection of protein binding*[22]

Lieber *et al.*[22] used the single-crystal boron-doped (*p*-type) SiNWs (Si nanowires) to prepare NW nanosensors for protein detection. The nanosensors were fashioned by flow-aligning[23] SiNWs on SiO_2 substrates and then making electrical contacts to the NW ends with electron-beam lithography.[24–26] In order to explore biomolecular sensors, Lieber *et al.* functionalized SiNWs with biotin and studied the well-characterized ligand-receptor binding of biotin-streptavidin (Fig. 9.19(a)). The conductance measurements show that the conductance of biotin-modified SiNWs increases rapidly to a constant value upon addition of a 250 nM streptavidin

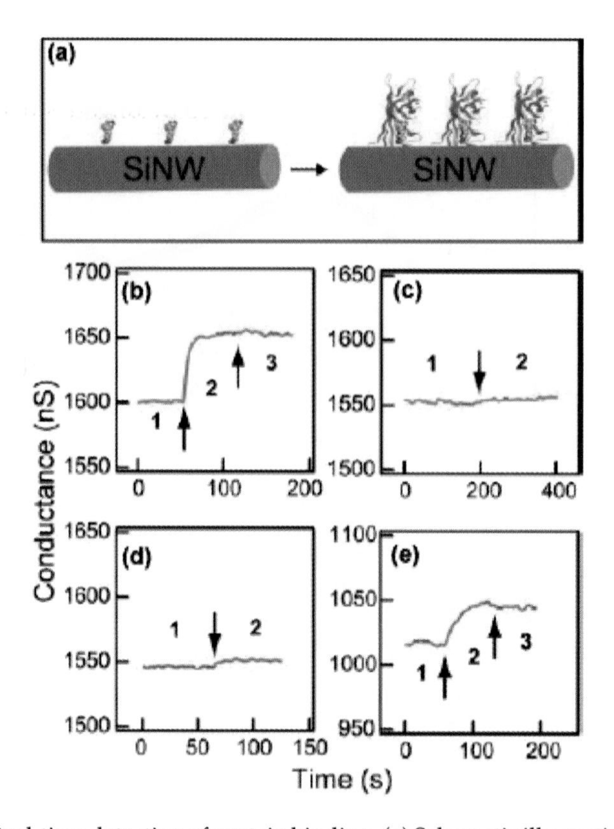

Fig. 9.19. Real-time detection of protein binding. (a) Schematic illustrating a biotin-modified SiNW (left) and subsequent binding of streptavidin to the SiNW surface (right). The SiNW and streptavidin are drawn approximately to scale. (b) Plot of conductance versus time for a biotin-modified SiNW, where region 1 corresponds to the buffer solution, region 2 corresponds to the addition of 250 nM streptavidin, and region 3 corresponds to the pure buffer solution. (c) Conductance versus time for an unmodified SiNW; regions 1 and 2 are the same as in (b). (d) Conductance versus time for a biotin-modified SiNW, where region 1 corresponds to the buffer solution and region 2 to the addition of a 250 nM streptavidin solution that was preincubated with 4 equivalents *d*-biotin. (e) Conductance versus time for a biotin-modified SiNW, where region 1 corresponds to the buffer solution, region 2 corresponds to the addition of 25 ppm streptavidin, and region 3 corresponds to the pure buffer solution. Arrows mark the points when the solutions were changed.[22]

and that this conductance value remains a constant after the addition of the pure buffer solution (Fig. 9.19(b)). The increase of conductance is caused by binding of negatively-charged streptavidin to the *p*-type SiNW surface because streptavidin is negatively-charged at the pH of the researcher's measurements. The conductance does not decrease with addition of a pure buffer. This is because the dissociation constant (Ka $\sim 10^{-15}$ M) and the corresponding dissociation rate for biotin-streptavidin are small.

In order to confirm that the observed conductance changes are due to the specific binding of streptavidin to the biotin ligand, Lieber *et al.* conducted the following experiments. When streptavidin solution was added to an unmodified SiNW, the conductance did not change (Fig. 9.19(c)). The second experiment showed that addition of streptavidin solution, in which the biotin-binding sites were blocked by reaction with 4 equivalents of *d*-biotin, did not show the conductance change of biotin-modified SiNWs (Fig. 9.19(d)). These results demonstrate there is little non-specific binding of protein to either the bare or biotin-modified SiNW surface.

Lieber *et al.* found that this nanosensor can detect streptavidin bind-down to a concentration of at least 10 pM (Fig. 9.19(e)). This detection level is substantially lower than the nanomolar range for detection of single molecules. However, biotin-streptavidin binding interaction is essentially irreversible. Therefore, real-time monitoring of varying protein concentrations cannot be realized. To explore the issue, Lieber *et al.* studied the reversible binding of monoclonal antibiotin (*m*-antibiotin) with biotin-modified SiNWs. The results are in Fig. 9.20. Figure 9.20 presents the change of conductance with time for a biotin-modified SiNW. From this figure, it is clear that a conductance drop after addition of *m*-antibiotin occurs, and

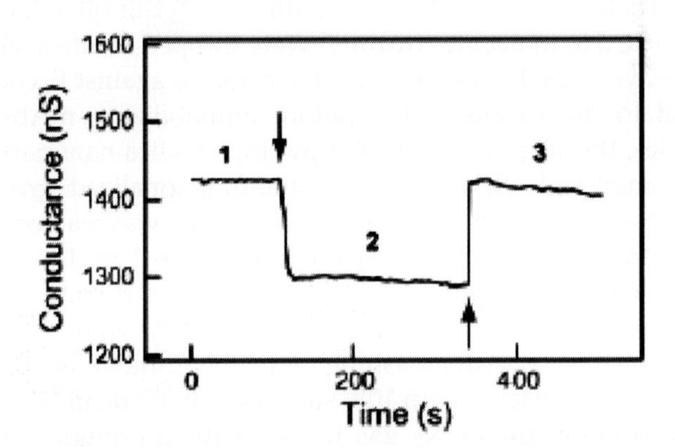

Fig. 9.20. Real-time detection of reversible protein binding for a biotin-modified SiNW, where region 1 corresponds to buffer solution, region 2 corresponds to the addition of $\sim 3\,\mu$m *m*-antibiotin (460 μg/ml), and region 3 corresponds to flow of pure buffer solution.[22]

upon addition of the pure buffer solution, the conductance increases to about the original value. The decrease in conductance upon *m*-antibiotin addition demonstrates that the *m*-antibiotin with a positive charge at pH = 7 binds to the SiNW. Upon addition of the pure buffer solution, the increase of conductance is associated with *m*-antibiotin dissociation.

9.2.2.5. *Bio-conjugated nanoparticles for rapid detection of single bacterial cell*[27]

Using fluorescent-bioconjugated silica nanoparticles, Tan *et al.*[27] have developed an ultrasensitive method for rapid detection of single bacterial cell quantization. Each nanoparticle encapsulates thousands of fluorescent dye molecules in a silicon matrix. Therefore, high fluorescent signal amplification was achieved when the antibody-conjugated and dye-doped silica nanoparticles bound to antigens on the surface of the bacteria. For a bacterium, there are many surface antigens available for specific recognition by using antibody-conjugated nanoparticles. Therefore thousands of nanoparticles can bind to each bacterium, thereby producing a greatly amplified signal. Using these nanoparticles, Tan *et al.* have developed an array tool, enabling the detection of one bacterium call per given sample in < 20 minutes.

The preparation process of fluorescent-bioconjugated silica nanoparticles are described as follows by Tan *et al.*[27] Firstly, RuBpy-doped silica nanoparticles were synthesized with a water-in-oil microemulsion method.[28] The dye molecules (RuBpy) were trapped inside a silica matrix to form the dye-doped nanoparticles. Then, mAbs against E. coli O157:H7 was bound to the nanoparticles. Before immobilizing mAbs onto the nanoparticles, the surfaces of the RuBpy-doped silica nanoparticles were chemically modified. To form the amine-functionalized group on the nanoparticle surfaces, 32 mg of silica nanoparticles was reacted with 20 ml of 1% trimethoxysilyl-propyldiethylenetriamine in 1 mM acetic acid for 30 minutes at room temperature, with continuous stirring. The amine-functionalized nanoparticles were obtained. They were washed three times in distilled water. After washing with N, N-dimethylformamide, the nanoparticles were reacted with 10% succinic anhydride in N, N-dimethyl formamide solution under N_2 gas for 6 h with continuous stirring. As a result, carboxyl groups were formed onto the silica nanoparticle surface for conjugation of antibodies. In an alternative nanoparticle synthesis method, carboxylated nanoparticles would be directly produced by adding a carboxylated siliane, N-(trimethoxysilylpropy)-ethylenediamine, during the postwashing of the silica nanoparticles. With stage at

4°C, the chemically-modified RuBpy-doped, silica-coated nanoparticles were viable for several months. After water-wash, the carboxylated nanoparticles were activated by using 5 ml of 100 mg/ml 1-ethy-3-3 (3-dimethylaminopropyl) carbodiimide hydrochloride and 5 ml of 100 mg/ml of N-hydroxy-succimimide in a 2-morpholino ethanesulfonic acid (Mes) buffer (pH 6.8), for 25 minutes at room temperature with continuous stirring. After being water-washed, nanoparticles were dispersed in 10 ml of 0.1 M PBS (pH 7.3). To covalently immobilize mAbs against *E.* coli O157 onto the nanoparticle surface, 5 ml of 0–1 μg/ml nanoparticles was reached with 2 ml of 5 μg/ml antibody for *E.* coli O157 for 2 ~ 4 h at room temperature with continuous stirring to form the resultant antibody-conjugated nanoparticles, followed by washing with a PBS buffer. Finally, the antibody-conjugated nanoparticles were reacted with 1% BSA and washed in 0.1 μ PBS (pH 7.3) before use, so that the effects of non-specific binding in the subsequent immunoarray were reduced.

The RuBpy-doped silica nanoparticles have high photostability. To verify this, Tan *et al.* compared pure RuBpy dye molecules with RuBpy-doped silica nanoparticles. Both of them were continuously irradiated with 450 nm of light for 1000 s. The fluorescence intensity of the pure dye molecules was reduced by 81%, whereas that of the nanoparticles remained constant. After biochemical modification of the nanoparticle surface, mAbs against the O-antigen of E. coli O157:H7 were covalently immobilized onto the nanoparticles. When stored at 4°C, the antibody-nanoparticle conjugates are viable for antigen recognition for up to 4 weeks, whereas when stored at −20°C, the nanoparticles are stable for several months.

Antibody-bioconjugated and RuBpy-doped silica nanoparticles exhibit high signal amplification when the antibody-conjugated nanoparticles bound to antigens on the surface of the bacteria. The mAb was highly selective for E. coli O157:H7 in the immunoassay because the antibody-conjugated nanoparticles specifically associated with the E. coli O157:H7 cell surface. The scanning electron microscope image of the E. coli O157:H7 cell after incubation with the nanoparticles shows that there were thousands of antibody-conjugated nanoparticles bound to a single bacterium, providing significant fluorescent signal amplification as compared with a single dye molecule. Therefore, the fluorescent image of E. coli O157:H7 after incubation with antibody-conjugated nanoparticles can be easily seen. After 20 minutes of continuous excitation, the fluorescence intensity remained constant. In solution-based experiments for bacteria detection, the result showed that signal amplification by the antibody-conjugated nanoparticles is > 1000 times greater than that produced with the dye molecule-labeled antibody. This result indicates that the high fluorescence signal enhancement by the nanoparticle-based antibody provides the foundation for the rapid detection of a single bacterium in solution samples. Tan *et al.* indicates that by using these antibody bioconjugated RuBpy-doped silica nanoparticle, one bacterium cell per given solution sample can be detected in less than 20 minutes with a spectrofluorometer.

9.2.2.6. *Near-infrared optical sensors based on single-walled carbon nanotubes*[28]

Carbon nanotubes fluorescence is in a region of the near infrared where human tissue and biological fluids are particularly transparent to their emission. Barone *et al.*[28] used carbon nanotubes as near-infrared optical sensors to measure blood glucose of patients by modulation fluorescent emission. In the following, we will introduce their work. The carbon nanotube optical sensor was prepared as follows. An ultrasonicated and purified solution of HiPCO (high pressure CO decomposition) carbon nanotubes was suspended according to a recently developed protocol except that 2 wt.% sodium cholate surfactant in TRIS (pH 7.4) buffered solution was used. The solution was dialysed against a surfactant-free buffer for 20 h in the presence of glucose oxidase (1 : 200 ratio with carbon atoms) to replace the surfactant with an immobilized, porous layer of protein at the surface. When single-walled carbon nanotubes are suspended in this manner, potassium ferricyanide, $K_3Fe(CN)_6$, irreversibly adsorbed on the surface. As a result, Barone *et al.* found that the adsorption of $K_3Fe(CN)_6$ shifted the Fermi levels into the valence bands, or acted to quench the emission after photo-excitation. The former mechanism is experimentally evident as an irreversible diminution of absorption transition that is mirrored in the band gap fluorescence of the nanotube. They observe that at 37°C, the emission fluorescence from the (6, 5) nanotube at 785 nm excitation decreases with increasing amounts of ferrocyanide, and the fluorescence decreases as increasing amounts of ferricyanide are added to the TRIS-buffered (pH 7.4) bath with the effect saturating at 225 nm and an attenuation of 83.3%. The ferrocyanide reduction product, ferrocyanide, reduces the signal by only 27.4% under identical conditions. This means that when ferricyanide is gradually reduced to ferrocyanide, the fluorescence from carbon nanotubes becomes gradually strong.

The suspension of glucose oxidase-suspended SWNT (single-walled carbon nanotube) after complete functionalization with $K_3Fe(CN)_6$ was loaded into a sealed 200 μm diameter dialysis capillary (13 KDa molecular weight cut-off) where the target analyte is free to diffuse across the capillary boundary which the nanotube sensing complex is retained. This capillary can be used as near-infrared optical sensors to measure blood glucose because it has several features. (i) This capillary is easily imaged in the near-infrared through a human epidermal tissue sample. (ii) Glucose oxidase at the surfaces of carbon nanotubes can catalyse the reaction of β-n-glucose to the D-glucose-1, 5- lactone with a H_2O_2 co-product. The H_2O_2 at 37°C and pH 7.4 can make partial reduction of the $Fe(CN)_6^{3-}$. Namely, ferricyanide is reduced to become ferrocyanide. At this time, the fluorescence from carbon nanotubes at excitation becomes strong. In other words, the fluorescence can respond to the glucose concentration because with a higher glucose concentration, more H_2O_2 is produced and more ferricyanide is reduced, resulting in fluorescence inhancement. Barone *et al.* proved experimentally this relation. Figure 9.21(a) gives the fluorescence relative intensity as the function of time. It can be observed that the fluorescence increases gradually after the addition of 1.4 mM, 2.4 mM and 4.2 mM glucose. Figure 9.21(b) shows the change of relative intensity with increasing glucose concentration. Clearly, the fluorescence relative intensity increases with the increase of the glucose concentration. Therefore, when the capillary is implanted into human tissue, the blood glucose enters the capillary to produce H_2O_2 under catalysis of glucose oxidase at the carbon nanotube surfaces and thus reduce ferricyanide at the surfaces, resulting in changes in the fluorescence intensity. Thus, we can measure the fluorescence relative intensity

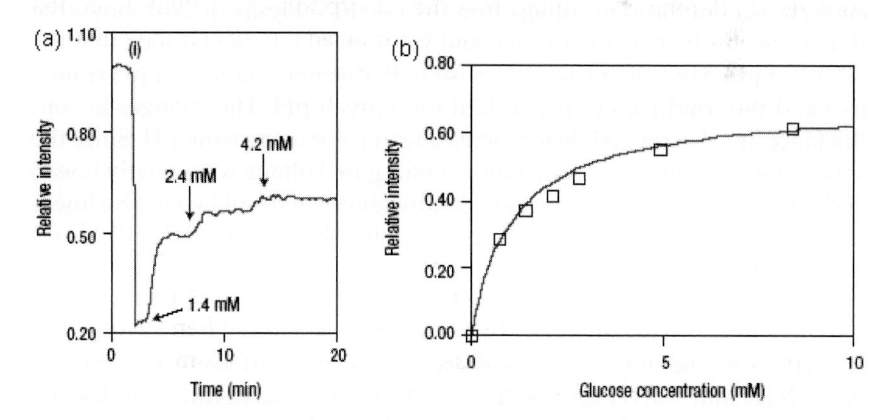

Fig. 9.21. (a) The fluorescence response to the addition of 1.4 mm, 2.4 mm and 4.2 mm glucose after the ferricyanide surface reaction at 37°C and pH buffered at 7.4. (b) The response function relates the normalized intensity to the local glucose concentration in the range of blood glucose detection with a type I absorption isotherm.[28]

and obtain the blood glucose concentration because the near-infrared fluorescence from the carbon nanotubes can penetrate human tissue.

9.2.3. Chemical sensors

Nanowire nanosensors for highly sensitive and selective detection of chemical species have been studied by many researchers. Here we will introduce the pH nanosensor reported by Lieber *et al.*[22] Lieber *et al.* connected a planar field effect transistor (FET) into a pH chemical nanosensor by modifying the gate oxide (without gate electrode) with molecular receptors for the analyte. Such chemically sensitive field effect transistors have an attractive feature such that binding of receptors with the analyte can be monitored by a direct change in the related electrical property. The structure of this pH nanosensor is that a silicon NW (SiNW) solid state FET, whose conductance is modulated by an applied gate, is transformed into a nanosensor by modifying the silicon oxide surface of the Si nanowire with 3-aminopropyltriethoxysilane (APTEs) to provide a surface that can undergo protonation and deprotonation, where a change in the surface charge can chemically gate the SiNW (Fig. 9.22(a)). The SiNWs are the single-crystal boron-doped (*p*-type) SiNWs. The preparation process of devices is the same as that described in 9.2.2.4. In solid state FET (insets, Fig. 9.22(b)), the conductance (dI/dV) measured in air at $V = 0$ as a function of time was stable for a given gate voltage and showed a stepwise increase with discrete changes of the gate voltage from 10 to − 10 V. Figure 9.22(b) shows the change of conductance with time and solution pH. It can be seen that the SiNW conductance increases stepwise with discrete changes in pH from 2 to 9 and the conductance is constant for a given pH. The changes in conductance are also reversible for increasing and/or decreasing pH. Since the solid state FET plots of conductance versus gate voltage were nearly linear, and for the pH nanosensor, plots of conductance versus pH were also linear over the pH 2 to 9 range, it was thus suggested that modified SiNWs could function as nanoscale pH sensors.

These results were analyzed by Lieber *et al.* as follows. Covalently linking APTEs to the SiNW oxide surface results in a surface terminating in both $-NH_2$ and –SiOH groups (Fig. 9.22(a)), which have different dissociation constants pKa (29, 30, 31). At low pH, the $-NH_2$ group is protonated to $-NH_3^+$ (30) and acts as a positive gate, which depletes hole carriers in the *p*-type SiNW and decreases the conductance. At high pH, –SiOH is deprotonated to $-SiO^-$, which correspondingly causes an increase in conductance. The observed linear response can be attributed to an approximately linear change in the total surface charge density (versus pH) because of the

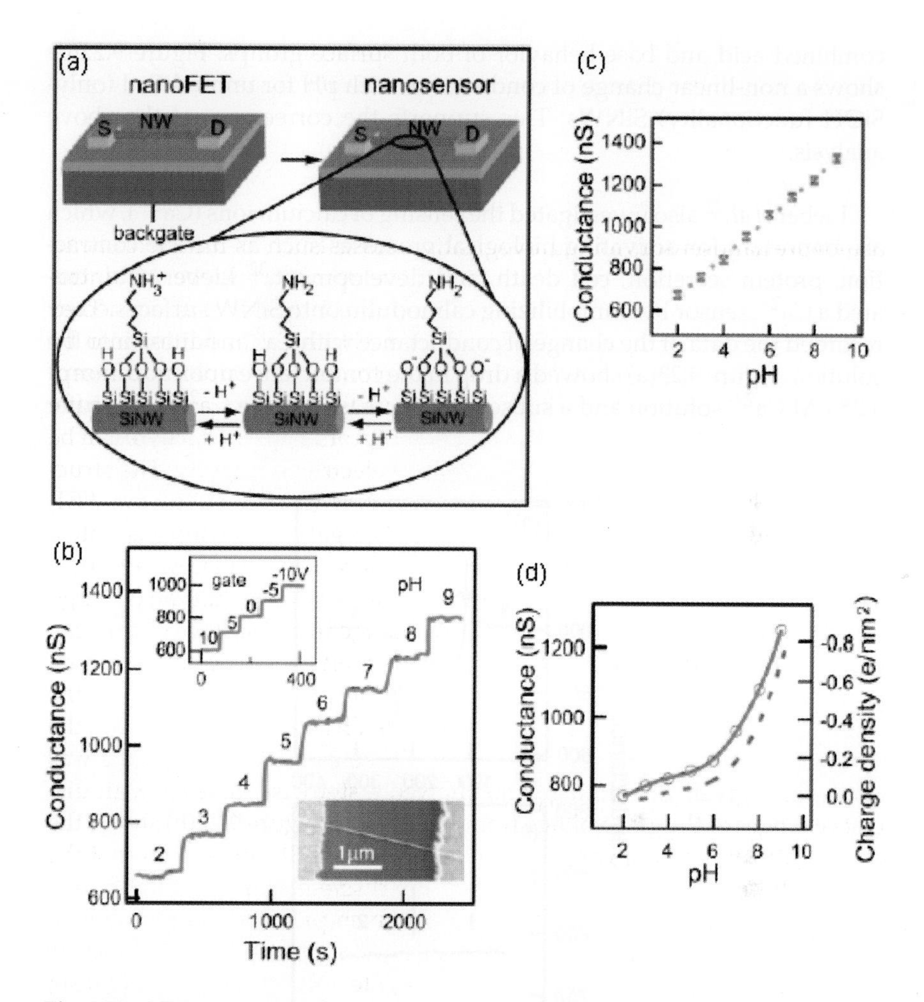

Fig. 9.22. NW nanosensor for pH detection. (a) Schematic illustrating the conversion of NW FET into NW nanosensors for pH sensing. The NW is contacted with two electrodes, a source (S) and drain (D), for measuring conductance. Zoom-in of the APTES-modified SiNW surface illustrates changes in the surface charge state with pH. (b) Real-time detection of the conductance for an APTES-modified SiNW for pHs from 2 to 9; the pH values are indicated on the conductance plot. (inset top) Plot of the time-dependent conductance of a SiNW FET as a function of the backgate voltage. (inset, bottom) Field-emission scanning electron microscopy image of a typical SiNW device (c) Plot of the conductance versus pH; the red points (error bars equal $\pm$ 1SD) are experimental data, and the dashed green line is the linear fit through this data. (d) The conductance of unmodified SiNW (red) versus pH. The dashed green curve is a plot of the surface charge density for silica as a function of pH (adapted from Ref. 22).

combined acid and base behavior of both surface groups. Figure 9.22(d) shows a non-linear change of conductance with pH for unmodified (only-SiOH functionality) SiNWs. This supports the correctness of the above analysis.

Lieber *et al.*[22] also investigated the sensing of calcium ions (Ca^{2+}), which are important for activating biological processes such as muscle contraction, protein secretion, cell death, and development.[32] Lieber *et al.* created a Ca^{2+} sensor by immobilizing calmodulin onto SiNW surfaces. They recorded the data of the change of conductance with Ca^{2+} addition into the solution. Figure 9.23(a) showed a drop in the conductance upon addition of a 25 μM Ca^{2+} solution and a subsequent increase when a Ca^{2+}-free buffer

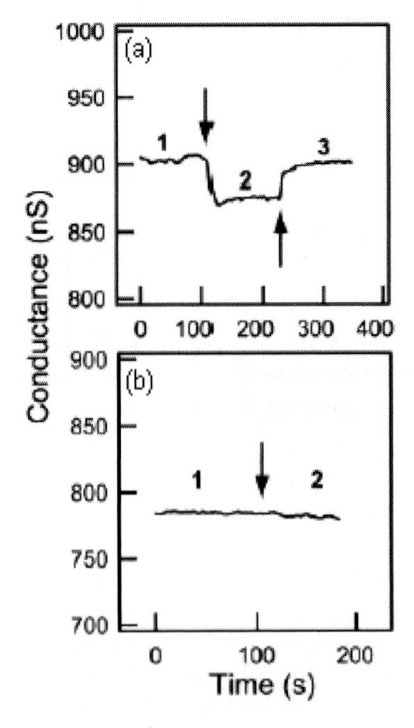

Fig. 9.23. Real-time detection of Ca^{2+} ions. (a) Plot of conductance versus time for a calmodulin-terminated SiNW, where region 1 corresponds to the buffer solution, region 2 corresponds to the addition of 25 μM Ca^{2+} solution, and region 3 corresponds to the pure buffer solution. (b) Conductance versus time for an unmodified SiNW; regions 1 and 2 are the same as in (a). Arrows mark the points when the solutions were changed. Calmodulin-modified NWs were prepared by placing a drop ($\sim$20 μl) of calmodulin solution (250 μg/ml) on SiNW for 1 h and then rinsing with water for three times.[22]

was subsequently followed through the device. For unmodified SiNW there was not a conductance change when Ca^{2+} was added. Thus, this demonstrates that the calmodulin receptor is essential for detection. In addition, the decrease in observed conductance in modified SiNWs is consistent with chemical gating by positive Ca^{2+}.

9.3 Field emission of carbon nanotubes and its application

Field emission of nanomaterials would be potential candidates for the use in cold cathodes, flat panel displays, electron guns, electron sources, and nanodevices. The field-emission characteristics of III-V and IV-VI compound wide band gap semiconductors have been widely investigated, such as GaN pyramids,[33] BN/GaN etched films,[34] AlN nanotubes,[35] SiC nanowires,[36] Cu_2S nanowires,[37] etc. Another class of wide band gap semiconductors worthy of investigation is that of chemically inert conductive oxides, such as indium, tin, molybdenum, and chromium oxides.[38,39] Carbon nanotubes (CNTs) are considered one of the most promising materials for field emission cathodes in vacuum electronic devices because of their high aspect ratio, high current-carrying capacity, robust structure and chemical inertness. Therefore, in the next paragraph, field emission of carbon nanotubes and its application will mainly be introduced.

Field emission characteristics from carbon nanotube (CNT) arrays are attracting much interest because CNT has many advantages which makes it the most promising material for the cold cathode in the next generation of high performance electronic devices, in which a large area electron emission with a low operation voltage is highly desired. But the field emission current of the CNT array is low and the threshold electric field is high as a result of screening effects. Therefore, researchers try to improve these properties using different methods. As we know, the surface treatment can induce a change in the field emission characteristics of CNT films, especially the threshold electric field and the emission site density. Several treatment methods have been put forward, such as mechanical surface treatment,[40,41] focused ion beam irradiation,[42,43] plasma exposure,[44–46] and laser irradiation.[47,48] Recently, Zhu *et al.*[49] presented the self-assembly of three-dimensional (3D) micropatterns on aligned carbon nanotube films through a water-spreading method. Chakrapani *et al.*[50] also reported two-dimensional contiguous cellular foams coming from dense carbon nanotube arrays as a result of capillary forces arising during the evaporation of liquids. These self-assembly structures could be promising when used to decrease the screening effect in the course of electron field emission and improve the field emission properties of CNT arrays. Here we will introduce the work on field emission properties of CNT arrays reported by Fan *et al.*[51] They put CNT arrays with chemical vapor deposited (CVD) into ethanol to make shrunk structures with separate nanotube "walls" for better field emission properties. Such structures decrease the screening effects and reduced the turn-on electric field at 10 $\mu A/cm^2$ from 1.68 to 1.23 $V/\mu m$. The field enhancement factor was calculated to increase by 23% according to Fowler–Nordheim (F–N) equation. The number of emission sites also increased and their distribution was more uniform.

Fan *et al.*[52] also investgated the effect of plasma treatment on field emission properties of CNT arrays. For three plasma treatment conditions of CNY arrays, they obtained three different samples (samples A, B and C). For sample A, only Ar plasma is used to etch the CNT array surface, while Ar/O_2 plasma with a volume ratio of 1 : 1 are used to etch the surface for sample B and C. Respectively, for the three samples, the vacuum pressure is 2 Pa, 1 Pa, and 2 Pa; the sputtering current is 80 mA, 40 mA, and 80 mA; and the treatment time is 300 s, 60 s, and 300 s. The surface of the original CNT arrays is flat and the carbon nanotubes on the surface are just disorderly tangled. After the plasma treatment, the surface is not flat anymore and a novel bundle structure is obtained. Furthermore, the morphologies of samples A, B and C are quite different. In sample A, no carbon nanotube can be found in each bundle. However, in the other two samples, the bundle contains both carbon nanotubes and amorphous carbon or carbon nanotubes only. The bundles are much sharper in sample C than those in sample B, and the distance between two bundles is closer in sample C.

The field emission properties of samples A, B and C before and after Ar/O_2 plasma etching are compared. After plasma treatment, not all the field emission properties of samples are improved. Sample A does not emit electrons even at the highest voltage of 1100 V. The field emission current of sample B is greatly decreased. A current density decreases by a factor of about 1.8 from 52 to 28 mA/cm^2 in an electric field of 5.5 $V/\mu m$. The threshold field, at which emission current density reaches 10 mA/cm^2, increases from 4.1 to 4.7 $V/\mu m$. Only in sample C, enhanced field emission ability is observed with the threshold field decreasing from 3.9 to 3.6 $V/\mu m$ and the largest emission current density achieved greatly increases from 50 to 107 mA/cm^2 in an electric field of 5.5 $V/\mu m$. Fan *et al.* attributed the enhanced field emission properties to the electrostatic screen effect being decreased.

Fan *et al.*[53] found that the field emission from the carbon nanotube can be affected by common carbon-containing residual gases in vacuum systems.

Exposure to CO and CO_2 at 10^{-5} Pa can reduce the emission current from 22% to 49%, depending on the specific partial pressure and exposure time. Such a reduction can be fully recorded by continuous emission under a high vacuum of 10^{-6} Pa. Exposure to CH_4, and C_2H_4 can increase the current, but the current stability is poor, and after the exposure, the current does not recover.

Fan *et al.*[54] investigated field emission properties of carbon nanotubes grown on silicon nanowire arrays. The silicon nanowires are highly single crystalline with lengths up to 100 μm and diameters ranging from 50 to 500 nm. The field emission properties of silicon nanowire arrays is very low. For the field emission properties of the carbon nanotube grown on silicon nanowire arrays, the low turn on the field (in which the current density is 10 μA/cm^2) and the threshold field (in which the current density is 10 mA/cm^2) are 2.3 and 3.2 V/μm, respectively. The field emission current densities become as large as 70 mA/cm^2 in an electric field of 5 V/μm.

In the following, several applications of the field effect (FE) from carbon nanotubes (CNTs) which were reported by Robertson[55] will be simply introduced.

A clear application of FE from CNTs is in electron guns for the next generation scanning electron microscopes (SEM) and transmission electron microscopes (TEM). In low voltage SEM, an electron source with a small energy spread-in is needed. A second critical parameter is the reduced brightness of the source, the current density per unit solid angle. The brightness influences the spatial resolution of an electron microscope and the amount of current in a certain probe size. Present day microscopes use Schottky emitters of doped Si or metal FE tips. A single MWNT FE source is found to have a factor of 30 higher brightness than existing electron sources and a small energy width of 0.25 eV. MWNTs are preferred to SWNTs because of their greater mechanical stiffness.

A second use of CNTs is in FE cathodes in high power microwave amplifiers such as klystrons for base stations. A klystron would have a CNT FE cathode with a microwave signal applied to the gate. This configuration is not possible with a conventional thermionic cathode, but the absence of heat from a CNT cold cathode allows a short cathode-gate distance to be used. Otherwise, the microwave signal must be introduced after a grid, as in a traveling wave tube, which makes the device much larger. This is another high-value niche where CNT sources are viable if they reach performance objectives.

A third possible application is in the electron source for miniature X-ray sources. Here, an electron beam from a CNT cathode is accelerated to strike a metal target to create X-rays. A small, high-brightness device with a pulsed electron beam allows real-time imaging. For this application, it is necessary to compete against existing X-ray source and three-dimensional tomographic imaging techniques.

For these applications, the CNTs should operate at the highest current density. In microwave and X-ray sources, a multi-nanotube source is used. Otherwise the field enhancement is reduced by screening from adjacent nanotubes.

A fourth application is in the FE displays (FEDs). This application has been a huge attraction because the total display market is $50 billion per annum. Now, the display market is dominated by active matrix liquid crystal displays (AMLCDs). The FED is a flat panel display in which the electron beam from many nanotube electron guns are controlled to create an image on a screen of phosphor pixels. The advantages of FEDs over LCDs are the higher video, and wider operating temperature range.

A fifth application is in the sadde-field ionization vacuum gauge with a carbon nanotube field emission cathode.[56] The gauge's cathode was prepared by adhering the as-grown carbon nanotube array to one end of the nickel rod. This nanotube cathode sadde-field gauge has the advantage of small electrode sizes, high sensitivity and low power consumption. The power consumption of the gauge is less than 8 mW. The gauge has a great potential in pressure measurement in the UHV/XHV (ultrahigh vacuum/extreme high vacuum) region and in sealed devices.

The FE applications likely to succeed are those where cost is not a key issue, such as electron guns for SEMs.

9.4 Light polarization

Some previous results showed that one-dimensional nanotubes or nanowires have different light absorption ability for light with different polarization directions.[57,58]

In an optical study, some researchers found that some nanowires or nanotubes, such as carbon nanotues (CNTs), InP nanowires etc. demonstrated giant polarization anisotropy. Namely, the light was not absorbed when the polarization direction was perpendicular to the axis of the nanowire or the nanotube, but the light was greatly absorbed when the polarization direction was parallel to the axis of the nanowire or the nanotube. Fan *et al.*[59] show a new finding that the CNT array can be self-assembled into yarns by simply pulling from super-aligned CNY arrays. Then, they made the CNT polarizer, which was implemented by parallel aligning a large multitude of yarns on a glass plate, as shown in Fig. 24(a). When they observed the polarized light source through this polarizer, with rotation of the polarizer to alter the angle, θ, the angle formed between the laser polarization direction and the polarization direction of the polarizer. The transmittance light intensity, I, exhibited a periodic ($\cos^2\theta$) dependence on the angle, θ, (Figs. 9.24(a)–9.24(d)). They proved experimentally that this polarizer can work in the ultraviolet region. This is related to the CNT yarns of the

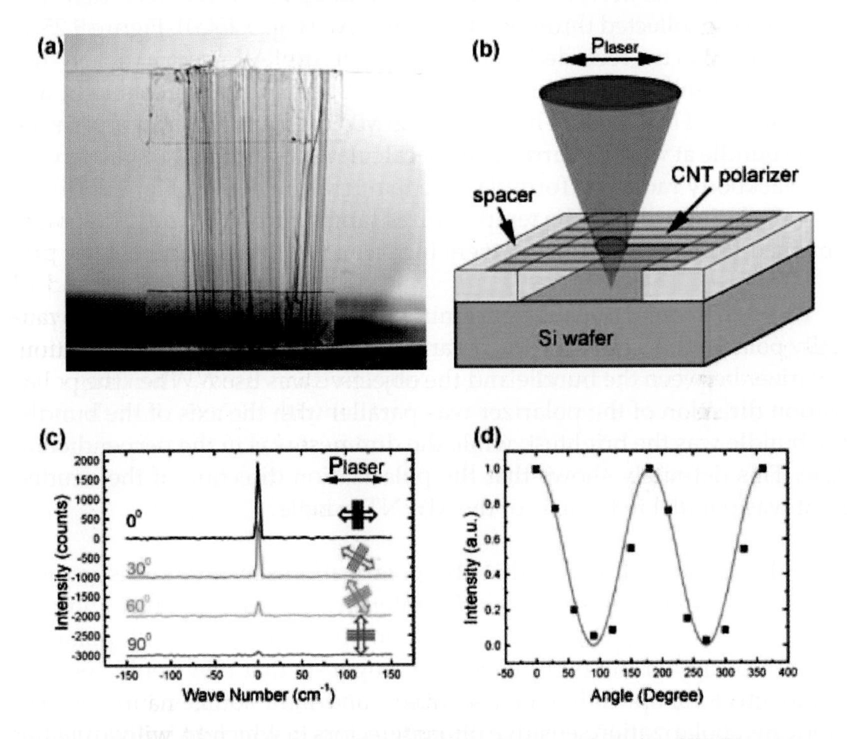

Fig. 9.24. CNT polarizer. (a) Parallel aligning a large multitude of CNT yarns on a glass plate; (b) the schema of polarization property measurement of the polarizer; (c) the change of the laser intensity with the θ. The solid arrow is the laser polarization direction. The hollow arrow is the polarization direction of the polarizer, θ is the angle formed between the laser polarization direction and the polarization direction of the polarizer. The abscissa is the wavenumber relative to 325 nm. The ordinate is the photon count. (d) Normalized laser intensity I/I_{max} versus θ.

polarizers, having the 10 nm diameter. The CNT polarizer could even be used for ultraviolet light of wavelength in the region of tens of nanometers.

To investigate the resistance heating light emission of multi-walled carbon nanotube (MWNT) bundles, Fan *et al.*[60] incorporated a MWNT bundle into a two terminal device on a silicon substrate with a thick oxide layer, on which a 40 nm thick Au/Ti electrode was deposited using lithography. An Au as-grown MWNT bundle was transferred from the MWNT array to electrodes under an optical microscope with a three-dimensional manipular. To make a good contact, they directly pressed the Au wires on the MWNT bundle and bonded them to the gold electrodes by an ultrasonic wire bonder. The device was placed in a vacuum chamber, which was pumped to a vacuum of 1×10^{-4} Pa, then the current was applied and the light emission spectra of the MWNT bundle at different currents were recorded. The spectra were collected through a 10 × objective (Fig. 9.25(a)). Figure 9.25(b) shows the spectra recorded at different currents. All spectra closely followed the plank blackbody radiation distribution, indicating that light was really induced by the joule heating of the MWNT bundle. The temperature of the bundle at various current can be calculated by fitting the spectra with the blackbody radiation formula. The temperature data in Fig. 9.25(b) are the calculating values. The temperatures ranging from 900 to 1200 K were obtained from the spectra and were increasing with the current.

Fan *et al.*[60] found that the light emitted from the MWNT bundle was partially polarized. To examine the polarization of the emitted light, a rotating polarizer between the bundle and the objective was used. When the polarization direction of the polarizer was parallel with the axis of the bundle, the bundle was the brightest, while the dimmest was in the perpendicular case. This definitely shows that the polarization direction of the emitted light was parallel to the axis of the MWNT bundle.

Lieber *et al.*[61] observed that the polarization-sensitive measurement revealed a striking anisotropy in PL intensity recorded parallel and perpendicular to the long axis of a InP nanowire. The extreme PL polarization anisotropy of these InP nanowires suggests that they could serve as photodetectors, optically gated switches, and light sources. Liber *et al.*[60] fabricated polarization sensitive photodetectors in which an individual InP nanowire serves as the device element (Fig. 9.26(a)). Nanowires were dispersed onto silicon substrates, and electrical contacts were defined at the nanowire ends with the use of electron beam lithography and followed by thermal evaporation of the metal electrodes. The nanowire devices were then placed in the epifluorescence microsope used for the PL image, and

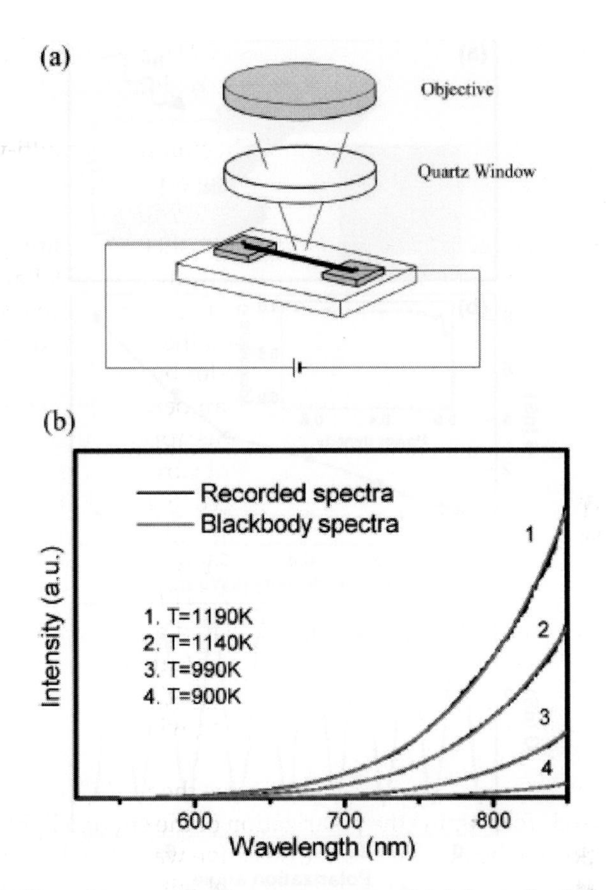

Fig. 9.25. CNT spectra recorded by spectrometer. (a) Schematic diagram of light emitting and spectra recording. (b) Light emission spectra of MWNT bundle at different currents. The fitted temperatures of four curves are given on the left.[60]

the change in conductance (G) of nanowires was measured as a function of the laser intensity and polarization.

In general, the conductance of individual InP nanowires increased by two to three orders of magnitude with increasing excitation power density. The photoconductivity (PC) is reproducible and reversible with respect to the change in excitation power. The PL also shows a striking polarization anisotropy with parallel excitation, producing G that is over an order of magnitude larger than the perpendicular excitation (Fig. 9.26(b)). The photoconductivity anisotropy ratio, $\sigma = (G_{\parallel} - G_{\perp})/(G_{\parallel} + G_{\perp})$, where $G_{\parallel}(G_{\perp})$ is the conductance with parallel (perpendicular) excitation, is 0.96

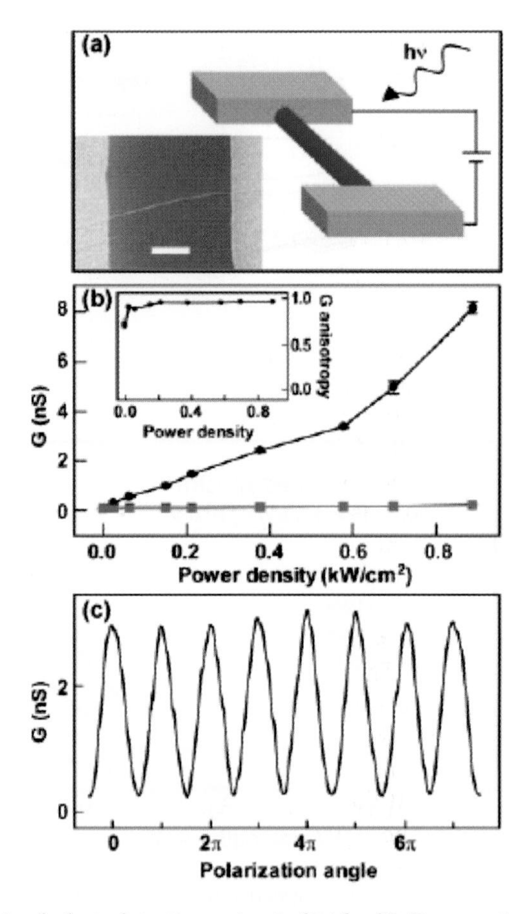

Fig. 9.26. Polarized photodetection using individual InP nanowires. (a) Schematic depicting the use of a nanowire as a photodetector by measuring the change in PC as a function of incident light intensity and polarization. Inset: field-emission scanning electron microscopy image of a 20 nm diameter nanowire and contact electrodes for PC measurements. Scale bar is 2 μm. Nanowires were first dispersed in ethanol and then deposited onto silicon substrates (600 nm oxide, 1 to 10 ohm cm resistivity). Electrical contacts to the wires were defined using electron beam lithography, and Ni/In/Au contact electrodes were thermally evaporated. (b) Conductance, G, versus excitation power density. We observe the PC response when the illumination is polarized parallel (black) and perpendicular (red) to the wire. Inset: PC anisotropy, σ, versus excitation power calculated from (b). The measured anisotropy for the shown device is 0.96. (c) Conductance versus polarization angle as the polarization was manually rotated while measuring the PC. All PC measurements were done at room temperature. Current collected at drain electrode was measured using standard lock-in techniques, with an excitation voltage of 50 mV at 31 Hz. No gate voltage was applied. An excitation wavelength of 514.5 nm was used for these measurements.[61]

for the figure (Fig. 9.26(b), inset). This is in excellent agreement with the polarization ratio measured from PL. The reproducibility of the PC polarization response is explicitly seen in plots of conductance, recorded as the excitation polarization vector is continuously rotated (Fig. 9.26(c)).

9.5 Light-bulb filaments made of carbon nanotube yarns[62]

While attempting to pull out a bundle of carbon nanotubes (CNTs) from a CNT array several hundred micrometers high and grown on a silicon substrate, Fan *et al.*[62] obtained instead a continuous yarn of CNTs. They estimated that an array area of roughly one cm^2 can generate 10 m of yarn. Also, they found that continuous yarns can only be drawn out from super-aligned arrays in which the CNTs are aligned parallel to one another and are held together by Van der Waals interactions to form bundles. The yarns usually appear as thin ribbons composed of parallel threads that have diameters in the range of several hundreds of nanometers, with the width of the yarn roughly depending on the number of threads in the yarn.

Fan *et al.* constructed a light bulb filament by winding a CNT yarn between two metal leads. The filament emits incandescent light when a DC voltage is applied in a vacuum of 5×10^{-3} Pa. After emitting light for 3 h at 70 V, the conductivity of the filament increases by 13% and the tenside strength changes from 1 mN to 6.4 mN. This indicates that some welding effect may be occurring at the weak connection points of the CNT during light emission, because these points have higher resistivity and as a result, a higher temperature when a current is applied.

9.6 Electronic and optoelectronic nanoscale devices

Nanowires and nanotubes carry charge and excitons efficiently, and are therefore potentially ideal building blocks for nanoscale electronics and optoelectronics.[62,63] Carbon nanotubes (CNTs) have already been exploited in devices such as the field-effect[65–68] and single-electron transistors.[69,70]

CNTs can carry the highest current density of any metal, 10^9–10^{10} A/cm^3, over 1000 times that for Cu, before failing as a result of electromigration.[55] It is well-known that the feature sizes in Si integrated circuits are continually reducing. This forces the interconnects between each transistor to carry increasingly large current densities. The vertical interconnects as vias could be replaced by CNTs and thus satisfy this requirement. This application is relatively easy, particularly since it requires vertical nanotube growth. Multiple CNTs in parallel can be used to lower the overall resistance. However, the greatest demand is not for vias, but for horizontal interconnects. The horizontally-directed nanotube growth can be done, but not with high reliability and yield.

The first generation field effect transistor (FETs) of CNTs were made by dispersing SWNTs (single-walled carbon nanotubes) on SiO_2-covered Si wafers and then making contacts. The Si substrate is used as the bottom gate electrode. This is not suitable for a real device, as the gate would be common to all FETs on the wafer. Robertson indicates that the second generation FET requires a separate 'top gate' over each individual nanotube. Wind *et al.*[67] have achieved FETs with very good performance figures. Javey *et al.*[68] subsequently used a ZrO_2 gate to obtain the best gate turn-on performance of any nanotube device.

There are two disadvantages for using CNTs as components of FETs. One is that they are expensive and the second huge disadvantage is that the practical utility of CNT components for building electronic circuits is limited, as it is not yet possible to selectively grow semiconducting or metallic nanotubes.[71,72] Therefore, many researchers began to use other nanowires or nanotubes to construct FETs. Lieber *et al.*[73] reported the assembly of functional nanoscale devices from InP nanowires, the electrical properties of which are controlled by selective doping. Firstly, we introduce the construction and electrical transport of nanoscale InP field-effect transistors. In order to construct these FETs, they prepared single-crystal Inp nanowires with *n*- and *p*-type doping by laser-assisted catalytic growth (lCG).[74] To confirm the presence and type of dopants in nanowires, they have performed gate-dependent, two-dimensional transport measurements on individual wires. The figure 9.27 shows the typical gate-dependent current–voltage (I–V) curves obtained from individual Te- and Zn-doped nanowires, respectively. The curves are linear or nearly linear at $V_g = 0$, indicating that the metal electrodes make ohmic contact to the nanowires. For Te-doped nanowires, the transport data show an increase in conductance for $V_g > 0$, whereas the conductance decreases for $V_g < 0$. These data clearly indicate that Te-doped InP nanowires are *n*-type. However, for Zn-doped nanowires, opposite changes occur in conductance with variation in V_g: for $V_g > 0$, conductance decreases and for $V_g < 0$, conductance increases (Fig. 9.27(b)). These results indicate that Zn-doped InP nanowires are *p*-type.

Figures 9.27(a) and 9.27(b) demonstrate gate effects in *p*-type and *n*-type InP nanowire cases. V_g can be used to completely deplete electrons and holes in *n*- and *p*-type nanowires such that the conductance becomes immeasurably small. For example, the conductance of the nanowires can be

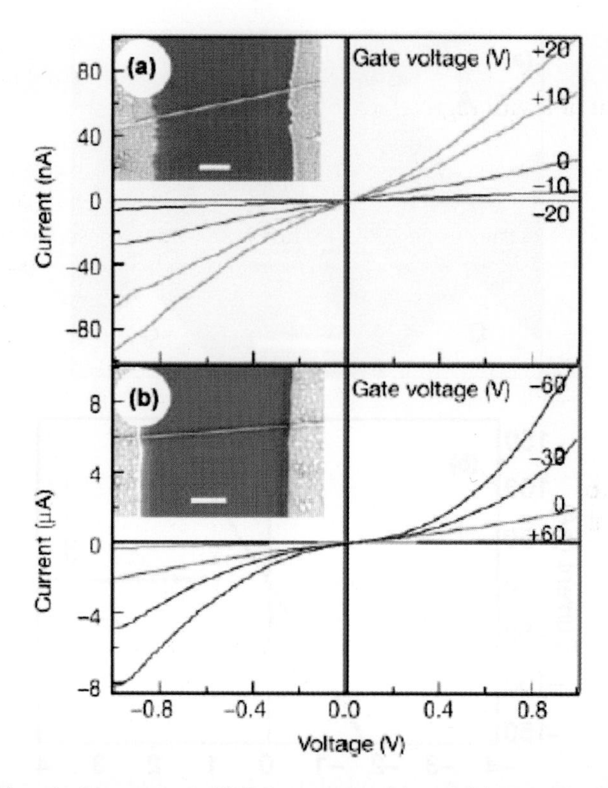

Fig. 9.27. Electrical transport of InP nanowires. (a) and (b) Gate-dependent $I \pm V$ behavior for Te- and Zn-doped NWs, respectively. Insets show the nanowire measured with two-terminal Ni/In/Au contact electrodes. Scale bars are 1 mm. The diameter of the nanowire in (b) is 47 nm, while that in (c) is 45 nm. Gate voltages used in the measurements are indicated on the corresponding I–V curves (right side). Data were recorded at room temperature.[73]

switched from a conducting (on) to an insulating (off) when $V_g \leq -20$ V. Thus it functions as a FET. This gate-dependent behavior is similar to that of metal-oxide-semicondutor FETs and recent semiconducting nanotube FETs.

In the following, we will introduce the research results on the electrical transport behavior of *p-n* junctions reported by Lieber *et al.*[73] These junctions were made by sequential deposition of dilute solutions of *n*- and *p*-type InP nanowires with intermediate drying. Each *p-n* junction was formed by crossing one *n*-type and one *p*-type nanowire, as shown in Fig. 9.28(a). The transport data recorded on the individual nanowires

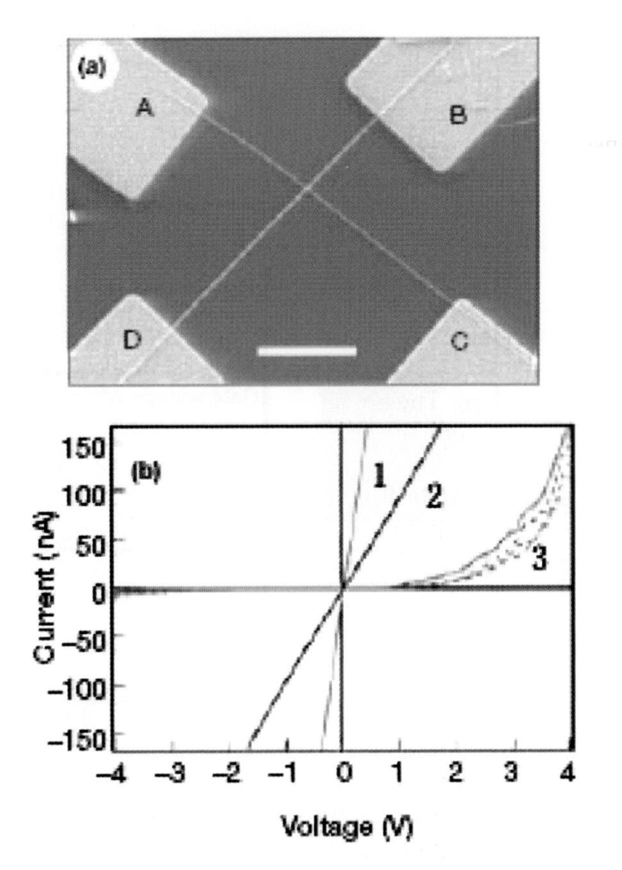

Fig. 9.28. Crossed nanowire junctions and electrical properties. (a) FE-SEM image of a typical crossed nanowire device with Ni/In/Au contact electrodes. The four arms are designed as A, B, C, D. Scale bar is 2 μm. The diameters of the nanowires are 29 nm (AC) and 40 nm (BD). (b) Curves 1 and 2 correspond to I–V for the individual n- and p-type nanowire components, respectively. Curve 3 corresponds to I–V for the p-n junction.[73]

(AC, BD) show linear or nearly linear I–V behavior. These results indicate that the metal electrodes make ohmic or nearly ohmic contact to the nanowires and hence will not make significant contributions to the I–V measurements across junctions. Lieber *et al.* found that crossed nanowires make reasonably good electrical contact with each other. This suggests that functional devices should be possible. Therefore, they have explored p-n junctions made from crossed p- and n-type InP nanowires. Typical I–V data recorded on a crossed p-n junction (Fig. 9.28(b)) shows linear or nearly linearly I–V for the individual n- and p-type nanowire components (curves

1 and 2 respectively). This indicates that they are chemically-contacted, and there exists current rectification across the *p-n* junction (curve 3). That means that little current flows in reverse bias, whereas there is a sharp current onset in forward bias. This diode-like behavior is similar to bulk semiconductor *p-n* junctions, which form the bias for many critical electric and optoelectric devices. For a standard *p-n* junction, rectification arises from the potential barrier at the interface between *p*- and *n*-type materials. In the InP nanowire *p-n* junction case, due to a thin oxide at the *p-n* junction interface, when either type of junction is forward biased, the barrier is reduced and a relatively large current can flow through the junction. On the other hand, only a small current can flow in reverse bias as the barrier is further increased. The above data show that Lieber *et al.* assembled crossed-wire *p-n* junctions that exhibit rectificating behavior.

Lieber *et al.* have made the smallest light-emitting diodes, in which the InP *p-n* junctions emit light strongly. While studying these nanowires, they studied the photoluminescence (PL) and electroluminescence (EL) from crossed nanowire *p-n* junctions. The experimental data show that the light originates from the nanowire *p-n* junction. The EL spectrum shows a maximum intensity around 820 nm, which is substantially blue-shifted relative to the bulk band gap of InP (925 nm). This suggests that these shifts may be due in part to the quantum confinement of the excitons. The electron to photon quantum efficiency of these initial devices is relatively low, at about 0.001%. The efficiency is comparable to that of early bulk InP LEDs ($\sim 0.002\%$). Lieber *et al.* attributed the low quantum efficiency to non-radiative recombination via surface states.

Lieber *et al.* pointed out that the electric-field-directed assembly can be used to create lightly integrated device arrays from nanowire building blocks.[73]

Bibliography

1. A. D. Kolmakov, Y. X. Zhang, G. S. Cheng, and M. T. Moskovits, *Adv. Mater.* **15**, 997 (2003).

2. E. Comini, G. Faglia, G. Sberveglieri, D. Calestani, L. Zanotti, and M. Zha, *Sens. Actuators* **B111–112**, 2 (2005).

3. J. Tamaki, N. Nagaishi, Y. Teraoka, N. Miara, and Y. Yamazoe, *Surf. Sci.* **221**, 283 (1999).

4. F. Solymosi and J. Kiss, *J. Catal.* **41**, 202 (1976).

5. G. Ghiotti, A. Chiorino, W. X. Pan, and L. Marchese, *Sens. Actuators* **B7**, 691 (1992).
6. H. Idriss and E. G. Seebauer, *J. Mol. Catal. A: Chem.* **152**, 201 (2000).
7. F. Q. Sun, W. P. Cai, Y. Li, L. C. Jia, and F. Lu, *Adv. Mater.* **17**, 2872 (2005).
8. Such sensors cannot be constructed from 3D colloidal crystals for three obvious reasons: (i) if there are too many layers, the 3D colloidal crystal cannot be lifted off from the flat substrate; (ii) even if colloidal crystals with a few layers can be transferred onto ceramic tubes, the porous films will produce many breaks during the heat-treatment process; (iii) the number of the layers cannot be controlled, and therefore the thickness of the gas-sensing film will not be controllable.
9. T. Sumida, Y. Wada, T. Kitamura, and S. Yanagida, *Langmuir* **18**, 3886 (2002).
10. X. J. Huang, F. L. Meng, Z. X. Pi, W. H. Xu, and J. H. Li, *Sens. Actuators* **B99**, 444 (2004).
11. S. C. Chang, in *Proc. of the Int. Meeting on Chemical Sensors*, Elsevier, Amsterdam, 1983, pp. 78–83.
12. Y. Zhao, Z. C. Feng, and Y. Liang, *Sens. Actuators* **B56**, 224 (1999).
13. Z. Ying, Q. Wan, Z. T. Song, and S. L. Feng, *Nanotechnology* **15**, 1682 (2004).
14. F. Patolsky, G. Zheng, O. Hayden, M. Lakadamyali, X. W. Zhuang, and C. M. Lieber, *PNAS* **101**, 14017 (2004).
15. S. Jin, D. M. Wang, M. C. McAlpine, R. S. Friedman, Y. Wu, and C. M. Lieber, *Nano Lett.* **4**, 915 (2004).
16. Y. Cui, Q. Wei, H. Park, and C. M. Lieber, *Science* **293**, 1289 (2001).
17. B. Ilic, Y. Yang, and H. G. Craighead, *Appl. Phys. Lett.* **85**, 2604 (2004).
18. S. H. Chung, A. Hoffmann, S. D. Bader, C. Liu, B. Kay, L. Makowski, and L. Chen, *Appl. Phys. Lett.* **85**, 2971 (2004).
19. Q. A. Pankhurst, J. Connolly, S. K. Jones, and J. Dobson, *J. Phys.* **D36**, R167 (2003).
20. M. I. Shliomis, *Sov. Phys. Usp.* **17**, 153 (1974).
21. J. Connolly and T. G. St. Pierre, *J. Magn. Magn. Mater.* **225**, 156 (2001).
22. Y. Cui, Q. Q. Wei, H. K. Park, and C. M. Lieber, *Science* **293**, 1289 (2001).

23. V. Huang, X. Duan, Q. Wei, and C. M. Lieber, *Science* **291**, 891 (2001).

24. Y. Cui, X. Duan, J. Hu, and C. M. Lieber, *Science* **104**, 5213 (2000).

25. Y. Cui and C. M. Lieber, *Science* **291**, 851 (2001).

26. SiNWs with diameters of either 10 or 20 nm were suspended in ethanol flow aligned on oxidized Si substrates (1 to 10 ohm cm, 600 nm oxide; Silicon Sense), and contact leads (50 nm Al or Ti + 100 nm Au) were defined with electron-beam lithography. The separation between contacts was typically 2 to 4 μm. The conductance of SiNW devices as a function of time was determined directly with a computerized apparatus with lock-in amplifier (Stanford Research, SR 830); a 31 Hz sine wave with 30 mV amplitude at zero DC bias was used in most measurements. The conductances of the SiNW devices were between 500 and 2000 ns (resistance: 2 megohms to 500 kilohms). This relatively small range testifies to the good control of doping in our NWs.

27. X. J. Zhao, L. R. Hilliard, S. J. Mechery, Y. P. Wang, R. P. Bagwl, S. G. Jin, W. H. Tan, *PNAS* **101**, 15027 (2004).

28. P. W. Barone, S. H. Baik, D. A. Heller, and M. S. Strano, *Nature Materials* **4**, 86 (2005).

29. R. K. Iler, *The Chemistry of Silicon*, Wiley, New York, 1979.

30. D. V. Vezenov, A. Noy, L. F. Rozsnyai, and C. M. Lieber, *J. Am. Chem. Soc.* **119**, 2006 (1997).

31. G. H. Bolt, *J. Phys. Chem.* **61**, 116 (1957).

32. C. B. Klee and T. C. Vanaman, *Adv. Protein Chem.* **35**, 213 (1982).

33. B. L. Ward, O. H. Nam, J. D. Hartman, S. L. English, B. L. McCArson, R. Schlesser, Z. Sitar, R. F. Davis, and R. J. Nemanicha, *J. Appl. Phys.* **84**, 5238 (1998).

34. C. Kimura, T. Yamamoto, T. Hori, and T. Suginoa, *Appl. Phys. Lett.* **79**, 4533 (2001).

35. V. N. Tondare, C. Balasubramanian, S. V. Shenda, D. S. Joag, V. P. Godbole, and S. V. Bhoraskara, *Appl. Phys. Lett.* **80**, 4813 (2002).

36. Z. Pan, H. L. Lai, F. C. K. Au, X. Duan, W. Zhou, W. Shi, N. Wang, C. S. Lee, N. B. Wang, S. T. Lee, and S. Xie, *Adv. Mater.* (Weinheim, Germany) **12**, 1182 (2000).

37. J. Chen, S. Z. Deng, N. S. Xu, S. Wang, X. Wen, S. Yang, C. Yang, J. Wang, and W. Ge, *Appl. Phys. Lett.* **80**, 3620 (2002).

38. Y. B. Li, Y. Bando, D. Golberg, and K. Kurashima, *Appl. Phys. Lett.* **81**, 5048 (2002).

39. H. B. Jia, Y. Zhang, X. H. Chem, J. Shu, X. H. Luo, Z. S. Zhang, and D. P. Yu, *Appl. Phys. Lett.* **82**, 4146 (2003).

40. T. J. Vink, M. Gillies, J. C. Kerege, and H. W. J. J. van de Laar, *Appl. Phys. Lett.* **83**, 3552 (2003).

41. K. B. Kim, Y. H. Song, C. S. Huwang, C. H. Chung, J. H. Lee, I. S. Choi, and J. H. Park, *J. Vac. Sci. Technol.* **22**, 1331 (2004).

42. A. Sawada, M. Iriguchi, W. J. Zhao, C. Ochiai, and M. Takai, in *Proceedings of the 14th International Vacuum Microelectronics Conference*, University of California, Davis, California, 2001, p. 29.

43. D. H. Kim, C. D. Kim, and H. R. Lee, *Carbon* **42**, 1807 (2004).

44. S. H. Choi, J. H. Han, T. Y. Lee, J. B. Yoo, C. Y. Park, T. W. Jung, S. G. Yu, W. K. Yi, H. J. Kim, N. S. Lee, and J. M. S. Kim, in *Proceedings of the 14th International Vacuum Microelectronics Conference*, University of California, Davis, California, 2001, p. 35.

45. Y. Kanazawa, T. Oyama, K. Murakami, and M. Takai, *J. Vac. Sci. Technol.* **B22**, 1342 (2004).

46. Y. M. Liu, L. Liu, P. Liu, L. M. Sheng, and S. S. Fan, *Diam. Relat. Mater.* **13**, 1609 (2004).

47. W. J. Zhao, A. Sawada, and M. Takai, *Appl. Phys. Part 1* **41**, 4314 (2002).

48. M. Takai, W. J. Zhao, A. Sawada, A. Hosono, and S. Okuda, *SID Digest*, Society for Information Display, Baltimore, USA, 2003, p. 794.

49. H. Liu, S. H. Li, J. Zhai, H. J. Li, Q. S. Zheng, L. Jiang, and D. B. Zhu, *Angew. Chem. Int. Ed.* **43**, 1146 (2004).

50. N. Chakrapani, B. Q. Wei, A. Carrilo, P. M. Ajayan, and R. S. Kane, *PNAS* **101**, 4009 (2004).

51. L. M. Sheng, M. Liu, P. Liu, Y. Wei, L. Liu, and S. S. Fan, *Appl. Sur. Sci.* **250**, 9 (2005).

52. Y. M. Liu, L. Liu, P. Liu, L. M. Sheng, and S. S. Fan, *Diamond & Related Materials* **13**, 1609 (2004).

53. L. M. Sheng, P. Liu, Y. M. Liu, L. Qian, Y. S. Huang, L. Liu, and S. S. Fan *J. Vac. Sci. Technol.* **A21**, 1202 (2003).

54. Y. M. Liu and S. S. Fan, *Solid State Commun.* **133**, 131 (2005).

55. J. Robertson, *Materials Today*, **October**, 46 (2004).

56. L. M. Sheng, P. Liu, Y. Wei, L. Liu, J. Qi, and S. S. Fan, *Diamond & Related Materials* **14**, 1695 (2005).

57. J. F. Wang, M. S. Gudiksen, X. F. Duan, Y. Cui, and C. M. Lieber, *Science* **293**, 1455 (2001).

58. Z. M. Li, Z. K. Tang, H. J. Liu, N. Wang, C. T. Chan, R. Saito, S. Okada, G. D. Li, J. S. Chen, N. Nagasawa, and S. Tsuda, *Phys. Rev. Lett.* **87**, 1274011 (2001).

59. K. L. Jiang, Q. Q. Li, and S. S. Fan, *Physics* **32**, 506 (2003) (in Chinese).

60. P. Li, K. L. Jiang, M. Liu, Q. Q. Li, and S. S. Fan, *Appl. Phys. Lett.* **82**, 1763 (2003).

61. J. F. Wang, M. S. Gudiksen, X. F. Duan, Y. Cui, and C. M. Liber, *Science* **293**, 1455 (2001).

62. K. L. Jiang, A. Q. Li, and S. S. Fan, *Nature* **419**, 801 (2002).

63. J. Hu, T. W. Odom, and C. M. Lieber, *Acc. Chem. Res.* **32**, 435 (1999).

64. C. Dekker, *Phys. Today* **52**, 22 (1999).

65. S. J. Tans, R. M. Verschueren, and C. Dekker, *Nature* **393**, 49 (1998).

66. R. Martel, T. Schmidt, H. R. Shea, T. Hertel, and P. Avouris, *Appl. Phys. Lett.* **73**, 2447 (1998).

67. S. J. Wind, J. Appenzeller, R. Martel, V. Derycke, and P. Avouris, *Appl. Phys. Lett.* **80**, 3817 (2002); P. Avouris, J. Appenzeller, R. Martel, and S. J. Wind, *Proc. IEEE* **191**, 1772 (2003).

68. A. Javey, H. Kim, M. Brink, Q. Wang, A. Ural, J. Guo, P. McIntyre, P. McEuen, M. Lundstrom, and H. J. Dai, *Nat. Mater.* **1**, 241 (2002); A. Javey, J. Guo, D. B. Farmer, Q. Wang, D. W. Wang, R. G. Gordon, M. Lundstrom, and H. J. Dai, *Nano Lett.* **4**, 447 (2004).

69. S. J. Tans, M. H. Devoret, H. J. Dai, A. Thess, R. E. Smalley, L. J. Geerligs, and C. Dekker, *Nature* **386**, 474 (1997).

70. M. Bockrath, D. H. Cobden, P. L. McEuen, N. G. Chopra, A. Zettl, A. Thess, and R. E. Smalley, *Science* **275**, 1922 (1997).

71. T. W. Qdom, J. L. Huang, P. Kim, and C. M. Lieber, *Nature* **391**, 62 (1998).

72. J. W. G. Wildoer, L. C. Venema, A. G. Rinzler, R. E. Smalley, and C. Dekker, **391**, 59 (1998).

73. X. F. Duan, Y. Huang, Y. Cui, J. F. Wang, and C. M. Lieber, *Nature* **409**, 66 (2001).

74. M. S. Gudiksen, L. J. Lauhon, J. Wang, D. C. Smith, C. M. Lieber, *Nature* **415**, 617 (2002).

Index